Modern Robotics

Modern Robotics

Edited by
Lauren Barrett

MURPHY & MOORE
www.murphy-moorepublishing.com

Published by Murphy & Moore Publishing,
1 Rockefeller Plaza,
New York City, NY 10020, USA

ISBN: 978-1-63987-373-9

Cataloging-in-Publication Data

Modern robotics / edited by Lauren Barrett.
p. cm.
Includes bibliographical references and index.
ISBN 978-1-63987-373-9
1. Robotics. 2. Automation. 3. Machine theory. 4. Robots. I. Barrett, Lauren.
TJ211 .M63 2022
629.892--dc23

For information on all Murphy & Moore Publications
visit our website at www.murphy-moorepublishing.com

MURPHY & MOORE

Contents

Preface

Robotics is a branch of engineering which is concerned with the conception, design, construction and operation of robots. It is an interdisciplinary field overlapping with artificial intelligence, electronics, mechatronics, computer science nanotechnology and bioengineering. Robots are designed to replicate human action. Today, robots are being used for domestic, military and commercial purposes. Many robots that are being designed are inspired by natural designs, thus contributing to the field of bio-inspired robotics. Much of the research in robotics is directed at the design and manufacture of new types of robots. A new innovation in robot design is evolutionary robots. This is based on the use of evolutionary computation to develop improved robots. The aim of this book is to present researches that have transformed the field of robotics aided its advancement. The topics covered in this extensive book deal with the core areas of robotics. It will help the readers in keeping pace with the rapid changes in this field.

This book is a comprehensive compilation of works of different researchers from varied parts of the world. It includes valuable experiences of the researchers with the sole objective of providing the readers (learners) with a proper knowledge of the concerned field. This book will be beneficial in evoking inspiration and enhancing the knowledge of the interested readers.

In the end, I would like to extend my heartiest thanks to the authors who worked with great determination on their chapters. I also appreciate the publisher's support in the course of the book. I would also like to deeply acknowledge my family who stood by me as a source of inspiration during the project.

Editor

Base Detection Research of Drilling Robot System by using Visual Inspection

Zhanxi Wang, [1,2] Jing Bai, [1] Xiaoyu Zhang, [1] Xiansheng Qin, [1] Xiaoqun Tan, [1] and Yali Zhao [2]

[1] *School of Mechanical Engineering, Northwestern Polytechnical University, Xi'an 710072, China*

[2] *Xinlian College, Henan Normal University, Zhengzhou 451400, China*

Correspondence should be addressed to Zhanxi Wang; zxwang@nwpu.edu.cn

Academic Editor: L. Fortuna

This paper expounds the principle and method of calibration and base detection by using the visual measurement system for detection and correction of installation error between workpiece and the robot drilling system. This includes the use of Cognex Insight 5403 high precision industrial camera, a light source, and the KEYENCE coaxial IL-300 laser displacement sensor. The three-base holes method and two-base holes method are proposed to analyze the transfer relation between the basic coordinate system of the actual hole drilling robot and the basic coordinate system of the theoretical hole drilling robot. The corresponding vision coordinates calibration and the base detection experiments are examined and the data indicate that the result of base detection is close to the correct value.

1. Introduction

Industrial robotic applications are becoming widely used in the aviation sector [1] because of lower investment, increasing automation, table working performance, and good accessibility. From the beginning of twenty-first Century, American GEMCOR, EI (ElectroimPact), Italian COMAU, and German BROETJE-Automation have been committed to the design and development of drilling robot systems, and their drilling robot system production has been used in aircraft manufacturing enterprises, such as the F-16, F-22 vertical wall, F-35 aircraft wing panel, and A380 wing panel [2–4]. S H Bi developed a drilling robot system, which uses an industrial camera to establish the relationship between the workpiece and the robot coordinate system [5]. During the drilling process of the movable robot, the moving device locks floatingly when the drilling equipment is in the working position; then the robot will arrive at the designated location following the programmed workpiece coordinate system [6]. Due to errors caused by the orientation of the motion device or the installation of the workpiece, differences in the relative location of the workpiece and the robot drilling system will occur, which will result in the offset of the actual arrival position [4, 7].

To simplify this problem, working pieces are assumed to be in a theoretical position, and all installation errors can be concentrated on the robot coordinate system [8, 9]. The fundamental theory of base detection can be illustrated, first, by the actual position being checked through measuring the datum holes on the workpiece when the drilling robot system arrives at the designated location and, second, as the matrix between the theoretical robot coordinate system and the actual robot coordinate system being deduced to a generally correct position of the holes involved in programming.

2. Calibration of Base Detection System

In this paper, we have developed a drilling end effector system and installed it on a Kuka KR-500 robot as the experiment platform, as shown in Figure 1.

The calibration of the visual coordinate system is composed of two parts: the position calibration and the direction calibration [10, 11]. Position calibration is used to obtain the 2D coordinates of the robot Tool Center Point (TCP) in the

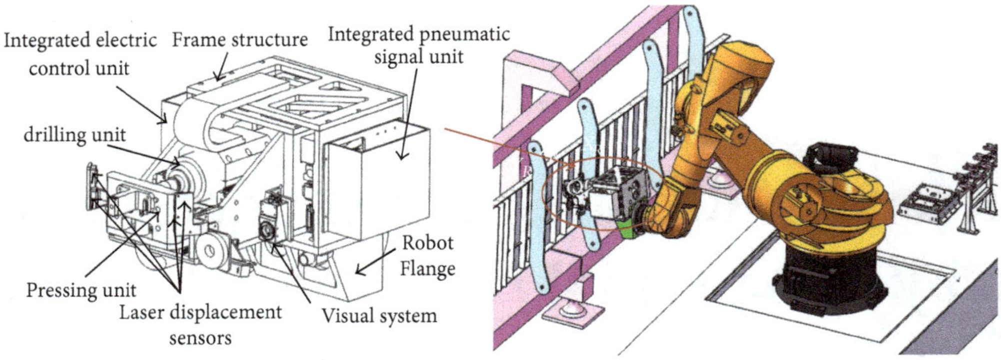

FIGURE 1: The specific structure of drilling end effector and robot system.

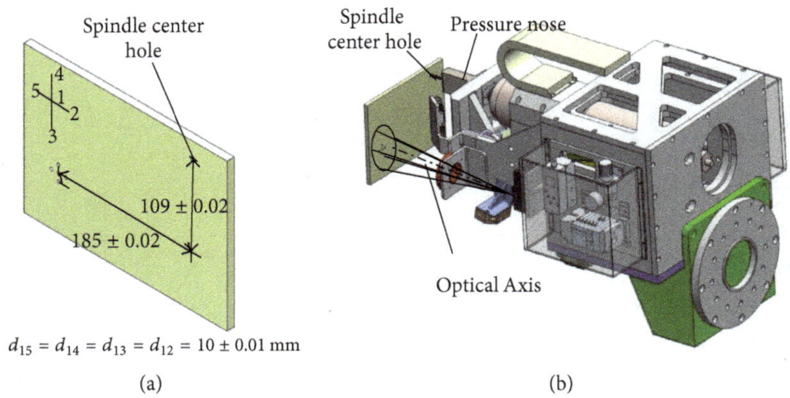

FIGURE 2: Visual system calibration plate.

visual coordinate system, and the readings of the four laser displacement sensors, where the distance between pressure nose tip and the workpiece surface is 45 mm.

Direction calibration is used to obtain the rotation angle between the visual coordinate system and the tool coordinate system around the X-axis of tool coordinate system, and the laser emission direction of laser displacement sensor [12, 13]. Because the vision coordinate system is fixed on base detection system, the calibration of the visual coordinate system described is consistent with the calibration of base detection system.

2.1. Position Calibration. A special calibration board has been designed for position calibration where several holes are finished with certain relative positions.

As shown in Figure 2, the standard cutter mounted on the spindle is sheathed in the upper right hole of calibration plate. This hole is referred to as the spindle center hole. The remaining five holes in the calibration plate can be captured by adjusting the focal length of camera. According to the relative positions of the spindle center hole and the remaining holes (Figure 2(a)), the coordinates of the spindle center hole can be calculated in visual coordinates.

According to the image of spindle center hole taken by industrial cameras, the position in default coordinate system can be calculated by the contour detection and image segmentation algorithm [14, 15], as illustrated in Figure 3.

The coordinate of spindle center hole in visual coordinate system is given by

$$\begin{bmatrix} dx \\ dy \end{bmatrix} = \overrightarrow{OP1} + \frac{\overrightarrow{P1P4}}{|P1P4|} \bullet l_1 + \frac{\overrightarrow{P1P2}}{|P1P2|} \bullet l_2, \qquad (1)$$

where $\overrightarrow{OP1}$ is the vector from O to $P1$, $\overrightarrow{P1P4}$ is vector from $P1$ to $P4$, $\overrightarrow{P1P2}$ is the vector from $P2$ to $P1$, l_1 is the vertical distance from TCP to O, and l_2 is the radial distance from TCP to O.

As shown in Figure 3, (dx, dy) can be obtained by substituting $l_1 = 185$ mm and $l_2 = 109$ mm into (1).

The laser displacement sensor should be installed on the right position to ensure that the laser is projected near the reference hole. Manual teaching robot is used to keep the front end of the pressure nose at a certain distance (programming safety height) from the workpiece surface. The reading of the laser displacement sensor s_{z0} is recorded as the zero position of the laser displacement sensor by this time.

2.2. Direction Calibration. The direction calibration of the vision coordinate system is mainly to obtain the direction of the optical axis of the industrial camera and laser displacement sensor. Calibration method is demonstrated in Figure 4

The calibration plate is fixed to the test bench, and the coordinates of the hole P on the calibration board in the

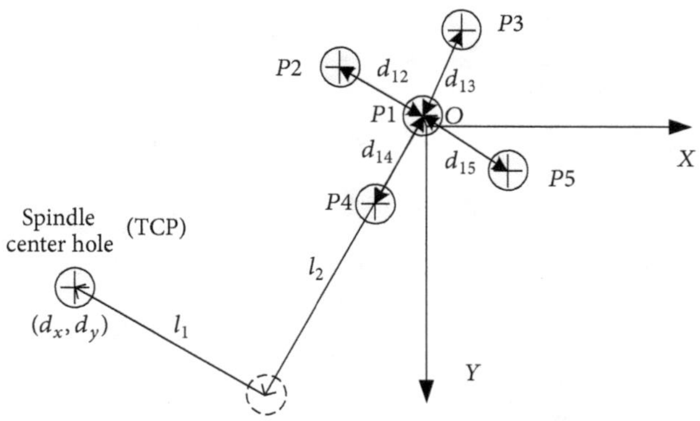

FIGURE 3: Position of spindle center hole (TCP).

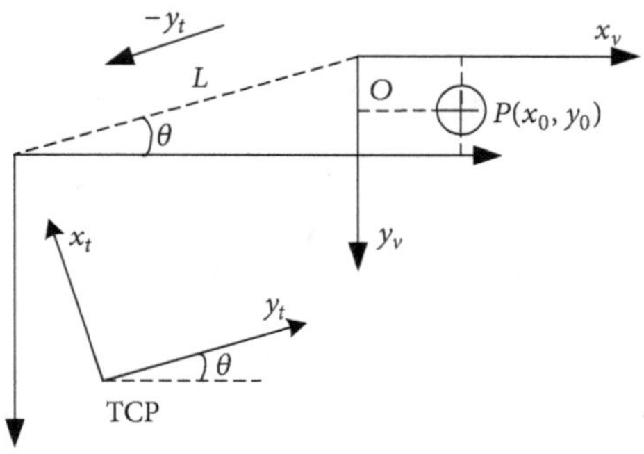

FIGURE 4: Direction calibration of industrial camera.

camera coordinate system (x_0, y_0) are recorded. Then, the industrial robot is moved a certain distance of L along the $-Y$ direction of tool coordinate system and set while the calibration plate is remeasured with the industrial camera. If both the X- and Y-axes of the visual coordinate system are in accordance with the $+Y$ direction and the $-X$ direction of the tool coordinate system, then the X coordinate value of hole P will increase L while the Y coordinates remain unchanged.

If there is a certain angle θ between the $+X$-axis of visual coordinate system and $+Y$-axis of tool coordinate system (positive for clockwise, negative for counterclockwise), the coordinates of remeasured hole A will be replaced with $(x_0 + L \times \cos(\theta), y_0 - L \times \sin(\theta))$. Then angle θ can be calculated by the coordinate values (x, y) obtained from the actual measurements.

To calibrate the direction of laser displacement sensor, a series of planes that have a certain distance paralleled to the pressure foot plane should be measured, so that the plane angle between laser and pressure foot plane θ_z can be easily obtained according to the distance between two adjacent planes and the reading of the laser displacement sensor, as shown in Figure 5.

$$\theta_z = \arcsin\left(\frac{d}{s}\right). \quad (2)$$

3. Base Detection Theory

Two methods are used for base detection in this paper, including three-base holes detection and two-base holes detection. Three-base holes detection is based on both their actual and theoretical locations to analyze the transfer relation between basic coordinate system of the actual hole drilling robot and the basic coordinate system of the theoretical hole drilling robot.

On account of the actual and theoretical position of one base hole, the actual position, actual normal, theoretical position, and theoretical normal of the other base hole, the two-base hole detection can be used to calculate the transfer relation.

As shown in Figure 6, the theoretical coordinates of three-base holes P_1, P_2, and P_3 in robot base coordinate system are given by

$$P_1 = \begin{bmatrix} P_{1x} \\ P_{1y} \\ P_{1z} \end{bmatrix}, \quad P_2 = \begin{bmatrix} P_{2x} \\ P_{2y} \\ P_{2z} \end{bmatrix}, \quad P_3 = \begin{bmatrix} P_{3x} \\ P_{3y} \\ P_{3z} \end{bmatrix}. \quad (3)$$

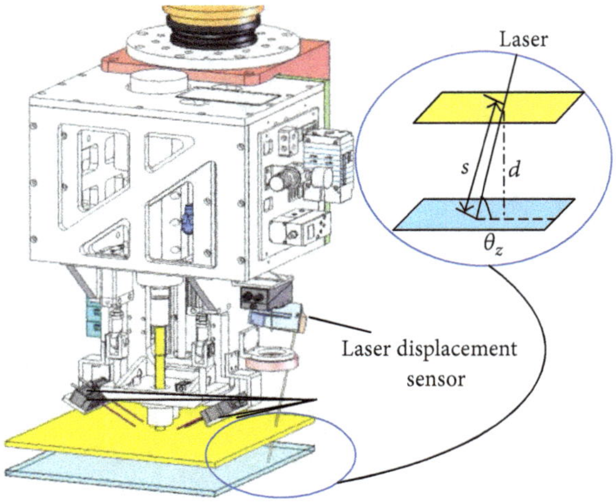

FIGURE 5: Direction calibration of laser displacement sensor.

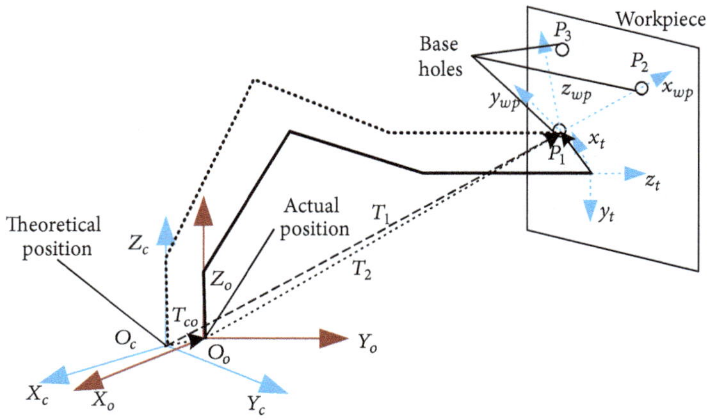

FIGURE 6: Base detection theorem of three-base holes.

Due to the error of workpiece installation and drilling robot orientation, the actual coordinates of three datum holes can be described as

$$P_1^o = \begin{bmatrix} P'_{1x} \\ P'_{1y} \\ P'_{1z} \end{bmatrix},$$
$$P_2^o = \begin{bmatrix} P'_{2x} \\ P'_{2y} \\ P'_{2z} \end{bmatrix}, \qquad (4)$$
$$P_3^o = \begin{bmatrix} P'_{3x} \\ P'_{3y} \\ P'_{3z} \end{bmatrix}.$$

3.1. Three-Base Hole Theorem. Direct calculation and least square matching are used to serve three-base hole theorem.

Direct Calculation. Coordinate system $P1\text{-}XYZ$ is established by taking P_1 for origin, the direction of $\overrightarrow{P_1P_2}$ for X-axis, and the normal plane that contains P_1, P_2, and P_3 as the Z-axis (as shown in Figure 6). Theoretically, the transfer matrix from the robot-based coordinate to coordinate $P_1\text{-}XYZ$ is

$$T_1 = \begin{bmatrix} n_1 & o_1 & a_1 & p_1 \\ 0 & 0 & 0 & 1 \end{bmatrix}. \qquad (5)$$

The transfer matrix from the actual robot-based coordinate $O_c\text{-}X_cY_cZ_c$ to actual workpiece coordinate is written as

$$T_2 = \begin{bmatrix} n_2 & o_2 & a_2 & p_2 \\ 0 & 0 & 0 & 1 \end{bmatrix}, \qquad (6)$$

where

$$p_1 = \overrightarrow{P_1}$$
$$n_1 = \frac{\overrightarrow{P_2} - \overrightarrow{P_1}}{\left|\overrightarrow{P_2} - \overrightarrow{P_1}\right|}$$

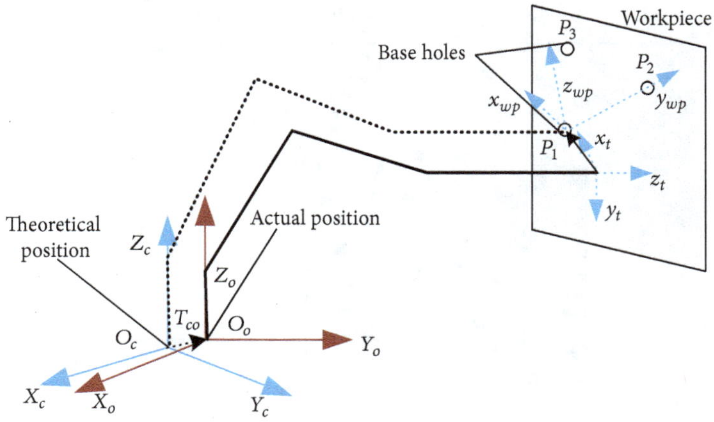

FIGURE 7: Base detection theorem of two-base holes.

TABLE 1: Industrial camera's raw data of position calibration.

P1	P2	P3	P4	P5
(872.5, 594.5)	(437.25, 344.5)	(1120.5, 158.25)	(623.5, 1028.5)	(1307.25, 842)

$$a_1 = \frac{\left(\vec{P_2} - \vec{P_1}\right) \times \left(\vec{P_3} - \vec{P_1}\right)}{\left|\vec{P_2} - \vec{P_1}\right| \bullet \left|\left(\vec{P_3} - \vec{P_1}\right)\right|}$$
$$o_1 = a_1 \times n_1. \tag{7}$$

We can obtain T_2 by substituting $P_1{}^o$ by P_1, $P_2{}^o$ by P_2, and $P_3{}^o$ by P_3.

$$T_{co} \bullet T_2 = T_1; \qquad T_{co} = T_1 \bullet T_2^{-1}. \tag{8}$$

The least square solution of transformation matrix can be calculated according to $\{P_1, P_2, P_3\}$ and $\{P_1^0, P_2^0, P_3^0\}$ [14, 15].

3.2. Two-Base Hole Theorem. As shown in Figure 7, two-base point detection is used to obtain the transfer matrix by knowing the actual position of point P_1^o, the actual normal direction A_1^o, B_1^o, theoretical position P_1, theoretical normal direction A_1, B_1, actual position P_2^o, and theoretical position of another point P_2.

The principle is shown in Figure 7.

The workpiece coordinate system is established by the method shown in Figure 7, where P_1 is the origin, the direction of $\overrightarrow{P_1P_2}$ is represented by the Y-axis, the normal vector of the plane that contains P_1 P_2 P_3 is the X-axis, and the Z-axis is determined by the right hand rule. The definitions of T_1 and T_2 are those as demonstrated in (5) and (6).

$$p_1 = \vec{P_1}$$
$$n_1 = \left[\sin B_1 \cos C_1 \quad -\sin C_1 \quad \cos B_1 \cos C_1\right]^T$$
$$o_1 = \frac{\vec{P_2} - \vec{P_1}}{\left|\vec{P_2} - \vec{P_1}\right|} \tag{9}$$
$$a_1 = n_1 \times o_1,$$

where B_1, C_1 are Euler angle of ZYX that can be detected in normal direction.

We can obtain T_2 by substituting $P_1{}^o$ by P_1, $P_2{}^o$ by P_2, and $P_3{}^o$ by P_3.

$$T_{co} \bullet T_2 = T_1; \qquad T_{co} = T_1 \bullet T_2^{-1}. \tag{10}$$

4. Base Detection Experiment

4.1. Calibration Experiment of the Base Detection System. First, the position calibration of the industrial camera and the laser displacement sensor is carried out. According to the installation method of the calibration plate shown in Figure 2, the image collected by Insight Explorer software is shown in Figure 8. Coordinates in the default camera coordinate system (units: pixels) are shown in Table 1.

The relation between the pixels and the distance (1 mm = 50.11 pixel) is obtained by using the distance between the five calibration holes, d_{15}, d_{14}, d_{13}, d_{12} = 10 ± 0.01 mm.

TABLE 2: Base hole's coordinate in camera coordinate system.

$P1$	$P2$	$P3$	$P4$	$P5$
(1.4468, −0.1098)	(−7.2392, 5.0988)	(6.396, −8.8157)	(−3.5223, 0.5513)	(10.1229, 4.8294)

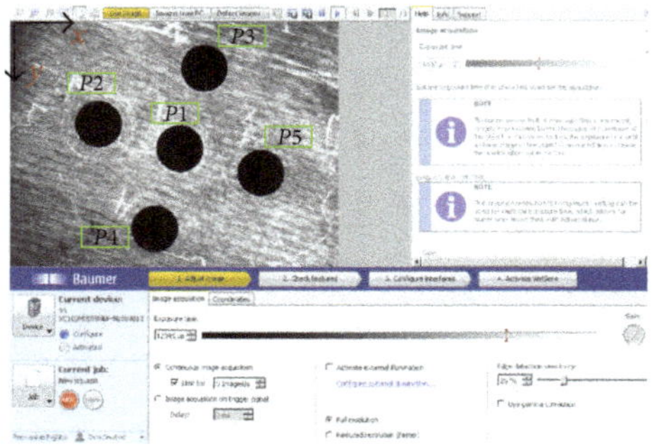

FIGURE 8: Image of calibration plate.

FIGURE 9: Direction calibration of industrial camera.

After moving the default camera coordinate system to the central point of view, the coordinates of the calibration hole in the camera coordinate system are obtained and shown in Table 2.

According to (1), the coordinate of the spindle hole in the visual coordinate system (unit: mm) is given as (−186.5838, 106.1815).

Figure 9 illustrates the direction calibration of the industrial camera. The calibration board is adjusted to the normal level after it is fixed to the test fixture in two different positions, making sure that the robot tool axis is perpendicular to the test fixture. Then the coordinates of $P1$, $P2$, and $P3$ are measured in the default camera coordinate (pixels), as shown in Table 3.

When the calibration plate is placed in position 1, the coordinate system in the robot base coordinates is (1354.51, −2622.831494.66, −90.08, −0.55, −1.57), and when in position 2 the tool coordinate position in the robot base coordinate system is (1545.86, −2665.15, 1461.93, −89.4, 0.3, −1.56); when the drilling robot moves a certain distance in the $+x$ direction of the tool coordinate system, the coordinates of holes $P1$, $P2$, and $P3$ (pixel) are remeasured. The results are shown in Table 4.

After moving, the tool coordinate in the robot base coordinate system is (1354.68, −2622.78, 1499.15, −90.07, −0.55, −1.57) in position 1 and (1534.36, −2665.26, 1462.24, −89.4, 0.3, −1.56) in position 2.

According to the relative position of the calibration holes $P1$, $P2$, and $P3$ on the calibration board, the corresponding relationship between pixels and the distance can be obtained:

1 mm = 47.741 pixel in position 1 and 1 mm = 50.2792 pixel in position 2.

It can be seen from Table 5 that when the moving distances of robot are 4.493 mm and 11.5047 mm, the error between the actual value and the theoretical value is less than 0.06 mm, and the rotation angle of the tool coordinate system in Z direction can be neglected.

The calibration method of laser displacement sensor is used to make the end effector tool perpendicular to the calibration plate with the normal adjustment function. Readings of sensors in Z direction are shown in Table 6 with different height gauge added onto the calibration plate.

Table 3: Data before direction calibration.

	$P1$	$P2$	$P3$
Position 1	(947.25, 700.25)	(470.25, 686.75)	(959, 224.75)
Position 2	(1350.75, 878.25)	(847.75, 866.25)	(1365.5, 375.5)

Table 4: Data after direction calibration.

	$P1$	$P2$	$P3$
Position 1	(949.75, 915.5)	(473.5, 902.5)	(961.25, 439.25)
Position 2	(771.25, 875.25)	(268.75, 863.75)	(785.5, 372.5)

Table 5: The difference of coordinates before and after moving.

	$P1$	$P2$	$P3$	Displacement of robot
Position 1	(0.0524, 4.5087)	(0.0576, 4.5191)	(0.0471, 4.493)	$+X$: 4.493 mm
Position 2	(−11.5320, −0.0596)	(−11.5221, −0.0498)	(−11.5420, −0.0543)	$+Y$: 11.5047 mm

Table 6: Readings of sensors in Z direction.

Height of gauge (mm)	0	20	30	40	50	60	70
Readings of sensors in Z direction (mm)	−60.4	−81.2	−91.6	−102	−112.4	−122.8	−133.2

Table 7: Values before changing coordinate system.

	Coordinate of industrial camera	Value of laser displacement Sensor	Gesture of robot (X, Y, Z, A, B, C) X/mm, Y/mm, Z/mm, A/°, B/°, C/°
$P1$	(855.5, 573.5)	−36.5	(185.09, −2609.33, 1689.50, −89.99, 0.026, 0.0259)
$P2$	(881.75, 585.00)	−36.3	(215.49, −2609.37, 1689.62, −89.99, 0.021, 0.020)
$P3$	(889.5, 590.75)	−36.3	(185.57, −2609.41, 1659.88, −90.00, 0.021, 0.024)

Table 8: Coordinates of base holes in robot coordinate system before changing coordinate system.

	$P1$	$P2$	$P3$
X	183.894	213.73	183.64
Y	−2608.07	−2608.31	−2608.34
Z	1690.07	1689.94	1660.08

The plane angle of the laser displacement sensor and pressure foot is given as $\arcsin(10/10.4)$.

The sensor reading is $S_{z0} = -34.8$ mm where the height between the front face of pressure nose and calibration plate $d = 45$ mm.

4.2. Base Detection Experiment of Robot Drilling System. This study has determined that the three-point detection theorem is best suited to verify the correctness of the benchmark detection principle. The experiment field is shown in Figure 10. The calibration board is fixed on the test fixture and in order to reduce the coordinate acquisition errors, the normal adjustment function is used to control the verticality of tool and calibration plate under 0.1 degrees before industrial the camera takes images.

The drilling robot is taught manually to make the industrial camera lens center aligned to the base hole; then the laser tracker is used to verify whether the end effector spindle is perpendicular to the calibration plate. The three holes on the calibration board are measured and the coordinates of the corresponding cameras and the laser displacement sensors recorded (unit: pixel). The readings in the robot teaching device (unit: mm) are shown in Table 7.

The manual work piece coordinate system is set to [12, −5, −12, −0.3, −0.05, −0.01], the robot pose is controlled to repeat positioning to the gesture in Table 8 (in work piece

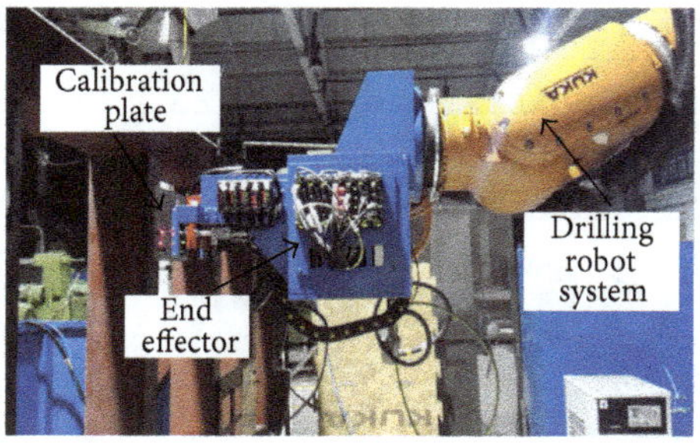

Figure 10: Base detection experiment of robot drilling system.

Table 9: Values after changing coordinate system.

Actual value	Coordinate of industrial camera	Value of laser displacement sensor
$P1$	(620.75, 396.75)	−39.1
$P2$	(637.75, 396.5)	−39.2
$P3$	(648.25, 406.75)	−39.0

Table 10: Coordinates of base holes in robot coordinate system after changing coordinate system.

	$P1$	$P2$	$P3$
X	189.10	219.124	188.82
Y	−2605.55	−2605.49	−2605.72
Z	1694.05	1694.18	1664.03

coodinates), and the board holes are recalibrated to achieve the pixel coordinates shown in Table 9.

The three coordinates of $P1$, $P2$, and $P3$ are converted to the robot base coordinate system, and the coordinates of three points are obtained, as shown in Table 10.

According to the three-base hole detection principle, compensation matrix is given as [13.70174, −6.99429, −12.833, −0.36, −0.08, −0.0188] by using direct method, and the compensation matrix is given as [12.064, −6.9798, −12.064, −0.36082, −0.02936, −0.01886] by using least square matching method.

By comparison, it can be seen that both two methods can find accurate solutions, but the least square matching method is closer to the true value. This proves that the detection principle is feasible. Deviation caused by the robot's positioning error and various conversion errors can be reduced by robot positioning accuracy compensation.

5. Conclusion

Improving the positioning accuracy of the drilling robot is significant in improving the quality hole drilling. The deviation of the relative position between the workpiece and the robot will affect the positioning accuracy of the drilling holes. In order to identify the actual relative position of the workpiece and the robot, the base detection should be done before robot drilling preparation. This study has discussed calibration using the base detection system, and the base detection technology principle, and then we designed the calibration experiment and base detection experiment to verify base detection methods. The data indicate that the result of base detection is close to the correct value and that the principle is feasible.

Conflicts of Interest

The authors declare that there are no conflicts of interest regarding the publication of this paper.

Acknowledgments

This work is partially supported by the National Natural Science Foundation of China (Grant no. 51505380), Shaanxi Science and Technology Innovation Project (Grant nos. 2016KTZDGY06-01, 2016KTZDGY4-12), and the "111 Project" of China (Grant no. B13044).

References

[1] A. Olabi, R. Béarée, O. Gibaru, and M. Damak, "Feedrate planning for machining with industrial six-axis robots," *Control Engineering Practice*, vol. 18, no. 5, pp. 471–482, 2010.

[2] J. Atkinson, J. Hartmann, S. Jones, and P. Gleeson, "Robotic drilling system for 737 aileron," in *Proceedings of the SAE 2007 AeroTech Congress and Exhibition*, pp. 1–3821, SAE Technical Papers, Los Angeles, Calif, USA, 2007.

[3] R. DeVlieg and E. Feikert, "One-up assembly with robots," Training, 2008. p. 09-30, 2013.

[4] R. Devlieg, K. Sitton, E. Feikert, and J. Inman, "ONCE (ONe-sided Cell End effector) robotic drilling system," *SAE Technical Papers*, 2002.

[5] S. Bi and J. Liang, "Robotic drilling system for titanium structures," *The International Journal of Advanced Manufacturing Technology*, vol. 54, no. 5-8, pp. 767–774, 2011.

[6] Y. Gao, D. Wu, Y. Dong, X. Ma, and K. Chen, "The method of aiming towards the normal direction for robotic drilling," *International Journal of Precision Engineering and Manufacturing*, vol. 18, no. 6, pp. 787–794, 2017.

[7] H. Chen, T. Fuhlbrigge, S. Choi, J. Wang, and X. Li, "Practical industrial robot zero offset calibration," in *Proceedings of the 4th IEEE Conference on Automation Science and Engineering*, pp. 516–521, Washington, DC, USA, August 2008.

[8] A. Nubiola and I. A. Bonev, "Absolute calibration of an ABB IRB 1600 robot using a laser tracker," *Robotics and Computer-Integrated Manufacturing*, vol. 29, no. 1, pp. 236–245, 2013.

[9] J. Wang, H. Zhang, and T. Fuhlbrigge, "Improving machining accuracy with robot deformation compensation," in *Proceedings of the 2009 IEEE/RSJ International Conference on Intelligent Robots and Systems, IROS 2009*, pp. 3826–3831, usa, October 2009.

[10] S. Garnier, K. Subrin, and K. Waiyagan, "Modelling of Robotic Drilling," in *Proceedings of the 16th CIRP Conference on Modelling of Machining Operations, CIRP CMMO 2017*, pp. 416–421, France, June 2017.

[11] B. Mei, W. Zhu, K. Yuan, and Y. Ke, "Robot base frame calibration with a 2D vision system for mobile robotic drilling," *The International Journal of Advanced Manufacturing Technology*, vol. 80, no. 9-12, pp. 1903–1917, 2015.

[12] A. Frommknecht, J. Kuehnle, I. Effenberger, and S. Pidan, "Multi-sensor measurement system for robotic drilling," *Robotics and Computer-Integrated Manufacturing*, vol. 47, pp. 4–10, 2017.

[13] Y. Liu, Z. Shi, P. Yuan, D. Chen, M. Lin, and Z. Li, "A visual positioning and measurement system for robotic drilling," in *Proceedings of the 14th IEEE International Workshop on Advanced Motion Control, AMC 2016*, pp. 461–466, New Zealand, April 2016.

[14] P. Arbeláez, M. Maire, C. Fowlkes, and J. Malik, "Contour detection and hierarchical image segmentation," *IEEE Transactions on Pattern Analysis and Machine Intelligence*, vol. 33, no. 5, pp. 898–916, 2011.

[15] P. Arena, A. Basile, M. Bucolo, and L. Fortuna, "An object oriented segmentation on analog cnn chip," *IEEE Transactions on Circuits and Systems I: Fundamental Theory and Applications*, vol. 50, no. 7, pp. 837–846, 2003.

2

Design of a New Nonlinear Stiffness Compliant Actuator and its Error Compensation Method

Shaobin Lan and Zhibin Song

Key Laboratory of Mechanism Theory and Equipment Design of Ministry of Education, Tianjin University, Tianjin 300072, China

Correspondence should be addressed to Zhibin Song; songzhibin@tju.edu.cn

Academic Editor: Huosheng Hu

Compliant actuators are more advantageous than stiff actuators in some circumstances, for example, unstructured environment robots and rehabilitation robots. Compliant actuators are more adaptive and safe. Constant stiffness compliant actuators have some limitations in impedance and bandwidth. Variable stiffness actuators improve their performance owing to introducing an extra motor to tune the stiffness of the actuators. However, they also have some limitations such as the bulky structure and heavy weight. It was also found that there are some waste functions existing in the current variable stiffness actuators and that the fully decoupled position control and stiffness tune are not necessary, because there exist some regular phenomena during most circumstances of human interaction with the robots which are "low load, low stiffness and high load, high stiffness". In this paper, a design method for nonlinear stiffness compliant actuator was proposed which performed the predefined deflection-torque trajectory of the regular phenomenon. A roller and a cantilever which has special curve profile constitute the basic mechanical structure of the nonlinear stiffness compliant actuators. An error compensation method was also proposed to analyze the stiffness of elastic structure. The simulation results proved that the proposed method was effective in designing a predefined nonlinear stiffness compliant actuator.

1. Introduction

Traditional actuators used in robotics or machines are stiff actuators which could obtain good position control and bandwidth. However, since gear boxes are usually used in stiff actuators, the backdrivability is always poor, which is not suitable in some circumstances, for example, human-machine interaction, bioinspired robots and unstructured environment robots, and rehabilitation robots. The compliant actuator using an elastic component between the actuator and load was first proposed by Pratt in MIT [1], which is also called series elastic actuator (SEA). This kind of passive compliant actuator can be looked at as a pure force source, which is able to perform perfect force control.

From the early SEA with a linear spring to the later one with a torsional spring [1–3], then to the constant stiffness SEA optimized via finite element methods [4], and finally to the latest SEA with variable stiffness [5], passive compliant actuators were greatly improved owing to more compact structure and wider bandwidth of force control [6, 7]. The variable stiffness SEA has attracted lots of interests and developed very quickly recently. Summarizing these proposed ideas, they could be divided into three groups to obtain the property of variable stiffness [8]:

(1) Change transmission between the load and elastic structures. AwAS [9] implemented the adjustable stiffness of its actuator via changing the effective length of linkages. AwAS-II [10], CompAct-VSA [11], and VSTA [12] varied their stiffness of actuators through adjusting the fulcrum position.

(2) Change the preload of the spring. FSJ [13], BAVS [14], and BAFSA [15] changed the stiffness of actuators via changing the contact position between the roller and cam mechanism with compressing the springs.

(3) Change the physical structure of springs. Jack Spring [16] and others obtained the variable stiffness actuator through changing the number of effective spring coins.

In general, most of variable stiffness actuators use two actuator units: one for position control and the other for the stiffness regulation, which makes the position control and stiffness regulation independent. Of course, variable stiffness actuator is flexible to adjust the stiffness whatever the position of rotor is. However, it also leads to the bulky and complicated mechanical structure [17]. On the other hand, decoupled position control and stiffness regulation in most of variable stiffness actuators are not used sufficiently.

Few researchers studied nonlinear stiffness SEA without an additional motor. Migliore et al. developed a nonlinear stiffness SEA with antagonistic nonlinear elastic springs [18]. Palli et al. also designed a kind of nonlinear stiffness SEA [19]. The nonlinear stiffness mimics that of the biological joints [20]. It was noted there was a common property among these researches that the stiffness of actuators increases with the load increasing. It could be summarized as "low load, low stiffness and high load, high stiffness," which is also common in human activity. Human limbs will stiffen to lift a heavy object but soften to clean up with a whisk.

This functionality is useful for human to obtain dexterous performance with low load and stability with high load [21]. To obtain the better running performance, researchers in MIT Biomimetic Lab developed MIT Cheetah robot with quadratic elastic stiffness in robot's ankle joint [22]. Omer et al. proposed an adjustable stiffness joint for a bipedal robot where two linear springs with different stiffness were adopted and assembled in parallel with an offset. When lower load was given to the output link, the spring with lower stiffness was compressed and the stiffness of system is low. When the spring of low stiffness had been compressed fully, the spring with high stiffness was compressed and the stiffness of system became high [23]. Torrealba and Udelman proposed an adjustable compliant actuator by designing a particular cam shape. The stiffness can be adjusted by compressing the floating spring, and a group of torque-deflection curves show that the stiffness increased with respect to the deflection [15]. Kilic et al. proposed a wrapping cam mechanism with a torsion spring to attain quadratic behavior [24]. Palli et al. utilized a compliant four linkages' mechanism as topology structure and optimized its dimensions of each bar using pseudo-rigid-body model to obtain the nonlinear relationship between position and stiffness of the actuator [19]. The different methods were used to generate nonlinear stiffness of compliant actuator. However, the feasible theoretical method to design the compliant mechanism which possesses the predefined nonlinear stiffness has not been given yet.

In this paper, a new design method of compliant actuators with predefined nonlinear stiffness trajectory of "low load, low stiffness and high load, high stiffness" was proposed. The basic structure of the proposed mechanism consisted of a roller, an elastic component, and a contact part. One joint of 3-DOF (degrees of freedom) rehabilitation robot for shoulder training was introduced as the compliant actuator with defined nonlinear stiffness. Theory deduction and resolution process using software were described in detail. The simulation and experiments were also conducted.

2. Mechanical Design of the Compliant Actuator

In this part, one compliant joint of a rehabilitation robot for shoulder complex was introduced which used the proposed

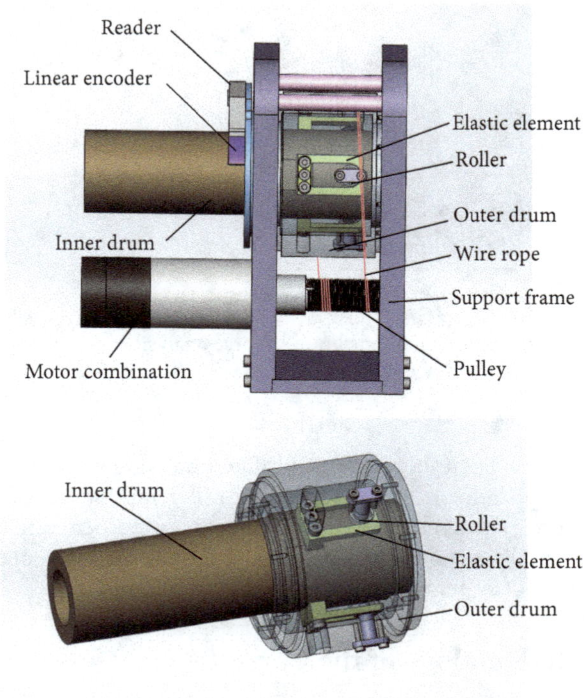

Figure 1: CAD model of the first joint of rehabilitation robot.

nonlinear elastic elements. This robot was designed with 3 DOF (RRP configuration) by International Center of Advanced Mechanisms and Robotics (iCAMAR) of Tianjin University. All of these joints installed the nonlinear elastic elements and the mechanical design of the first joint was introduced as an example which is shown in Figure 1. The compliant joint consists of an inner drum, an outer drum, elastic elements, rollers, a linear magnetic encoder, wire rope, a pulley, motor combination (including a motor, gear box, and an encoder), and support frame. The inner drum connects with external loads and the outer drum is driven by a BLDC motor (Maxon) via wire rope and the pulley. To implement bidirectional compliant actuation, each compliant mechanism consists of two symmetrically arranged elastic elements and a roller. There are three compliant mechanisms distributed on the surface of inner drum with 120 degrees among them. The elastic elements connect with the inner drum and the rollers are installed on the outer drum via flanges. Bearings are used as rollers in order to decrease friction. A linear encoder and a reader which constitute linear magnetic encoder are installed on the outer drum and inner drum, respectively, so that the relative motion between them can be detected precisely. The equilibrium status is that the roller is located in the center line of two symmetrical elastic elements.

The key issue of the design of a nonlinear stiffness compliant actuator is to design elastic mechanisms with nonlinear stiffness. In this paper, a very basic and common elastic structure was adopted as the elastic element, namely, cantilever structure, which is easy to design and accurate to calculate. The main idea of topology design is to propose an elastic element composed of the elastic part and the contact

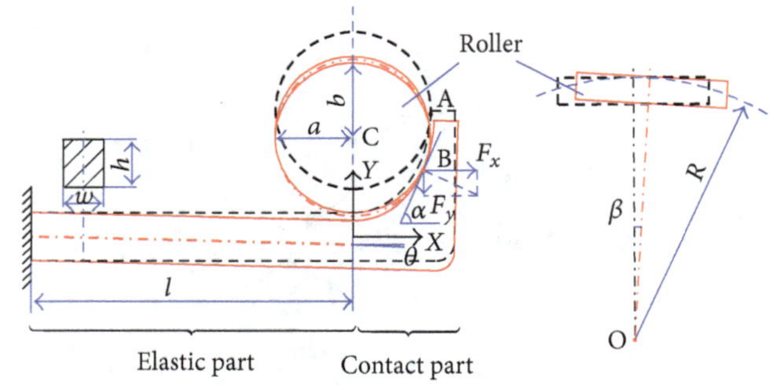

FIGURE 2: The schematic diagram of force analysis of the proposed mechanism.

part which has a special curve profile contacting with a roller. The impedance in the direction of roller motion could be changed when the slope of the tangent plane between the roller and the contact part varied, and then the mechanical stiffness of system changed.

3. Methodology of the Curve Profile Design

The core of designing the nonlinear stiffness compliant actuator is the nonlinear elastic mechanism. As we investigated from the current literature, there is no method to obtain the mechanism whose stiffness matches the predefined nonlinear stiffness trajectory. In this paper, we proposed a simple mechanism to achieve it, which consists of an elastic element and a roller. The elastic element consists of two parts: the elastic part and the contact part. The elastic part is a uniform cantilever which will generate deflection and rotation when it is given a certain external force. The contact part has a specific curve profile.

Figure 2 shows the schematic diagram of the proposed basic nonlinear stiffness mechanism. According to this figure, the elastic element was compressed by the roller when force was exerted on the curve profile of the contact part. The stiffness would be zero when the roller moved on the tangent plane at the contact point and generated contact force was zero. Alternatively, the stiffness would be maximum when the roller moved perpendicularly to the tangent plane at the contact point. Between both of extremity situations, due to the deflection and deflection angle of the cantilever, the angle between the direction of roller motion and the tangent plane would be within 0–90 degrees. The relationship between the roller motion and components of contact force along the roller motion could be adjusted with the slope of contact point, so that the stiffness could be changed.

The force analysis in the mechanism is also shown in Figure 2 where a roller compresses the elastic element. Solid red line shows the current status of the roller and the elastic element and dotted black line shows the last status. Point A represents the contact position between the roller and the elastic element in past status. Point B shows the current contact position of them. C is the center of ellipse in vertical view when the roller rotates around the point O. From the axial direction of the outer drum, the roller which is installed on a rotational component moves along the arc of radius R. The rest of the symbols and their description during force and motion analysis of the proposed mechanism are shown in the Abbreviations.

To simplify analysis of the deflection of the elastic element, the elastic element was analyzed by divided it into two parts. One is a simple cantilever and the other is variable cross section cantilever. Generally speaking, the deflection of elastic element mainly happened on the elastic part and the deflection of the contact part is very little. So firstly analyze the deflection of elastic element by assuming that the contact part is rigid body. Then calculate the curve profile of contact part on the elastic element and conduct the simulation via ANSYS software. Finally, compare the desired torque-deflection trajectory with simulation results, and a method of error compensation was proposed to decrease the error of obtained torque-deflection caused by the assumption of rigidity of the contact part.

3.1. Analysis of Deflection of Elastic Element Assuming That the Contact Part Is Rigid Body. In this research, the elastic element is required to deform within a small deflection, so that the mechanical system can be of good elastic performance. Therefore, the deflection of the elastic element can be described using differential equations of the deflection curve. In Figure 2, the elastic element yields the contact force $F(F_x, F_y)$ to bear the pressure exerted by the roller. Due to the contact part being viewed as a rigid body, to simplify force analysis, the components of contact forces along x-axis and y-axis were moved to the extreme of neutral plane of the elastic part. Therefore, the equivalent torque can be generated owing to the forces transformation:

$$M_0 = -\left(F_y x + y F_y \tan \alpha\right). \quad (1)$$

According to the differential equations of the deflection curve, the deflection angle and deflection of the extreme of elastic component can be described as the following equations:

$$\begin{aligned} \theta &= -\frac{1}{EI}\left[\left(F_y x + y F_y \tan \alpha\right) l + F_y \frac{l^2}{2}\right], \\ v &= -\frac{1}{EI}\left[\left(F_y x + y F_y \tan \alpha\right) \frac{l^2}{2} + F_y \frac{l^3}{3}\right], \end{aligned} \quad (2)$$

where E represents Young's modulus of material and I expresses the second moment of area with respect to the y-axis, which is determined by the right hand rule.

The geometric relationships shown in Figure 2 can be described as follows:

$$\begin{aligned} x &= m_i \cos \theta - n_i \sin \theta, \\ y &= m_i \sin \theta + n_i \cos \theta + v. \end{aligned} \quad (3)$$

Equations (3) express the coordinates of contact point B before and after the elastic component deflection. Furthermore, contact point B is not only on the profile of the contact part but also on the profile of the roller. The constraints can be obtained by (4). γ represents the angle of the parametric equations of the ellipse

$$x - x_c = a\cos\gamma,$$
$$y_c - y = b\sin\gamma. \quad (4)$$

The slope of contact point B before the deformation of the elastic element can be expressed by (5). Furthermore, the slope of contact point B after the deformation of the elastic element can be described by taking the derivative of (4), which is shown by (6). Consider

$$\frac{n_{i-1} - n_i}{m_{i-1} - m_i} = \tan(a - \theta), \quad (5)$$

$$\frac{b}{a}\cot\gamma = \tan\alpha. \quad (6)$$

The length of the semimajor axis and the semiminor axis of ellipse can be described by (7). At the same time, the coordinates of the center of the roller can be calculated via (8) when the roller moves along the circle around its axis:

$$a = r,$$
$$b = r\cos\beta, \quad (7)$$

$$x_c = 0,$$
$$y_c = y_0 - R\sin\beta. \quad (8)$$

Assume that an external load applied to the system is τ and reaction forces between the roller and the contact part are generated to equilibrate the system. The relationship between τ and F_y can be obtained via

$$\tau = F_y R. \quad (9)$$

3.2. Design Process of the Nonlinear Stiffness Elastic Element Using MATLAB Software. For a predefined deflection-torque relationship of "low load, low stiffness and high load, high stiffness" which cannot be described using a mathematical function, a numerical method was proposed to solve the adequate curve profile by using MATLAB software. The following part shows the procedure of process for solution of the following available point on the curve profile of the contact part:

Initialization

For $i = 1$: totalNum (total step numbers)

Obtain rotational angel and torque
Resolve equations using *fsolve* function
Get next contact point (m_i, n_i) and save it
$m_{i-1} = m_i, n_{i-1} = n_i$;

End

An interval (iteration step) which affects the accuracy of curve profile was chosen firstly to obtain the number of steps. According to (5), the initial point must be determined. The coordinates of contact point before deformation would be also derived by the rotational angle, torque, and the previous point. When the iteration was over, the resolved points were saved to draw the curve profile.

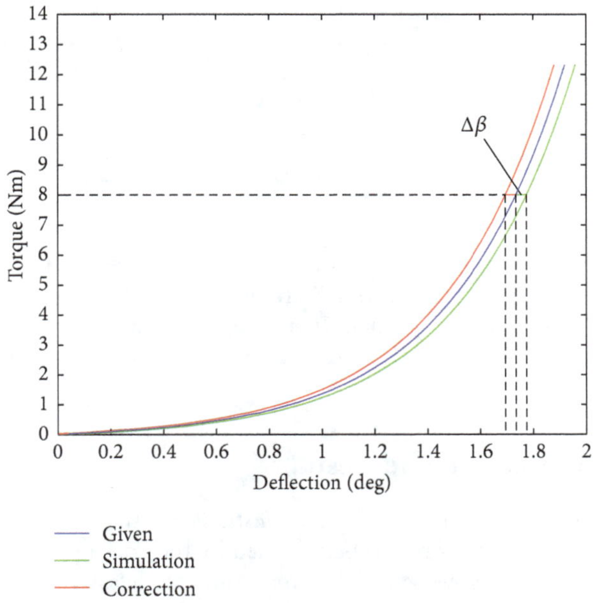

FIGURE 3: Errors analysis of theoretical calculation and simulation results.

3.3. A Method of Error Correction to Compensate the Error of Obtained Torque-Deflection. Since the combination of a cantilever and contact part with a curve profile was used as elastic element of nonlinear elastic system, it is inevitable to refer to the elastic deflection calculation of variable cross section. However, it is difficult to describe this deflection precisely with some algebraic equations.

A simple resolution was proposed to describe the relationship of the deflection and torque in this case. Firstly, analyze the deflection and deflection angle of the cantilever while assuming the contact part is rigid body so that the preliminary shape of the curve profile of the contact part could be derived as mentioned above. According to the simulation results, the derived trajectory of torque-deflection is always a little larger than that derived via theory calculation where the same torque was set. It is not difficult to analyze the reason that the contact part also deflects tinily in horizontal and vertical directions. So some errors should be generated via calculating the geometric equations. Meanwhile, the coordinates of the curve profile of undeformed elastic element were derived via those of deformed elastic element, which makes the derived torque-deflection trajectory of simulation always lower than that derived from the theoretical calculation.

Errors analysis of theoretical calculation and simulation results is shown in Figure 3. The blue curve shows the desired relationship between the torque and deflection and the green curve shows the simulation results. From this figure, it is easy to find that the simulation results are always lower compared to desired trajectory. From different view of this relationship, it would be perfect that the desired trajectory is green curve. So if this curve was set as the desired one, a pseudo desired trajectory should be given firstly such as the desired blue curve for obtaining the green simulated trajectory. Due to the tiny deflection between the desired trajectory and simulation trajectory, there exists linear relationship between the errors

Table 1: The result of optimization.

Design variable	Optimized value (mm)
l	24.4
w	3.4
h	4.9

of pseudo desired trajectory and the errors of desired trajectory. The identical errors $\Delta\beta$ are determined to get the red pseudo desired trajectory. Therefore, the simulation results are likely close to the desired trajectory which is shown as blue curve.

4. Simulation and Results

4.1. Optimization Design of the Elastic Part. The stiffness of compliant actuators can be obtained by the curve profile of contact part. However, the inherent stiffness of elastic part has an effect on the range of the stiffness of compliant actuator. It is essential to determine the inherent stiffness firstly. When the inherent stiffness of elastic part is high, the range of the compliant actuator is large. However, the high stiffness generally induces large stress in the elastic element owing to the fact that the inherent stiffness is not fully involved. The inherent stiffness of the elastic part depends on its dimension, and the shape is necessary to be optimized to acquire the appropriate stiffness.

Take initial inherent stiffness of the extreme of the elastic part as optimal object. Its value is higher than the required value which can be calculated via the desired torque-deflection trajectory. The design and optimization of details of contact part were implemented. When the derived shape of the elastic part is available in terms of allowable stress and assembly conditions, the inherent stiffness would be determined. Otherwise, the next inherent stiffness and optimization process would be repeated for the available shape of the elastic part.

Three groups of elastic elements are installed on a plane and each bears the maximum forces 12 Nm within 1.9 degrees in our proposed compliant robotic joint. The dimensions of the cantilever are design variables and the stress is the status variable. The chosen material is 50 CrVA (elasticity modulus 2.06 Gpa, poisson ratio: 0.29, and yield strength: 1.32 Gpa).

To obtain the optimized structure of the elastic element, the genetic algorithm was used under the constraint conditions of design variables and status variables by using the commercial software ANSYS. The assembly constraint conditions in our proposed structure are 23 mm < length < 25 mm, 3 mm < width < 3.5 mm and 3 mm < height < 5 mm, and stress < 900 Mpa. The obtained stiffness of the extreme of the elastic part is 1400 N/mm. Table 1 shows the details of dimensions.

4.2. The Thickness of the Contact Part. The thickness of the contact part is related to the simulation results of nonlinear stiffness trajectory. Figure 4 shows the thickness of contact part. The thicker the contact part is, the more accurate the

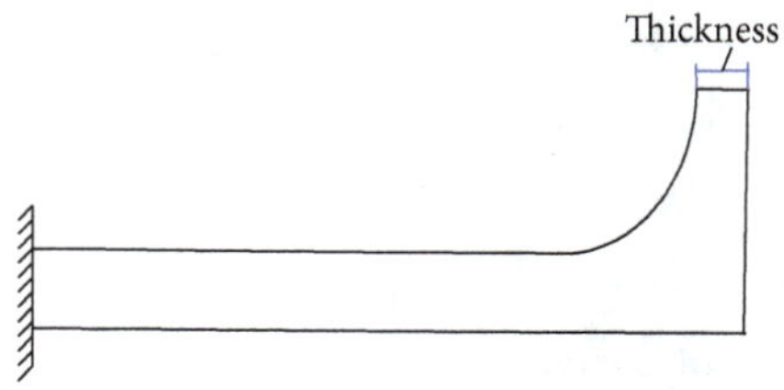

Figure 4: The thickness of the contact part in the elastic element.

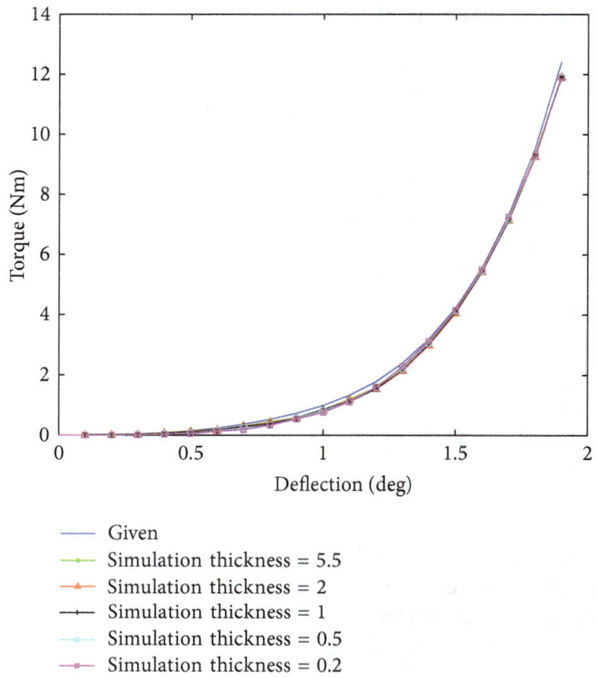

Figure 5: The torque-deflection trajectory with different thicknesses.

simulation results of nonlinear stiffness would be, because the deflection of the contact part behaves more approximately to a rigid body. It is also the indicator that whether it is necessary to consider the deflection of contact part during designing such a nonlinear stiffness elastic mechanism. The simulation results are shown in Figure 5, which includes the variation of thickness along x-axis from 0.2 mm to 5.5 mm. From this figure, there is tiny difference for different thicknesses of contact part in the elastic element, which indicates that it is not necessary to consider the influence of the thickness of contact part.

4.3. Error Compensation Using Pseudo Desired Trajectory. For the circumstance of high precision requirement to stiffness characteristic, the error compensation has to be considered, which is induced by the assumption of rigid body of contact part. The method of error compensation is mentioned in Section 3.3, and a pseudo desired trajectory is proposed. Firstly, the uncompensated torque-deflection trajectory is shown in Figure 6. The desired torque-deflection trajectory is exponential format following $\tau = (3/20)(10)^{\beta} - 3/20$, which is shown in blue curve. The pink triangle points are

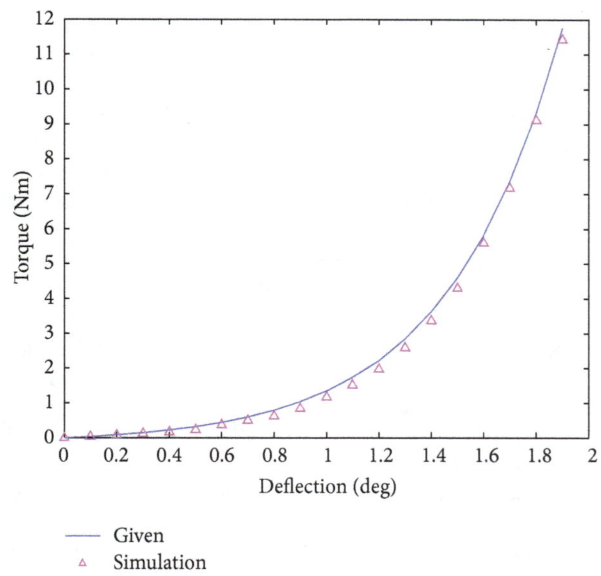

FIGURE 6: Uncompensated torque-deflection trajectory.

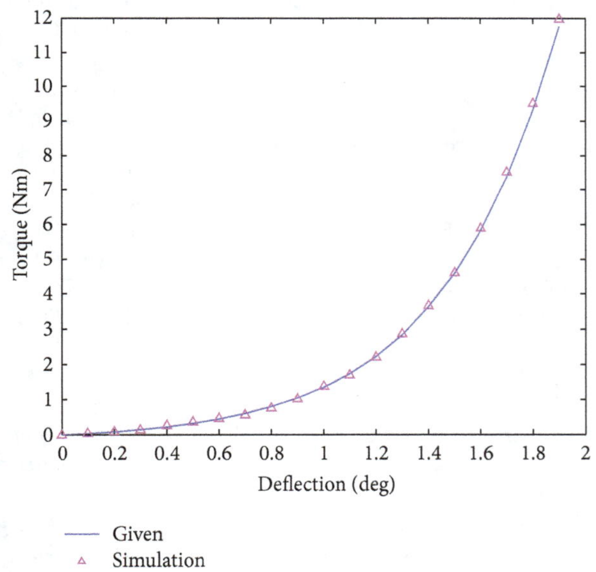

FIGURE 7: Compensated torque-deflection trajectory.

the simulation results derived via ANSYS software. From this figure, the simulation results are regularly lower than the desired trajectory which verifies previous reason analysis of error generation.

The compensated torque-deflection trajectory with the same desired trajectory as Figure 6 is shown in Figure 7, which adapts the pseudo desired trajectory. The simulation results show that compensated torque-deflection trajectory is closer to the desired trajectory than uncompensated one. For comparison between the simulation result using the error compensation method and those not using the error

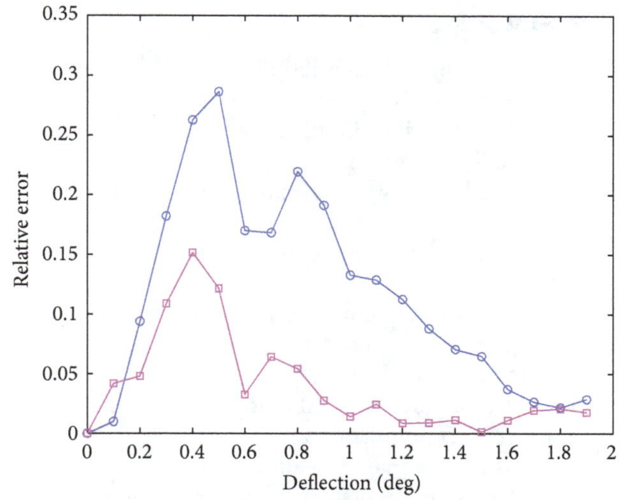

FIGURE 8: Relative errors using and not using compensation method.

compensation method, relative errors are calculated via the following equation:

$$e_r = \frac{|\tau_s - \tau|}{\tau}, \tag{10}$$

where τ_s is the torque derived via simulation and τ is the theoretical torque.

Figure 8 shows the blue line is the relative errors without compensation between the simulation results and the desired trajectory and the pink line shows the relative errors using the error compensation method. From this figure, the maximum errors of no compensation reach nearly 0.29. Meanwhile, the errors in the part of small deflection are larger than those in the part of large deflection. With the increasing of torque acting on the elastic element, the relative errors of torque decreased quickly. The relative errors using the error compensation method are lower than those not using the compensation method, and the maximum error derived from the compensation method is 0.15, only half of that without compensation.

5. Conclusions

In this paper, a new design methodology of compliant mechanisms whose stiffness matches the predefined deflection-torque trajectory of "low load, low stiffness and high load and high stiffness" and an error compensation method were proposed. The assumption that the contact part of elastic element was viewed as rigid body is almost effective in getting the predefined nonlinear stiffness trajectory mentioned above. For obtaining more accurate trajectory, an error compensation method was proposed. According to the simulation results, the proposed method is highly effective in decreasing the relative errors. One joint of the rehabilitation robot for shoulder complex developed by ourselves was illustrated. This research achievement will be useful to design new compliant actuators used in bioinspired robotics, rehabilitation robotics, and so on.

Abbreviations

l: The length of cantilever
r: The radius of roller
R: The radius of circular trajectory of the roller movement
a: The length of the semimajor axis of ellipse C
b: The length of the semiminor axis of ellipse C
w: The width of cross section of the cantilever
h: The height of cross section of the cantilever
α: The angle between the slope plane at the contact point and $x$axis
β: The rotational angle of roller motion in terms of the rotatory component where the elastic element was installed
v: Deflection of the end of the elastic part
θ: Deflection angle of the end of the elastic part
m_{i-1}, n_{i-1}: The coordinates of point A before the deformation of the elastic element
m_i, n_i: The coordinates of point B before the deformation of the elastic element
x, y: The coordinates of point B after the deformation of the elastic element
x_c, y_c: The current coordinates of the roller center
x_0, y_0: The initial coordinates of the roller center
F_x, F_y: The component force of contact forces between the roller and the elastic element
τ: The external torque.

Competing Interests

The authors declare that they have no competing interests.

Acknowledgments

The financial support from the Natural Science Foundation of China (Project no. 51475322) and the Programme of Introducing Talents of Discipline to Universities ("111 Program") under Grant no. B16034 and Open Fund of Key Laboratory of Mechanism Theory and Equipment Design of Ministry of Education of China is greatly acknowledged.

References

[1] G. A. Pratt and M. M. Williamson, "Series elastic actuators," in *Proceedings of the IEEE/RSJ International Conference on Intelligent Robots and Systems*, vol. 1, pp. 399–406, August 1995.

[2] J. F. Veneman, R. Ekkelenkamp, R. Kruidhof, F. Van der Helm, and H. Van der Kooij, "Design of a series elastic-and bowden cable-based actuation system for use as torque-actuator in exoskeleton-type training," in *Proceedings of the IEEE 9th International Conference on Rehabilitation Robotics (ICORR '05)*, pp. 496–499, 2005.

[3] C. Lagoda, A. C. Schouten, A. H. A. Stienen, E. E. G. Hekman, and H. Van der Kooij, "Design of an electric series elastic actuated joint for robotic gait rehabilitation training," in *Proceedings of the 3rd IEEE RAS and EMBS International Conference on Biomedical Robotics and Biomechatronics (BioRob '10)*, pp. 21–26, Tokyo, Japan, September 2010.

[4] G. Carpino, D. Accoto, F. Sergi, N. L. Tagliamonte, and E. Guglielmelli, "A novel compact torsional spring for series elastic actuators for assistive wearable robots," *Journal of Mechanical Design, Transactions of the ASME*, vol. 134, no. 12, Article ID 121002, 2012.

[5] G. Grioli, S. Wolf, M. Garabini et al., "Variable stiffness actuators: the user's point of view," *International Journal of Robotics Research*, vol. 34, no. 6, pp. 727–743, 2015.

[6] V. R. Ham, T. G. Sugar, B. Vanderborght, K. W. Hollander, and D. Lefeber, "Compliant actuator designs: review of actuators with passive adjustable compliance/controllable stiffness for robotic applications," *IEEE Robotics and Automation Magazine*, vol. 16, no. 3, pp. 81–94, 2009.

[7] H. Vallery, J. Veneman, E. van Asseldonk, R. Ekkelenkamp, M. Buss, and H. van Der Kooij, "Compliant actuation of rehabilitation robots," *IEEE Robotics & Automation Magazine*, vol. 15, no. 3, pp. 60–69, 2008.

[8] B. Vanderborght, A. Albu-Schaeffer, A. Bicchi et al., "Variable impedance actuators: a review," *Robotics and Autonomous Systems*, vol. 61, no. 12, pp. 1601–1614, 2013.

[9] A. Jafari, N. G. Tsagarakis, and D. G. Caldwell, "A novel intrinsically energy efficient actuator with adjustable stiffness (AwAS)," *IEEE/ASME Transactions on Mechatronics*, vol. 18, no. 1, pp. 355–365, 2013.

[10] A. Jafari, N. G. Tsagarakis, and D. G. Caldwell, "AwAS-II: a new actuator with adjustable stiffness based on the novel principle of adaptable pivot point and variable lever ratio," in *Proceedings of the IEEE International Conference on Robotics and Automation (ICRA '11)*, pp. 4638–4643, Shanghai, China, May 2011.

[11] N. G. Tsagarakis, I. Sardellitti, and D. G. Caldwell, "A new variable stiffness actuator (CompAct-VSA): design and modelling," in *Proceedings of the 2011 IEEE/RSJ International Conference on Intelligent Robots and Systems: Celebrating 50 Years of Robotics (IROS '11)*, pp. 378–383, IEEE, San Francisco, Calif, USA, September 2011.

[12] C. Pew and G. K. Klute, "Design of lower limb prosthesis transverse plane adaptor with variable stiffness," *Journal of Medical Devices, Transactions of the ASME*, vol. 9, no. 3, Article ID 035001, 2015.

[13] S. Wolf, O. Eiberger, and G. Hirzinger, "The DLR FSJ: energy based design of a variable stiffness joint," in *Proceedings of the IEEE International Conference on Robotics and Automation (ICRA '11)*, pp. 5082–5089, May 2011.

[14] W. Friedl, H. Höppner, F. Petit, and G. Hirzinger, "Wrist and forearm rotation of the DLR hand arm system: mechanical design, shape analysis and experimental validation," in *Proceedings of the IEEE/RSJ International Conference on Intelligent Robots and Systems (IROS '11)*, pp. 1836–1842, San Francisco, Calif, USA, September 2011.

[15] R. R. Torrealba and S. B. Udelman, "Design of cam shape for maximum stiffness variability on a novel compliant actuator using differential evolution," *Mechanism and Machine Theory*, vol. 95, pp. 114–124, 2016.

[16] K. W. Hollander, T. G. Sugar, and D. E. Herring, "Adjustable robotic tendon using a 'jack spring'," in *Proceedings of the IEEE 9th International Conference on Rehabilitation Robotics (ICORR '05)*, pp. 113–118, July 2005.

[17] N. L. Tagliamonte, F. Sergi, D. Accoto, G. Carpino, and E. Guglielmelli, "Double actuation architectures for rendering

variable impedance in compliant robots: a review," *Mechatronics*, vol. 22, no. 8, pp. 1187–1203, 2012.

[18] S. A. Migliore, E. A. Brown, and S. P. DeWeerth, "Novel nonlinear elastic actuators for passively controlling robotic joint compliance," *Journal of Mechanical Design*, vol. 129, no. 4, pp. 406–412, 2007.

[19] G. Palli, G. Berselli, C. Melchiorri, and G. Vassura, "Design of a variable stiffness actuator based on flexures," *Journal of Mechanisms and Robotics*, vol. 3, no. 3, Article ID 034501, 2011.

[20] S. Rapoport, J. Mizrahi, E. Kimmel, O. Verbitsky, and E. Isakov, "Constant and variable stiffness and damping of the leg joints in human hopping," *Journal of Biomechanical Engineering*, vol. 125, no. 4, pp. 507–514, 2003.

[21] C. B. Churchill, D. W. Shahan, S. P. Smith, A. C. Keefe, and G. P. McKnight, "Dynamically variable negative stiffness structures," *Science Advances*, vol. 2, no. 2, Article ID e1500778, 2016.

[22] S. Seok, A. Wang, M. Y. Chuah et al., "Design principles for energy-efficient legged locomotion and implementation on the MIT Cheetah robot," *IEEE/ASME Transactions on Mechatronics*, vol. 20, no. 3, pp. 1117–1129, 2015.

[23] A. Omer, R. Ghorbani, K. Hashimoto, H.-O. Lim, and A. Takanishi, "A novel design for adjustable stiffness artificial tendon for the ankle joint of a bipedal robot: modeling & simulation," *Machines*, vol. 4, no. 1, article 1, 2016.

[24] M. Kilic, Y. Yazicioglu, and D. F. Kurtulus, "Synthesis of a torsional spring mechanism with mechanically adjustable stiffness using wrapping cams," *Mechanism and Machine Theory*, vol. 57, pp. 27–39, 2012.

3

Nonlinear Friction and Dynamical Identification for a Robot Manipulator with Improved Cuckoo Search Algorithm

Li Ding,[1,2] Xingcheng Li,[1] Qilin Li,[1] and Yuan Chao[1]

[1]*College of Mechanical Engineering, Jiangsu University of Technology, Changzhou 213001, China*
[2]*Department of Industrial and System Engineering, The Hong Kong Polytechnic University, Kowloon 999077, Hong Kong*

Correspondence should be addressed to Li Ding; nuaadli@163.com

Academic Editor: Yangmin Li

This paper concerns the problem of dynamical identification for an industrial robot manipulator and presents an identification procedure based on an improved cuckoo search algorithm. Firstly, a dynamical model of a 6-DOF industrial serial robot has been derived. And a nonlinear friction model is added to describe the friction characteristic at motion reversal. Secondly, we use a cuckoo search algorithm to identify the unknown parameters. To enhance the performance of the original algorithm, both chaotic operator and emotion operator are employed to help the algorithm jump out of local optimum. Then, the proposed algorithm has been implemented on the first three joints of the ER-16 robot manipulator through an identification experiment. The results show that (1) the proposed algorithm has higher identification accuracy over the cuckoo search algorithm or particle swarm optimization algorithm and (2) compared to linear friction model the nonlinear model can describe the friction characteristic of joints better.

1. Introduction

Modern industry is increasingly oriented towards the production of small batches of a large variety of products, asking for flexibility and automation in the manufacturing systems [1, 2]. Additionally, the increasing quality standards, international competition, and economic reasons put higher requirements on reliability and accuracy and especially on the speed of production processes. In this context, industrial serial robot manipulators have become an indispensable means of automation to increase productivity and flexibility of production units. Robots are programmed by teaching the sequence of the attitude and position which are necessary to execute the desired task. To reach a sufficient accuracy, this teaching is mostly done on-site and relies on a good repeatability, rather than on a good absolute accuracy. To improve the operation accuracy, a precise dynamical model is essential for accurate offline programming.

In the academic aspect, a typical manipulator identification procedure consists of dynamic modelling, excitation trajectory design, data collection, signal preprocess, parameter identification, and model validation [3]. When a priori knowledge is available about the robot system, parametric models can be derived based on the laws of physics and mechanics resulting in a set of differential equations. The unknown dynamical parameters have a physical meaning and can be identified by several approaches. Atkeson et al. [4] used the least square method (LS) to implement the load estimation of dynamical parameters on a PUMA600 robot. According to weighted least squares method (WLS), Gautier and Poignet [5] proposed a dynamical identification approach only from the torque data, without other sensors. Grotjahn et al. [6] used the two-step method to execute the friction and rigid body identification of robot dynamics. Considering the effect of measurement noise, Olsen and Petersen [7] used the maximum likelihood estimating (MLE) method for parameters identification of an industrial robot with a statistical framework. Recently, some novel dynamical identification methods for robot manipulators have been reported using intelligence algorithms. For instance, Bingül and Karahan [8] integrated the particle swarm optimization (PSO) algorithm with LS algorithm to estimate the dynamical parameters of Staubli RX-60 robots. In the identification experiment, the velocity and acceleration are measured by three high-speed cameras, and the joint torques are measured by six load cell sensors. Without velocity and acceleration sensors, Ding et al. [9] used the motor current and joint positions to calculate the torques, velocity, and acceleration of joints and proposed an artificial bee colony algorithm (ABC) to obtain the unknown dynamical parameters. Nevertheless, when handling complicated and high-dimensional parameters identification problems, the flaw that the premature convergence can make those intelligence computation algorithms stuck in a local optimum.

As known, the friction is a major source of disturbances affecting motion quality. Therefore, it must be included as an additional component in robot modelling. In robot identification applications, a model including Coulomb and viscous friction is frequently applied [10, 11]. With such linear model, the parameters estimation is significantly simplified. However, this friction model is not capable of describing the experimentally measured friction characteristic, especially the static model at joint reversal. Aiming at these problems, we add a nonlinear friction model into the dynamical model of a 6-DOF industrial serial manipulator and use an improved cuckoo search (ICS) algorithm for dynamical identification. The idea of the method is to measure the positions and gravitational torques of different joints through designing Fourier series as excitation trajectories. The collected values are used to calculate the dynamical parameters based on ICS. In ICS, considering the outstanding performance of chaotic operator and emotion operator, these improved operators are used to enhance the performance of the classical CS. And the comparison of three different identification methods, CS, PSO, and ICS, illustrated the superiority of our proposed algorithm in the application of dynamical identification.

This article is organized as follows. The dynamical model with a nonlinear friction model of a robot manipulator is given in Section 2. For the unknown dynamical parameters, the improved cuckoo search algorithm is introduced to realize the parameters estimation in Section 3. Then, the design of excitation trajectory, data collection, and preprocessing are presented in Section 4. Later, an ER-16 robot is used as a test platform for identification experiment, and the results are analyzed in Section 5. In addition, the superiority of linear and nonlinear friction model has been compared through the model validation in Section 6. Finally, Section 7 discussed the key findings and prospective research target.

2. Dynamic Modelling

According to the literature [12], a n-DOF serial manipulator is described as a kinematic chain of several rigid bodies. Hence, we can utilize the Newton-Euler method to deduce the dynamical model of the manipulator:

$$\boldsymbol{\tau} = \boldsymbol{\tau}_f + \mathbf{M}(\mathbf{q})\ddot{\mathbf{q}} + \mathbf{C}(\mathbf{q}, \dot{\mathbf{q}}) + \mathbf{G}(\mathbf{q}), \quad (1)$$

where joint torque $\boldsymbol{\tau}$, joint position $\mathbf{q}$, joint velocity $\dot{\mathbf{q}}$, and joint acceleration $\ddot{\mathbf{q}}$ are n-dimensional vector. $\boldsymbol{\tau}_f$ denotes the n-dimensional joint friction vector. $\mathbf{M}(\mathbf{q})$ represents $n \times n$ inertial matrix, $\mathbf{C}(\mathbf{q}, \dot{\mathbf{q}})$ is a n-dimensional vector including Coriolis and centrifugal forces, and $\mathbf{G}(\mathbf{q})$ is n-dimensional gravity vector.

Equation (1) except for friction torques can be rewritten as a linear form if using the modified Newton-Euler parameters [13] or the barycentric parameters [11]:

$$\boldsymbol{\tau}_s = \boldsymbol{\Phi}_s(\mathbf{q}, \dot{\mathbf{q}}, \ddot{\mathbf{q}})\,\boldsymbol{\theta}_s, \quad (2)$$

where $\boldsymbol{\Phi}_s$ only contains the motion data, which can be treated as $n\times 10n$ identification matrix or observation matrix. $\boldsymbol{\theta}_s$ is the barycentric parameter vector. This conversion vastly reduces the complexity of parameters identification. In addition, the dynamical parameters of link i are governed by the form

$$\boldsymbol{\theta}_s^i = \left[I_{xxi}, I_{xyi}, I_{xzi}, I_{yyi}, I_{yzi}, I_{zzi}, m_i r_{xi}, m_i r_{yi}, m_i r_{zi}, m_i\right]^T, \quad (3)$$

where $I_{\zeta\zeta i}$ ($\zeta = x, y, z$) is the inertial tensor of the link i. Similarly, m_i is the mass of the link i and $m_i r_{\zeta i}$ is the inertia moment.

Generally, the identification matrix $\boldsymbol{\Phi}_s$ in (2) is not full of rank, that is, not all dynamical parameters give contribution to the joint torques. In the literature [14], some methods like case-by-case analysis or singular value decomposition are adopted to eliminate the redundant parameters. And the barycentric parameter vector $\boldsymbol{\theta}_s$ can be replaced by a vector of minimal barycentric parameters $\boldsymbol{\theta} \in \mathbf{R}^p$ with $p < 13n$. Hence, (2) can be transformed into another form

$$\boldsymbol{\tau} = \boldsymbol{\Phi}(\mathbf{q}, \dot{\mathbf{q}}, \ddot{\mathbf{q}})\,\boldsymbol{\theta}, \quad (4)$$

where $\boldsymbol{\Phi}$ denotes the $n\times p$ observation or identification matrix and p is the number of minimal barycentric parameters.

Except for the dynamical parameters in (2), there also exists friction torques and extra torques caused by inertias of motor rotors. In general, the inertias of motor rotors are provided by manufacturers, and corresponding torques can easily be compensated to the dynamical model. As for joint friction model, it is regarded as a complex nonlinear model. To simplify the model, the Coulomb and viscous friction were used to describe the friction model. But the researchers found [15] that the friction torques of some joints exceeded their full speed range, and the simple friction model could not cover the characteristics, especially at motion reversal. A better description of the joint friction characteristics may be based on the following nonlinear equations:

$$\boldsymbol{\tau}_f = f_0 + f_c \operatorname{sign}(\dot{\mathbf{q}}_i) + f_v \dot{\mathbf{q}}_i + f_a \operatorname{atan}(f_b \dot{\mathbf{q}}_i), \quad (5)$$

where f_0 is the zero drift error of friction torque, f_c is the Coulomb friction coefficient, f_v is the viscous friction coefficient, and f_a and f_b are the experiential friction coefficients. It should be noted that this model has a discontinuity at zero velocity.

In summary, using the nonlinear friction model yields the whole dynamical model of the 6-DOF robot as

$$\boldsymbol{\tau} = \boldsymbol{\Phi}(\mathbf{q}, \dot{\mathbf{q}}, \ddot{\mathbf{q}})\,\boldsymbol{\theta} + \boldsymbol{\tau}_f, \quad (6)$$

where $\boldsymbol{\tau}_f$ denotes the $5n$ friction torques vector and $5n$ is the number of friction coefficients. Obviously, the classical least square method could not solve the above nonlinear equation. Hence, applying an intelligence algorithm for solving this problem may be a feasible method.

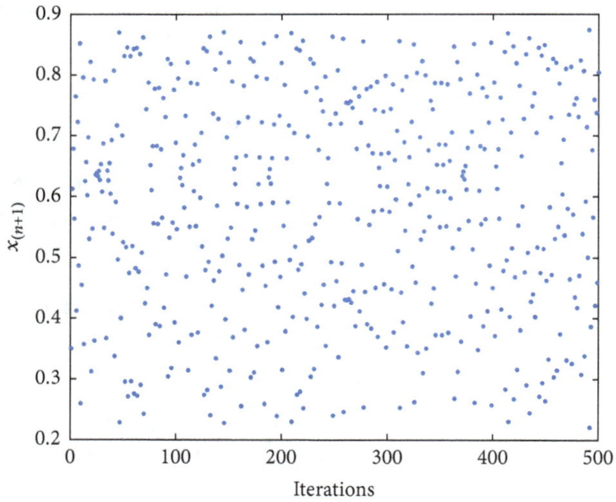

FIGURE 1: A tent map.

3. Identification Algorithm

3.1. Introduction to Chaos Theory. Chaos theory is epitomised by the so-called "buttery" detailed by Lorenz [16]. He discovered that tiny changes in an initial state would make a radically different final result and typically rendering long-term prediction impossible. Chaotic map has ergodic and stochastic properties, which is regarded as a bounded nonlinear system with deterministic dynamical behavior. Moreover, it has a very sensitive dependence on initial conditions. There are many forms about chaotic map, such as tent map, Gauss map, logistic map, and tent map [17]. Considering the high robustness and stability of tent map, we choose it to generate the chaotic sequence. The description of the tent map is written as

$$x_{n+1} = \begin{cases} 2\gamma x_n & 0 \le x_n \le 0.5 \\ 2\gamma(1 - x_n) & 0.5 < x_n \le 1, \end{cases} \quad (7)$$

where n is iterations and γ denotes a positive number. The value of x_n is updated with the initial condition $x_n \in (0, 1)$. A tent map for $\gamma = 0.65$ after 500 iterations is shown in Figure 1. As the value of n increases, x_{n+1} gets a new value. In this paper, the parameter γ is chosen as 0.6 after many trials.

3.2. Basic Principles of CS. The Cuckoo search algorithm (CS) is a new metaheuristic algorithm formulated by Professor Yang and Deb in 2009 [18]. It is a swarm intelligence type of algorithm inspired by the behaviors of cuckoos, consisting of locating nests, laying eggs, and the Lévy flight. With a simple structure and a few control parameters, CS has been widely used in practical engineering optimization problems. There are three basic rules in CS. (1) Each cuckoo chooses a random nest to lay an egg once. (2) Only the best nests with a high quality of eggs would be retained to the next generations. (3) A host cuckoo has a probability of $P_a \in [0, 1]$ to detect the egg of another cuckoo in its own nest. Through the switching coefficient P_a, CS can subtly combine a global search and a local search.

The local search can be described as the following formulation, which gains an insight into CS search mechanisms:

$$x_i^{t+1} = x_i^t + \alpha s \otimes H(P_a - \varepsilon) \otimes \left(x_j^t - x_k^t\right), \quad (8)$$

where x_j^t and x_k^t are two different random solutions, $H(\cdot)$ represents a Heaviside function, the symbol "⊗" represents point-to-point multiplication, ε is a random parameter which satisfied a uniform distribution, s is step size, and α is a scaling coefficient over zero.

For the global search, the process is controlled by a Lévy flight behavior. The mathematical formula for this behavior is described as follows:

$$\begin{aligned} x_i^{t+1} &= x_i^t + \alpha L(s, \lambda) \\ L(s, \lambda) &= \frac{\lambda \Gamma(\lambda) \sin(\pi\lambda/2)}{\pi} \frac{1}{s^{1+\lambda}}, \end{aligned} \quad (9)$$

where $L(s, \lambda)$ is a Lévy flight function that complicates integration, λ represents the power coefficient, and $\Gamma(\lambda) = \int_0^\infty e^{-t} t^{\lambda-1} dt$.

The basic steps of CS are presented as follows.

Step 1 (parameters initialization). The initial positions of nests are randomly generated from

$$x_{ij} = lb_j + \text{rand}(0, 1)\left(ub_j - lb_j\right), \quad (10)$$

where $i \in \{1, \ldots, FN\}$, $j \in \{1, \ldots, D\}$, FN is the position size, D is the dimensionality of unknown dynamical parameters, and lb_j and ub_j are the lower and upper limits in the solution space.

Step 2 (objective value calculated). In this paper, CS searched for the dynamical identification by minimizing the objective function written as follows:

$$\begin{aligned} F_i = \frac{1}{N} \sum_{i=1}^{N} \Big(& \delta_1 \left\| \tau_{1i} - \tau_{p1i} \right\| + \delta_2 \left\| \tau_{2i} - \tau_{p2i} \right\| \\ & + \delta_3 \left\| \tau_{3i} - \tau_{p3i} \right\| \Big), \end{aligned} \quad (11)$$

where F_i is the objective function, N is the data length, and $\tau_{\xi i}$ $(\xi = 1, 2, 3)$ denotes joint torques measured from the first three joints. Similarly, $\tau_{p\xi i}$ $(\xi = 1, 2, 3)$ denotes predicted torques calculated by the identified model. δ_ξ $(\xi = 1, 2, 3)$ is a weight coefficient between 0 and 1.

Step 3. Through (8)-(9), the nest positions can be updated. After generating a group of new nest positions, the superior positions are retained to the next generation by comparing the objective value between the original positions and the postupdate nest positions.

Step 4. A random number of uniform distributions $r \in [0, 1]$ is compared with P_a. If $r > P_a$, then the located nest is changed randomly. Otherwise, the nest remains unchanged.

Step 5. The termination condition of the CS is determined. If the condition is satisfied, then the optimal solution becomes an output. Otherwise, Step 3 is repeated.

3.3. Improved Operator. In general, the parameter α in Lévy flight is the key factor to affect the convergence of CS. Due to the infinite variance and mean, the classical CS algorithm may have a premature search process. To overcome this problem, the tent map is used to generate a chaotic sequence for parameter α. In that case, the algorithm searches the new position in the neighborhood of the current optimal position. Meanwhile, a new emotional acceptance criterion is used to prevent the algorithm from getting trapped into local optima.

In the ICS, two cuckoos' emotions (positive and negative) can be described as follows:

$$e = -k \ln \left| \frac{S\{F(x_i) - F(x_j)\}}{S_0} \right| \tag{12}$$

$$\text{IF } \alpha < e \text{ Then positive Else negative,}$$

where e represents the function of cuckoo's emotion, k is a constant, S is the stimulus function, S_0 is a stimulus threshold, and F is the objective function. It should be noted that k is selected as 1 and $S = e^x$.

4. Design of Exciting Trajectories and Data Preprocess

It is essential to consider these conditions before designing an identification experiment, that is, (1) whether the excitation trajectory is sufficient to provide fast and accurate parameters estimation and (2) whether the processing of the experimental data is simple and yields stable and accurate results. In fact, the imprecise modelling and measurement noise are assumed to be an additive frequently distributed zero-mean stochastic disturbance, causing bias errors and uncertainty in the dynamical identification. To reduce the effect of the disturbance, an appropriate excitation trajectory must be designed carefully.

The choice of parameterization for the excitation trajectory is a very important issue. It directly determines the number of parameters in the optimization problem and the effort needed to calculate velocity and acceleration from the joints positions measurements. The first considerations on finding excitation trajectory for the dynamical identification of manipulator were proposed by Armstrong [19]. Then, Grotjahn and Daemi [20] proposed an interpolated trajectory consisting of two parts, part I overcame the given boundary conditions and part II overcame the homogeneous boundary conditions. Furthermore, Gautier [21] used fifth-order polynomials to obtain smooth joint trajectories to be executed by the manipulator, and the polynomial coefficients are fixed by imposing continuity constraints between the trajectory segments. In this paper, we adopt a finite Fourier series which was proposed by Swevers et al. [22] as excitation trajectories. The trajectory for joint i of a manipulator is designed as

$$q_i(t) = q_{i,0} + \sum_{k=1}^{N} a_{i,k} \sin(k\omega_f t) + \sum_{k=1}^{N} b_{i,k} \cos(k\omega_f t), \tag{13}$$

where $q_{i,0}$ is the offset term and ω_f is the fundamental pulsation of the Fourier series. This Fourier series specifies a periodic function with period $T_f = 2\pi/\omega_f$. Each Fourier series contains $2N + 1$ parameters, and $a_{i,k}$, $b_{i,k}$ are the amplitudes of the trigonometric functions. It should be noted that although this trajectory is hard to implement on an industrial setup, this parameterization has several advantages. It is possible to calculate the velocity and acceleration in the frequency domain, which avoids phase distortions.

The antinoise ability and convergence speed of an identification experiment are related to the constraints of the excitation trajectory. It should be noted that the configurations of measurements must correspond to a good conditional simplified identification matrix, since the corresponding input/output represents some limitations. In literature [23], the constraints of an excitation trajectory are given as

$$\begin{gathered} \min \operatorname{cond}(\mathbf{\Phi}) \\ q_{\min} \le q(\beta) \le q_{\max}, \\ |\dot{q}(\beta)| \le \dot{q}_{\max} \\ |\ddot{q}(\beta)| \le \ddot{q}_{\max} \\ w(q(\beta)) \subset W_o \\ \boldsymbol{\tau}_{\min} \le \mathbf{\Phi}(q(\beta), \dot{q}(\beta), \ddot{q}(\beta))\theta \le \boldsymbol{\tau}_{\max}, \end{gathered} \tag{14}$$

where $q_{\min}$ and $q_{\max}$ are the lower and upper of the joint positions, $\dot{q}_{\max}$ and $\ddot{q}_{\max}$ are the upper of velocity and acceleration, β is optimal trajectory parameters, W_o is the available workspace of robot, and $\boldsymbol{\tau}_{\max}$ is the maximum joint torque.

When the robot joints repeatedly track the excitation trajectories with the PID controllers, motor current and joints positions can be sampled in the time domain. The motor current can be transformed into joint torques with a simple torque constant. Since there are measurement noises in experiments, it is necessary to preprocess the collection data before identification. In order to remove outliers and attenuate the effect of interference signal, a five-point triple smoothing method is adopted to smooth the raw data according to the following equation [24]:

$$\begin{aligned} \overline{y}_1 &= \frac{1}{70}[69y_1 + 4(y_2 + y_4) - 6y_3 - y_5] \\ \overline{y}_2 &= \frac{1}{35}[2(y_1 + y_5) + 27y_2 + 12y_3 - 8y_4] \\ \overline{y}_i &= \frac{1}{35}[-3(y_{i-2} + y_{i+2}) + 12(y_{i-1} + y_{i+1}) + 17y_i] \\ \overline{y}_{m-1} &= \frac{1}{35}[2(y_{m-4} + y_m) - 8y_{m-3} + 12y_{m-2} + 27y_{m-1}] \\ \overline{y}_m &= \frac{1}{70}[-y_{m-4} + 4(y_{m-3} + y_{m-1}) - 6y_{m-2} + 69y_m], \end{aligned} \tag{15}$$

where $i = 3, \ldots, m-2$, $Y = [y_1, y_2, \ldots, y_m]$ is the sampled raw data, and $\overline{Y} = [\overline{y}_1, \ldots, \overline{y}_m]^T$ is the data for identification after preprocessing. The more the number of using (15) is, the smoother the curves will be. It should be noted that excessive use (15) to smooth the raw data can lead to the error of the parameters identification enlarging.

Due to the shortage of speed sensor and acceleration sensor, the velocity and acceleration of joints cannot be measured directly. The calculation of the joint velocities and acceleration can be performed by analytical differentiation of the measured angles [25]. The velocity and acceleration are given as follows:

$$\dot{q}_i(t) = \sum_{k=1}^{N} a_{i,k} k\omega_f \cos(k\omega_f t) - \sum_{k=1}^{N} b_{i,k} k\omega_f \sin(k\omega_f t)$$

$$\ddot{q}_i(t) = \sum_{k=1}^{N} - a_{i,k} k^2 \omega_f^{\,2} \sin(k\omega_f t) - \sum_{k=1}^{N} b_{i,k} k^2 \omega_f^{\,2} \cos(k\omega_f t), \tag{16}$$

For the analytical differentiation, the measured encoder readings are first approximated with the least square method, as a finite sum of trigonometric functions. This approximation corresponds to frequency domain differentiation combined with frequency domain windowing. Firstly, the discrete Fourier transform of the preprocessing encoder readings is calculated and the excited frequency curves are selected by frequency domain windowing. Then, the selected curves are multiplied with the frequency response of a pure single and double differentiator, that is, multiplied with $j\omega$ and $-\omega^2$, with ω being the frequency in radians per second. The obtained frequency spectra are then transformed back into time domain using the inverse discrete Fourier transform, yielding joint velocity and acceleration. The application of a Fourier series gives a trajectory which is continuously differentiable up to any order. And it can avoid the excitation of unknown dynamic effects. It is clear that the finite Fourier series has several advantages over the classical excitation trajectories.

5. Parameters Identification

To test the effectiveness and viability of our proposed method, the identification procedure was implemented on the first three joints of the ER-16 6-DOF industrial robot manipulator without payload, as shown in Figure 2. It should be noted that the values of the dynamical parameters of the first three joints are much bigger than those of the other three joints. It is reasonable to ignore the effect of the torques caused by the 4~6 joints [26]. The motion constraints of the ER-16 are given in Table 1.

We use 5th order Fourier series as the excitation trajectories in the experiment. The fundament pulsation is 0.05 Hz, resulting in a period of 25 s. The data is sampled with 1 Hz. The excitation trajectories with 0.25 Hz bandwidth are shown in Figure 3, containing 11 optimal parameters in each joint. These parameters are listed in Appendix A. Figure 4 shows the trajectory of the effect center point of the ER-16 in the workspace.

TABLE 1: Motion constraints of robot.

Parameters	Joint	Min	Max
	1	−180	180
$q/(°)$	2	−60	60
	3	−170	80
	1	−145	145
$\dot{q}/(°\cdot s^{-1})$	2	−105	105
	3	−170	170
	1	−50	50
$\ddot{q}/(°\cdot s^{-2})$	2	−40	40
	3	−65	65
	1	-	1200
$\tau/(N\cdot m)$	2	-	1000
	3	-	600

FIGURE 2: ER-16 6-DOF robot manipulator.

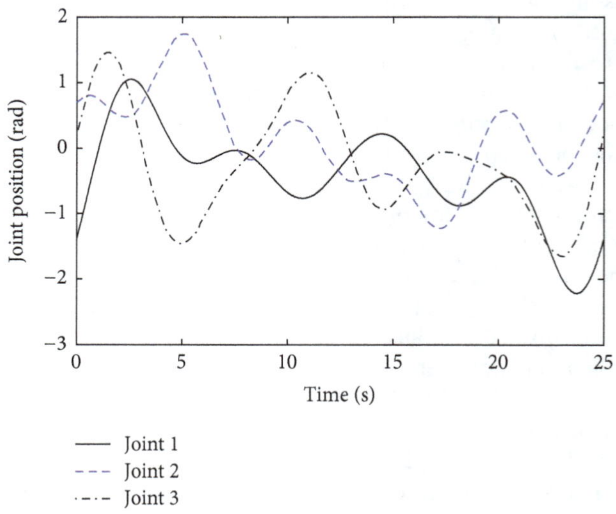

FIGURE 3: Optimized exciting trajectories in joint space.

Identification procedures discussed above are carried out with ICS algorithm in MATLAB 2016b programming environment on an Intel Core i7-3770 PC running Windows 7. No commercial tools are used. In order to test the performance of the proposed algorithm, the classical CS algorithm and PSO algorithm [27] are employed as the comparison. The initial parameters of the three algorithms are listed in Table 2. All the algorithms are run three times and the maximal iteration

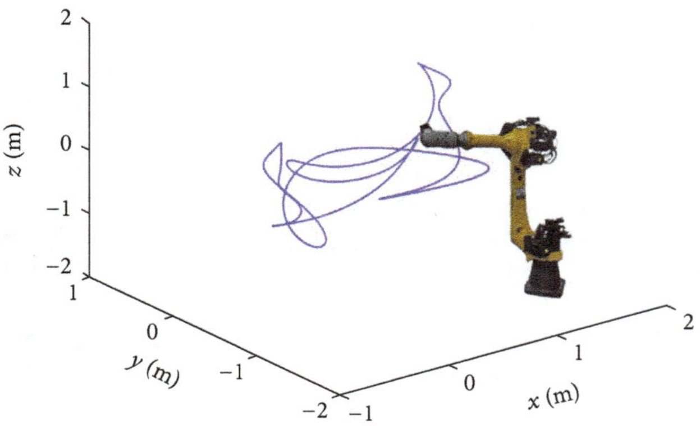

FIGURE 4: 3D excitation trajectory in work space.

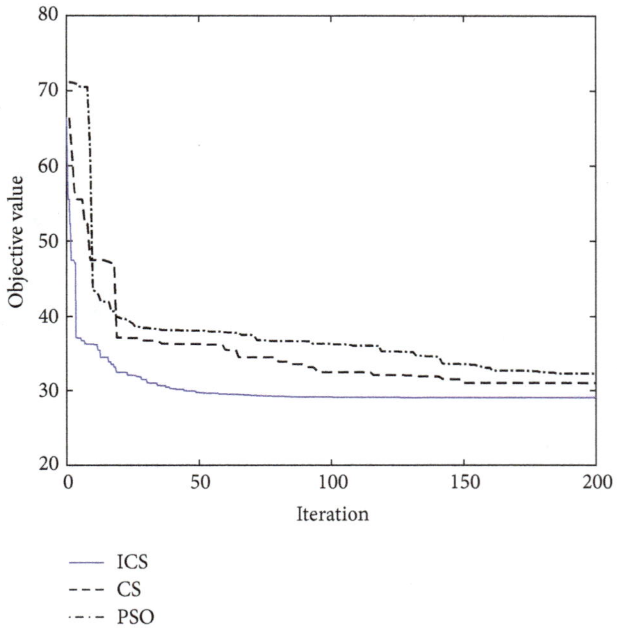

FIGURE 5: Evolutionary curves of ICS.

is 200. The best results would be retained. Theoretically, those results are large probability global optimum. Additionally, the search scope of the unknown parameters is also listed in Appendix B.

Figure 5 shows the evolution curves of the above three algorithms regarding (11). The figure demonstrates that the objective value increases as the generation iterates with time, gradually converging to an optimal result. Compared with CS and PSO, ICS achieves a better result with the higher objective value after 50 iterations. The optimal objective value of ICS is 29.0753, whereas those of CS and PSO are 30.2513 and 31.8865, respectively. Obviously, ICS has found the optimum with the objective value equal to 29.0753 which is a 4.04% or 9.67% increase with respect to the CS or PSO. It can be seen from the result that the chaotic operator can not only avoid the traditional CS being trapped in a local optimum but also improve its robustness and efficiency.

The dynamical model of the first three joints contains 15 barycentric parameters and 15 friction parameters. These parameters identified by our proposed algorithm are listed in Table 3.

Figure 6 compares the measured torques for the excitation trajectory with the predicted torques based on the identified dynamical parameters. The results show that the identified data by applying the three algorithms have the same trend as the measured data. Nonetheless, it turns out that the predicted torques generated by ICS approximate the actual test torques best.

TABLE 2: The initial parameters of the three algorithms.

Algorithm	Parameters
CS	Population size = 25, $P_a = 0.4$, $\alpha = 0.25$, $\lambda = 1.5$
ICS	Population size = 25, $P_a = 0.4$, $\alpha = 0.25$, $\lambda = 1.5$
PSO	Population size = 25, inertia weight = 0.5, acceleration coefficients = 0.2

TABLE 3: Dynamical parameters of ER-16.

Parameter	Value
I_{zz1} (kg·m^2)	51.3846
I_{xx2} (kg·m^2)	−24.4871
I_{xy2} (kg·m^2)	1.1452
I_{xz2} (kg·m^2)	−0.3699
I_{yz2} (kg·m^2)	0.4574
I_{zz2} (kg·m^2)	44.5863
m_2r_{x2} (kg·m)	32.1287
m_2r_{y2} (kg·m)	5.4040
I_{xx3} (kg·m^2)	3.1263
I_{xy3} (kg·m^2)	1.9361
I_{xz3} (kg·m^2)	0.4849
I_{yz3} (kg·m^2)	0.6900
I_{zz3} (kg·m^2)	11.7725
m_3r_{x3} (kg·m)	5.1317
m_3r_{y3} (kg·m)	10.3145
f_{01} (N·m)	0.7044
f_{c1} (N·m)	3.9877
f_{v1} (Nm·s/rad)	8.0000
f_{a1} (N·m)	32.3929
f_{b1}	6.9975
f_{02} (N·m)	53.2922
f_{c2} (N·m)	−2.8909
f_{v2} (Nm·s/rad)	−70.0174
f_{a2} (N·m)	70.0243
f_{b2}	19.7363
f_{03} (N·m)	−3.4499
f_{c3} (N·m)	30.6465
f_{v3} (Nm·s/rad)	4.3297
f_{a3} (N·m)	−2.7533
f_{b3}	5.0000

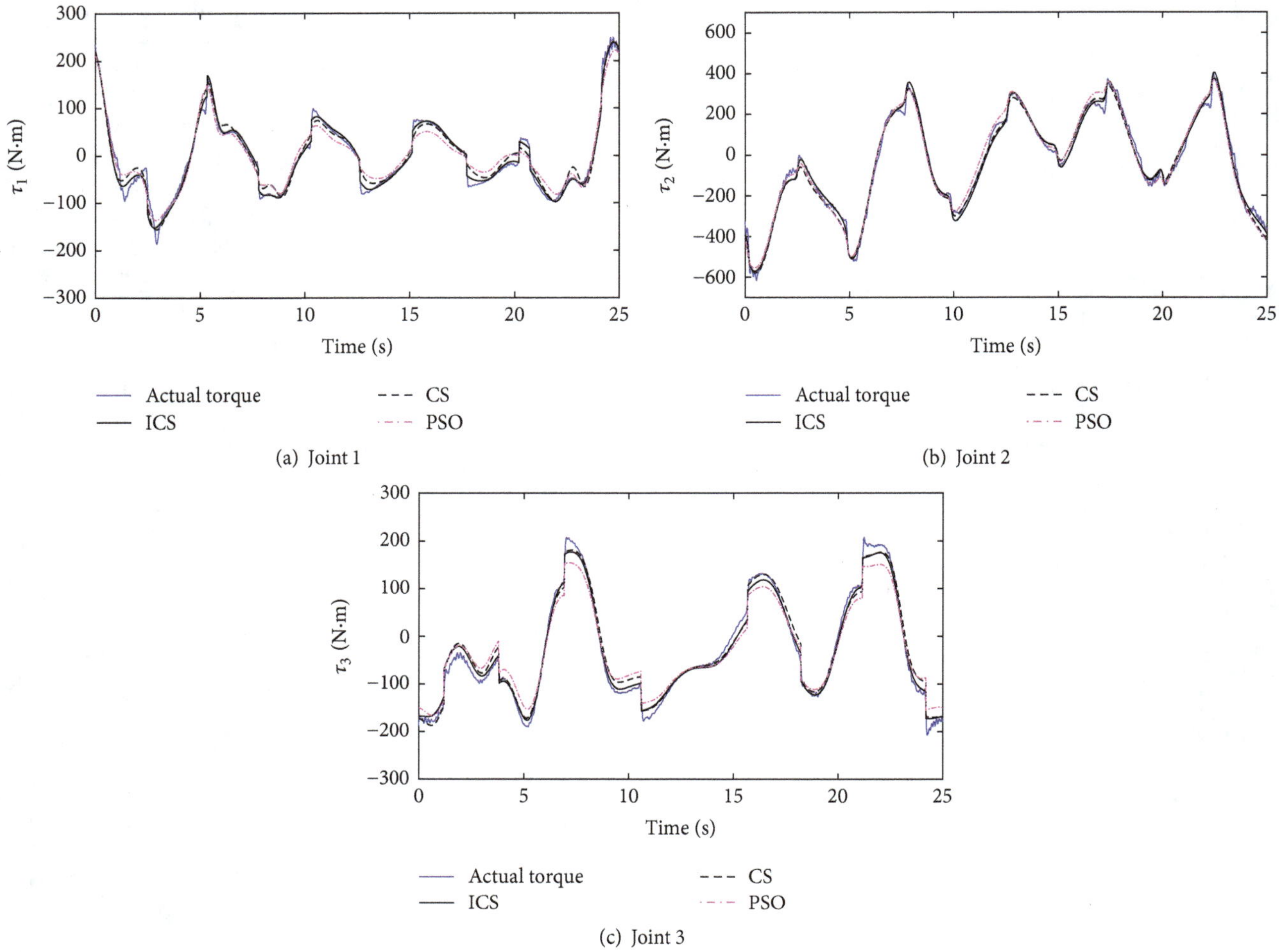

FIGURE 6: Comparison of the measured torques and predicted torques.

To verify the precision of the identified model by the above three algorithms, the correlation coefficient between the measured torques τ_i and predicted torques τ_{mi}, defined as the normalized cross-covariance function, is used to estimate how well the identified model can reproduce the measured torques, and the function is defined as

$$\rho = \frac{\sum_{i=1}^{N} (\tau_i - \overline{\tau}) (\tau_{mi} - \overline{\tau}_m)}{\sqrt{\sum_{i=1}^{N} (\tau_i - \overline{\tau})^2 \sum_{i=1}^{N} (\tau_{mi} - \overline{\tau}_m)^2}}, \tag{17}$$

where $\overline{\tau} = (1/N) \sum_{i=1}^{N} \tau_i$ and $\overline{\tau}_m = (1/N) \sum_{i=1}^{N} \overline{\tau}_{mi}$. The closer the correlation coefficient is to unity, the better the identified model is. While the coefficient is close to zero, the identified model is poor. From Table 4, we can clearly see that the correlation coefficient of the identification results produced by ICS is better than those of other algorithms. It indicates that the identified procedure based on ICS has a higher identification precision due to the strong search ability of ICS.

TABLE 4: Comparison of the correlation coefficient.

Joint	PSO	CS	ICS
1	97.22%	98.03%	99.07%
2	99.10%	99.27%	99.50%
3	99.18%	99.28%	99.67%

6. Validation Experiment

According to literature [9], the peaks in the prediction error would abruptly increase when the joints reverse. And the following linear friction model, which includes viscous and Coulomb friction only, is unable to capture the complex dynamical friction behavior. With the linear friction model, the dynamical model of the robot is linear and the unknown parameters can be identified by the least square method or intelligence algorithms.

$$\boldsymbol{\tau}_f = f_c \operatorname{sign}(\dot{\mathbf{q}}_i) + f_v \dot{\mathbf{q}}_i, \tag{18}$$

In order to validate the accuracy of the dynamical model with the nonlinear friction model, we used the linear friction model as a comparison. Similarly, the dynamical parameters of the two models are identified by the ICS algorithm. The excitation trajectories and simulation conditions are set the same as those in Section 5. The comparison of the measured torques and the predicted torques based on different friction models is given in Figure 7.

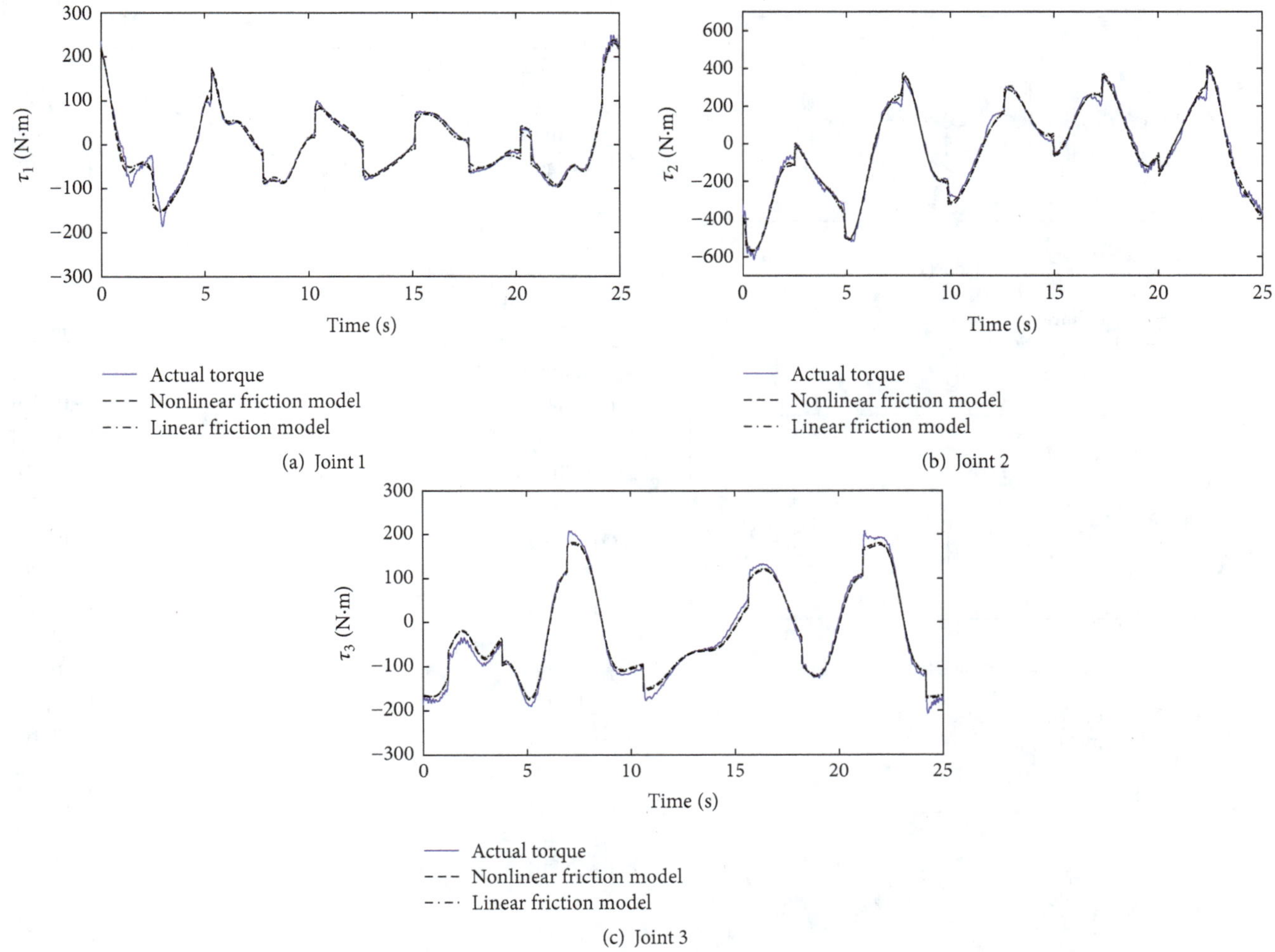

FIGURE 7: Comparison of the different friction models.

As we have seen, both the predicted torques from linear and nonlinear dynamical models can match the measured torques well. Nonetheless, the nonlinear model describes the friction behavior better for the points of joint inversion, obviously in Figure 8. The nonlinear friction model can reduce the error peak, which is beneficial to the design of the controllers.

7. Conclusions

In this paper, an improved cuckoo search algorithm has been proposed to solve the dynamical parameters identification for a robot manipulator with nonlinear friction property. The dynamical model has been established by Newton-Euler method and processed into a linear form. Then, a nonlinear friction model is added to the joint model for realizable friction compensation at motion reversal. Based on the experimental data collected from identification experiment, we use a novel identification algorithm ICS including CS methods, the chaotic operator, and emotion operator to identify the unknown parameters of the robot model. Compared to other two identification algorithms, CS and PSO, the model generated by using our proposed algorithm matches the actual torques better. What is more, ICS has a fast convergence speed and strong search ability. Furthermore, a linear friction model is used as a comparison to test the effect of the nonlinear friction model in describing the friction characteristic. The results show that the nonlinear friction model can restrain the saltation at motion reversal, resulting in higher identification accuracy. In the future, we will attempt to study the controller design based on the identification dynamical model.

Appendix

A. Optimal Trajectory Parameters

The optimal trajectory parameters for the 5th order Fourier series are listed as follows:

$$\beta = \begin{bmatrix} -0.42 & -0.21 & -0.17 & -0.25 & -0.25 & -0.08 & 0.43 & 0.62 & 0.21 & 0.49 & 0.10 \\ 0.15 & 0.45 & -0.12 & -0.35 & 0.12 & 0.44 & 0.67 & 0.01 & 0.04 & 0.01 & -0.12 \\ -0.24 & -0.24 & 0.23 & 0.10 & -0.09 & 0.21 & 0.21 & 0.03 & 0.93 & -0.23 & -0.34 \end{bmatrix}. \tag{A.1}$$

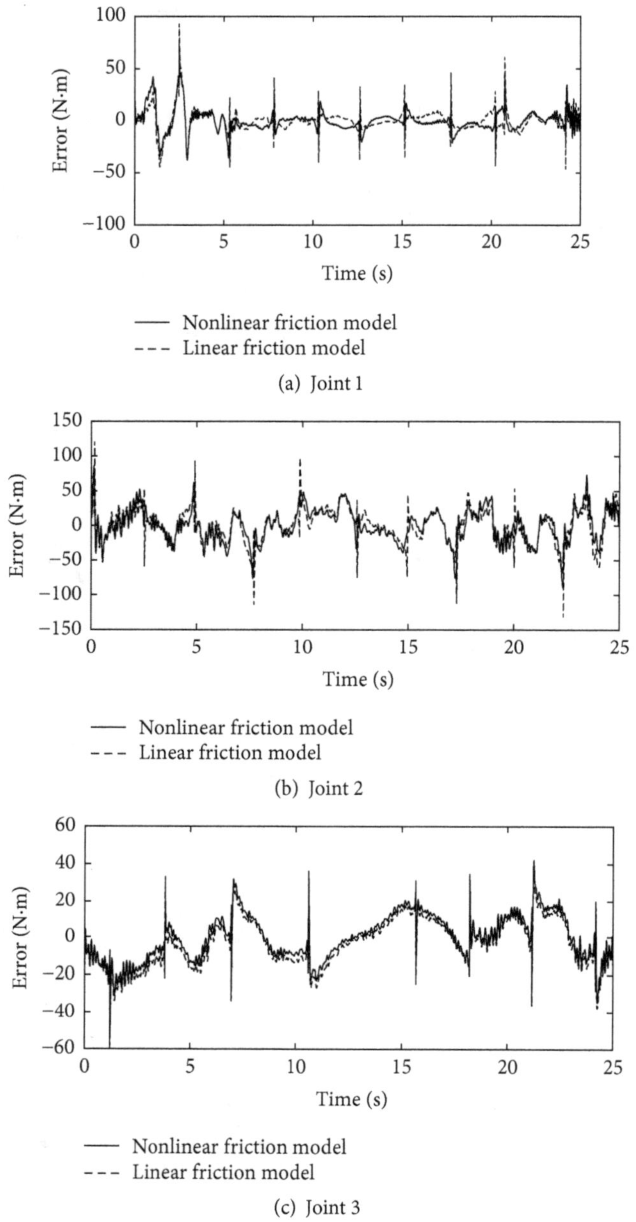

FIGURE 8: Comparison of linear and nonlinear models.

B. Search Scope of Dynamical Parameters

The search scope of the unknown parameters is listed in Table 5.

TABLE 5: Search scope of dynamical parameters.

Parameter	Scope
I_{zz1} (kg·m^2)	[0, 80]
I_{xx2} (kg·m^2)	[−40, 40]
I_{xy2} (kg·m^2)	[−10, 10]
I_{xz2} (kg·m^2)	[−5, 5]
I_{yz2} (kg·m^2)	[−3, 3]
I_{zz2} (kg·m^2)	[−10, 10]
m_2r_{x2} (kg·m)	[0, 100]
m_2r_{y2} (kg·m)	[−20, 20]
I_{xx3} (kg·m^2)	[−10, 10]
I_{xy3} (kg·m^2)	[−10, 10]
I_{xz3} (kg·m^2)	[−10, 10]
I_{yz3} (kg·m^2)	[−10, 10]
I_{zz3} (kg·m^2)	[−30, 30]
m_3r_{x3} (kg·m)	[−10, 10]
m_3r_{y3} (kg·m)	[−20, 20]
f_{01} (N·m)	[−10, 10]
f_{c1} (N·m)	[−10, 10]
f_{v1} (Nm·s/rad)	[−10, 10]
f_{a1} (N·m)	[−30, 30]
f_{b1}	[−10, 10]
f_{02} (N·m)	[−100, 100]
f_{c2} (N·m)	[−10, 10]
f_{v2} (Nm·s/rad)	[−150, 150]
f_{a2} (N·m)	[−100, 100]
f_{b2}	[−50, 50]
f_{03} (N·m)	[−20, 20]
f_{c3} (N·m)	[−50, 50]
f_{v3} (Nm·s/rad)	[−10, 10]
f_{a3} (N·m)	[−10, 10]
f_{b3}	[−10, 10]

Conflicts of Interest

The authors declare that there are no conflicts of interest regarding the publication of this paper.

Acknowledgments

This work was partially supported by the Foundation Research Project of Jiangsu Province (the Natural Science Fund no. BK20170315), Changzhou Sci&Tech Program of China (Grant no. CJ20179017), and Jiangsu Proactive of Production, Study and Research (BY2016030-03).

References

[1] S. Chen, T. Zhang, and Y. Zou, "Fuzzy-sliding mode force control research on robotic machining," *Journal of Robotics*, vol. 2017, pp. 1–8, 2017.

[2] S. Ramabalan, R. Saravanan, and C. Balamurugan, "Multi-objective dynamic optimal trajectory planning of robot manipulators in the presence of obstacles," *The International Journal of Advanced Manufacturing Technology*, vol. 41, no. 5-6, pp. 580–594, 2009.

[3] J. Wu, J. Wang, and Z. You, "An overview of dynamic parameter identification of robots," *Robotics and Computer-Integrated Manufacturing*, vol. 26, no. 5, pp. 414–419, 2010.

[4] C. G. Atkeson, C. H. An, and J. M. Hollerbach, "Estimation of inertial parameters of manipulator loads and links," *International Journal of Robotics Research*, vol. 5, no. 3, pp. 101–119, 1986.

[5] M. Gautier and P. Poignet, "Extended Kalman filtering and weighted least squares dynamic identification of robot," *Control Engineering Practice*, vol. 9, no. 12, pp. 1361–1372, 2001.

[6] M. Grotjahn, M. Daemi, and B. Heimann, "Friction and rigid body identification of robot dynamics," *International Journal of Solids and Structures*, vol. 38, no. 10–13, pp. 1889–1902, 2001.

[7] M. M. Olsen and H. G. Petersen, "A new method for estimating parameters of a dynamic robot model," *IEEE Transactions on Robotics and Automation*, vol. 17, no. 1, pp. 95–100, 2001.

[8] Z. Bingül and O. Karahan, "Dynamic identification of Staubli RX-60 robot using PSO and LS methods," *Expert Systems with Applications*, vol. 38, no. 4, pp. 4136–4149, 2011.

[9] L. Ding, H. Wu, Y. Yao, and Y. Yang, "Dynamic model identification for 6-DOF industrial robots," *Journal of Robotics*, vol. 2015, Article ID 471478, pp. 1–9, 2015.

[10] J. Swevers, W. Verdonck, and J. De Schutter, "Dynamic model identification for industrial robots," *IEEE Control Systems Magazine*, vol. 27, no. 5, pp. 58–71, 2007.

[11] N. Vuong, A. H. Marcelo, Y. Li, and S. Lim, "Improved dynamic identification of robotic manipulators in the linear region of dynamic friction," *IFAC Proceedings Volumes*, vol. 42, no. 16, pp. 167–172, 2009.

[12] S. Kim, J.-P. Kim, and J. Ryu, "Adaptive energy-bounding approach for robustly stable interaction control of impedance-controlled industrial robot with uncertain environments," *IEEE/ASME Transactions on Mechatronics*, vol. 19, no. 4, pp. 1195–1205, 2014.

[13] A. Calanca, L. M. Capisani, A. Ferrara, and L. Magnani, "MIMO closed loop identification of an industrial robot," *IEEE Transactions on Control Systems Technology*, vol. 19, no. 5, pp. 1214–1224, 2011.

[14] Z. Qin, L. Baron, and L. Birglen, "A new approach to the dynamic parameter identification of robotic manipulators," *Robotica*, vol. 28, no. 4, pp. 539–547, 2010.

[15] W. Xi Q, B. Chen, L. Ding et al., "Dynamic identification for robot manipulator with nonlinear friction mode," *Transactions of the Chinese Society for Agricultural Machinery*, vol. 48, no. 2, pp. 393–399, 2017.

[16] E. N. Lorenz, "Deterministic nonperiodic flow," *Journal of the Atmospheric Sciences*, vol. 20, no. 2, pp. 130–141, 1963.

[17] X. Liu and M. Fu, "Cuckoo search algorithm based on frog leaping local search and chaos theory," *Applied Mathematics and Computation*, vol. 266, pp. 1083–1092, 2015.

[18] X.-S. Yang and S. Deb, "Cuckoo search via Lévy flights," in *Proceedings of the World Congress on Nature and Biologically Inspired Computing (NABIC '09)*, pp. 210–214, Coimbatore, India, December 2009.

[19] B. Armstrong, "On finding exciting trajectories for identification experiments involving systems with nonlinear dynamics," *International Journal of Robotics Research*, vol. 8, no. 6, pp. 28–48, 1989.

[20] M. Grotjahn and M. Daemi, "Practical experiences with LS methods for the identification of robot dynamics," in *Proceedings of the 2nd ECPD International Conference on Advanced Robotics*, pp. 535–540, 1996.

[21] M. Gautier, "Dynamic identification of robots with power model," in *IEEE International Conference on Robotics and Automation*, pp. 1922–1927, 3, 1997.

[22] J. Swevers, C. Ganseman, D. B. Tükel et al., "Optimal robot excitation and identification," *IEEE Transactions on Robotics and Automation*, vol. 13, no. 5, pp. 730–740, 1997.

[23] L. Ding, T. H. Wu, Y. Yao et al., "Parameters identification for industrial robots based on WLS-ABC algorithm," *South China University of Technology*, vol. 44, no. 5, pp. 90–95, 2016.

[24] L. Ding, H. T. Wu, and Y. Yao, "Chaotic artificial bee colony algorithm for system identification of a small-scale unmanned helicopter," *International Journal of Aerospace Engineering*, vol. 2015, Article ID 801874, 11 pages, 2015.

[25] J. Swevers, C. Ganseman, J. De Schutter, and H. Van Brussel, "Experimental robot identification using optimised periodic trajectories," *Mechanical Systems and Signal Processing*, vol. 10, no. 5, pp. 561–577, 1996.

[26] Y. Ding D, B. Chen et al., "An identification method of industrial robots dynamical parameters," *South China University of Technology (Natural Science Edition)*, vol. 43, no. 3, pp. 49–56, 2015.

[27] C.-M. Huang, C.-J. Huang, and M.-L. Wang, "A particle swarm optimization to identifying the ARMAX model for short-term load forecasting," *IEEE Transactions on Power Systems*, vol. 20, no. 2, pp. 1126–1133, 2005.

4

Design of Connectivity Preserving Flocking using Control Lyapunov Function

Bayu Erfianto,[1,2] Riyanto T. Bambang,[2] Hilwadi Hindersah,[2] and Intan Muchtadi-Alamsyah[3]

[1]*School of Computing, Telkom University, Bandung, Indonesia*
[2]*School of Electrical Engineering & Informatics, Institut Teknologi Bandung, Bandung, Indonesia*
[3]*Faculty of Mathematical Sciences, Institut Teknologi Bandung, Bandung, Indonesia*

Correspondence should be addressed to Bayu Erfianto; bayu.erfianto@gmail.com

Academic Editor: Shahram Payandeh

This paper investigates cooperative flocking control design with connectivity preserving mechanism. During flocking, interagent distance is measured to determine communication topology of the flocks. Then, cooperative flocking motion is built based on cooperative artificial potential field with connectivity preserving mechanism to achieve the common flocking objective. The flocking control input is then obtained by deriving cooperative artificial potential field using control Lyapunov function. As a result, we prove that our flocking protocol establishes group stabilization and the communication topology of multiagent flocking is always connected.

1. Introduction

Flocking in nature is a coordinated movement of group of living entities. Recently, flocking behavior has been adopted in a multiagent system to resemble collective movement of living system, for example, collective movement of group of multi-robot system and autonomous vehicles. Those multiple agents together form a cooperative system to solve complex problems in distributed manner that is impossible to be carried out by a single agent.

With respect to the literature of multiagent system, many theories denote that flocking model is built using three basic heuristic rules, that is, cohesion, separation, and velocity alignment, which is initiated for the first time by [1]. These three rules describe how an individual agent moves according to position and velocity of the nearest agents in flocks. Flocks always move together with velocity converging to the same value and guaranteeing no collision among agents.

In multiagent robot, common flocking rule can further be described technically as centroid attraction, collision avoidance, and velocity alignment, respectively. Centroid attraction rule tries to pull the agent to stay close to the nearby agents in the flocks if the agents are moving away. Meanwhile, collision avoidance rule forces the agent to avoid collision with nearby agents in the flock, while velocity synchronization rule will drive the agents to synchronize their velocities while moving in the flock. The most notable flocking controllers are based on the works of Tanner et al. [2, 3] and Olfati-Saber [4]. Olfati-Saber [4] does not use flock centering rule to build flocking formation; instead of that he defines lattice formation to keep interagent distance.

The synchronization of movement in flocking can be carried out through local or global communication, with explicit communication and without communication (stigmergic) as summarized in [5]. Erfianto et al. [6] have extended Reynold flocking protocol with explicit communication using broadcast gossip protocol, where the explicit communication of flocking is simulated in Netlogo multiagent simulator. By equipping mobile robot agent with communication capability, it means that flocking protocol is then restricted by communication capability. Previous research works on flocking algorithm are mostly based on the assumption that the network topology is always connected without any mechanism to achieve this circumstance.

Network communication is mainly used to exchange information state between an agent and the nearby agents. Regarding this matter, the problem is that the initially connected network can not guarantee connectivity over the time during flock movement. Once disconnected, flock movement can be separated and that causes collective flocking goal not being achieved. Thus, it is important to prevent separation of the flock into multiple separate groups when the network is being disconnected. Regarding this issue, many kinds of research have been conducted to design flocking control input (flocking protocol) to ensure that the flocks achieve the collective goal while preserving connectedness during the movement. Connectivity preserving flocking of multiagent mobile robot then becomes a new trending area, and various flocking protocols have been proposed including both centralized and decentralized approaches, for example, presented in [7–9].

Since connectivity of agents in flocking formation depends on its communication range, then the communication topology among agents can be connected or disconnected affected by the movement of an agent in flocks. The changes of signal power may also affect to the actual interagent distance since the distance is correspondingly perceived from measured signal power. Regarding this issue, a connectivity preserving mechanism is required to maintain connectivity during the flock movement. Fang et al. in [10] further divide connectivity preservation into two main categories, that is, conservative connectivity preservation and flexible connectivity preservation. In conservative connectivity preservation, the approach is to preserve all existing links; thus the network topology is strictly fixed, as presented in [7, 11, 12]. In flexible connectivity preservation approaches, the underlying communication topology is allowed to switch into different topologies, which means that edges are allowed to be added or removed from a node as long as the global graph topology is always connected. It is done by maintaining the spanning tree of the underlying network. Regarding these two approaches, therefore, the challenge is to design cooperative control law where it depends on the assumption that the underlying communication topology is connected over the time during the whole flocking evolution.

Formation of flocking is commonly used in the various practical implementation of flocking protocol, such as in coverage control using multiagent mobile robot, a formation of multiple UAV, and cooperative mobile sensor networks, as well as cooperative mobile robot, for a surveillance system. Therefore, the design of formation flocking needs to consider navigation capability along with separation, cohesion, velocity alignment, and connectivity of a multiagent network.

In this paper, our work is to preserve network connectivity by using an artificial potential function, where flocking protocol can also perform collision avoidance and obstacle avoidance function. The artificial potential function is then considered as control Lyapunov function candidate to be derived as continuous flocking control input. We also define the algebraic connectivity gradient controller and blend this gradient controller into flocking control input. This is the major difference of our work from [7, 9, 13, 14]. As a result, we conclude that, given the initially connected network, our flocking protocol can make all agents converge to a common velocity, collisions between agents, as well as between agent and obstacle, can be avoided, and the connectivity of underlying network can be preserved; thus the desired stable flocking behavior can be achieved.

This paper is organized as follows. Section 1 is the introduction of this paper. Section 2 of this paper is about problem definition to design connectivity preserving flocking. Section 3 discusses the design of artificial potential function that will be involved in control Lyapunov function. Flocking controller design is presented in Section 4, while the simulation to verify the controller is presented in Section 5.

2. Problem Formulation and Assumption

Let us consider a group of points of mobile robots that consists of N robot agents moving in 2D space. The dynamic of mobile robot agent is modeled as

$$\begin{aligned} \dot{q}_i &= p_i, \\ \dot{p}_i &= u_i, \end{aligned} \tag{1}$$

where $q = [q_1^T, q_2^T, \ldots, q_n^T]^T \in \mathbb{R}^2$ denotes position vector, where $q_{ij} = \|q_i - q_j\|$ is the Euclidean distance between two agents, which is clearly $q_{ij} \geq 0$. Let $p_i \in \mathbb{R}^2$ be the velocity vector and $u_i \in \mathbb{R}^2$ be the control input of agent i, respectively.

Assumption 1. We assume that each agent obtains its position from sensors. Each agent is also equipped with a wireless communication network; therefore information state can be exchanged within the neighboring agents via a communication network.

Regarding Assumption 1, interaction among agents can further be modeled by bidirectional communication network, that is, undirected graph topology.

Definition 2 (flocking topology). Topology of a flock of robots is a dynamic undirected topology, denoted as $\mathcal{G} = (\mathcal{V}, \mathcal{E})$, that consists of

(i) node or vertices $\mathcal{V} = \{1, \ldots, n\}$;

(ii) set of edges or time varying communication links $\mathcal{E} = \{i, j \in \mathcal{V} \times \mathcal{V} \mid q_{ij} \leq R\}$ in which each undirected edge represents the communication between a pair of agents limited by communication range R;

(iii) set of neighbors $\mathcal{N}_i = \{j \in \mathcal{V} \mid (i, j) \in \mathcal{E}\}$ with symmetric property $i \in \mathcal{N}_j \leftrightarrow j \in \mathcal{N}_i$.

Without loss of generality, each agent is strictly not self-communicating agent $(i, j) \notin \mathcal{E}$. Regarding Definition 2, interagent communication can then be constructed by adjacency matrix $A \in \mathbb{R}^{N \times N}$, with element

$$a_{ij} = \begin{cases} 1 & \text{if } q_{ij} \leq R \\ 0 & \text{otherwise,} \end{cases} \tag{2}$$

where q_{ij} is the Euclidean distance between two agents, and each agent is supposed to have the same communication radius R. Some fundamentals of graph theory used in the paper, including algebraic graph theory, can be found in the main reference [15].

From the adjacency matrix A, the degree matrix is defined as $D = \text{diag}(a_{ij}) = \sum_{j=1}^{N} a_{ij}$. Laplacian matrix of $\mathcal{G}$ can be further obtained by subtracting adjacency matrix from degree matrix, such that $L = D - A$, where Laplacian matrix L is positive semidefinite. Let the eigenvalues of L be $0 = \lambda_1 \leq \lambda_2 \leq \cdots \leq \lambda_N$ which is nondecreasing order with corresponding eigenvectors $\{v_1, v_2, \ldots, v_n\}$. We denote that the remarkable properties of Laplacian matrix are as follows.

Lemma 3 (see [15]). *Given the undirected graph* $\mathcal{G}$

(i) $L(\mathcal{G})$ *is always symmetric and positive semidefinite and satisfies sum of square property*

$$x^T L x = \frac{1}{2}\sum_{i=1}^{N} a_{ij}\left(x_i - x_j\right)^2, \tag{3}$$

where $x_i \in \mathbb{R}^N$;

(ii) $\lambda_1 = 0$; *thus* $\lambda_2 > 0$ *if graph* $\mathcal{G}(t)$ *is connected; then* λ_2 *is referred to algebraic connectivity of* $\mathcal{G}(t)$;

(iii) $L(\mathcal{G})$ *has a single zero eigenvalue and the corresponding eigenvector is* $\mathbf{1}_N$.

This paper will focus on the design of flocking protocol that can preserve connectivity in terms of algebraic graph representation.

We define the stable flocking protocol as the necessary condition to preserve connectivity of the network all the time.

Definition 4 (stable flocking protocol). Group of robot agents is said to flock when all agents approach the same velocity, with no collisions among them, such that

$$\begin{aligned} \left\|q_i - q_j\right\| &\leq R, \\ p_i &= p_j, \\ \lambda_2 &> 0, \end{aligned} \tag{4}$$

where λ_2 is the second smallest eigenvalue of Laplacian matrix L.

The connectivity maintenance of flocking topology is approximately determined by the distance between agents in the flock. The movement of agents that might cause the connectivity can be unavailable during flocking motion. Therefore, our problem is dealing with designing a cooperative flocking protocol using only local information to drive all robot agents to achieve the same velocity, avoiding collision with other agents in flock, avoiding static obstacle, driving the agents to the goal position, and guaranteeing that the network is always connected during the flocking, indicated by $\lambda_2(L) > 0$.

3. Potential Function Design

This section introduces the design of artificial potential functions for the purpose of target attraction, obstacle avoidance, and interagent attraction and repulsion. Since first used in [16], the artificial potential function has been commonly used to solve the problems of motion planning in mobile robotic applications. Theoretically, the artificial potential function generates repulsive fields close to obstacles and attractive fields close target. In artificial potential fields, a noncollided motion is determined by how much the robot is attracted by other agents and repelled by either the obstacle or other agents. We add centroid function and attract-repel function to avoid local minima caused by goal attraction and obstacle function. The potential function method has several advantages, and the most important one is being easy to implement.

3.1. Environmental Potential Field. Consider the multiagent robots move in configuration workspace, which is defined as follows.

Definition 5 (workspace). Suppose a point of robot moves in the workspace $\mathcal{W} \triangleq \{q \in \mathbb{R}^2 \mid q_{ij} \leq s^2\}$; that is, the Euclidean distance between two points does not exceed the area of s^2.

We define the free space, where there is no obstacle within the area.

Definition 6 (free space). Free space $\mathcal{F}$ is defined as $\mathcal{F} \triangleq \mathcal{W} \bigcup_{k=1}^{m} O_k$, where a robot agent could move safely without colliding with any obstacles.

Given the position of robot agents in stack vector $q = [q_1^T, \ldots, q_N^T]^T \in \mathbb{R}^2$, we assume all information of the environment in the $\mathcal{W}$ is known by each robot agent. A target or goal position of agent i is defined as $q_t \in \mathcal{F}$, which is in the free area of configuration workspace. To attract the robot agent to reach the desired goal position, the attraction-to-target potential function $\mathcal{T}$ is defined as follows.

Definition 7 (target attraction function). $\mathcal{T}_i : \mathcal{F} \to [0, \infty)$ is a metric that measures how close a robot agent is to the desired goal position. Target or goal attraction function is defined as

$$\mathcal{T}_i = \frac{1}{2}\left\|q_i - q_t\right\|^2. \tag{5}$$

Let $O_k \in \mathbb{R}^2$ be kth obstacle in the workspace and $O_{i,k} \in \mathbb{R}^2$ be the kth obstacle that has to be avoided by agent i.

Definition 8 (obstacle). O_k is a kth disc obstacle that exists in workspace $\mathcal{W}$ with center $q_o \in \mathbb{R}^2$ and radius r_k:

$$O_k = \left\{q \in \mathbb{R}^2 \mid \left\|q - q_o\right\| \leq r_k\right\} \tag{6}$$

such that the distance of any point to the center of obstacle does not exceed r_k.

Using the definitions above, all robot agents have to avoid obstacles that block the robot route toward the desired

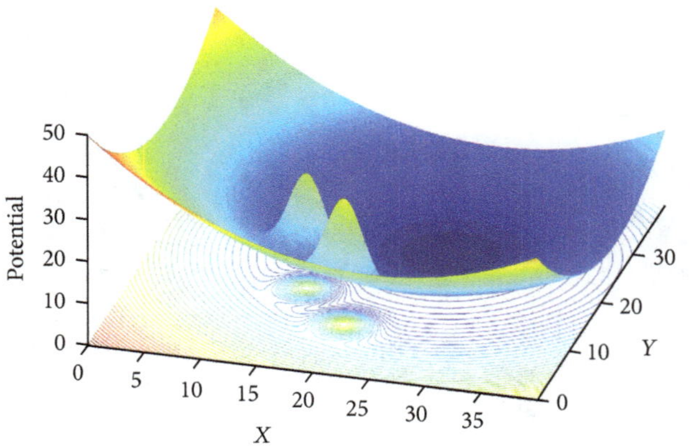

FIGURE 1: Example of environment model used in this paper. Goal position is represented by the lowest potential value; two hills represent different size of obstacles.

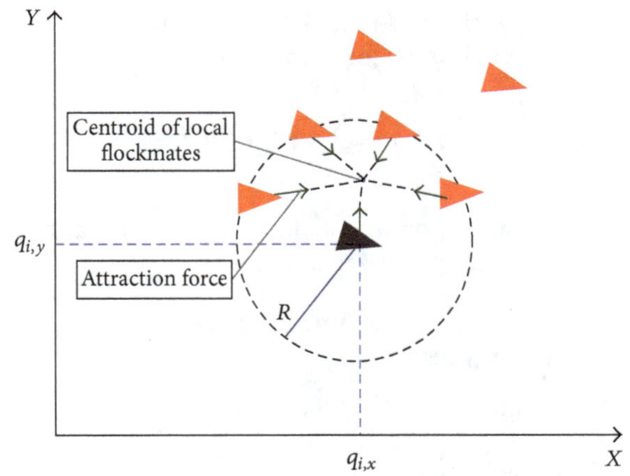

FIGURE 2: Illustration of an agent disk radius and centroid attraction among the flockmates.

position. Therefore, each robot has to execute an obstacle avoidance function that is constructed by measuring the distance between the robot agent and the obstacle center $q_o \in \mathscr{W}$. The obstacle avoidance function computed by ith agent is defined as follows.

Definition 9 (obstacle avoidance function). Robot agent i could avoid kth obstacle with the following potential function:

$$\mathscr{O}_{i,k} = \exp\left(-\frac{(1/2)\left\|q_i - q_o\right\|^2}{r_k}\right) r_k, \tag{7}$$

where $q_o \in \mathbb{R}^2$ is the center of obstacle and r_k is radius of obstacle k, such that robot agent i could avoid obstacle O_k if none of them intersect each other within $\|q_i - q_o\| > r_k$.

Using the definitions above, the environmental potential field is composed of goal attractiveness and obstacle repulsiveness, and the result is depicted in Figure 1. This figure illustrates two different sizes of obstacles indicated by two hills with high potential value, while the lowest potential, the darkest valley at the center of the basin, illustrates the desired position.

3.2. Interagent Potential Function. We assume that the sensing capability of an agent equals its communication radius. Suppose there exists pair interaction between robot agents i and j via wireless communication network. The interaction between paired agents is limited by communication range, where the coverage of robot agent i resembles a disk with a certain radius R.

Definition 10 (agent disc radius). Let $\mathscr{B}(q_i, R)$ be a disk with coverage radius R, and $q_i = (x_i, y_i) \in \mathbb{R}^2$ is position of agent i a (x, y) plane:

(i) Open disk radius is an area covered by $\mathscr{B}(q_i, R)$ if $\{q_j \in \mathbb{R}^2 \mid \|q_i - q_j\| < R\}$.

(ii) Close disk radius is an area covered by $\mathscr{B}(q_i, R)$ if $\{q_j \in \mathbb{R}^2 \mid \|q_i - q_j\| \leq R\}$.

3.3. Attraction to Centroid. The idea of flock centroid is initiated in [1]. Action to attract flock member to the center of a flock, or flock centering, makes the agents move in a direction toward the centroid. By means of flock centering, the flocking algorithm provides a mechanism to make sure that agents are attracted to each other and not moving away. This concept is further known as a part of the well-known one of three heuristic flocking rules initiated by Reynolds [1]. Illustration of agent's disk radius and the center of the flock can be seen in Figure 2. This figure also explains how the interaction among agents in the flock occurs.

The long-range attraction occurs between agents in the flocks. To make sure that robot agents in the flocks are attracted to each other, as well as attracted by the center of the flock, an attractive function is constructed by measuring the distance between individual agents and the center of the flocks. The flock centering attraction function is simply given by

$$\mathscr{C}_i(q_i) = \sum_{i=1}^{N}\left\|q_i - \frac{1}{N}\sum_{i=1}^{N} q_i\right\|^2. \tag{8}$$

3.4. Interagent Attraction Repulsion. This paper uses the definition of closed disk coverage, where two interagent distances may occur within $\|q_i - q_j\| \leq R$. This condition is further coded into communication edge $(i, j) \in \mathscr{E}$. Based on this interaction, the neighborhood of an agent i is determined by $\mathscr{N}_i = \{j \in \mathscr{V} \mid (i, j) \in \mathscr{E}\}$. Using the definition of $\mathscr{N}_i$, the interagent collision avoidance function is further defined.

Definition 11 (interagent collision avoidance). Collision avoidance function is used to keep $\|q_i - q_j\| \leq R$ such that communication can be preserved and no agents collided during the flock movement.

Interagent collision avoidance is implemented by attract-repel function. The long-range attraction occurs between two agents in the flocks. It is to make sure that the agents in the flock are attracted by the following function:

$$\Psi_{ij}\left(q_{ij}\right) = \sum_{j\in\mathcal{N}_i} \frac{c_a}{2}\left\|q_i - q_j\right\|^2, \tag{9}$$

where c_a is attraction constant. Properties of attraction function can be characterized as follows:

(i) Attractive potential function can be viewed as potential energy between agents in the flock.

(ii) Let $\Psi_{ij}(q_{ij})$ be the attraction function; then

(a) $\Psi_{ij}(q_{ij}) \rightarrow 0$, as agent $\|q_i - q_j\| \rightarrow \infty$; that is, j moves away from agent i.

(b) $\Psi_{ij}(q_{ij})$ is differentiable along $q_{ij} > 0$ such that $\nabla_{q_i}\Psi_{ij}(q_{ij}) > 0$.

Interagent collision avoidance is implemented by short range repulsive function, which is formulated from the distance between agents. The objective of this function is to repel neighboring agents if they are too close; therefore collision can be avoided. The short range repulsive potential function is defined as

$$\Phi\left(q_{ij}\right) = \sum_{j\in\mathcal{N}_i} c_r \exp\left(-\frac{(1/2)\left\|q_i - q_j\right\|^2}{R}\right)R, \tag{10}$$

where c_r is repulsion constant. The properties of repulsive function can be characterized as follows:

(i) Repulsive potential function can be viewed as potential energy between agents in the flock within the range $[0, R)$ such that no collision occurs between agents $j \in \mathcal{N}_i$.

(ii) Let $\Phi_{ij}(q_{ij})$ be the repulsive function, such that

(a) $\Phi_{ij}(q_{ij}) \rightarrow \infty$, as $\|q_i - q_j\| \rightarrow 0$;

(b) $\Phi_{ij}(q_{ij})$ is differentiable along $q_{ij} > 0$, such that $\nabla_i\Phi_{ij}(q_{ij}) > 0$.

We combine attraction and repulsion potential function to generate attract-repel function as

$$\begin{aligned}\mathcal{R}_{ij}\left(q_{ij}\right) &= \Phi_{ij}\left(q_{ij}\right) + \Psi_{ij}\left(q_{ij}\right)\\ &= \sum_{j\in\mathcal{N}_i}\frac{c_a}{2}\left\|q_i - q_j\right\|^2\\ &\quad + \sum_{j\in\mathcal{N}_i} c_r \exp\left(-\frac{(1/2)\left\|q_i - q_j\right\|^2}{R}\right)R.\end{aligned} \tag{11}$$

The attractive-repulsive potential function may guarantee that agent j never moves away from $\mathcal{N}_i$. The plot of attract-repel function is depicted in Figure 3.

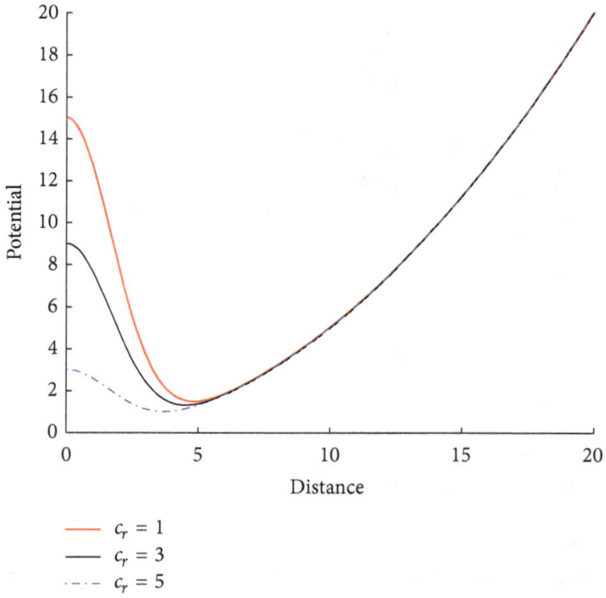

FIGURE 3: Interagent attract-repel potential function with various repulsive constant c_r, where $c_a = 1$ and $R = 3$.

4. Flocking Controller Design

This section describes the design of flocking controller by utilizing control Lyapunov function. We use the direct method of Lyapunov to obtain the flocking control input which is derived from overall artificial potential function that has been defined in the previous section. The reader may refer to [17–19] for the detail of control Lyapunov function for multiagent system.

4.1. Control Lyapunov Function. Regarding [1, 4, 7], flocking control law, which is also known as flocking protocol, is composed of negative gradient of potential force and consensus velocity of flock members. The force is obtained from the gradient of potential fields. Generally, the objective of flocking control is to steer the flock such that velocity between agents i and j is approximately the same.

We extend the control law of formation control problem in [14] into flocking problem. First of all, consider a multiagent system dynamics

$$\begin{aligned}\dot{q}_i &= p_i,\\ \dot{p}_i &= u_i,\end{aligned} \tag{12}$$

where $q_i \in \mathbb{R}^2$ is position of agent in two-dimensional space, p_i is agent velocity, and u_i in control input to be designed. Our intention is to use control Lyapunov function to obtain control input u_i, which is derived from the potential function. We recall the sum of potential functions from (5), (7), (8), and (11),

$$\begin{aligned}&\mathcal{P}_i\left(q_i\right)\\ &= \sum_{i=1}^{N}\left(\mathcal{C}_i\left(q_i\right) + \sum_{j\in\mathcal{N}_i}\mathcal{R}_{ij}\left(q_{ij}\right) + \mathcal{T}_i\left(q_i\right) + \mathcal{O}_{i,k}\left(q_i\right)\right).\end{aligned} \tag{13}$$

Definition 12. A control Lyapunov function is continuously positive-definite function; that is, $V(x)$ is positive except for $x = 0$, such that

$$\dot{V}(x,u) = \nabla V(x) f(x,u) < 0. \tag{14}$$

Regarding Definition 12, the candidate of Lyapunov function is then defined by combining all potential energy obtained from potential function, with kinetic energy as the difference between velocity of agent i and agent j,

$$V_i = \mathscr{P}_i(q_i) + \frac{1}{2}\sum_{j\in\mathscr{N}_i}(p_i - p_j)^2. \tag{15}$$

Candidate of Lyapunov function for all agents can then be written as

$$V = \sum_{i=1}^{N} V_i = \sum_{i=1}^{N}\Bigg(\mathscr{C}_i(q_i) + \sum_{j\in\mathscr{N}_i}\mathscr{R}_i(q_i) + \mathscr{T}_i(q_i) + \mathscr{O}_{i,k}(q_i) + \frac{1}{2} \cdot \sum_{j\in\mathscr{N}_i}(p_i - p_j)^2\Bigg). \tag{16}$$

To extract the control input from candidate of Lyapunov function, we differentiate all potential function components in V, such that flocking control input for system (12) is obtained from $\dot{V}(q_i)$. We consider the assumption described in previous section to generate the following theorem.

Theorem 13. *The N holonomic agents use dynamics in system (12). The stable flocking control input u_i can be obtained from the gradient of (16). The flocking protocol is stable such that each agent moves in the flock, the obstacle can be avoided, and there is no interagent collision, and finally agents stop at a defined goal position.*

To prove the above theorem, we consider (16). It is obvious that $V_i(0) = 0$, and V_i is positive-definite function and continuous when

$$\sum_{i=1}^{N}\mathscr{C}_i(q) > 0,$$
$$\sum_{i=1}^{N}\sum_{j\in\mathscr{N}_i}\mathscr{R}_{ij}(q) > 0, \tag{17}$$
$$q_i \in \mathbb{R}^2.$$

Since the force is negative gradient along the trajectories of q_i, then we can write

$$\dot{V} = \sum_{i=1}^{N}\frac{\partial V_i}{\partial q_i}\frac{\partial q_i}{\partial t} = \sum_{i=1}^{N}\frac{\partial V_i}{\partial q_i}\cdot\dot{q}_i = \sum_{i=1}^{N}\frac{\partial V_i}{\partial q_i}p_i. \tag{18}$$

The velocity is the form of negative force that equals

$$p_i = -\sum_{i=1}^{N}\frac{\partial V_i}{\partial q_i}. \tag{19}$$

We then check the definiteness of Lyapunov function, such that

$$\dot{V} = -\sum_{i=1}^{N}\left(\frac{\partial V_i}{\partial q_i}\cdot\frac{\partial V}{\partial q_i}\right) = -\sum_{i=1}^{N}p_i^T p_i \le 0. \tag{20}$$

It is clear that since (20) is negative semidefinite, then control input u_i is Lyapunov stable. Therefore, the flocking control input u_i can then be obtained from the gradient of Lyapunov function $\dot{V}(q_i, p_i)$ as follows:

$$\begin{aligned} u_i &= -\nabla_q\mathscr{C}_i(q) - \sum_{j\in\mathscr{N}_i}\nabla_q\mathscr{R}_{ij}(q) - \nabla_q\mathscr{T}_i(q) \\ &\quad - \nabla_q\mathscr{O}_{i,k}(q) + \sum_{j\in\mathscr{N}_i}(p_i - p_j) \\ &= -\left(q_i - \sum_{i=1}^{N}\overline{q}_i\right) + \exp^{-(q_i-q_o)/r_s}(q_i - q_o) \\ &\quad - (q_i - q_t) + \left(c_a - c_r\exp^{-(q_i-q_j)/R}\right)(q_i - q_j) \\ &\quad + \sum_{j\in\mathscr{N}_i}(p_i - p_j). \end{aligned} \tag{21}$$

The notation $\overline{q} = (1/N)\sum_{i=1}^{N} q_i$ is the center of the flock. Thus, we can substitute flocking control input in system (12) with (21). It can be observed that the last term of control input is velocity consensus.

4.2. Convergence of Velocity. In this subsection, we present the analysis of velocity convergence which is the necessary condition of stable flocking controller as in Definition 4. We begin with the stationary of the center of the flocks, which is to attract all flock members during the movement. Now let us consider the following lemma.

Lemma 14. *Using system (12) and control input in (21), the centroid of the flocks is stationary $\forall t > 0$.*

Proof. Let $\overline{q} = (1/N)\sum_{i=1}^{N} q_i$ be the center of the flocks. The time derivative of $\overline{q}$ can further be defined as

$$\dot{\overline{q}} = \frac{1}{N}\sum_{i=1}^{N}\dot{q}_i. \tag{22}$$

Let us consider the system $u_i = \dot{q}_i$ and let $\dot{q}_i = \nabla_{q_i}\mathscr{R}_i(q) = -(c_a - c_r\exp^{-(q_i-q_j)/R})(q_i - q_j)$ be the attract-repel force. Let $f(-a) = -(c_r\exp^{-(q_i-q_j)/R})(q_i - q_j)$; then $f(-a)$ is odd function such that $f(-a) = -f(a)$. Due to the symmetry of odd function, (22) can be written as

$$\dot{\overline{q}} = \frac{1}{N}\sum_{i=1}^{N}\sum_{j\in\mathscr{N}_i}(f(-a) + f(a)) = 0. \tag{23}$$

Let the invariant set of equilibrium (or stationary) points be $\Omega_e = \{q \mid \dot{\bar{q}} = 0\}$. It implies $\dot{q}_i = \dot{\bar{q}}$, which means that all individual agents' velocity equals the velocity of center of the flocks. □

We begin with following theorem.

Theorem 15. *Robot agent that uses control input (21) eventually can reach the same velocity during flocking and collision can be avoided and achieve the desired goal.*

Proof. To prove this theorem, we rewrite the sum of potential function $\mathcal{P}_i$ defined in (13):

$$\mathcal{P}_i(q_i) = \sum_{i=1}^{N}\left(\mathcal{C}_i(q_i) + \sum_{j\in\mathcal{N}_i}\mathcal{R}_i(q_i) + \mathcal{T}_i(q_i) + \mathcal{O}_k(q_i)\right). \tag{24}$$

The candidate of Lyapunov function is then defined, which consists of artificial potential function and velocity of each agent.

$$V_i = \frac{1}{2}\left(\sum_{i=1}^{N}\mathcal{P}_i(q_i)^2 + p_i^2\right). \tag{25}$$

The derivative of Lyapunov function along q_i is

$$\dot{V}_i = \sum_{i=1}^{N}\dot{q}_i\left(\nabla_{q_i}\mathcal{C}_i(q) + \nabla_{q_i}\mathcal{T}_i(q) + \nabla_{q_i}\mathcal{O}_i(q) + \sum_{j\in\mathcal{N}_i}\nabla_{q_i}\mathcal{R}_{ij}(q)\right) + \sum_{i=1}^{N}p_i^T\dot{p}_i. \tag{26}$$

By replacing $\dot{p}$ in (26) with u_i in (21), we obtain

$$\begin{aligned}\dot{V}_i &= \sum_{i=1}^{N}\dot{q}_i\left(\nabla_{q_i}\mathcal{C}_i(q) + \nabla_{q_i}\mathcal{T}_i(q) + \nabla_{q_i}\mathcal{O}_i(q) + \sum_{j\in\mathcal{N}_i}\nabla_q\mathcal{R}_{ij}(q)\right) + \sum_{i=1}^{N}p_i^T u_i = \sum_{i=1}^{N}p_i^T\Bigg[\Bigg(\nabla_{q_i}\mathcal{C}_i(q) \\ &\quad + \nabla_{q_i}\mathcal{T}_i(q) + \nabla_{q_i}\mathcal{O}_i(q) + \sum_{j\in\mathcal{N}_i}\nabla_q\mathcal{R}_{ij}(q)\Bigg) + \sum_{i=1}^{N}u_i\Bigg] = \sum_{i=1}^{N}p_i^T\Bigg[\Bigg(\nabla_{q_i}\mathcal{C}_i(q) + \nabla_{q_i}\mathcal{T}_i(q) \\ &\quad + \nabla_{q_i}\mathcal{O}_i(q) + \sum_{j\in\mathcal{N}_i}\nabla_q\mathcal{R}_{ij}(q)\Bigg) - \sum_{i=1}^{N}\nabla_{q_i}\mathcal{C}_i(q) - \nabla_{q_i}\mathcal{T}_i(q) - \nabla_{q_i}\mathcal{O}_i(q) - \sum_{j\in\mathcal{N}_i}\nabla_q\mathcal{R}_{ij}(q) \\ &\quad - \sum_{j\in\mathcal{N}_i}(p_i - p_j)\Bigg] = -p_i^T\sum_{j\in\mathcal{N}_i}(p_i - p_j).\end{aligned} \tag{27}$$

The neighborhood $j \in \mathcal{N}_i$ is represented by adjacency element a_{ij}; then (27) can be written as

$$\dot{V} = -\sum_{i=1}^{N}\left(p_i^T\sum_{j\in\mathcal{N}_i}a_{ij}(p_i - p_j)\right) = -p^T(L\otimes I)p \le 0, \tag{28}$$

where L is the Laplacian matrix and I is the identity matrix. Therefore it means that when $-p^T(L\otimes I)p = 0$, the agent velocity $p_1 = p_2 = \cdots = p_N$, $i = 1, 2, \ldots, N$. □

4.3. Algebraic Connectivity Gradient Controller. In this section, we add the term of gradient of algebraic connectivity along q_i into u_i in (21). We redefine the adjacency matrix A as follows:

$$a_{ij} = \begin{cases}\exp^{-((1/2)\|q_i - q_j\|^2/R)}R, & \|q_i - q_j\| \le R \\ 0, & \text{otherwise.}\end{cases} \tag{29}$$

We can also write the second smallest eigenvalue of Laplacian matrix $\lambda_2 = v_2^T L v_2$, where v_2 is corresponding eigenvector of λ_2. Due to the properties of Laplacian matrix, the connectivity is guaranteed if $\lambda_2(L) > 0$. Suppose we define $(0, 1] \in \epsilon$ to be the lower bound of λ_2. It is clear that when λ_2 goes to ϵ, it means that the graph connectivity is weak. Regarding this condition, we consider adding the value of algebraic connectivity as the additional term in flocking protocol, which is to make sure that $\lambda_2 \ge \epsilon$.

Let us define the following theorem.

Theorem 16. *Suppose $\varphi(q_i) = \coth^2(\lambda_2(q_i) - \epsilon)$ is the short range function of λ_2 along q_i. Then $\varphi(q_i)$ is differentiable along q_{ij}. Therefore, its gradient exists along q_{ij}, and $\varphi(\lambda_2 - \epsilon)$ approaches a constant value as λ_2 increases.*

Proof. To proof this theorem, we follow [20]. We can also write

$$\inf_{v\in\mathbf{1}^\perp}\lambda_2 = v^T L v = \inf_{v\in\mathbf{1}^\perp}\sum_{i=1}^{n}\sum_{j\in\mathcal{N}_i}a_{ij}(v_i - v_j)^2, \tag{30}$$

where v_i is corresponding eigenvector of λ_2. Suppose we want to define the gradient controller with algebraic connectivity function. The control input can be defined by taking the derivative of λ_2 along q_i as follows:

$$u_i^\lambda = \frac{\partial\varphi(\lambda_2(q_i) - \epsilon)}{\partial q_i} = \frac{\partial\varphi(\lambda_2(q_i) - \epsilon)}{\partial\lambda_2}\frac{\partial\lambda_2}{\partial q_i}. \tag{31}$$

The solution of first term is partial derivative of $\varphi(\lambda_2(q_i))$ on λ_2,

$$\frac{\partial\varphi(p)}{\partial\lambda_2} = \operatorname{csch}^2(\lambda_2 - \epsilon). \tag{32}$$

The next step is to find the solution of second term of partial derivative of $\partial\lambda_2/\partial q_i$ in (31). We use [20] to find $\partial\lambda_2/\partial q_i$.

Lemma 6 (see [20]). *Given any positively weighted graph $\mathcal{G}$, λ_2 is a nondecreasing function of each weight a_{ij}.*

Second smallest eigenvalue can be written in the form of

$$\lambda_2 = \min_{x \perp 1} \frac{x^T L x}{x^T x} = \min_{v \perp 1} \frac{\sum_{(i,j)\in\mathscr{E}} a_{ij} \left(v_i - v_j\right)^2}{v^T v}. \tag{33}$$

As a consequence, therefore, the partial derivative of λ_2 can also be written as follows:

$$\frac{\partial \lambda_2}{\partial q_i} = v^T \frac{\partial L}{\partial q_i} v. \tag{34}$$

By replacing $x^T L x$ with $\sum_{(i,j)\in\mathscr{E}} a_{ij}(v_i - v_j)^2$ we obtain

$$v^T \frac{\partial L}{\partial q_i} v = \sum_{(i,j)\in\mathscr{E}} \frac{\partial a_{ij}}{\partial q_i} \left(v_i - v_j\right)^2. \tag{35}$$

Therefore, the gradient of $\varphi(\lambda_2 - \epsilon)$ exists along q_i, and this value asymptotically approaches a constant value as λ_2 increases. □

We further define the gradient of algebraic connectivity as control input as follows:

$$u_i^{\lambda} = -\operatorname{csch}^2\left(\lambda_2 - \epsilon\right) \sum_{j\in\mathscr{N}_i} a_{ij} \left(v_i - v_j\right)^2 \left(q_i - q_j\right). \tag{36}$$

We define the following nonnegative energy function as the candidate of Lyapunov function:

$$V = \coth\left(\lambda_2 - \epsilon\right). \tag{37}$$

With respect to [20], $\nabla_{q_i} V = \dot{q}_i = u_i$. Thus, the derivative of candidate of Lyapunov function is

$$\dot{V}(q) = \nabla_{q_i} V(q)\, \dot{q} = \dot{q}^T \dot{q} = -\sum_{i=1}^{N} p_i^T p_i. \tag{38}$$

Equation (36) can then be added into the previous flocking protocol as a new term in (21),

$$\begin{aligned} u_i = {} & -\nabla_{q_i} \mathscr{C}_i(q) - \nabla_{q_i} \mathscr{T}_i(q) + \nabla_{q_i} \mathscr{O}_i(q) \\ & + \sum_{j\in\mathscr{N}_i} \nabla_q \mathscr{R}_{ij}(q) - \sum_{j\in\mathscr{N}_i} \left(p_i - p_j\right) \\ & - \operatorname{csch}^2\left(\lambda_2 - \epsilon\right) \\ & \cdot \sum_{j\in\mathscr{N}_i} \exp^{-((1/2)\|q_i - q_j\|^2/R)} R \left(v_i - v_j\right)^2 \left(q_i - q_j\right). \end{aligned} \tag{39}$$

To analyze the stability of flocking controller after u_i^{λ} addition, we take the derivative in (28) with respect to q_i and add it with the derivative of Lyapunov function in (38) with respect to q_i, such that

$$\begin{aligned} \dot{V}(q) &= -\sum_{i=1}^{N} \left(p_i^T \sum_{j\in\mathscr{N}_i} a_{ij} \left(p_i - p_j\right)\right) - \sum_{i=1}^{N} p_i^T p_i \\ &= -2 \sum_{i=1}^{N} \left(p^T (L \otimes I)\, p\right) = -2 \left(p^T (L \otimes I)\, p\right) \\ &\le 0. \end{aligned} \tag{40}$$

---

initialization all variables and constants
while $t < t_{\text{end}}$ **do**
  capture position q;
  for $i = 1$ to N **do**
    compute $\nabla_{q_i} \mathscr{T}_i$, $\nabla_{q_i} \mathscr{O}_{i,k}$;
    if $q_{ij} \le R$ **then**
      build $[a_{ij}]$;
      compute $\sum_{i=1}^{N} (\nabla_{q_i} \mathscr{R}_i + \nabla_{q_i} \mathscr{C}_i)$;
    end
  end
  compute L, λ_2, p_i;
  compute u_i
  update q_i;
  update u_i;
end

Algorithm 1: Main algorithm.

It means that adding gradient of algebraic connectivity to the previous flocking control input remains guarantees the stability in the sense of Lyapunov, connectivity can be maintained, collision between agents as well as with obstacle can be avoided, and the agent can reach the common desired position.

5. Simulation of Multiagent Connectivity Preserving Flocking

This section demonstrates the simulation results of connectivity preserving flocking in multiagent system using the stable control input obtained from the previous section. The simulation verifies flocking protocol, which is derived using control Lyapunov function. The common algorithm to simulate multiagent robot with connectivity preserving flocking is depicted in Algorithm 1. For simulation purpose, the different flocking scenarios are also presented: workspace without obstacle, with single obstacle, and with multiple obstacles. In all scenarios, the point mass mobile robot agents move from the initial position to the goal position while avoiding each other.

All scenarios in simulation are conducted in Matlab. The initial position of each robot agent is generated randomly with initial connected topology at $X = 1$ to 5, $Y = 1$ to 5. The size of obstacles is varied from $r_k = 1$ to $r_k = 3$. For the purpose of simulation, we use the same communication radius $R = 5$. The simulation scenarios are determined as follows:

(1) multiagent flocking without obstacle,

(2) multiagent flocking with single obstacle,

(3) multiagent flocking with multiple obstacles,

(4) multiagent flocking with algebraic connectivity controller.

(1) Multiagent Flocking without Obstacle. In the first scenario, we present the result of simulation of connectivity preserving

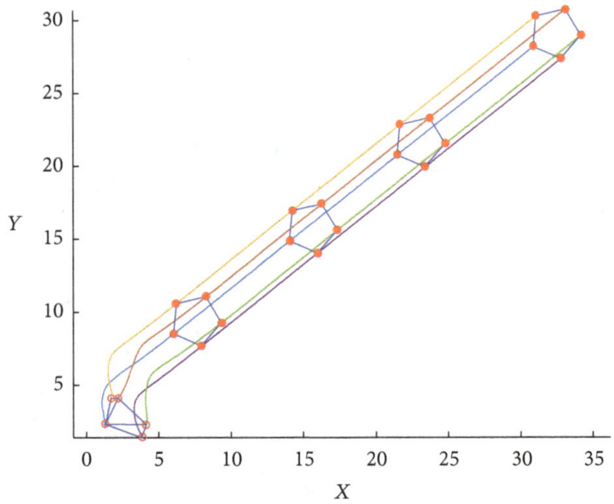

FIGURE 4: Multiagent flocking trajectories without obstacle.

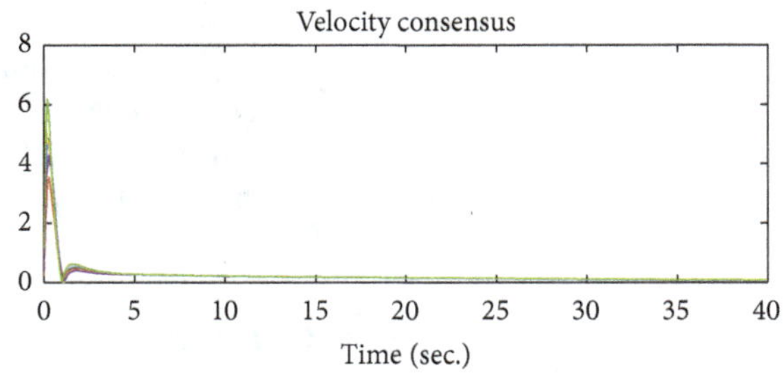

FIGURE 5: Multiagent velocity trajectory in no obstacle scenario.

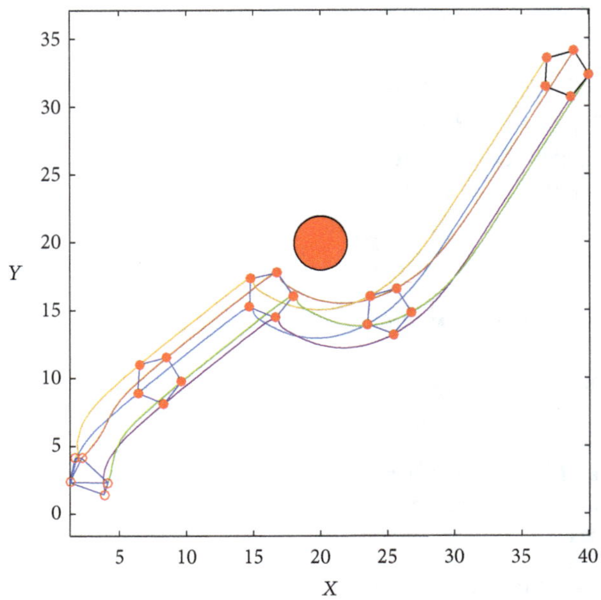

FIGURE 6: Multiagent robot trajectories with single obstacle.

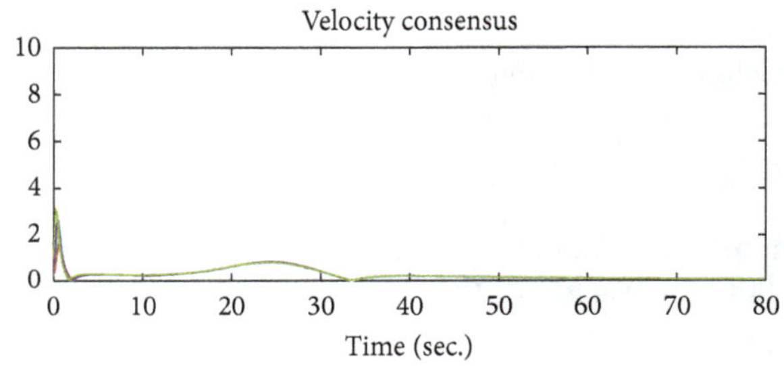

FIGURE 7: Multiagent velocity trajectory in the scenario with obstacle.

flocking without obstacle in the configuration workspace. We simulate five points of robots that represent robot agent, randomly starting from initial condition around $XY = (0, 0)$ up to $(5, 5)$ in coordinate plane, and target position is $XY = (35, 35)$. We plot the trajectory of robot during simulation and the result is depicted in Figure 4.

From simulation, it can be seen that the multiagent graph is always connected during flocking evolution. The convergence of agent's velocity is determined by consensus term as described in control input in (21). In Figure 5 it can be seen that the velocity of agents converges to the same value when consensus is used in the connectivity preserving flocking protocol.

(2) Multiagent Flocking with Single Obstacle. We plot the trajectory of a robot during simulation as depicted in Figure 6. The multiagent graph is always connected during flocking evolution. In this simulation, we can also see that velocity of agents still converges to the same value, as depicted in Figure 7. A slightly different result is obtained when obstacle is present in the workspace. Agents can reach velocity convergence in the beginning. However, when detecting the obstacles, the velocity is a bit varying before it finally converges.

(3) Multiagent Flocking with Multiple Obstacles. In this scenario, multiple sizes of multiple obstacles are present in the workspace. We assume that three multiple sizes of obstacles represent a small, medium, and large size of obstacle. For this purpose, we use two different subscenarios: first is simulation with five robot agents, and the second one is with ten robot agents.

The robot trajectories during simulation of multiagent with multiple size obstacles can be seen in Figure 8. We also plot the velocity convergence of flock during simulation and the results are depicted in Figure 9. In Figure 10, ten robot agents are simulated to perform flocking and reach the goal position. This is to see how the flocking control input works with large number of agents. Ten agents can still move in the flock while avoiding three different sizes of obstacle and finally reach the desired position.

(4) Multiagent Flocking with Algebraic Connectivity Controller. In this scenario, we use (39) to preserve connectivity by using gradient of algebraic connectivity. During simulation, the values of algebraic connectivity, which indicates the measure of connectedness of multiagent network, are plotted and depicted in Figure 13. From the beginning of simulation, it can be seen that the value of algebraic connectivity remains >0, meaning that the flocking controller still guarantees

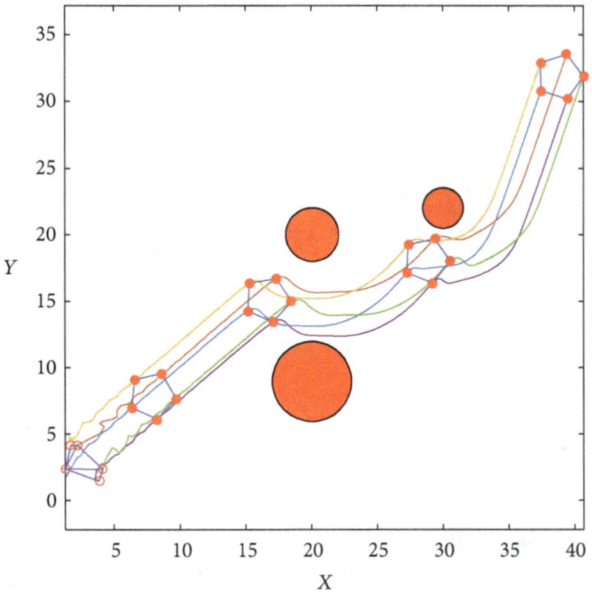

FIGURE 8: Multiagent robot trajectories with multiple size multiple obstacles.

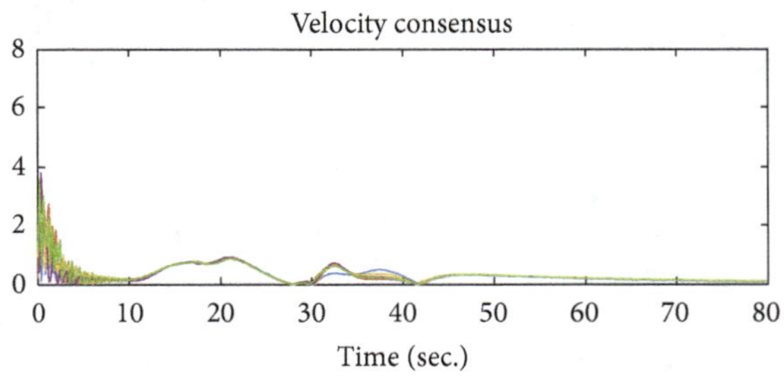

FIGURE 9: Multiagent velocity convergence in multiple size multiple obstacles.

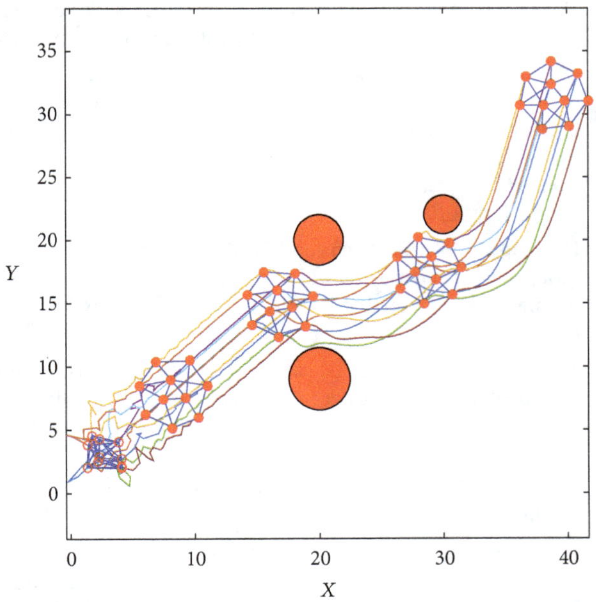

FIGURE 10: Multiagent robot trajectories of 10 agents with multiple obstacles.

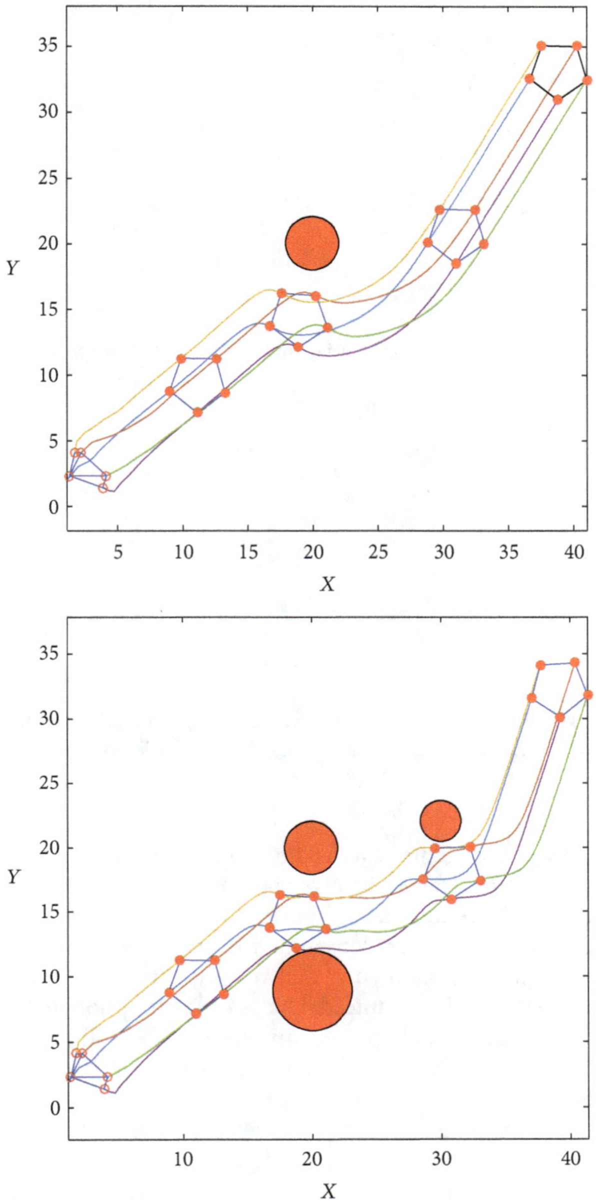

FIGURE 11: Multiagent robot trajectories with single and multiple obstacles with algebraic connectivity controller.

connectivity while avoiding obstacle as illustrated by robot trajectories shown in Figure 11.

We also compare the velocity convergence where the gradient of λ_2 is added into the flocking control input, and the plot is depicted in Figure 12. By adding λ_2 gradient controller, velocity alignment in flocking protocol is smooth and faster compared to flocking controller without λ_2 gradient controller.

6. Conclusion

In this paper, the control Lyapunov function is successfully used to construct flocking control input with connectivity preservation in a group of networks of mobile robot agents.

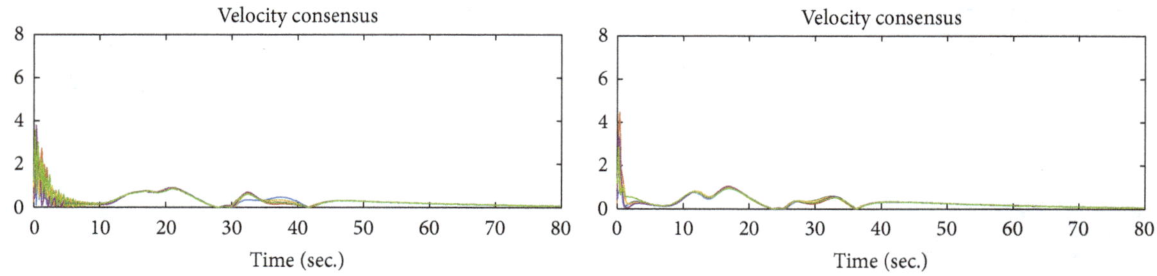

(a) (b)

FIGURE 12: Multiagent velocity with multiple obstacles scenario. (a) Without λ_2 gradient. (b) With λ_2 gradient.

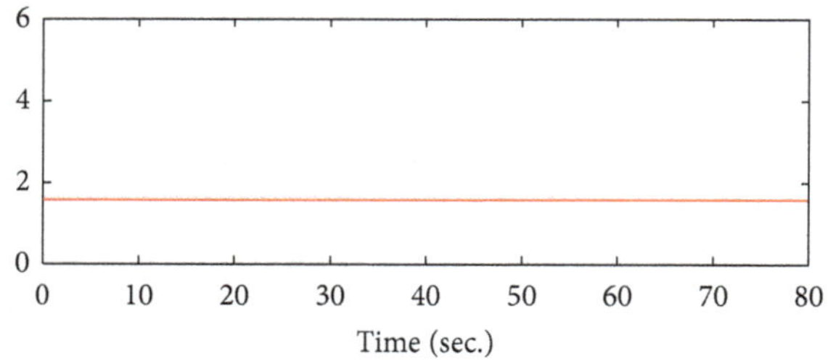

FIGURE 13: Algebraic connectivity value during simulation of multiagent velocity convergence in multiple size multiple obstacles.

We model the agents as a dynamic point mass, and the proposed control strategy involves potential force and velocity consensus. Several potential forces are designed to steer the neighbor agents to predetermined distance while avoiding collision among them such that connectedness of network topologies can be guaranteed. Under the assumption that the initial flocking topology is connected, the overall flocking topology is preserved to be connected while obstacle and collision avoidance are also guaranteed.

Simulation results verify the new proposed control input with gradient of algebraic connectivity. From the multiagent flocking point of view, it implies that the overall controller could maintain connectivity of agents, forming cohesive multiple agents, and matching velocities. In order to guarantee goal achievement of multiagent flocking, connectivity is still preserved by keeping the second smallest eigenvalue of flocking topology $\lambda_2 > 0$.

Competing Interests

The authors declare that there are no competing interests regarding the publication of this article.

References

[1] C. W. Reynolds, "Flocks, herds and schools: a distributed behavioral model," in *Proceedings of the 14th Annual Conference on Computer Graphics and Interactive Techniques (SIGGRAPH '87)*, vol. 21, pp. 25–34, ACM, Anaheim, Calif, USA, 1987.

[2] H. G. Tanner, A. Jadbabaie, and G. J. Pappas, "Stable flocking of mobile agents, part i: fixed topology," in *Proceedings of the 42nd IEEE Conference on Decision and Control*, vol. 2, pp. 2010–2015, IEEE, Maui, Hawaii, USA, 2003.

[3] H. G. Tanner, A. Jadbabaie, and G. J. Pappas, "Stable flocking of mobile agents part II: dynamic topology," in *Proceedings of the 42nd IEEE Conference on Decision and Control*, vol. 2, pp. 2016–2021, IEEE, 2003.

[4] R. Olfati-Saber, "Flocking for multi-agent dynamic systems: algorithms and theory," *IEEE Transactions on Automatic Control*, vol. 51, no. 3, pp. 401–420, 2006.

[5] J. Penders, "Robot swarming applications," in *Liber Amicorum ter gelegenheid van de 60e verjaardag van Prof.dr. H. Jaap van den Herik*, J. Donkers, Ed., pp. 227–234, Maastricht ICT Competence Center, 2007.

[6] B. Erfianto, B. R. Trilaksono, and H. Hindersah, "Signal strength-based adjacency matrix and its eigenvalue in mobile robotic networks," in *Proceedings of the International Conference of Information and Communication Technology (ICoICT '13)*, pp. 247–252, Bandung, Indonesia, March 2013.

[7] M. Zavlanos and G. Pappas, "Flocking while preserving connectivity," in *Proceedings of the IEEE Conference on Decision and Control (CDC '07)*, pp. 2919–2924, New Orleans, La, USA, December 2007.

[8] M. M. Zavlanos, H. G. Tanner, A. Jadbabaie, and G. J. Pappas, "Hybrid control for connectivity preserving flocking," *IEEE Transactions on Automatic Control*, vol. 54, no. 12, pp. 2869–2875, 2009.

[9] Y. Mao, L. Dou, H. Fang, H. Liu, and H. Cao, "Connectivity-preserving flocking of multi-agent systems with application towheeled mobile robots," in *Proceedings of the 29th Chinese Control Conference (CCC '10)*, pp. 4494–4500, IEEE, Beijing, China, July 2010.

[10] H. Fang, Y. Wei, J. Chen, and B. Xin, "Flocking of second-order multiagent systems with connectivity preservation based on algebraic connectivity estimation," *IEEE Transactions on Cybernetics*, 2016.

[11] M. Ji and M. Egerstedt, "A graph-theoretic characterization of controllability for multi-agent systems," in *Proceedings of the American Control Conference (ACC '07)*, pp. 4588–4593, IEEE, New York, NY, USA, July 2007.

[12] M. Ji and M. B. Egerstedt, "Distributed coordination control of multiagent systems while preserving connectedness," *IEEE Transactions on Robotics*, vol. 23, no. 4, pp. 693–703, 2007.

[13] R. K. Williams and G. S. Sukhatme, "Topology-constrained flocking in locally interacting mobile networks," in *Proceedings of the IEEE International Conference on Robotics and Automation (ICRA '13)*, pp. 2002–2007, Karlsruhe, Germany, May 2013.

[14] B. N. Sharma, J. Vanualailai, and U. Chand, "Flocking of multi-agents in constrained environments," *European Journal of Pure and Applied Mathematics*, vol. 2, no. 3, pp. 401–425, 2009.

[15] C. Godsil and G. F. Royle, *Algebraic Graph Theory*, vol. 207 of *Graduate Texts in Mathematics*, Springer, New York, NY, USA, 2001.

[16] E. Rimon and D. E. Koditschek, "Exact robot navigation using artificial potential functions," *IEEE Transactions on Robotics and Automation*, vol. 8, no. 5, pp. 501–518, 1992.

[17] P. Ögren, M. Egerstedt, and X. Hu, "A control Lyapunov function approach to multiagent coordination," *IEEE Transactions on Robotics and Automation*, vol. 18, no. 5, pp. 847–851, 2002.

[18] S. Ge, "Lyapunov design. Control Systems, Robotics And Automation, XIII," 2007.

[19] B. Sharma, J. Vanualailai, and J. Raj, "Obstacle and collision avoidance control laws of a swarm of boids," *International Journal of Mathematical, Computational, Statistical, Natural and Physical Engineering*, vol. 8, no. 2, pp. 253–258, 2014.

[20] P. Yang, R. A. Freeman, G. J. Gordon, K. M. Lynch, S. S. Srinivasa, and R. Sukthankar, "Decentralized estimation and control of graph connectivity for mobile sensor networks," *Automatica*, vol. 46, no. 2, pp. 390–396, 2010.

5

A Robust Control for an Aerial Robot Quadrotor under Wind Gusts

Li Ding[1,2] and Zhenwei Wang[1]

[1]College of Mechanical Engineering, Jiangsu University of Technology, Changzhou 213001, China
[2]Department of Industrial and System Engineering, The Hong Kong Polytechnic University, 999077 Hung Hom, Hong Kong

Correspondence should be addressed to Li Ding; nuaadli@163.com

Academic Editor: L. Fortuna

A robust flight controller based on linear active disturbance rejection control (LADRC) is proposed for stability control of an aerial robot quadrotor under wind gusts. The nonlinear dynamical model of the quadrotor, considering the wind disturbance, is firstly established through Newton-Euler method. Subsequently, a robust LADRC technique is proposed to design the controllers for the inner loop and outer loop of the aircraft. In this control scheme, the linear extended state observer (LESO) serves as a compensator which can effectively reject the wind gusts. Then, a method of parameter tuning is introduced to obtain the optimized control performance. Finally, the effectiveness and advantages of the proposed controller are demonstrated through series of simulation case.

1. Introduction

Quadrotor is the most popular configuration of multirotor unmanned aerial robot, extensively for civilian applications [1, 2]. In recent years, significant growth of interest is witnessed towards the academic research of the quadrotor. This is possibly attributed to their certain features, such as simple mechanical structure, hovering, and agile maneuverability. With the increasing requirement of autonomous flight under different conditions, control of the quadrotor is an important challenge. Complex rotation and translation operations are also commonly encountered in the mathematical model, resulting in complex equations that have to be taken into account when designing control laws. On the other hand, the position control and attitude control of the quadrotor are extremely sensitive to external disturbance, such as wind gusts. Particularly, robustness issues may be critical for the control of quadrotor since they are subjected to the complicated dynamics and external disturbances. Several proper control methodologies for the quadrotor have been reported in the literatures, such as proportion integration differentiation (PID) control [3], backstepping control [4], sliding mode control [5], and model predictive control [6]. Nevertheless, there are a few researches about the robust control of the quadrotor, especially under the wind gusts.

In order to weaken the effect of wind gusts, some authors have proposed complicated control techniques to achieve stability. In the literatures [7, 8], adaptive controllers of different structures were verified by simulations and experiments on test bench. For control of a multirotor aircraft, Salazar et al. [9] proposed a novel mathematical model of the wind gusts and designed a simple nonlinear control law to resist the wind disturbance. The effect of wind disturbance was discussed towards the issue of (unmanned air vehicle) UAV path planning in the literatures [10, 11]. From the analysis, compensation control for the aircraft may be an effective method. Dong et al. [12] designed a flight controller with disturbance observer (DOB) to guarantee the high-performance trajectory tracking of a quadrotor. The DOB-based controller is robust to wind gusts without the use of high control gain or extensive computational power.

Our team has proposed in previous articles, like [13], the use of LADRC to control a small-scale unmanned helicopter. This control strategy is rooted in active disturbance rejection control (ADRC) which was firstly proposed by Han [14]. LADRC is a simplified implementation of the ADRC, in which bandwidth is the only adjustable parameter of the control performance. A previous contribution from our team presented the parameter tuning towards the LADRC based on the artificial bee colony algorithm [15]. Detailed background and principles on the ABC can be found in the literatures [16, 17], and the authors analyzed the performance of the ABC on optimization problem.

The content is organized in the following manner: Section 2 presents the mathematical model of the quadrotor under the wind gusts. In Section 3, the robust controller is described using the LADRC strategy. And the proof of the stability analysis for the proposed controller is given in Section 4. A method of parameter tuning is applied in Section 5, to optimize the performance of our proposed controller. The control law with wind gusts is simulated and illustrated in Section 6. Finally, our conclusions and contributions are discussed in Section 7.

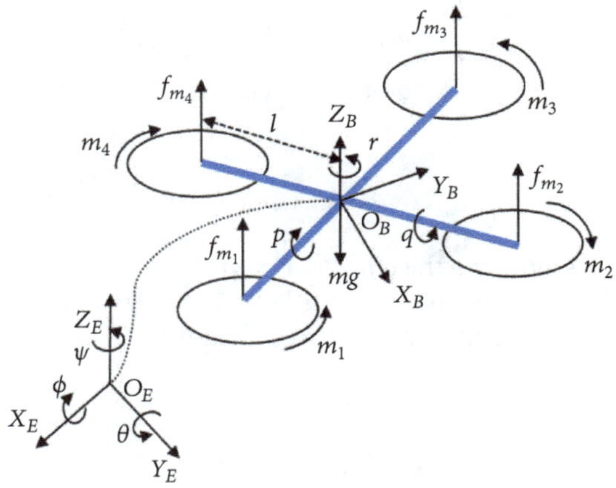

Figure 1: Free body diagram of quadrotor.

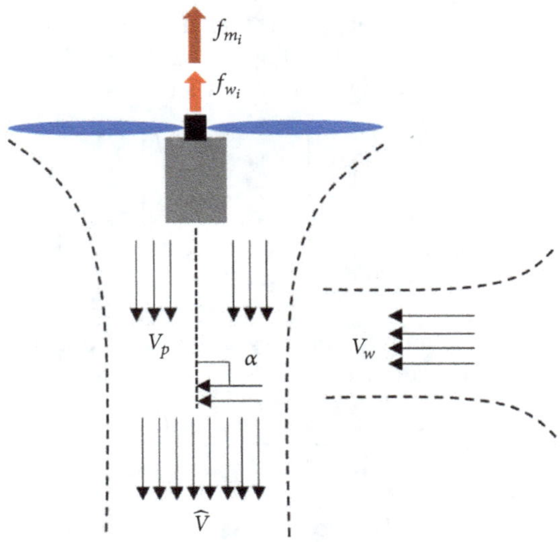

Figure 2: Analysis of the main and lateral thrusts.

2. Mathematical Model

The free body diagram and coordinate systems for the quadrotor are given in Figure 1. It is controlled by the angular speeds of four electric motors. Each motor produces a thrust and a torque m_i $(i = 1, 2, 3, 4)$, whose combination generates the main thrust input U_1, pitch input U_2, roll input U_3, and yaw input U_4. Two reference frames are subjected to the quadrotor, namely, the earth-fixed frame $\{O_E X_E Y_E Z_E\}$ and the body-fixed frame $\{O_B X_B Y_B Z_B\}$. The position (x, y, z), attitude (pitch θ, roll ϕ and yaw ψ), and motion can be achieved through the two frames. According to Newton-Euler method. The six-degrees-of-freedom dynamics of the quadrotor are modeled as [18]

$$\begin{aligned}
\ddot{x} &= \frac{U_1 (c\psi s\theta c\phi + s\psi s\phi)}{m} \\
\ddot{y} &= \frac{U_1 (s\psi s\theta c\phi - c\psi s\phi)}{m} \\
\ddot{z} &= \frac{U_1 c\phi c\theta}{m} - g \\
\ddot{\phi} &= \frac{U_2}{I_{xx}} + \frac{\dot{\theta}\dot{\psi}(I_{yy} - I_{zz})}{I_{xx}} \\
\ddot{\theta} &= \frac{U_3}{I_{yy}} + \frac{\dot{\phi}\dot{\psi}(I_{zz} - I_{xx})}{I_{yy}} \\
\ddot{\psi} &= \frac{U_4}{I_{zz}} + \frac{\dot{\phi}\dot{\theta}(I_{xx} - I_{yy})}{I_{zz}}
\end{aligned} \tag{1}$$

where I_{xx}, I_{yy}, and I_{zz} are the moment of inertia, m is the mass, and g is the gravitational acceleration. And $s\alpha$ and $c\alpha$ are the abbreviations for $\cos\alpha$ and $\sin\alpha$.

The thrust of each propeller is calculated by $f_{m_i} = k_t \omega_i^2$ $(i = 1, 2, 3, 4)$ with k_t being the thrust coefficient and ω_i being the angular speeds. The relationship between the four inputs and angular speed is expressed by

$$\begin{bmatrix} U_1 \\ U_2 \\ U_3 \\ U_4 \end{bmatrix} = \begin{bmatrix} k_t & k_t & k_t & k_t \\ 0 & -k_t l & 0 & k_t l \\ -k_t l & 0 & k_t l & 0 \\ -k_m & k_m & -k_m & k_m \end{bmatrix} \begin{bmatrix} \omega_1^2 \\ \omega_2^2 \\ \omega_3^2 \\ \omega_4^2 \end{bmatrix} \tag{2}$$

where l denotes the distance from the center of mass to the center of the propeller and k_m is termed as antitorque coefficient.

In the outdoor flight, the quadrotor is generally exposed to lateral wind gusts. The aircraft will be yawed or overturned if it is subjected to the lateral wind gusts. Stating precisely, this leads to additional lateral air flow acting on the propeller, as shown in Figure 2. Hence, the total thrust $f_{T_i} = f_{m_i} + f_{w_i}$ $(i = 1, 2, 3, 4)$ could be described as [19]

$$f_{T_i} = 2\rho A\widehat{V}V_p \tag{3}$$

where ρ is the air density and A is the propeller area. V_p is the induced wind speed of the propeller and $\widehat{V}$ is the total wind induced speed of the rotor. The relationship between them is given by

$$\widehat{V} = \left[\left(V_w \cos\alpha + V_p\right)^2 + \left(V_w \sin\alpha\right)^2\right]^{1/2} \tag{4}$$

where α represents the angle between the propeller axis and the lateral wind gusts. Since the latter is perpendicular to the former, $\alpha = 90°$. At this moment, the additional lateral forces are expressed by

$$f_{w_i} = 2\rho A V_p^2 \left(1 + \frac{V_w^2}{V_p^2}\right)^{1/2} - f_{m_i} \tag{5}$$

Moreover, the aerodynamics drag is defined as $m_{drag} = \rho A V_w^2/2 = k_{drag} V_w^2$, where $k_{drag} > 0$ is a constant depending on the parameter ρ, shape of the blade, and other factors. When the aircraft is disturbed by the lateral wind gusts, the additional torques acting on the propellers are represented by

$$\begin{bmatrix} m_{w_\phi} \\ m_{w_\theta} \\ m_{w_\psi} \end{bmatrix} = \begin{bmatrix} \left(f_{w_4} - f_{w_2}\right) l \\ \left(f_{w_3} - f_{w_1}\right) l \\ \sum_{i=1}^{4} m_{drag_i} \end{bmatrix} \tag{6}$$

Using (1), (5), and (6), the integrated quadrotor dynamical model can be rewritten as

$$\begin{aligned}
\ddot{x} &= \frac{U_1 \left(c\psi s\theta c\phi + s\psi s\phi\right)}{m} + \sum_{i=1}^{4} \frac{f_{w_i}}{m} \\
\ddot{y} &= \frac{U_1 \left(s\psi s\theta c\phi - c\psi s\phi\right)}{m} + \sum_{i=1}^{4} \frac{f_{w_i}}{m} \\
\ddot{z} &= \frac{U_1 c\phi c\theta}{m} - g + \sum_{i=1}^{4} \frac{f_{w_i}}{m} \\
\ddot{\phi} &= \frac{U_2}{I_{xx}} + \frac{\dot{\theta}\dot{\psi}\left(I_{yy} - I_{zz}\right)}{I_{xx}} + \frac{m_{w_\phi}}{I_{xx}} \\
\ddot{\theta} &= \frac{U_3}{I_{yy}} + \frac{\dot{\phi}\dot{\psi}\left(I_{zz} - I_{xx}\right)}{I_{yy}} + \frac{m_{w_\theta}}{I_{yy}} \\
\ddot{\psi} &= \frac{U_4}{I_{zz}} + \frac{\dot{\phi}\dot{\theta}\left(I_{xx} - I_{yy}\right)}{I_{zz}} + \frac{m_{w_\psi}}{I_{zz}}
\end{aligned} \tag{7}$$

3. Robust Controller Design

According to (7), the input and output of each control channel present second-order derivative relation. Thus, we use the second-order system in the following form for describing the LADRC:

$$\begin{aligned}
\ddot{x}_s &= f_s\left(x_s, \dot{x}_s, w_s, t\right) + b(t) u \\
y_s &= x_s
\end{aligned} \tag{8}$$

where u and y_s denote input and output, respectively. f_s is the external disturbance. x_s is state variables and $b(t)$ is the control function.

Suppose b_0 is the approximate value of $b(t)$, and (8) is written equivalently as

$$\begin{aligned}
\ddot{x}_s &= f_s\left(x_s, \dot{x}_s, w_s, t\right) + b_0 u = f_s + b_0 u \\
y_s &= x_s
\end{aligned} \tag{9}$$

By rewriting the above equation in the form of differential equation, one has

$$\begin{aligned}
\dot{x}_{s_1} &= x_{s_2} \\
\dot{x}_{s_2} &= x_{s_3} + b_0 u \\
\dot{x}_{s_3} &= h \\
y_s &= x_{s_1}
\end{aligned} \tag{10}$$

with $x_{s_3} = f_s$ added as an augmented state and $h = \dot{f}_s$ as unknown disturbance. Now f_s can be estimated through a state observer based on a state space model:

$$\begin{aligned}
\dot{\mathbf{x}} &= \mathbf{A}\mathbf{x} + \mathbf{B}\mathbf{u} + \mathbf{E}\mathbf{h} \\
\mathbf{y} &= \mathbf{C}\mathbf{x}
\end{aligned} \tag{11}$$

where

$$\begin{aligned}
\mathbf{A} &= \begin{bmatrix} 0 & 1 & 0 \\ 0 & 0 & 1 \\ 0 & 0 & 0 \end{bmatrix}, \\
B &= \begin{bmatrix} 0 \\ b_0 \\ 0 \end{bmatrix}, \\
\mathbf{C} &= \begin{bmatrix} 1 \\ 0 \\ 0 \end{bmatrix}, \\
\mathbf{E} &= \begin{bmatrix} 0 \\ 0 \\ 1 \end{bmatrix}
\end{aligned} \tag{12}$$

The LESO of (11) can be constructed as

$$\begin{aligned}
\dot{\mathbf{z}} &= \mathbf{A}\mathbf{z} + \mathbf{B}\mathbf{u} + \mathbf{L}\left(\mathbf{y} - \hat{\mathbf{y}}\right) \\
\hat{\mathbf{y}} &= \mathbf{C}\mathbf{z}
\end{aligned} \tag{13}$$

where $\mathbf{L}$ is the observer gain. It is calculated as [17]

$$\mathbf{L} = \begin{bmatrix} l_1 \\ l_2 \\ l_3 \end{bmatrix} = \begin{bmatrix} \alpha_1 \omega_o \\ \alpha_2 {\omega_o}^2 \\ \alpha_3 {\omega_o}^3 \end{bmatrix} \tag{14}$$

where $\omega_o > 0$ is the observer bandwidth. It results in the characteristic polynomial of (13) being

$$\lambda(s) = s^3 + \alpha_1 s^2 + \alpha_2 s + \alpha_3 = (s + \omega_o)^3 \quad (15)$$

From the polynomial, one obtains

$$\begin{aligned} \alpha_1 &= 3, \\ \alpha_2 &= 3, \\ \alpha_3 &= 1 \end{aligned} \quad (16)$$

The observer can track the state variables with well-tuned observer bandwidth and yield $z_1(t) \longrightarrow \hat{y}(t)$, $z_2(t) \longrightarrow \dot{\hat{y}}(t)$, and $z_3(t) \longrightarrow f_s$.

With the estimate of f_s, the LADRC can actively compensate for effect of the external disturbance in real time as

$$u = \frac{u_0 - z_3}{b_0} \quad (17)$$

The closed-loop control system is converted into a unit double integral plant. And the control law is [20, 21]

$$\ddot{y} = f - z_3 + u_0 \approx u_0 = \lambda_1 (r_e - z_1) - \lambda_2 z_2 \quad (18)$$

where r_e is the reference signal.

The transfer function of the second-order plant is given as

$$G_2 = \frac{\lambda_1}{s^2 + \lambda_2 s + \lambda_1} \quad (19)$$

Here, the gains can be calculated as [17]

$$\begin{aligned} \lambda_1 &= \omega_c^2, \\ \lambda_2 &= 2\omega_c \end{aligned} \quad (20)$$

4. Stability Analysis

Define the error as

$$\mathbf{e} = \mathbf{x} - \mathbf{z} \quad (21)$$

Subtracting (13) from (11), the error equation is obtained as

$$\dot{\mathbf{e}} = \mathbf{A}_e \mathbf{e} + \mathbf{E}\mathbf{h} \quad (22)$$

where

$$\mathbf{A}_e = \mathbf{A} - \mathbf{L}\mathbf{C} = \begin{bmatrix} -l_1 & 1 & 0 \\ -l_2 & 0 & 1 \\ -l_3 & 0 & 0 \end{bmatrix} \quad (23)$$

Obviously, the LESO is bounded-input bounded-output (BIBO) stable if the roots of the characteristic polynomial, namely (15), are all in the left half plane and $\mathbf{h}$ is bounded.

Theorem 1. *The design of LADRC from (13) to (18) yields a BIBO stable closed-loop system if the LESO and the control law for the unit double integral plant are stable.*

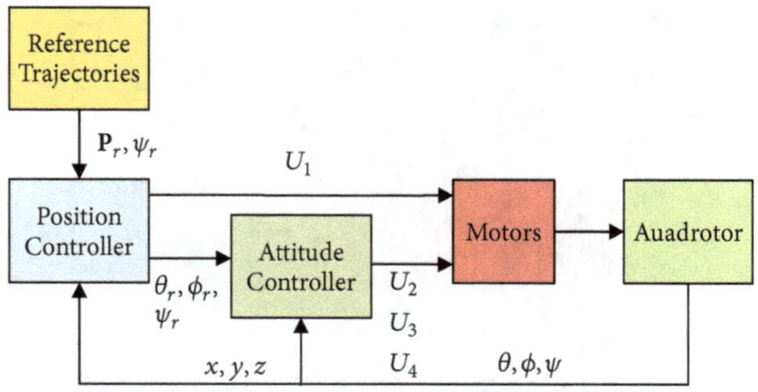

FIGURE 3: Control system architecture of the quadrotor.

Proof. According to (17) and (18), one has

$$\begin{aligned} u &= \frac{1}{b_0} \begin{bmatrix} -\lambda_1 & -\lambda_2 & -1 \end{bmatrix} \begin{bmatrix} z_1 - r_e \\ z_2 \\ z_3 \end{bmatrix} = \mathbf{Q}\mathbf{z} - \frac{1}{b_0} \begin{bmatrix} r_e \\ 0 \\ 0 \end{bmatrix} \\ &= \mathbf{Q}\mathbf{z} - \mathbf{G} \end{aligned} \quad (24)$$

where

$$\mathbf{Q} = \frac{1}{b_0} \begin{bmatrix} -\lambda_1 & -\lambda_2 & -1 \end{bmatrix} \quad (25)$$

The closed-loop control system is rewritten as

$$\dot{\mathbf{x}} = \mathbf{A}\mathbf{x} + \mathbf{B}\mathbf{u} + \mathbf{E}\mathbf{h} = \mathbf{A}\mathbf{x} + \mathbf{B}\mathbf{Q}\mathbf{z} - \mathbf{B}\mathbf{G} + \mathbf{E}\mathbf{h} \quad (26)$$

$$\begin{aligned} \dot{\mathbf{z}} &= \mathbf{A}\mathbf{z} + \mathbf{B}\mathbf{u} + \mathbf{L}(\mathbf{y} - \hat{\mathbf{y}}) \\ &= \mathbf{L}\mathbf{C}\mathbf{x} + (\mathbf{A} - \mathbf{L}\mathbf{C} + \mathbf{B}\mathbf{Q})\mathbf{z} - \mathbf{B}\mathbf{G} \end{aligned} \quad (27)$$

Using the state space equation to represent the closed-loop control system, one gets

$$\begin{bmatrix} \dot{\mathbf{x}} \\ \dot{\mathbf{z}} \end{bmatrix} = \begin{bmatrix} \mathbf{A} & \mathbf{B}\mathbf{Q} \\ \mathbf{L}\mathbf{C} & \mathbf{A} - \mathbf{L}\mathbf{C} + \mathbf{B}\mathbf{Q} \end{bmatrix} \begin{bmatrix} \mathbf{x} \\ \mathbf{z} \end{bmatrix} + \begin{bmatrix} -\mathbf{B} & \mathbf{E} \\ -\mathbf{B} & \mathbf{0} \end{bmatrix} \begin{bmatrix} \mathbf{G} \\ \mathbf{h} \end{bmatrix} \quad (28)$$

Similarly, the control system is BIBO stable if the eigenvalues of (28) are in the left half plane. Since the reference signal r_e is always bounded, with the condition that the disturbance f_s satisfies differentiable, $\mathbf{h}$ is bounded. □

5. Parameter Tuning

According to the previous analysis, the control structure of the quadrotor is divided into two loops, i.e., inner loop and outer loop. The block diagram of such a structure is schematically shown in Figure 3. In the outer loop, the position vector $\mathbf{P}_r$ and yaw angle ψ_r are chosen as the reference signals. In the inner loop, the reference signal $(\theta_r, \phi_r, \psi_r)$ comes from the outer loop. Moreover, the four inputs generated by the two loops are converted into the angular speeds of motors.

The evaluation function is crucial for the parameter tuning. Consequently, an improved integral of time multiplied by absolute error (IITAE) is employed as performance index.

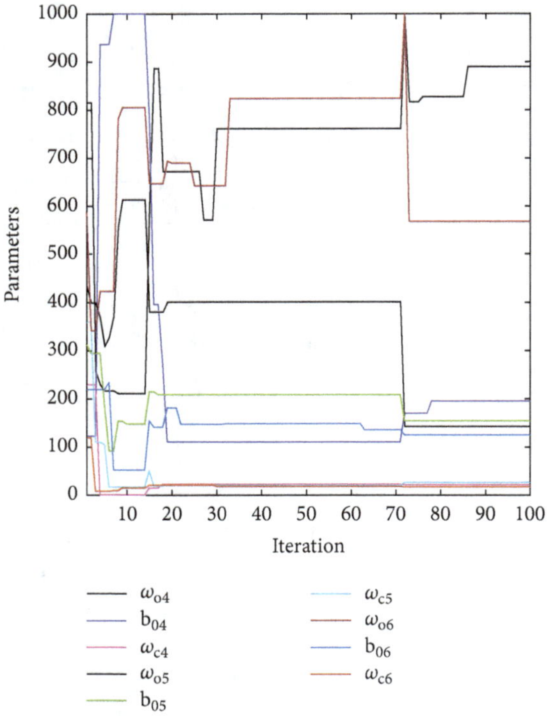

FIGURE 4: Evolving curves of the control parameters.

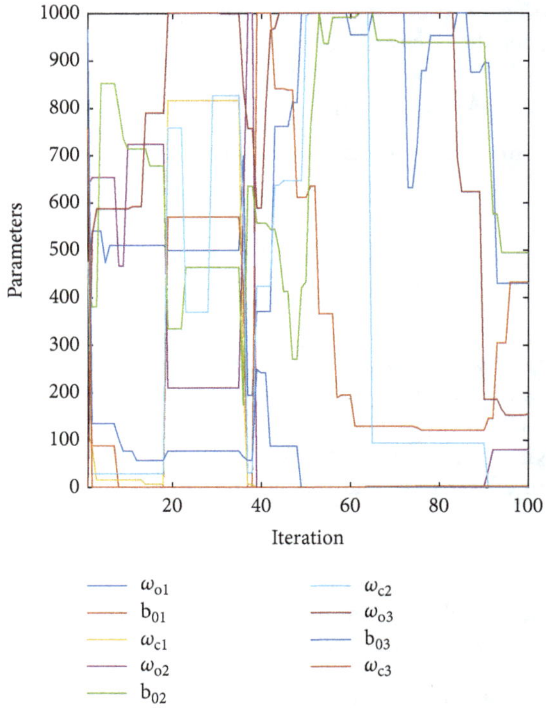

FIGURE 6: Tuned parameters of outer loop controller.

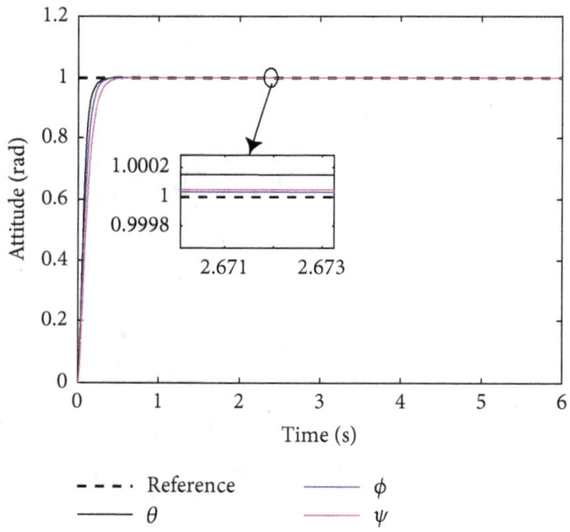

FIGURE 5: Response of the attitude.

TABLE 1: Physical parameters.

Item	Quantity
L (m)	0.2053
m (kg)	1.923
I_{xx} (kg·m^2)	0.094
I_{yy} (kg·m^2)	0.094
I_{zz} (kg·m^2)	0.086

The evaluation function of the performance index is defined as

$$F(t) = \alpha_1 \int_0^\infty t\,|e(t)|\,dt + \alpha_2 \int_0^\infty (u(t))^2\,dt + \alpha_3 t_s + \alpha_4 o \tag{29}$$

where $e(t)$ is the error signal in the time domain, t_s is the settling time, o_s is the overshoot, and α_1, α_2, α_3, and α_4 are the weight coefficients. Note that the parameter tuning of the LADRC is conducted by the ABC algorithm, and the exhaustive process is reported in the literature [13], and here repeating is unnecessary.

As taking the inner loop as an instance, the reference signal is set to 1 rad, 1 rad, 1 rad. There are three second-order LADRC in the inner loop, i.e., θ-LADRC, ϕ-LADRC, and ψ-LADRC. The physical parameters of the quadrotor are listed in Table 1. We use the ABC algorithm to tune the unknown control parameters under a unit step response case. The number of the iterations is 100. In addition, the simulation parameters of the wind gusts are given as ρ = 1.293 g/L, A = 0.0016 m^2, V_P = 2 m/s, and V_w = 6 m/s. And a white noise with the amplitude of 0.2 rad is added to the measured ports. The simulation time lasts for 6s. Figure 4 displays the evolving curves of the control parameters, which indicates the process of the parameter tuning by ABC. The attitude response is shown in Figure 5. It is obvious that the LADRC has a satisfactory performance of wind resistance, and the steady state error is less than 2%. Furthermore, the responses of the three attitude channels are fast and accurate.

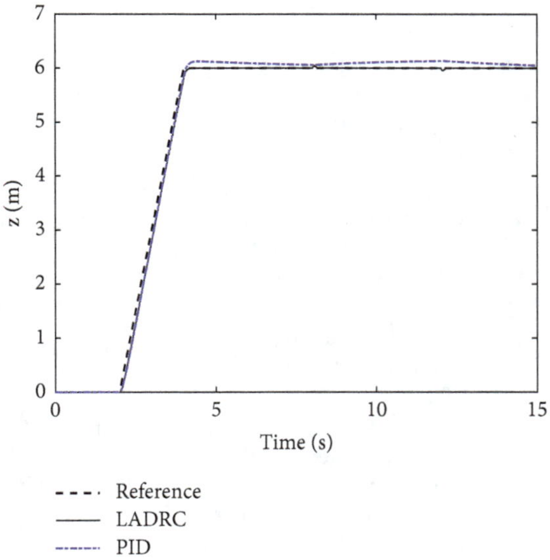

(a) $V_w = 4\,\mathrm{m/s}$

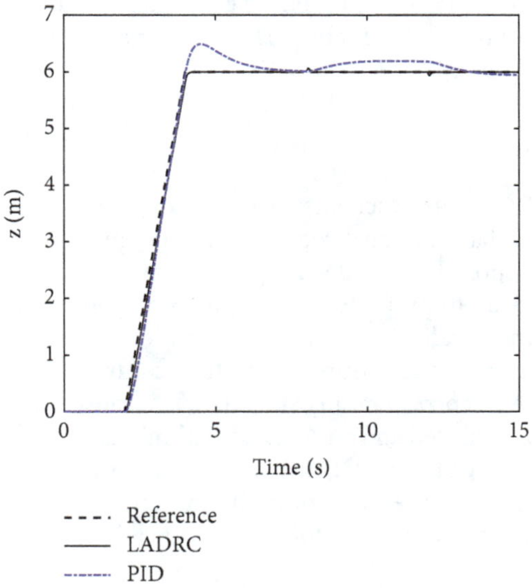

(b) $V_w = 10\,\mathrm{m/s}$

FIGURE 7: Position response under different wind gusts.

(a) Time=0s

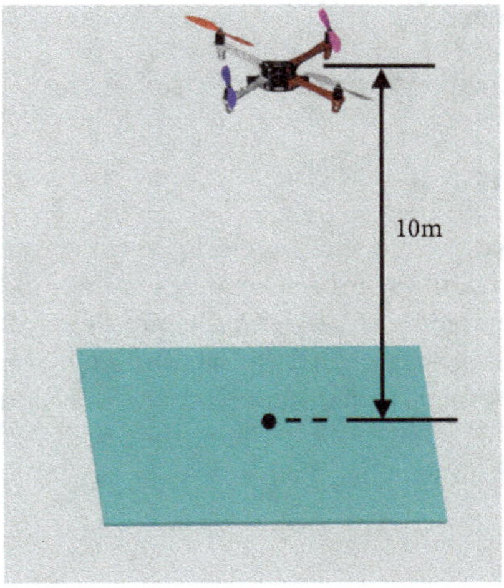

(b) Time=6s

FIGURE 8: Hovering control in Simscape environment.

6. Simulation Results

In order to validate the robustness and wind resistance of the proposed control scheme for stabilizing the quadrotor at trajectory tracking, the simulation is conducted in MATLAB 2017a programming environment on an Intel Core i7-4720 PC running Windows 10. Similarly, random wind gusts are added to the input of the outer loop.

As pointed out in Section 5, the control parameters of the outer loop controller are also tuned by the ABC algorithm. The tuned parameters are shown in Figure 6 using a unit step response case.

Moreover, we introduce a PID controller from the literature [22] as a comparison to test the robustness of our proposed controller. In the comparative simulation, the quadrotor follows a referenced trajectory of z direction as shown in Figure 7. Meanwhile, lateral wind gusts are applied with different wind speeds at a specific time period 8 s-12 s. The PID and LADRC are applied to stabilize the position of the aircraft, respectively. From the result, the response of LADRC is slightly faster than that of PID. As the wind speed increases, the tracking error becomes bigger. Nevertheless, it is obvious that the steady state error of LADRC is smaller than that of PID due to the LESO. In conclusion, the LADRC has a strong ability of wind disturbance compensation.

Lastly, a visual simulation is run in the MATALB/Simscape environment to display the trajectory tracking of the quadrotor. Note that the Simscape link utility creates a physical modeling XML file that represents the quadrotor's parts as bodies and maps the constraints between the parts into joints [23]. In this case, the task requires that the quadrotor takes off from the ground and flies in hover. The simulation time lasts for 8s. From the result in Figure 8, the quadrotor has a stable flight performance based on the proposed controller.

7. Conclusion

A new LADRC control scheme is given for the stability control of an aerial robot quadrotor under wind gusts in this article. The proposed controller roughly has two parts: the LESO part applied to properly estimate and compensate the wind disturbance leading to an attractive model-free feature and a PD part applied to ensure satisfactory control performance. Moreover, corresponding stability of the closed-loop control plant is analyzed and an ABC algorithm is applied for parameter tuning of the LADRC. Finally, all the simulation results show the proposed controller has a better ability of resisting wind gusts comparing to the traditional PID.

In the future, more advanced control strategies such as that reported in the literature [24] will be studied to control the movement of the quadrotor under wind gusts.

Conflicts of Interest

The authors declare that they have no conflicts of interest.

Acknowledgments

Helpful discussions with Professor Wu Hongtao from Nanjing University of Aeronautics and Astronautics on his guidance in dynamical modeling of aerial robots are gratefully acknowledged. This work was partially supported by the Foundation Research Project of Jiangsu Province (The Natural Science Fund no. BK20170315) and Changzhou Sci&Tech Program of China (Grant no. CJ20179017).

References

[1] R. Amin, L. Aijun, and S. Shamshirband, "A review of quadrotor UAV: Control methodologies and performance evaluation," *International Journal of Automation and Control Engineering*, vol. 10, no. 2, pp. 87–103, 2016.

[2] M. A. Tofigh, M. J. Mahjoob, and M. Ayati, "Dynamic modeling and nonlinear tracking control of a novel modified quadrotor," *International Journal of Robust and Nonlinear Control*, vol. 28, no. 2, pp. 552–567, 2018.

[3] S. An, S. Yuan, and H. Li, "Self-tuning of PID controllers design by adaptive interaction for quadrotor UAV," in *Proceedings of the 7th IEEE Chinese Guidance, Navigation and Control Conference, CGNCC 2016*, pp. 1547-1552, IEEE, Xiamen, China, August 2016.

[4] A. Das, F. Lewis, and K. Subbarao, "Backstepping approach for controlling a quadrotor using lagrange form dynamics," *Journal of Intelligent & Robotic Systems*, vol. 56, no. 1-2, pp. 127–151, 2009.

[5] D. Lee, H. J. Kim, and S. Sastry, "Feedback linearization vs. adaptive sliding mode control for a quadrotor helicopter," *International Journal of Control, Automation, and Systems*, vol. 7, no. 3, pp. 419–428, 2009.

[6] A. Chamseddine, Y. Zhang, C.-A. Rabbath, C. Fulford, and J. Apkarian, "Model reference adaptive fault tolerant control of a quadrotor UAV," in *Infotech@ Aerospace*, St. Louis, MO, USA, March 2011.

[7] F. Chen, F. Lu, B. Jiang, and G. Tao, "Adaptive compensation control of the quadrotor helicopter using quantum information technology and disturbance observer," *Journal of The Franklin Institute*, vol. 351, no. 1, pp. 442–455, 2014.

[8] S. Salazar, H. Romero, R. Lozano et al., "Modeling and real-time stabilization of an aircraft having eight rotors," *Journal of Intelligent and Robotic Systems*, vol. 54, no. 1-3, pp. 455–470, 2008.

[9] L. Ding, R. Ma, H. Wu, C. Feng, and Q. Li, "Yaw control of an unmanned aerial vehicle helicopter using linear active disturbance rejection control," *Proceedings of the Institution of Mechanical Engineers, Part I: Journal of Systems and Control Engineering*, vol. 231, no. 6, pp. 427–435, 2017.

[10] S. Jackson, J. Tisdale, M. Kamgarpour, B. Basso, and J. K. Hedrick, "Tracking controllers for small UAVs with wind disturbances: theory and flight results," in *Proceedings of the 47th IEEE Conference on Decision and Control (CDC '08)*, pp. 564–569, IEEE, Cancun, Mexico, December 2008.

[11] A. L. Jennings, R. Ordóñez, and N. Ceccarelli, "An ant colony optimization using training data applied to UAV way point path planning in wind," in *Proceedings of the 2008 IEEE Swarm Intelligence Symposium, SIS 2008*, USA, September 2008.

[12] W. Dong, G.-Y. Gu, X. Zhu, and H. Ding, "High-performance trajectory tracking control of a quadrotor with disturbance observer," *Sensors and Actuators A: Physical*, vol. 211, pp. 67–77, 2014.

[13] J. Song, Z. Gan, and J. Han, "Study of active disturbance rejection controller on filtering," *Control and Decision*, vol. 18, no. 1, pp. 110–112, 2003.

[14] D. Ma, Y. Xia, T. Li, and K. Chang, "Active disturbance rejection and predictive control strategy for a quadrotor helicopter," *IET Control Theory & Applications*, vol. 10, no. 17, pp. 2213–2222, 2016.

[15] D. Karaboga, "An idea based on honey bee swarm for numerical optimization," Technical report-tr06, Erciyes university, engineering faculty, computer engineering department, Kayseri, Turkey, 2005.

[16] L. Ding, H. Wu, Y. Yao, and Y. Yang, "Dynamic model identification for 6-DOF industrial robots," *Journal of Robotics*, vol. 2015, Article ID 471478, pp. 1–9, 2015.

[17] N. Michael, D. Mellinger, Q. Lindsey, and V. Kumar, "The GRASP multiple micro-UAV testbed," *IEEE Robotics and Automation Magazine*, vol. 17, no. 3, pp. 56–65, 2010.

[18] Y. Guo, B. Jiang, and Y. Zhang, "A novel robust attitude control for quadrotor aircraft subject to actuator faults and wind gusts," *IEEE/CAA Journal of Automatica Sinica*, vol. 5, no. 1, pp. 292–300, 2018.

[19] Z. Gao, "Scaling and bandwidth-parameterization based controller tuning," in *Proceedings of the American Control Conference*, pp. 4989–4996, Denver, Colo, USA, June 2003.

[20] Z. Gao, "Active disturbance rejection control: a paradigm shift in feedback control system design," in *Proceedings of the American Control Conference*, pp. 2399–2405, Minneapolis, Minn, USA, June 2006.

[21] Y. Liu, J. Liu, and S. Zhou, "Linear active disturbance rejection control for pressurized water reactor power," *Annals of Nuclear Energy*, vol. 111, pp. 22–30, 2018.

[22] W. Dong, G. Gu Y, X. Zhu et al., "Modeling and control of a quadrotor UAV with aerodynamic concepts," in *Proceedings of World Academy of Science, Engineering and Technology*, vol. 77, p. 437, World Academy of Science, Engineering and Technology (WASET), 2013.

[23] https://ww2.mathworks.cn/help/physmod/simscape/index.html?s_tid=srchtitle.

[24] L. Fortuna and G. Muscato, "A roll stabilization system for a monohull ship: modeling, identification, and adaptive control," *IEEE Transactions on Control Systems Technology*, vol. 4, no. 1, pp. 18–28, 1996.

6

Modeling and Analysis of Truck Mounted Concrete Pump Boom by Virtual Prototyping

Wu Ren,[1,2] Zhongwei Li,[1] Yanping Bi,[1] Shan Zhao,[1] Bo Peng,[3] and Liming Zhou[1]

[1]*School of Biomedical Engineering, Xinxiang Medical University, Xinxiang, Henan 453003, China*
[2]*State Key Laboratory for High Performance Complex Manufacturing, Central South University, Changsha, Hunan 410083, China*
[3]*Hunan Special Equipment Inspection and Testing Research Institute, Zhangjiajie Branch, Zhangjiajie, Hunan 427000, China*

Correspondence should be addressed to Wu Ren; renwu88@126.com

Academic Editor: Shahram Payandeh

By far there is lack of research on different working conditions between rigid and flexible dynamics of truck mounted concrete pump booms. First a 3D model has been established by using virtual prototyping technology of a 37 m long boom in Pro/Engineering software. Then the rigid body simulation model has been built. Next modal superimposition method is adopted to change the 4 rigid body booms into flexible ones. Kinematics law and dynamic characteristics of 4 common working conditions had been studied then. Next tip displacement and the first boom hydraulic cylinder force of the 4 working conditions between rigid and flexible models have been researched. Furthermore the first natural frequencies of the structure have been calculated. The results show that the frequency of the horizontal condition has the lowest of all and the roof condition has the largest of all. Besides the cylinder forces of the flexible model are larger than the corresponding rigid ones because of the flexible boom vibration. Finally an experiment has been done on a boom test rig which proved that the established simulation model is reasonable and the frequency results are correct. All of these provide design reference to mechanical manipulator as well as reducing product development cost of such mechanism.

1. Introduction

In recently years, CAD/CAE/CAM/PDM has been further developed including VP (virtual prototype) and PLM (product life-cycle manager) technology [1]. VPD (virtual product development) originated from the Boeing Company in 1990. The use of all digital paperless design made the cost decrease by 25% and manufacture cycle reduce by 50%. This application technology was called VP (virtual prototyping technology). Virtual prototyping relies on the technology of CAX (such as CAD, CAE, and CAM), DFX (such as design for assembly), and DFM (design for manufacture). Virtual product development has 3 main features: the first one is digitization [2–5]. Next is product life cycle. This is supported by CAD/CAE/CAM/PDM. The last one is network coordination. The development of modern Internet technology makes the network coordination become real. Until now more and more fields are using virtual prototyping technology like vehicle test, train system design, robot control, bridge construction, flexibility manufacturing, hazard identification, and so on. The use of this technology will greatly reduce the development costs and time [6–8]. The relationship between virtual prototyping and traditional CAD/CAE/CAM is shown in Figure 1.

The scientific and innovative contribution of the paper is the establishment of a truck mounted concrete pump boom model considering the boom flexibility and hydraulic actuator equivalent stiffness. In this paper flexible multibody theory has been used to get the dynamics characteristics of different working conditions. Thus the spray control and motion accuracy can be got during the working process. In the previous research the oil elasticity modulus was not considered in the calculating of hydraulic cylinder stiffness. So the flexibility of the cylinder is not accurate. In this paper the cylinder stiffness and boom flexibility are both considered. Thus the accurate boom kinematic and dynamics response could be obtained. The boom model is made up of multisection operating booms, connecting links, hydraulic

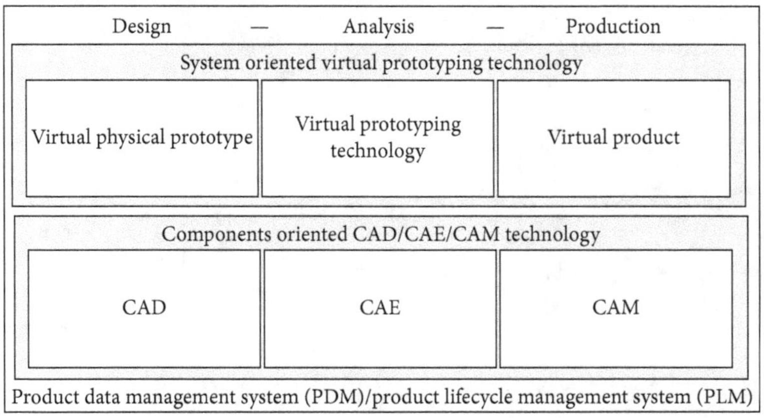

FIGURE 1: Virtual prototyping and traditional CAD/CAE/CAM comparison.

actuator cylinders, and joints. Usually the boom can be simplified as a slim beam and the load is mainly coming from self-weight and pump impact. Thus the boom system stiffness must be large and the self-weight should be light. Frequently boom system dynamics experiments will increase cost and waste time, which can not satisfy production demand. But virtual prototyping technology can solve this problem and reduce the product development cycle [9, 10].

Up to now, Slepniov et al. studied production, innovation, and service networks of companies all over the world [1]. They proposed the widening geographical and cognitive gaps between the networks becoming strategically important method in production research. Jiang et al. studied concrete pump outrigger's effect on the safety and stability of the system. The relationship between pipe length and outrigger area is obtained then [11]. Ling et al. did the simulation of a mobile pump boom motion process. The most dangerous working conditions of cylinders had been studied which offered theoretical basis for the optimization design of the hydraulic cylinders [12]. Wang et al. analyzed pouring process of a mobile concrete pump boom. Next they proposed intelligent pouring program of it. In the end the trajectory plan and virtual environments work had been simulated [13, 14]. Guo et al. established the dynamic equation of flexible booms according to robot theory and language equation which is help for the automatic pouring [15]. Chen and Sun studied the structural optimization of concrete pump displacing boom system by using BP neural network, finite element, and genetic algorithms [16]. Zhang et al. established a truck mounted concrete pump simulation model in ADAMS software and studied the force between the cylinder and link joints which provide design guide for such mechanism [17]. Guo et al. improved the automation pour process by using optimal control method of a mobile concrete pump boom [18]. Jiang built a concrete pump boom virtual prototyping model and carried out the simulation by multibody dynamics and finite element method [19]. Dai et al. used PD control method into concrete pump boom drive. By their study the boom can move in desired angle [20]. Finally they made the boom move with expected rotation angle and tip trajectory. Zhang et al. did control to a flexible boom driven by hydraulic cylinder and had researched the vibration depression of the system [21, 22]. The safety and diagnostic applied to concrete displacing boom are done by Ambrosio et al. Then Resta et al. established a 1:3 reduced model of a mobile concrete pump. Vibrations suppression, feedforward and feedback actions, disturbance estimator, and modal control were researched. Modal space control method is used to reduce the tip vibration and trajectory control [23–28]. Lenord et al. studied a flexible hydraulic mobile concrete pump truck. Linear and nonlinear models were simulated by damping optimization [29]. In the end a linear model was used to substitute the nonlinear model in order to reduced computational cost. Sun et al. proposed a closed-loop monitoring and open-loop control method to suppress the boom vibration [30].

Due to the fact that vibration is very important in the dynamics and kinematics analysis of the concrete pump boom, a rigid-flexible method is adopted in this paper to establish an accurate model of the boom. It is organized as follows. Firstly modal superimposition method is introduced and used. Then rigid and flexible body virtual prototyping models of a truck mounted concrete pump boom have been established. Next the tip displacement and force comparison of the two models in different working conditions are studied. Then natural frequency of the flexible model has been calculated and an experiment has been done on a boom test rig. In the end the conclusion has been discussed.

2. Modal Superimposition Method

The multiple degree-of-freedom system with proportional damping motion equation is as follows:

$$[M]\{\ddot{x}\} + [C]\{\dot{x}\} + [K]\{x\} = \{f(t)\}. \quad (1)$$

In (1),

$$[C] = \alpha[M] + \beta[K]. \quad (2)$$

Here in (2), α and β are proportional constant. $[C]$ is damping matrix and $[M]$ and $[K]$ are the mass matrix and stiffness matrix, respectively. {} represents vector and [] represents matrix. By the coordinate transformation, (3) is obtained

$$\{x\} = \sum_{r=1}^{N} q_r \{\varphi_r\}. \quad (3)$$

In (3), $\{\varphi_r\}$ is the r-order mode shapes. By replacing (3) with (1),

$$[M]\left(\sum_{r=1}^{N} \ddot{q}_r \{\varphi_r\}\right) + [C]\left(\sum_{r=1}^{N} \dot{q}_r \{\varphi_r\}\right) + [K]\left(\sum_{r=1}^{N} q_r \{\varphi_r\}\right) = \{f(t)\}. \quad (4)$$

Then (4) premultiplied $\{\varphi_s\}^T$

$$\{\varphi_s\}^T [M] \{\varphi_r\} = \begin{cases} 0, & r \neq s, \\ m_s, & r = s, \end{cases}$$
$$\{\varphi_s\}^T [K] \{\varphi_r\} = \begin{cases} 0, & r \neq s, \\ k_s, & r = s. \end{cases} \quad (5)$$

The following equation can be got:

$$\{\varphi_s\}^T [C] \{\varphi_r\} = \alpha \{\varphi_s\}^T [M] \{\varphi_r\} + \beta \{\varphi_s\}^T [K] \{\varphi_r\} = \begin{cases} 0, & r \neq s, \\ \alpha m_s + \beta k_s = c_s, & r = s. \end{cases} \quad (6)$$

Here m_s, k_s, c_s are the s-order modal mass, modal stiffness, and modal damping coefficient. Next (7) can be obtained

$$m_s \ddot{q}_s + c_s \dot{q}_s + k_s q_s = \{\varphi_s\}^T \{f(t)\}. \quad (7)$$

In (7) q_s is the s-order modal coordinates. Assuming that $\{f(t)\} = \{F\}e^{j\omega t}$ and $q_s = Q_s e^{j\omega t}$ by replacing them in (7) we can get

$$\left(-\omega^2 m_s + j\omega c_s + k_s\right) Q_s e^{j\omega t} = \{\varphi_s\}^T \{F\} e^{j\omega t}, \quad (8)$$

$$Q_s = \frac{\{\varphi_s\}^T \{F\}}{-\omega^2 m_s + j\omega c_s + k_s}. \quad (9)$$

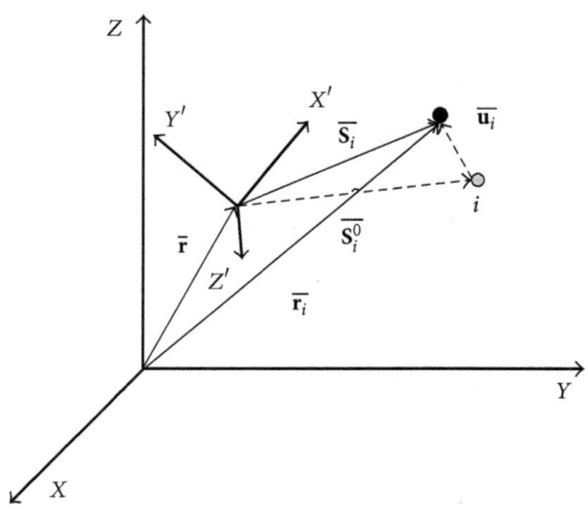

Figure 2: Flexible body coordinate system.

In (8) and (9), Q_s modal coordinates vector. Next the displacement response can be written as

$$\{x\} = \begin{Bmatrix} X_1 \\ X_2 \\ \vdots \\ X_N \end{Bmatrix} e^{j\omega t} = \sum_{r=1}^{N} q_r \{\varphi_r\} = \sum_{r=1}^{N} Q_r \{\varphi_r\} e^{j\omega t},$$
$$\{X\} = \begin{Bmatrix} X_1 \\ X_2 \\ \vdots \\ X_N \end{Bmatrix} = \sum_{r=1}^{N} Q_r \{\varphi_r\} = \sum_{r=1}^{N} \frac{\{\varphi_r\}^T \{F\} \{\varphi_r\}}{-\omega^2 m_r + j\omega c_r + k_r} \quad (10)$$
$$= \sum_{r=1}^{N} \frac{\{\varphi_r\}\{\varphi_r\}^T}{-\omega^2 m_r + j\omega c_r + k_r} \{F\}.$$

Modal superimposition method is based on FMBD (finite element multibody dynamics) and obtains flexible body dynamic stress, strain, and deformation through finite element analysis. Its calculation speed is fast and can calculate large-scale flexible bodies. So it is suitable for the flexibility analysis of construction machinery. A point i on the flexible body's three-dimensional coordinate can expressed as in Figure 2

$$\overline{\mathbf{r}}_i = \overline{\mathbf{r}} + \mathbf{A}\overline{\mathbf{S}}_i n = \overline{\mathbf{r}} + \mathbf{A}\left(\overline{\mathbf{S}}_i^0 n + \overline{\mathbf{u}}_i n\right). \quad (11)$$

In (11), $\mathbf{A}$ is the direction cosine matrix, $\overline{\mathbf{S}}_i^0$ is the point i not deformed vector, $\overline{\mathbf{u}}_i$ is the relative deformation, and i is the translational and rotational modal matrix as shown in

$$\mathbf{\Psi}^i = \left[\mathbf{\Psi}_t^{iT}, \mathbf{\Psi}_r^{iT}\right]. \quad (12)$$

The modal coordinate of $\overline{\mathbf{u}}_i$ is shown as

$$\mathbf{u}_i' = \mathbf{\Psi}_t^{i'} \overline{\mathbf{a}}. \quad (13)$$

FIGURE 3: Boom system CAD model.

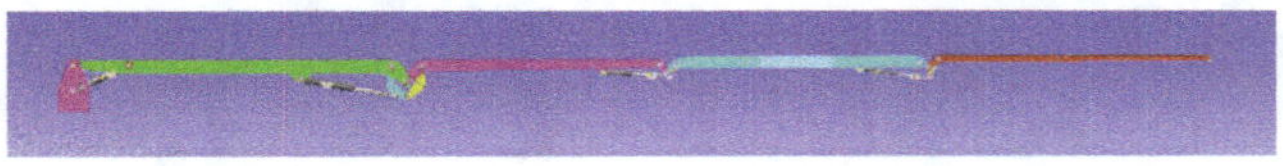

FIGURE 4: Boom system rigid model.

In (13), $\bar{\mathbf{a}}$ is the generalized coordinates of deformation; from (11) and (13), the following equation can be got:

$$\bar{\mathbf{r}}_i = \bar{\mathbf{r}} + \mathbf{A}\left(\bar{\mathbf{S}}_i^{0\prime} + \boldsymbol{\Psi}_t^{i\prime}\bar{\mathbf{a}}\right). \tag{14}$$

The displacement of the flexible body is shown as (15). According to modal information $\boldsymbol{\Psi}$ the corresponding velocity and acceleration equation of the flexible body can be obtained either

$$[\mathbf{u}] = \sum a_i [\boldsymbol{\varphi}]_i . \tag{15}$$

In (15), $[\mathbf{u}]$ is the sum of all nodes displacement; a_i is the modal factor; $[\boldsymbol{\varphi}]_i$ is the component modal. The flexible body deformation can be got by adding the nodes displacement.

The velocity and acceleration of point i can be obtained by derivation for (14) which is shown in

$$\begin{aligned} \dot{\bar{\mathbf{r}}}_i &= \dot{\bar{\mathbf{r}}} + \dot{\mathbf{A}}\left(\bar{\mathbf{S}}_i^{0\prime} + \boldsymbol{\Psi}_t^{i\prime}\bar{\mathbf{a}}\right) + \mathbf{A}\boldsymbol{\Psi}_t^{i\prime}\dot{\bar{\mathbf{a}}}, \\ \ddot{\bar{\mathbf{r}}}_i &= \ddot{\bar{\mathbf{r}}} + \ddot{\mathbf{A}}\left(\bar{\mathbf{S}}_i^{0\prime} + \boldsymbol{\Psi}_t^{i\prime}\bar{\mathbf{a}}\right) + 2\dot{\mathbf{A}}\boldsymbol{\Psi}_t^{i\prime}\dot{\bar{\mathbf{a}}} + \mathbf{A}\boldsymbol{\Psi}_t^{i\prime}\ddot{\bar{\mathbf{a}}}. \end{aligned} \tag{16}$$

3. Truck Mounted Concrete Pump Boom Modeling by Virtual Prototyping Technology

The established 3D model of a truck mounted concrete pump boom is shown in Figure 3. The system is built up by 4 booms, several links, hydraulic actuators, joints, and other components.

After importing the CAD model into RecurDyn software the virtual prototyping model with appropriate constraints is obtained too. RecurDyn is multibody system software of FunctionBay, Inc. It is an innovative Multibody Dynamics offering multibody dynamic capabilities with an integrated powerful FEM extension. With the finite element method combined with MFBD (Multiflexible Body Dynamics) the mechanical system simulation can provide more and more precise dynamic motion results including stress analysis for engineers. The flexible body contacts and nonlinear deformations can be solved in it too. So it is very useful in dealing with the long flexible boom here. The built rigid model is shown as Figure 4. Due to the fact that the booms are flexible large-scale motion components, the flexible influence should not be neglected. Then a flexible body truck mounted concrete pump boom model is established as shown in Figure 5. In it the four booms are modal flexible bodies calculated by finite element software ANSYS. The first boom flexible body is shown in Figure 6. The other three booms are similar to the first one. By importing them into RecurDyn the concrete pump boom flexible model has been built with constraints consistent with rigid body model. The control scheme of the large scale is used with PID regulator. And the model parameters are as in Table 1. The element mass and length are in Table 2.

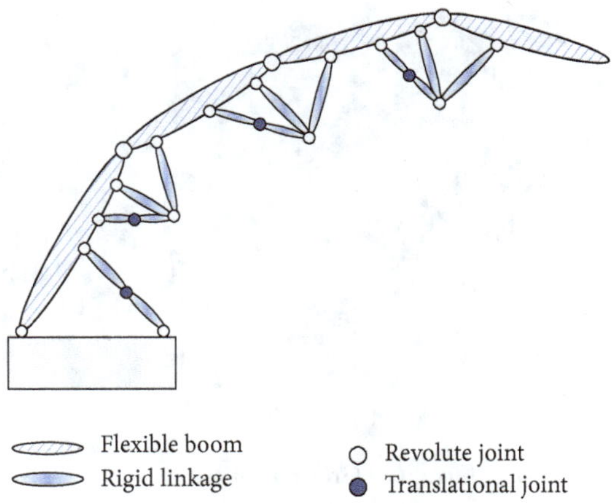

FIGURE 5: Flexible multibody boom topology sketch.

FIGURE 6: The first boom modal reduction flexible body.

TABLE 1: Parameters of boom system.

	Component	Number
Flexible body	Boom	4
Rigid body	Linkage	14
Joint	Bearing, linkage, boom	21
Translational joint	Hydraulic cylinder piston	4
Spring damper	Hydraulic cylinder	4

TABLE 2: Main parameters of each boom.

Boom	1	2	3	4
Length (mm)	3750	2830	2870	3200
Mass (kg)	88	38	32	18

4. Simulation and Experiment

4.1. Simulation Parameters. The general working conditions of concrete pump boom are horizontal condition, ground

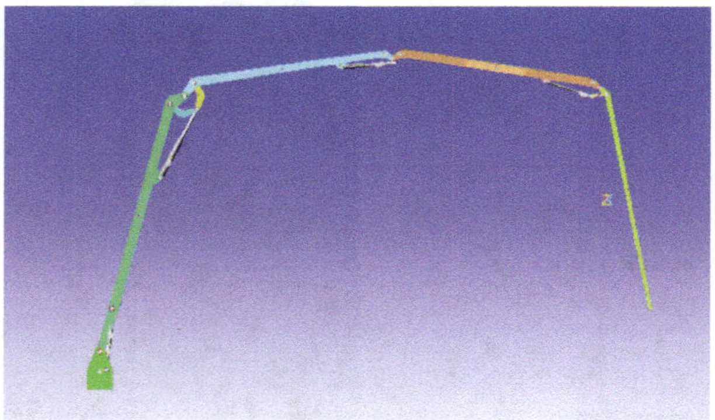

FIGURE 7: Ground condition.

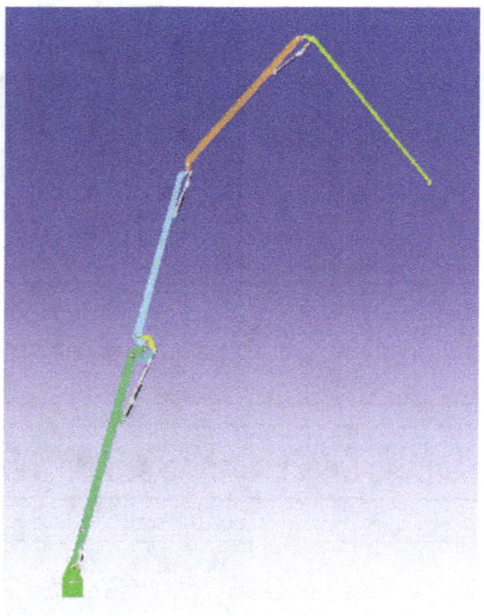

FIGURE 8: Wall condition.

TABLE 3: Each boom angle of working posture.

Posture	Boom 1	Boom 2	Boom 3	Boom 4
Horizontal	0°	0°	0°	0°
Ground	75°	15°	−15°	−75°
Wall	75°	45°	0°	−45°
Roof	75°	75°	45°	−45°

condition, wall condition, and roof condition. The first condition is simulating the long distance. Because this condition is the most dangerous working condition, the stress value and flexible effect are the largest of all. Next the second condition is used in the ground working process. And the third working condition is used in the wet concrete pumping of lower floors. The last simulation is about the upper floors. So these four simulations are typical working conditions of concrete pump boom. In the built model the tip displacements and the first hydraulic cylinder force are studied. The four simulation postures angles of the two models are shown in Figures 4, 7, 8, and 9. Table 3 is the posture angle of each working condition.

In order to get the truck mounted concrete pump boom kinematic rule and vibration characteristic, the motion simulation results of the above working conditions are studied.

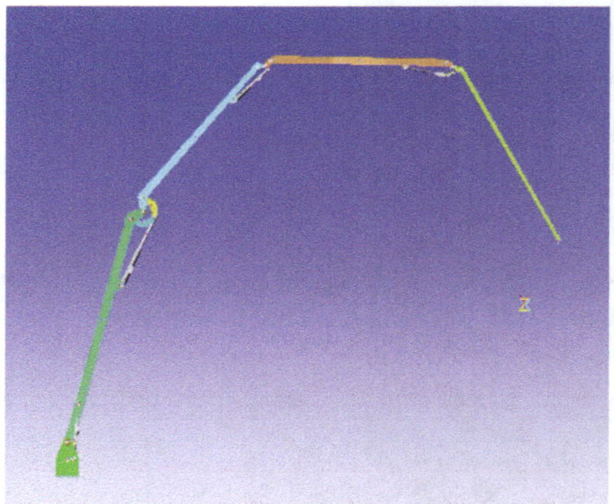

FIGURE 9: Roof condition.

Each boom is driven by its hydraulic actuator. The actuators motion is in Table 4. In every condition the first boom gets the motion at first. Then the other three booms successively move. During the simulation each segment mass, stiffness and oil stiffness are considered also. The time history of each cylinder during different working conditions is as in Figure 10. Because in the horizontal condition only the 1st cylinder has motion, the time history of the other three cylinders is zero.

4.2. Tip Displacement Simulation Results. The tip displacements of the rigid and flexible model between the four conditions are shown in Figure 11.

In Figure 11(a) we can find when the hydraulic cylinder motion stops the model immediately stops and there is no vibration of the rigid model. However the flexible model vibration will occur and last for a time due to the flexible deformation after the motion stops. It does downward movement when the motion occurs because of the gravity and then move upward. After several seconds the vibration disappears due to the structure damping. The other three working conditions seen in Figures 11(b), 11(c), and 11(d) are similar to horizontal one. But the first condition has the maximum vibration amplitude. This demonstrates that the first condition has the most flexibility during the four pouring process.

4.3. Hydraulic Cylinder Force Simulation Results. Figure 12 is the first hydraulic cylinder force between rigid model and flexible model. It can be seen that the corresponding flexible model force is larger than the rigid model at a certain degree. Along with the booms movement the force also increases or decreases periodically. After the motion termination the forces tend to constant value in the end. The maximum force of all the conditions of the flexible model is 72 kN in the horizontal working condition, which is 30 kN larger than the corresponding rigid model. This verifies that horizontal condition is the most dangerous of all. Therefore in design of similar large-scale motion manipulator the mechanical flexible influence should be considered.

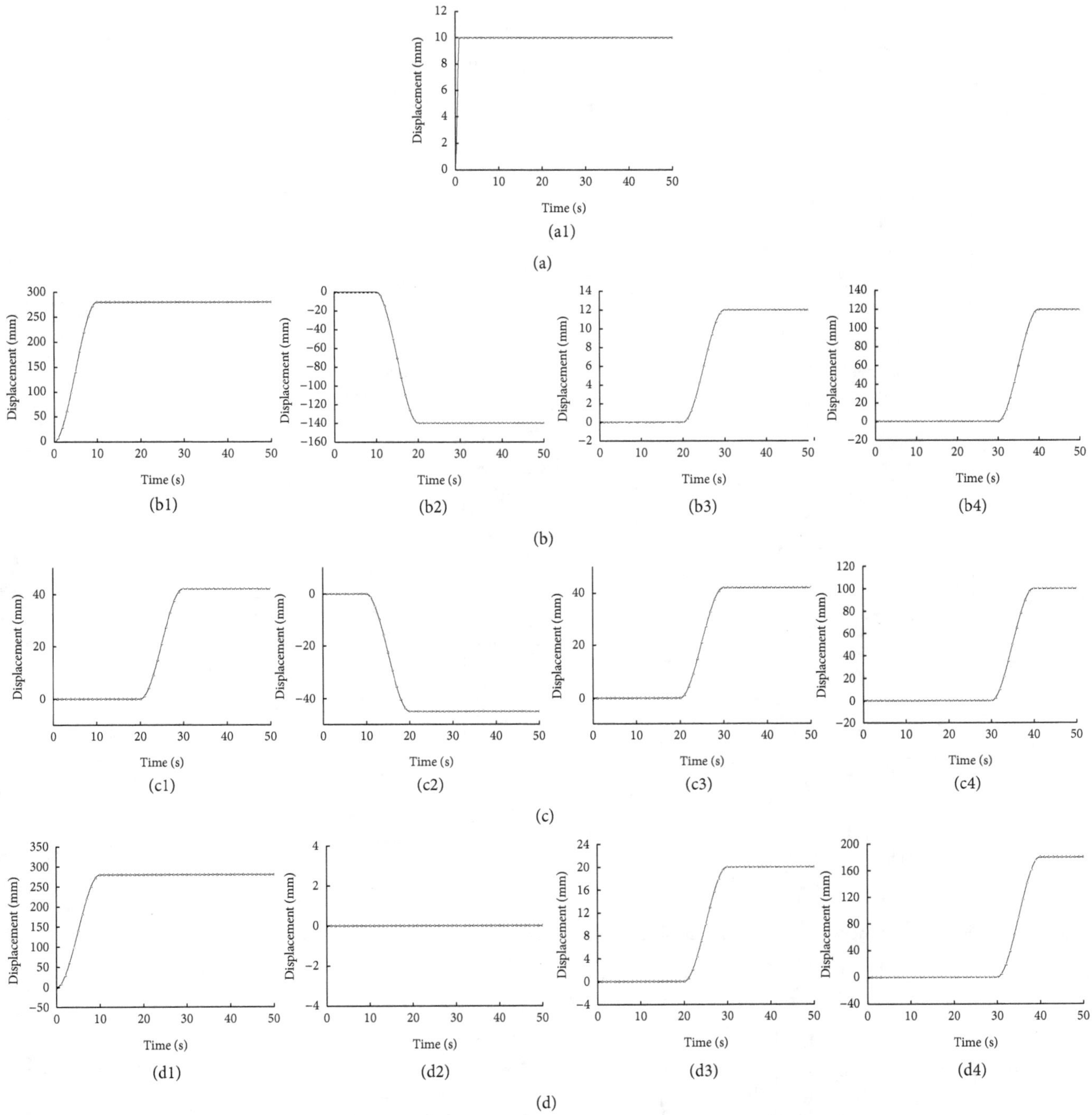

FIGURE 10: Time history of each cylinder during different working conditions ((a) horizontal condition; (b) ground condition; (c) wall condition; (d) roof condition; 1: cylinder 1; 2: cylinder 2; 3: cylinder 3; 4: cylinder 4).

TABLE 4: The cylinder motion function.

	First cylinder	Second cylinder	Third cylinder	Forth cylinder
Horizontal	Step (time, 0, 0, 1, 10)	0	0	0
Ground	Step (time, 0, 0, 10, 280)	Step (time, 10, 0, 20, −140)	Step (time, 20, 0, 30, 12)	Step (time, 30, 0, 40, 120)
Wall	Step (time, 0, 0, 10, 280)	Step (time, 10, 0, 20, −45)	Step (time, 20, 0, 30, 42)	Step (time, 30, 0, 40, 100)
Roof	Step (time, 0, 0, 10, 280)	Step (time, 10, 0, 20, 0)	Step (time, 20, 0, 30, 20)	Step (time, 30, 0, 40, 180)

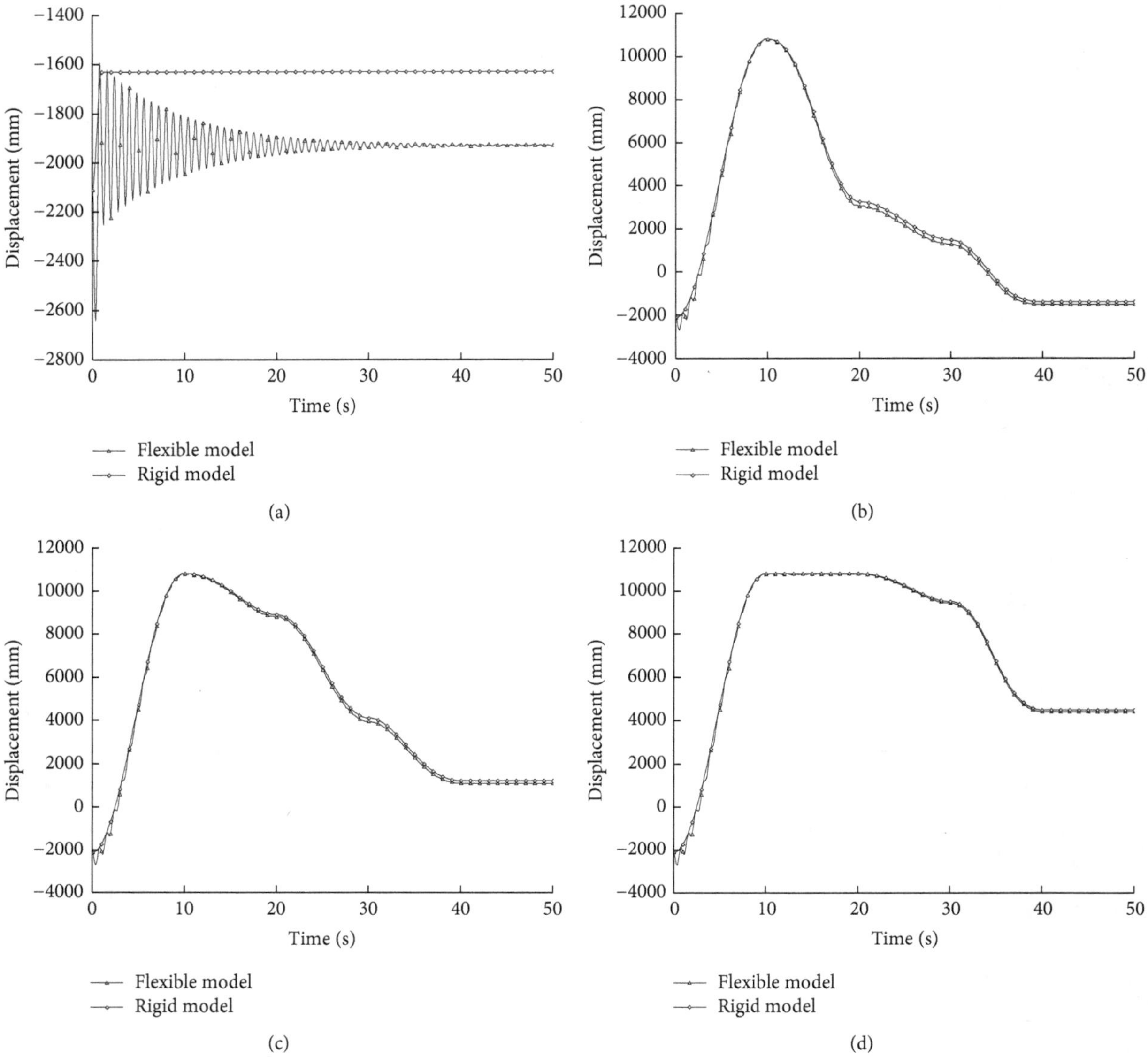

FIGURE 11: Tip displacements between rigid model and flexible model. (a) Horizontal condition; (b) ground condition; (c) wall condition; (d) roof condition.

4.4. Natural Frequency Analysis and Experiment. In this section, the first natural frequencies in the vertical plane have been studied. In order to verify that the simulation is correct a test has done on a boom test rig as shown in Figure 13. The test rig parameters are the same as simulation ones. The material is Q345B and is 13.46 m long. It mounts on a steel base with screw thread legs and casters. The thread legs are used to bear gravity and motion dynamics. The required instrument also includes DEWETRON dynamic data collection equipment, which has 24 channels, 8 accelerate channels, 2 CAN modules, and a CPU. Accelerometer strain sensor is also needed. The boom test rig is as shown in Figure 13.

By analyzing the obtained acceleration signal of the boom tip the modal flexible body natural frequency can be got by FFT transform. The corresponding frequency of simulation models is also given in Table 5.

TABLE 5: The first natural frequency of vertical posture between simulation and test.

	Horizontal	Ground	Wall	Roof
Simulation value	0.872	0.920	0.903	1.184
Test value	0.812	0.829	0.861	1.040

Figure 14 shows the first 10-second tip acceleration comparison of simulation and experiment in horizontal working condition.

According to Table 5, the obtained simulation results of the four working conditions are larger than the measured ones. This is because only the four booms are equivalent to flexible body and the other connecting links and hydraulic cylinder are still considered as rigid body. This increases the

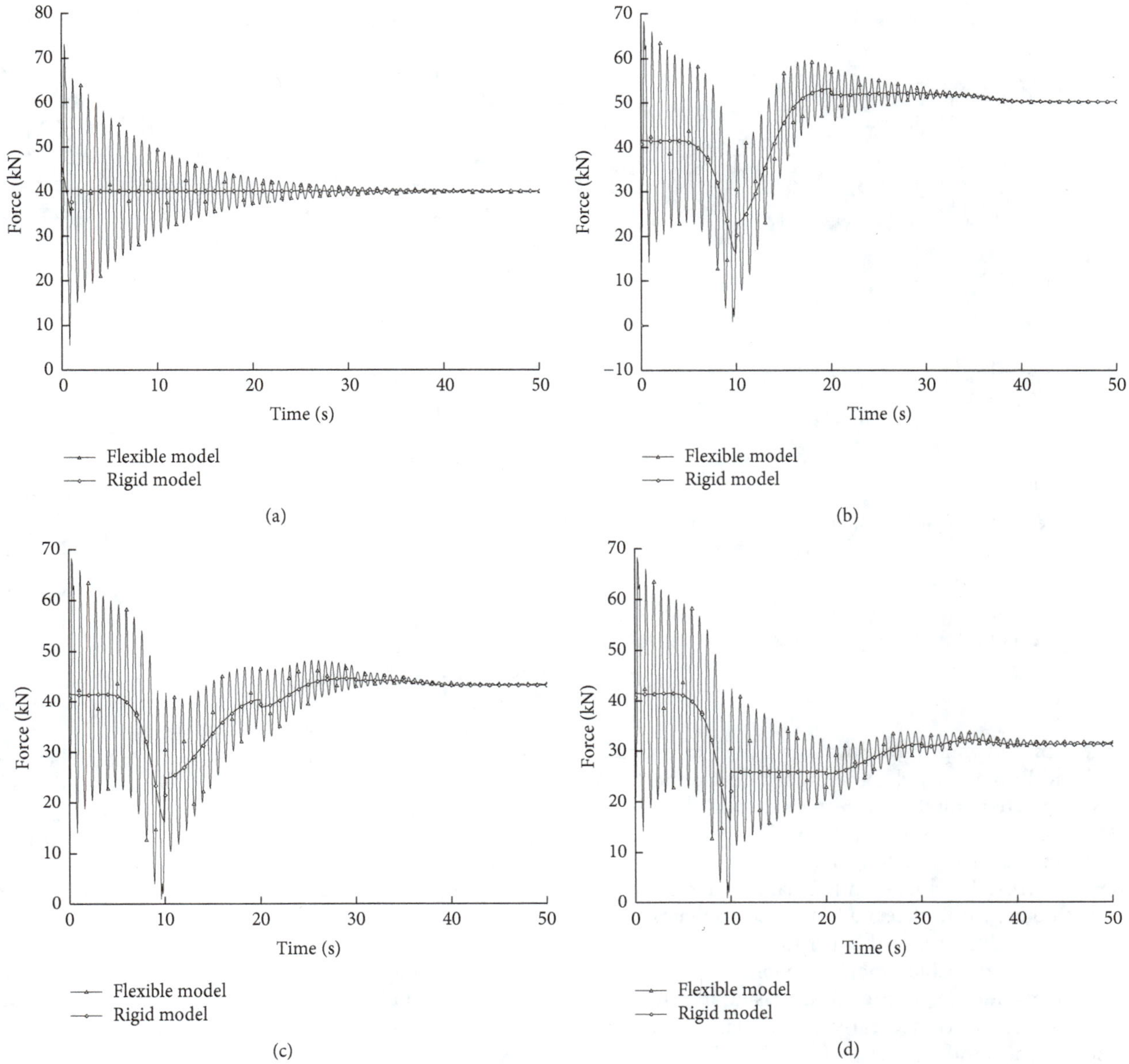

FIGURE 12: The first hydraulic cylinder force between rigid model and flexible model. (a) horizontal condition; (b) ground condition; (c) wall condition; (d) roof condition.

FIGURE 13: Boom test rig.

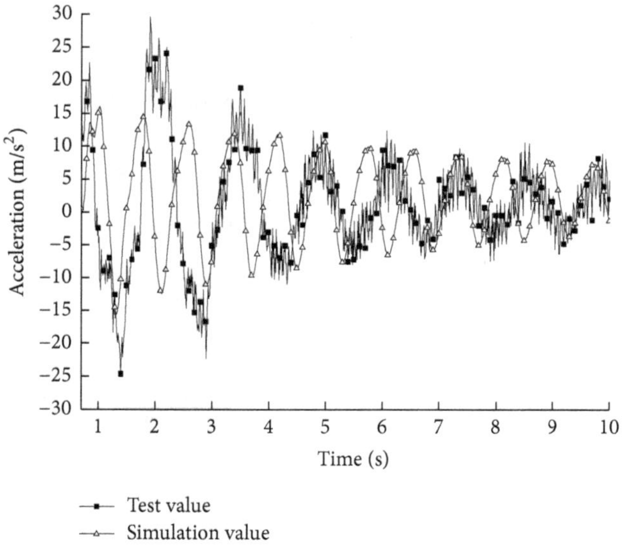

FIGURE 14: Acceleration signal of simulation and experiment ones.

natural frequency to a certain extent. If the other components are flexible the natural frequency will close to the test value.

5. Conclusion

The virtual prototyping technology is adopted to establish the rigid and flexible models of a large-scale motion truck mounted concrete pump boom in this study. In the flexible model modal superimposition method has been adopted. By comparing the two models, the tip displacement and cylinder force, the kinematic and dynamic characteristics have been studied. The flexible model tip displacements of the four working conditions are all larger than the rigid model. This is because the flexible model has vibration during the motion. The cylinder forces are all increasing largely in all the conditions and become nearly 2 times larger than the rigid model. The results show that the flexible analysis is necessary. Finally an experiment has been done on a test rig which proves that the proposed method is reasonable in this paper. The first natural frequencies in the vertical plane are researched then. The horizontal condition has the lowest frequency and the roof condition has the largest one. And the results are in good agreement too. All of these provide vibration control reference and fatigue analysis of such low-frequency manipulators. The virtual prototyping technology is convenient in reducing production design cost and time for engineers.

Conflicts of Interest

The authors declare that there are no conflicts of interest regarding the publication of this paper.

Acknowledgments

This work was supported by study on "Human Machine Dynamic Characteristics of Multi Degree of Freedom Rehabilitation Displacement Device" (172102310542), "Multi-DOF Rehabilitation Patient Transfer Device Dynamics Characteristic and Human-Machine Safety Comfort Research" (505140), "Human Comfort Study by Elevator Transportation Based on Multi-Body Dynamics" (no. 2015KYJH33), and the Natural Science Foundation of Henan Province (2014GGJS-096 and 152102310357); abnormal shear stress mediated expression of pcsk9 in atherosclerotic plaque cells (152102210339); effect of mechanical factors on protein expression in atherosclerotic plaque cells (15a180056).

References

[1] D. Slepniov, B. V. Wæhrens, and J. Johansen, "Production, innovation and service networks: complex interplay, evolution and coordination," *International Journal of Product Development*, vol. 17, no. 3-4, pp. 189–203, 2012.

[2] M. Karkee, B. L. Steward, A. G. Kelkar, and Z. T. Kemp II, "Modeling and real-time simulation architectures for virtual prototyping of off-road vehicles," *Virtual Reality*, vol. 15, no. 1, pp. 83–96, 2011.

[3] H. Li, N. K. Y. Chan, T. Huang, M. Skitmore, and J. Yang, "Virtual prototyping for planning bridge construction," *Automation in Construction*, vol. 27, pp. 1–10, 2012.

[4] C. K. Chun, H. Li, and M. Skitmore, "The use of virtual prototyping for hazard identification in the early design stage," *Construction Innovation*, vol. 12, no. 1, pp. 29–42, 2012.

[5] M. I. C. Dede, "Virtual prototyping for robot controllers," *International Journal of Design Engineering*, vol. 3, no. 3, pp. 276–288, 2010.

[6] M. Datar, I. Stanciulescu, and D. Negrut, "A co-simulation environment for high-fidelity virtual prototyping of vehicle systems," *International Journal of Vehicle Systems Modelling and Testing*, vol. 7, no. 1, pp. 54–72, 2012.

[7] A. Shahin and N. Rostamian, "Concurrent engineering and manufacturing flexibility: an integrated model," *International Journal of Productivity and Quality Management*, vol. 12, no. 2, pp. 121–140, 2013.

[8] K. H. Han, S. M. Bae, S. H. Choi, and G. Lee, "Parameter-driven rapid virtual prototyping of flexible manufacturing system," *International Journal of Mathematics and Computers in Simulation*, vol. 6, no. 4, pp. 387–396, 2012.

[9] M. H. Abidi, A. M. Tamimi, A. M. Ahmari, and E. S. Nasr, "Assessment and comparison of immersive virtual assembly training system," *International Journal of Rapid Manufacturing*, vol. 3, no. 4, pp. 266–283, 2013.

[10] L. Debao and L. Qiuhai, *Analysis of Experiments in Engineering Vibration*, Tsinghua University Press, Beijing, China, 2004.

[11] Y. Jiang, J. Li, and G. Zhang, "The effect of outrigger structure design on the stability of truck concrete pump," *Construction Machinery*, vol. 25, no. 3, pp. 67-68, 2005.

[12] Z. Ling, P. Fu, H. Liu et al., "Simulate research on the amplitude changing mechanism arm support of concrete pump vehicle," *Machinery Design and Manufacture*, vol. 31, no. 5, pp. 101–103, 2011.

[13] T. Wang, G. Wang, K. Liu, and S. Zhou, "Simulation control of concrete pump truck boom based on PSO and gravity compensation," in *Proceedings of the 2nd International Symposium on Intelligent Information Technology Application (IITA '08)*, pp. 960–963, Shanghai, China, December 2008.

[14] Z.-J. Gou, "The simulation of harmonic vibration response based on the rigid and flexible boom system of the ttruck-mounted concrete pump," in *Proceedings of the International Conference on Electronic and Mechanical Engineering and Information Technology (EMEIT '11)*, pp. 2404–2407, Harbin, China, August 2011.

[15] D.-M. Guo, K. Liu, D.-L. Xu, and J. Liu, "Effect of the vibration coupling pumped concrete fluid with conveying pipes on concrete pump truck," *Journal of Northeastern University*, vol. 31, no. 1, pp. 99–102, 2010.

[16] K. Chen and G. Sun, "Intelligent optimun design of boom frame of concrete pumping vehicle," *Journal of Wuhan University of Technology*, vol. 27, no. 2, pp. 244–246, 2003.

[17] S. Zhang, M. Feng, and P. Zhang, "Boom kinematic analysis of concrete pump truck based on ADAMS," *Construction Machinery*, vol. 21, no. 21, pp. 123–125, 2001.

[18] L. Guo, G. Zhang, and J. Li, "Approach to robotization of concrete boom truck for construction," *Machine Design and Manufacturing Engineering*, vol. 29, no. 2, pp. 6–8, 2000.

[19] T. Jiang, *Application study of virtual prototuping technology on concrete pump truck [Ph.D. thesis]*, Nanjing University of Technology, Nanjing, China, 2005.

[20] L. Dai, J. Wang, Y. Liu et al., "PD control of swiveling jib system of concrete delivery truck," *Journal of Northeastern University*, vol. 30, no. 11, pp. 1645–1648, 2009.

[21] C.-Y. Zhang and G.-C. Bai, "Extremum response surface method of reliability analysis on two-link flexible robot manipulator," *Journal of Central South University of Technology*, vol. 19, no. 1, pp. 101–107, 2012.

[22] G. Li and M. Wu, "Modeling and controlling of a flexible hydraulic manipulator," *Journal of Central South University of Technology*, vol. 12, no. 5, pp. 578–583, 2005.

[23] P. Ambrosio, G. Cazzulani, F. Resta, and F. Ripamonti, "An optimal vibration control logic for minimising fatigue damage in flexible structures," *Journal of Sound and Vibration*, vol. 333, no. 5, pp. 1269–1280, 2014.

[24] G. Cazzulani, S. Moschini, F. Resta, and F. Ripamonti, "A diagnostic logic for preventing structural failure in concrete displacing booms," *Automation in Construction*, vol. 35, pp. 499–506, 2013.

[25] G. Cazzulani, C. Ghielmetti, H. Giberti, F. Resta, and F. Ripamonti, "A test rig and numerical model for investigating truck-mounted concrete pumps," *Automation in Construction*, vol. 20, no. 8, pp. 1133–1142, 2011.

[26] G. Bagordo, G. Cazzulani, F. Resta, and F. Ripamonti, "A modal disturbance estimator for vibration suppression in nonlinear flexible structures," *Journal of Sound and Vibration*, vol. 330, no. 25, pp. 6061–6069, 2011.

[27] G. Cazzulani, F. Resta, and F. Ripamonti, "A feedback and feedforward vibration control for a concrete placing boom," *Journal of Vibration and Acoustics*, vol. 133, no. 5, Article ID 051002, pp. 1–8, 2011.

[28] F. Resta, F. Ripamonti, G. Cazzulani, and M. Ferrari, "Independent modal control for nonlinear flexible structures: an experimental test rig," *Journal of Sound and Vibration*, vol. 329, no. 8, pp. 961–972, 2010.

[29] O. Lenord, S. Fang, D. Franitza, and M. Miller, "Numerical linearisation method to efficiently optimise the oscillation damping of an interdisciplinary system model," *Multibody System Dynamics*, vol. 10, no. 2, pp. 201–217, 2003.

[30] X. Sun, H. Ye, and S. Fei, "A closed-loop detection and open-loop control strategy for booms of truck-mounted concrete pump," *Automation in Construction*, vol. 31, pp. 265–273, 2013.

7

The Encountered Problems and Solutions in the Development of Coal Mine Rescue Robot

Yong Wang,[1,2] **Peng Tian**,[2,3] **Yu Zhou**,[2] **and Qing Chen**[2]

[1] *College of Mechanical and Electrical Engineering, China University of Mining and Technology, 221000, China*
[2] *Xuzhou Products Quality Supervision and Inspection Center, 221000, China*
[3] *Institute of Environment and Surveying and Mapping, China University of Mining and Technology, 221000, China*

Correspondence should be addressed to Yong Wang; nicedone@163.com

Guest Editor: Saeed Khorashadizadeh

Based on the various problems in the process of coal robot applications, from three-dimensional analysis of the key technology, market demand, and policy regulations and standardization, the paper reviews the use of carbon fiber material, elastic material of traditional surface spraying and integral casting technology to realize lightweight robot, by using the dual redundant design to improve the reliability of the rescue robot CPU, through the functions of voice recognition obstacle sensor, gas sensor acquisition components, and distribution information to realize the underground rescue mission. In addition, virtual test technology, hardware in the loop simulation technology, and shrinkage model technology are proposed to solve the difficulties of hard field testing of rescue robots. These suggest that concrete measures have been pointed out to break the problem of the development of the coal mine relief robot.

1. Introduction

China has abundant coal resources and it is the largest coal-producing country of the world. The average depth of the mine is about 600 meters, and more than half are in the high gas area. Complex operating environment and lack of technology and management cause the fire, gas explosion, and other coal mines disasters. As Hemanth Reddy described in his article [1], "Underground mining is beset with numerous problems such as ground movement (fall of roof/sides), inundation, air blast, etc.; apart from gas explosions and dust explosions that are restricted to coal mines. Whatever may be the cause and type of accident or the extent of damage caused, it is a horrendous task for the rescue team to reach the trapped miners. The accident/irrespirable zone contains increased levels of harmful gases like carbon dioxide and carbon monoxide and explosive gases like methane apart from a deficiency of Oxygen. The entire gallery or roadway is filled with dust and smoke, or water in case of inundation, hindering the visibility of rescue personnel." How to seize the best time to implement rapid and effective rescue and how to avoid and reduce the casualties caused by secondary accidents are a hot social issue in recent years.

It has been more than ten years since the research of the mine rescue robots which were used to do the rescue work instead of human [2, 3]. The development of the rescue robot has achieved fruitful results. Great progress has been made in the key technical issues such as safety, communications, energy resources, control, lighting equipment, sensor technology, and physical design. But, so far, there is no robot capable of successfully completing the search and rescue mission. Through different cases, the author analyzes the bottlenecks in the development of the rescue robot deeply from three dimensions of market demand, policy license, and technology and puts forward some suggestions on the future development of the coal mine disaster rescue robot. The key to the development of mine disaster rescue robot is to find the problem; then the robot can be applied to the rescue work as soon as possible.

In this paper, authors have reviewed developments of the mine rescue robots and problems encountered in the past ten years. A prompt review paper on rescue robot is very useful,

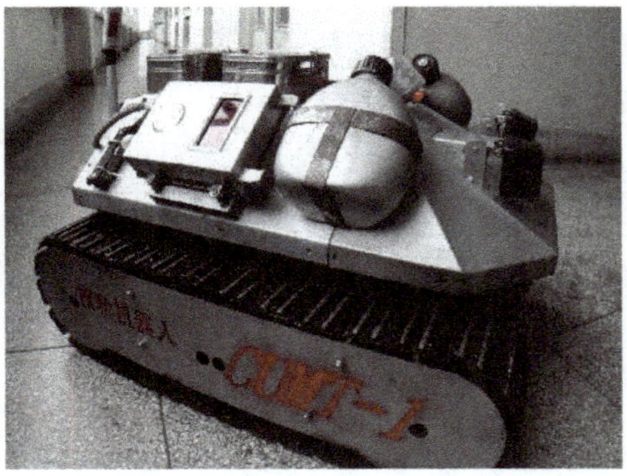

FIGURE 1: CUMT-1, the first generation of coal mine disaster rescue robots in China University of Mining and Technology.

FIGURE 2: V2 robot.

FIGURE 3: Robot of Western Australia Water Company.

not only to identify the current status of research but also to provide information to anyone in the field of the rescue robot. This paper is prepared based on the development of the mine rescue robot in the past ten years.

2. Review on Development Status

The systematic review on recent development in rescue robot has been done based on sets of their application in accident.

The world's first robot of mine disaster rescue RATLER was born in 1998 in the United States, a robot with a dangerous gas sensor, an infrared camera, a gyroscope, and a radio frequency signal transceiver. But due to the laggard equipment its remote control distance was only seventy-six meters. China's first coal mine disaster rescue robot CUMT-I was developed by China University of Mining and Technology in 2006, shown in Figure 1; exploration of this robot began in 2004, which was proposed by professor Ge Shirong who created the first R&D team in China [3]. CUMT-I had basic ability of remote control, video transmission, dangerous gas detection, and moving object. But neither RATLER nor CUMT-I had been applied to the actual rescue mission because of the lack of rescue requirements.

2.1. Review on Research and Application Status Overseas. The first and the only coal mine rescue robot used in the United States is the ANDROS Wolverine, nicknamed V2 [4–6], which is designed and manufactured by Remotec Company with entrustment of MSHA. As shown in Figure 2, V2 is about 50 inches high and weighs more than 1200 pounds and is driven by rubber tracks. The rescue robot is equipped with navigation and monitoring cameras, lighting equipment, night vision, two-way voice communication, and manipulator arm. Robot V2 can be remotely operated with a distance of up to 5000 feet; the important information in the mine can be transported through the fiber optic cable; the operators can access real-time information including video and the concentration of flammable and toxic gas. In the Sago Mine disaster in 2006 [7, 8], the robot V2 was used for the first aid, but the rescue failed with moving 790 meters into the mire. This robot of MSHA is the only one robot in the mining industry in the United States. Michael Knopp [5], President of Remotec Company, seemed pessimistic on the development of robots in the future; he thought the mine rescue robot had not been used deeply and widely because coal mining industry was not willing to pay for the robot; if there was no budget support, it will be difficult to develop.

Another use of the coal mine rescue robot was in the gas explosion accident at parker river mine on November 19, 2010 [9]. In the morning of 23, the rescue team sent an EOD robot into coal mine provided by New Zealand Defense Force, but it traveled only 550 m when short trouble occurred because of water in the mine after taking pictures of the mine [10–14]. In the morning of 24, the second robot was sent into the roadway for search and rescue. The third robot was a piping robot of Western Australia Water Service Company, which was designed for large pipeline operation. As shown in Figure 3, it was a robot with differential steering, equipped with camera and light and gas detection equipment and was controlled by fiber optic cable, with a maximum of 6 kilometers [14–16]. The second explosion in the 24th afternoon brought the rescue mission of the robot to an end, and all the robots were buried in abandoned alleys. In March of the following year, Western Australia Water Service Company sent another robot into the mine again; it took a lot of video data of the roadway after the explosion, to help the police to determine whether the mine can enter the recovery stage.

On February 25, 2016, near the city of Komi Republic of Russia Vorkuta [17, 18], a series of gas explosions happened in the north coal mine which was operated by Theo Kuta Mining Company; 36 people were killed. The second explosion killed all the trapped miners, including five rescue workers and one

Table 1: Robot products of Changzhou Research and Development Center with safety certification.

SN	Safety standard number	product name	Product model	End date	Safety standard number
1	MLE160001	5-explosion-proof robot device for mine.	ZR	/	normal
2	MLE160002	explosion-proof robot device for mine	ZR(A)	/	normal
3	MAB160422	The main control box of isolation and intrinsic safety robot for mine	ZR-K	/	normal

Figure 4: System of "Leader" robot.

Figure 5: CUMT-V, a coal mine disaster rescue robot with the Coal Mine Safety Certification.

miner; then the rescue stopped. On 29 February, the "leader" robots of Russian Special Force were transported to the mine; these robots were equipped with cameras, gas sensors, and other metering equipment, as shown in Figure 4. They can work in remote, dark, salt, fog, and dust areas to collect samples of air. The robot can be used to participate in the investigation instead of rescue workers to determine when to enter into the mine again [19, 20]. But the rescue only had the previous news and demonstration video of the robot; the rescue process is unknown.

2.2. Review on Research and Application Status Domestic. Domestic research and application of mine rescue robot are developing slowly. Since coal mine rescue robots CUMT-I, CUMT-III, and CUMT-III have been developed, the coal mine rescue robots from universities and units such as Beijing Institute of Technology, Shenyang Institute of Automation, Taiyuan University of Technology, and Xi'an University of Science and Technology appeared one by one. But only the CUMT-V robot developed by China University of Mining and Technology and the detection robot KQR48 developed by Tangshan Kai Cheng Group have obtained the Coal Mine Safety Certification (MA).

After two key projects of National Plan "863", "the coal mine environment detection and rescue robot system" and "development and application of the coal mine disaster rescue robot", the mine rescue robot CUMT-V developed by China University of Mining and Technology and Changzhou Scientific Research and Development Center Co., Ltd., obtained the Coal Mine Safety Certification (MA), as shown in Table 1; the end date "/" represents the certificate of safety sign of new products.

Tangshan Kai Cheng Group is the first company in China to create R&D and industrialization of coal mine disaster rescue robot, and the robot products certified by coal mine safety are shown in Table 2 [21]. Kaicheng company is also the first and only equipment manufacturer in China which obtained the Coal Mine Safety Certification (MA) and Explosion-Proof Certificates.

At present, China's coal mine rescue robot is still in the research stage of prototypes, and there are no reports of robots participating in the rescue mission.

On August 3, 2016, CUMT-V conducted an experiment of downhole rescue in Ta Shan Coal Mine of Datong Coal Mine Group, as shown in Figure 5. Three coal mine environmental detection robots were used in this experiment at Ta Shan Coal Mine, two coal mine rescue robots and a coal mine firefighting robot. The function and main performance of robots were verified, including underground walking, communication, and the successful opening of fire extinguishing equipment after the fire spot [22, 23].

2.3. Comprehensive Analysis of Development Status. Through the investigation and research above, the application status of coal mine rescue robot can be summarized as follows:

TABLE 2: Robot products of Kaicheng company with safety certification.

SN	Safety standard number	product name	Product model	End date	Safety standard number
1	MLE110002	Robot body.	KRZ I - BT	2020/12/1	effective
2	MLE110003	Mine robot device for mine.	KRZ I	2020/12/1	effective
3	MAJ110006	downhole robot flameproof display for mine.	KBX I - 12	2015/12/16	overdue
4	MAK120077	flameproof life detector for mine.	TBS12	2017/10/25	overdue
5	MLE120119	intrinsic safety robot for mine.	KQR24Z	/	normal
6	MLE120120	Detection robot for mine.	KQR48	/	normal
7	MAJ120275	console display of Ground output intrinsic safety robot.	KQR220Z-K	/	normal
8	MLE150001	Flameproof and intrinsically safe type underwater robot for mine.	KSJ24	/	normal
9	MLE170001	Flameproof and intrinsically safe type Inspection robot for mine.	KRXJ38	/	normal
10	MAB170513	Inspection robot device for mine.	ZRXJ127	/	normal

(1) The flame-proof design of V2 robots is in line with the relevant laws and regulations of the United States, but due to its size and hulking it was only about 790 feet in the mines and was disabled.

(2) There is flame-proof design for EOD robots of National Defense Force and a robot of Western Australia Water Service Company. The EOD robot did not take into account the humid environment, so short trouble appeared when driving 550 meters. Compared with the EOD robot, the robot of Western Australia Water Service Company has better waterproofing and sealing performance, so it entered the dangerous area. But rescue work of all two military robots and a water robot resulted in the second explosion.

In addition, a robot of Western Australia Water Service Company entered the mine disaster again in the following year for the analysis after disaster, but it was very difficult for this robot to avoid obstacles and the barrier in the narrow mine, with the adoption of four-wheel differential steering. On the other hand, its body was too heavy to be carried out, so the robot stayed in the park mine permanently after taking a lot of videos of the roadway after disaster.

(3) Russia's robots have no follow-up reports and no video data on the disaster scene.

(4) Robots have been developing steadily in China. The coal mine detection robot KQR48 developed by Tangshan Kai Cheng company and the coal mine rescue robot CUMT-V developed by China University of Mining and Technology have obtained the Coal Mine Safety Certification (MA) and obtained the permission to enter the mine, but they have not participated in the actual rescue activities yet.

So, the mine rescue robots are just of a preliminary use in the world, and at present it can only be used in very limited environment (fewer obstacles, less flooded, and short distance) and limited rescue phase. The work of rapid investigation and rescue after the accident have not achieved the desired effect. But even so the robots could be of great value to face the potential dangers of the second explosion in place of rescuers.

3. Problems and Countermeasures

Several countries have done much for the R&D of mine rescue robot in recent years, such as the United States, New Zealand, Russia, and China, with many problems appearing in this process. Then, the author will analyze these problems from three aspects of technology, policy, and market to try to find appropriate countermeasure.

3.1. The Key Technology. The improvement of technology is a very important aspect to promote the development of coal mine disaster rescue robot, especially the improvement of key technologies such as explosion-proof design, explosion-proof material, lightweight physical design, communication system [24], integration of electronic equipment, artificial intelligence, and test methods, which are problems that must be solved in the process of practical application of coal mine disaster rescue robot.

3.1.1. New Explosion-Proof Material and Lightweight Physical Design. At present the explosion-proof enclosure of the mine rescue robot is mainly composed of iron and steel, which is very heavy. Robot V2 in the United States is 1200 pounds [4], robot CUMT-V of China University of Mining and Technology is about 500 kg, and the weight of other robots is also similar to the former according to their size and structure. The bigger the robot, the bigger the drive motor, the more the drive circuit, and the more the power battery, so it is so heavy, once insufficient, short trouble or blocking occurs, three or four rescuers cannot move it, and it is difficult to go out from the mine again. That is why those four robots in the pike mine disasters in New Zealand are lost in the mine, causing huge losses [10]. Therefore, the new structural design and the lightweight of the robot are the most important work in the field of the rescue robots [25, 26]. Molyneaux Lance proposed methods to solve the difficulties of underground mine environments; a system level design is carried out in the paper of "Development of an Underground Mine Scout Robot" [27]. To operate in an explosive atmosphere, a purge and pressurisation system is applied to a fiber glass chassis, with intrinsic safety incorporated into the sensor design. The chassis is invertible, with a low center of gravity and a roll-axis pivot. This chassis design, in combination with spoked-wheels, allows traversal of the debris and rubble of a disaster site.

The explosion-proof and lightweight design can be explored from the following three aspects.

(1) Explosion-proof design with lightweight material, like carbon fiber, glass steel, conductive plastic, and other lightweight nonmetallic materials [28]: According to GB 3836.1, the material must meet the requirements of electrostatic surface and strength [29].The spraying robot can be referenced; when part of it is made of carbon fiber material, the weight is greatly reduced.

(2) The use of surface modification techniques: For traditional lightweight materials or based on traditional craft processes, it is possible to improve the surface electrical conductivity by special technique, to avoid the surface static electricity and to improve the impact resistance, for example, the application of elastomer materials for coating treatment, such as spray polyuria elastomer [30, 31].

(3) Integrated design: Try to use the pressure casting process on the robot shell [32]. At present, many mobile phones and laptop shells are made with casting process [33, 34]; as shown in Figure 6, the ThinkPad laptop shell of the carbon fiber with integrated structure is strong [35], compact, and lightweight; in the same way, it will be very beneficial for the lightweight research of robots.

3.1.2. Integration of Electronic Equipment. (1) High integration can simplify the circuit and improve the reliability and stability. On the other hand, smaller volume and smaller weight are good for lightweight research on robots.

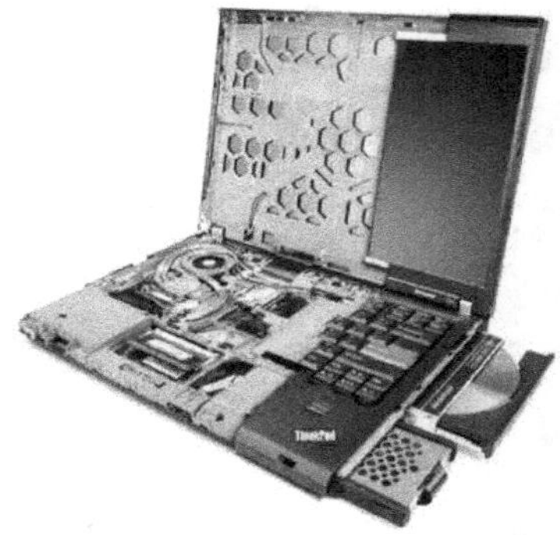

FIGURE 6: ThinkPad laptop shell of the carbon fiber with integrated structure.

(2) Reducing the number of connections by bus design: Bus design can improve the efficiency and reliability of system, and it is convenient to replace the various modules.

(3) Using reasonable heat dissipation design [36]: Due to the need of explosion-proof, the circuits are potted into a fully sealed housing; internal heat is hard to escape, so the structure should be designed to introduce the heat into the shell and try to use the shell as the cooling medium.

(4) Double redundancy design being adopted: The core device should adopt double redundancy design, especially the central processing unit, to avoid the failure of the whole robot caused by the crash [37].

3.1.3. Artificial Intelligence Specialization. The remote control is used in system for coal mine rescue robotic mainly, for several reasons [4, 15, 20, 22].

(1) Constraints of structure and security: To ensure the compliance of the requirements of the safety aspects of the design [38], many sensors cannot be applied in the mine rescue robot; for example, laser sensor and ultrasonic sensor are under strict control; robots are only equipped with cameras and dangerous gas sensors. Obviously, it is not enough for the rescue robot. It is necessary to develop special compound sensors for coal mine rescue robot to realize more rescue missions, such as intelligent navigational control and searching life signal.

(2) Rugged environment: The mine environment underground is extremely bad, dark, damp, dust, and black background, as shown in Figure 7. The camera is the most important sensor of remote control robot; it is difficult to learn and analyze of the image information extracted by the camera in this environment with artificial intelligence algorithm. It can only be controlled by artificial remote control [39, 40], and the robot will lose the capability of operation once it is separated from the remote control, and the rescue mission will fail.

(3) The unstructured environment of coal mines: At present, the artificial intelligence algorithm is still in primary stage; it can perform navigation work better in the structured environment compared to the unstructured environment.

Based on the above three reasons, navigation technology in the field of traditional robot research is difficult to be

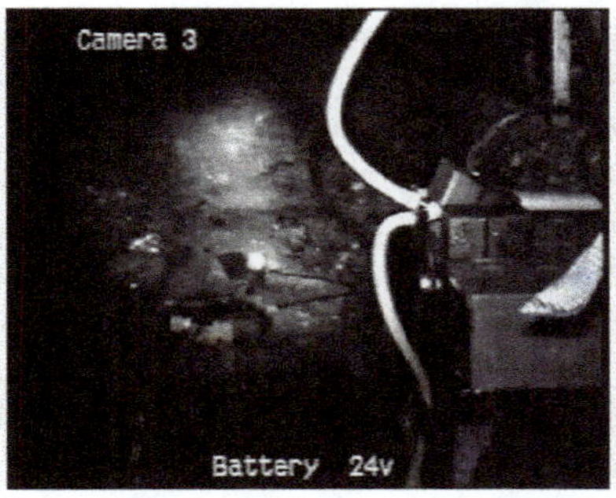

FIGURE 7: Mod robot and photos taken in environment of mine disaster.

applied in the coal mine rescue robot, but other artificial intelligence technologies following can be cited:

(1) Recognition and analysis of speech information [41, 42]: The robot can recognize all kinds of voice correctly by intelligent identification system and to judge the direction based on these signals, such as signal for help, including the human voice, tapping, and explosion under the noise environment of coal mine disasters.

(2) Image enhancement and obstacle recognition: The image in bad environment can be denoised and enhanced automatically so that the operator can identify or judge the surrounding environment, such as the distribution of obstacles. The ability of intelligent identification and measurement of the underground obstacles can assist the operator to make decisions about the obstacle avoidance of the robot [43–45].

(3) Analysis and identification of dangerous gas components [46]: Use modular gas sensors for recording the distribution and analysis of the gas components to identify the robot's environment risk. It can assist robot operator in judging robot environment risk in order to make the right judgment.

(4) Position navigation based on image features: According to the analysis of the position of the robot and determination of various borehole geometry, image characteristics in the environment after disaster, such as various kinds of markers, sign recognition vision system of robot, the operator can make an accurate navigation judgment.

3.1.4. Test Method and R&D Method. Like all other mechanical and electrical products, the R&D of the mine rescue robot cannot be apart from the development of the test methods. As a special robot, the opportunities of taking part in the rescue are few; in these conditions, how to carry out the test and development? The following methods can be used to help the robot to adapt to the environment in the disaster areas better.

(1) Virtual Test Technology. There are three types of virtual tests currently. The first is virtual testing technology based on virtual instrument and software, such as Labview. The second is the virtual testing technology based on software simulation [47, 48], software like Ansys, RecurDyn, Matlab, etc. The third is virtual reality based on virtual testing technology. A robot physical model similar to real robot is simulated in the computer. The robot can be controlled and operated with control programs running on the model of the robot, like Evolution Robotics ERSP, Microsoft Robotics Studio, OROCOS, Skilligent, URBI, Webots, Player, Stage, Gazebo, iRobot AWARE, Open JAUS, CLARAty, V-REP, Unity3D, and so on. Currently, there is virtual test of robot Gazebo, which is an important component of ROS in robot operating system, as shown in Figure 8. These tests can be used to test the functionality, maneuverability, and robustness of coal mine rescue robots.

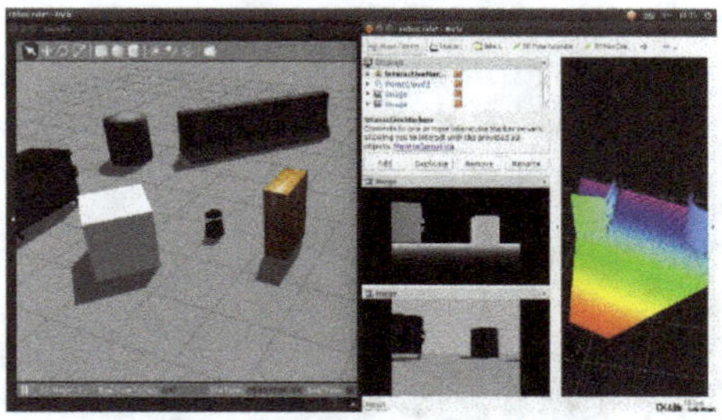

FIGURE 8: Virtual test of robot Gazebo.

(2) Semiactual Object Simulation Test. The semiactual object simulation is a technique of testing with connecting the object with the simulation model (such as mathematical simulation) on the computer [49–52]. The semiactual object simulation system of the driving system is shown in Figure 9. It is difficult to test with a real machine in a harsh environment where the mine rescue robot works; three coordinate test machines (temperature, humidity, and vibration) can be used to simulate the bad environment underground; combined with the virtual test of robot control system, the robot is simulated with semi-physical simulation to test the reliability of the robot controller and functionality.

(3) Test of Scaling Model. Test of scale model is mainly used in the case that the real machine cannot carry out the test or due to the larger cost of the experiment. For example, wind tunnel test of scale model can be used in the coal mine rescue robot with flying function; for EMC test of some large robots, the scale model can also be used [53].

(4) Iterative Development. Iterative development is generally used for development of systemic engineering, often found in the development of software [54, 55]. This method decomposes the entire development work into a series of short, fixed-length projects; each iteration includes the entire process from definition and requirement analysis to design, implementation, and testing. This method can simplify the complex problems gradually to refine the requirements according to the customer feedback, to improve the whole project solution through the iterative process continuously. The iterative development is shown in Figure 10. The product and development of coal mine rescue robot are complex systematic projects, which integrate the technology of machine, electricity, and software. At the beginning of the design, this work should be divided into many independent development units for independent development, like the drive unit, driving mechanism, control software, video transmission,

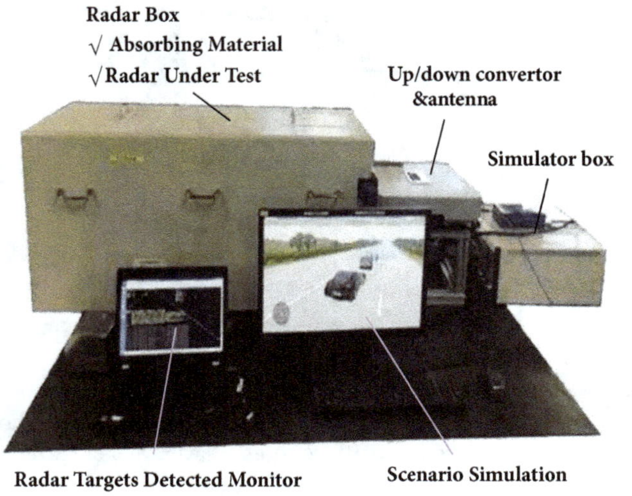

FIGURE 9: The semiactual object simulation system of the driving system.

etc., and then each unit is divided into several development stages again, like the interface design of drive unit, driver design, embedded software design, motor driver test, load test, and so on; each small project should also continue to iterate and improve and divide complex problems into small problems, to improve the entire life cycle of robot products step by step [56].

3.2. Standardization Problems

3.2.1. Security Permissions. The working environment of coal mine underground is special, and the environment after disaster is more complex, so the rescue work can be done only with permit of this industry. Robot and then how to obtain the franchise for the rescue work underground are the most confusing problems which have to be faced for every researchers of mine rescue robot.

The coal mine is strict with safety requirements for electrical equipment entering the mine [57, 58]; explosion-proof electrical equipment should be checked, with "the product certification", "the mining products safety approval", "the flame-proof certification", and safety performance before entering the mine; the flame-proof certification is issued by the qualified testing institution, and the mining products safety approval shall be issued by the national center for the safety of mineral products (the office of safety marks of mine products) [59]. It will take a long time with the process of testing and certification for the mine rescue robot before entering the coal mine. According to GB3836 standards stating that "the explosion environment is guarded against bursting the electrical equipment"[29] and international standards of IEC 60079 "electrical equipment used in explosive gas environment"[59], the safety problems of various components of coal mine rescue robot are analyzed and the corresponding suggestions are put forward in literature [47].

In order to solve problems of licensing issue of mine rescue robot, many enterprises put forward many measures; as mentioned earlier, Changzhou Research and Development Center Co., Ltd., and CITIC Heavy Industries Kaicheng Intelligent Equipment Company have adopted the new product safety sign certification prototype for their robot products, end date "/" in Tables 1 and 2, identified as security certificate for the new product; the "scope" of its "certification product details" says "this certificate is the new product safety certificate. it has certain limitation if just on the basis of the existing common standards and specifications for the product safety performance evaluation, there may be unknown risks, the holder should guide users to use and maintain correctly, to ensure the product used safely". In addition, regulations emphasize the certificate of quantity and number of the following products. But this method applies only to a limited number of new products and is not suitable for mass production.

3.2.2. Inspection, Evaluation, and Standardization. We must have a series of quality assurance measures, including the inspection, evaluation, and standardization of the quality of the robot products to promote the use of robots [60, 61].

The Ministry of Industry and Information Technology has been ahead in the inspection and evaluation of robots. The National Robot Detection and Assessment Center of National Development and Reform Commission has been approved and locates in Beijing. The establishing and operation of this public technology service platform are in the charge of China software center which can provide public technical service for robot industry [59, 62], and this platform has by approved by National Intelligent Robot Quality Supervision Test Center and Product Certification of Mechanical Equipment and Parts. Then, the Robot Detection and Evaluation Center is established in Shanghai, Chongqing, Guangzhou, and other regions.

Relative researches standardization works are now carried on actively by Standardization Working Group of special type of working robot in our country [63]. In July 2017, general technical requirements of robot for the ground search and rescue were launched; further testing, evaluation, and certification system would follow up with the development of standard work. It is believed that the coal mine disaster rescue robot will obtain the legal basis of entering the mine quickly with further advancement of standardization work.

3.2.3. Policies and Regulations. The ownership and underground operation of the coal mine rescue robots must be based on policies and regulations, for example, the method for the extraction and use of enterprise safe production costs [64–66]; this regulation was issued by the Treasury and the State Administration of Work Safety (SAWS); the coal production enterprises must improve the six systems spending, such as monitoring and control system for coal mine, personnel location system, emergency aversion, forced air self-help, communication, and avoiding, expenses for emergency rescue equipment communication, facility layout, and maintenance, and expenses for emergency evacuation and emergency shelter equipment configuration and emergency drill; the coal mine disaster rescue robot can be paid according to the underground detection and monitoring,

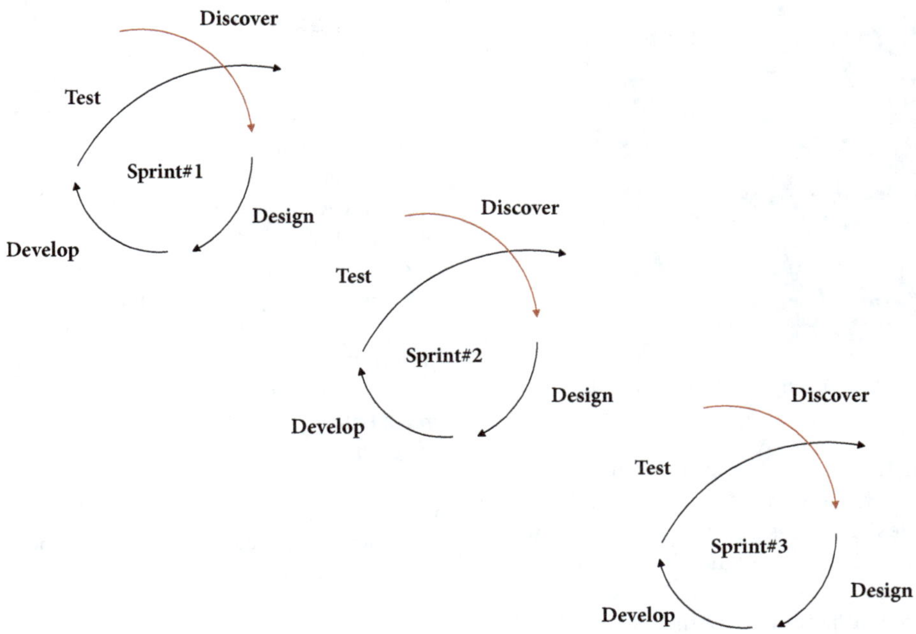

FIGURE 10: The iterative development.

communication and safety, emergency rescue technical equipment, and other items.

According to the safety regulations, such as the principle of classification explosive dangerous places, the mine explosion-proof electric locomotive must be cut off when the gas concentration exceeds 0.5%. It seems that current standards and specifications make it impossible for coal mine rescue robots to work underground. Literature [29] proposed that equipment of explosion-proof and intrinsic safety can be used for disaster rescue robot and improve the safety performance of the robot by the high-performance materials and special structure design, trying to solve the policy problem of mine disaster rescue robot from the standard and legal level.

It is worth mentioning that the relevant standards of spray painting robots have a great impact on the application of robots [67–70]. The environment is where spray painting work is poisonous, so the spraying robot is widely used in industry. During spraying, volatile organics (such as acetone, toluene, xylene, and ethyl ether) in the airtight room of spraying can form explosive environment. The explosive gas environment in the paint area is Level 1 danger zone area [71]. Due to the long term in the explosive gas environment, the spray painting robot work is dangerous similar to the work of the coal mine rescue robot. Nearly 30 years after multiple versions of revision to ensure that the spraying robot is able to be used under the special working environment, it provides a clear legal protection for robot manufacturer; it also provides a good demonstration effect for mine rescue robot.

According to literature [72], the cause of secondary gas explosion of coal mine is more complex [73], mainly because of the state of the coal mine disaster (comprehensive reasons, complicated reasons). It is impossible to prevent. Robots are not the major factor. Explosion-proof design is necessary [74, 75], but if grade of explosion-proof is overemphasized too much, it will be contrary to the original intention of the robot. The most important function of the coal mine disaster rescue robot is to collect the gas composition of the environment in the mine after the disaster and provide accurate analysis result to help the rescue workers in the background to make the correct judgment and decide whether to send rescue workers into mine to carry out rescue work; its effect is to reduce casualties in the process of rescue.

The excessive explosion-proof design will make the robot a huge volume and weight burden, which will bring a series of problems, like slow moving speed, poor ability of obstacle, short distance of operation, short battery life, etc.; these problems will make the robot unable to complete the detection and rescue mission. It is hoped that the relevant departments will give special regulations for the coal mine disaster rescue robot, so that this technology can truly benefit the mine and the vast coal mine workers.

3.3. Market Analysis and Recommendations

3.3.1. Analysis in Demand Side. The occurrence of coal mine disaster is not the normal event of coal mine, the use of the robot is seldom, and the iteration speed of R&D is bound to be affected by the frequency of use. In addition, the situation of the disaster is very different each time, which makes a great affection to the simulation experiment of the rescue robot. So even in the United States, where technology of robots is advanced, the development of coal mine rescue robots is very slow. Remotec is pessimistic about the market of coal mine rescue robots. The author holds that abnormal coal mine accidents caused the discontinuity of the market demand, which influence the development of robot's market.

The mine disaster is a threat to the safety of life and property of the broad masses of the people; once it occurs, the death toll is large and the impact is enormous; therefore, we

FIGURE 11: CITIC Kaicheng's firefighting robot.

should formulate policies actively to guide the demand of coal mining enterprises. On the other hand, with the development of The Times, people's demands for life safety and work environment are getting higher and higher; in addition to coal rescue robots, coal mining work robots will be used more and more to replace people in harsh environments [76, 77]; this growth in demand will be enough to drive demand for coal mine rescue robots.

3.3.2. Analysis in Purchasers. There are many robots with explosion-proof technology being used in the field of production, the use frequency of robots in the field of productive is very high, and the extensive use makes the technology of robots develop rapidly, like the firefighting robot, the spray paint robot (request explosion-proof), and so on. CITIC Heavy Industries has the whole product series of the firefighting robot. As shown in Figure 11, the fire robot series has succeeded in business, as the buyers are just the local governments. Similarly, the authors think the mine rescue robots will achieve success if the buyers are the new coal mining enterprise or the group with robot operation that can respond to the disaster quickly with the corresponding talent and the equipment support. It is beneficial for the rescue robot to play the biggest role.

3.3.3. Analysis in Operators and User. Like other robot products, the coal mine rescue robots need specialized teams for maintenance, training the operators, and calibrating the sensor regularly [78]. Therefore, the user of the rescue robot needs to establish a professional team to meet these needs. But for a coal mine it is difficult to establish such a professional team. It is more reasonable to build a robot rescue team based on the area since coal mining enterprises are relatively concentrated in one area. So the coalmine groups of the same area should adopt the form of cooperation; also, they can cooperate with the safety supervision, firefighting, and even specialized rescue companies to achieve the complete operation team of rescue robot.

4. Conclusion and Prospect

This review summarized existing problems and limitations of rescue robots in order to develop a reliable and practical mine rescue robot.

In conclusion, though a great deal of effort and practice have been done, there is still no successful case of rescue robots completing the rescue mission in the world. The scope of this paper has not been discussed in available review papers. The work remains a hard problem in the world; the bottleneck of this problem is not only about technology, but also related to market demand, policy license, and many other reasons. In this paper, the analysis is done from the theory and practice dimensions in the hope of attracting universal attention. The government support and the construction of legal system will ensure the achievement quickly and also will make enterprises and organizations have enough confidence in further research and development work, to promote the development of mine rescue robots faster and then realize the industrialization as soon as possible. The author hopes that the rescue robot can be applied to the underground rescue work early and successfully.

Conflicts of Interest

The authors declare that there are no conflicts of interest regarding the publication of this paper.

Acknowledgments

Yong Wang is working for a Natural Science Foundation Project of China [RMB 620,000]: "Research on the Mechanism Creation and Stability Control of Multi-terrain and Multi-motion Mine Rescue Robot".

References

[1] A. H. Reddy, B. Kalyan, and C. S. Murthy, "Mine Rescue Robot System – A Review," *Procedia Earth and Planetary Science*, vol. 11, pp. 457–462, 2015.

[2] S.-H. Qian, S.-R. Ge, Y.-S. Wang, and C.-Q. Liu, "Research status of the disaster rescue robot and its applications to the mine rescue," *Robot*, vol. 28, no. 2, pp. 350–354, 2006 (Chinese).

[3] W. Yong, "Current Status and Technical Problems in Research of Coal Mine Rescue Robot," *Coal Mine Machinery*, vol. 04, pp. 107–109, 2007.

[4] Y. Zhang, "Mine Safety and Health Administration (MSHA). Mine Rescue Robot[EB/OL]," https://arlweb.msha.gov/SagoMine/robotdetails.asp.

[5] Mining Technology, "Rescue Robots: The Search for Investment[EB/OL]," 2011, http://www.mining-technology.com/features/feature111373.

[6] United States Department of Labor, "Mine Safety and Health Administration(MSHA)[EB/OL]," https://www.msha.gov/about/history.

[7] Wikipedia., "Sago Mine disaster [EB/OL]," https://en.wikipedia.org/wiki/Sago_Mine_disaster.

[8] J. D. Mcateer, "The sago mine disaster. Buckhannon, 2006".

[9] Wikipedia, "Pike River Mine disaster [EB/OL]," https://en.wikipedia.org/wiki/Pike_River_Mine_disaster.

[10] "Robot broken in Pike River mine [EB/OL]," 2010, http://www.nzherald.co.nz/nz/news/article.cfm?c_id=1&objectid=10689488.

[11] Perth now, "WA sends rescue robot to NZ coal mine as rescue hopes fade [EB/OL]," 2010, http://www.perthnow.com.au/news/hopes-fading-for-29-trapped-miners-as-robot-breaks-down-at-pike-river-coal-mine/news-story/ae7b28164309d0d69b78d502543359af.

[12] In the News, "Robots deployed in Pike River mine[EB/OL]," https://www.sciencemediacentre.co.nz/2010/11/24/media-coverage-robots-deployed-in-pike-river-mine/.

[13] Science Media Center, "Experts on effectiveness of robots in mine rescue operations[EB/OL]," 2016, http://www.in-fonews.co.nz/default.cfm.

[14] Radio New Zealand, "Robots in the Pike River Mine-a timeline[EB/OL]," 2017, http://www.radionz.co.nz/news/national/329829/robots-in-the-pike-river-mine-a-timeline.

[15] Vimeo Pro, "PIKE RIVER MINE FOOTAGE[EB/OL]," https://vimeopro.com/user66181559/pike-river-mine/video/215922687.

[16] Laura Mills from the Greymouth Star, "Pike River charged for missing robot[EB/OL]," 2012, http://www.nzherald.co.nz/nz/news/article.cfm?c_id=1&objectid=10838786.

[17] Wikipedia, "Vorkuta mine disaster[EB/OL]," https://en.wikipedia.org/wiki/Vorkuta_mine_disaster.

[18] Everipedia, "Vorkuta mine disaster[EB/OL]," https://everipedia.org/wiki/Vorkuta_mine_disaster/.

[19] RT News, "Robots deployed in search operation following deadly Russian coal mine explosion (VIDEO)[EB/OL]," 2016, https://www.rt.com/news/334038-robots-vorkuta-coal-mine/.

[20] "Mining Products Safety Approval and Certification Center[EB/OL]," 2017, http://www.aqbz.org/Home/Index.aspx.

[21] G. E. Shi-Rong and Z. Hua, "Technical development status and tendency of rescue robot in dangerous environment," *Coal Science and Technology*, vol. 5, 8 pages, 2017.

[22] L. Dong-Xiao, "Application of robot technology to mine automation," *Coal Science and Technology*, vol. 5, pp. 62–64, 2007.

[23] Q. Xin-Yi, C. Dong-Xue, and Z. Shu-Jian, "Status of the Painting Robot and Its Application in Industry," *Modern Paint and Finishing*, vol. 8, pp. 53–55, 2016.

[24] J. Sun and F. Zhang, "A novel mine mobile communication system without blind zones," *Applied Mechanics and Materials*, vol. 58-60, pp. 249–254, 2011.

[25] S.-G. Niu, X. L. Wang, X. M. Jiang, and Z. Y. Hou, "New structural design of coal mine rescue robot," *Applied Mechanics and Materials*, vol. 470, pp. 650–653, 2014.

[26] C. Wang, W. Shi, Y. Li, and W. Cao, "Structural analysis of explosion proof mine pump base on the material of QT600-3," *Journal of Computational and Theoretical Nanoscience*, vol. 10, no. 12, pp. 2847–2852, 2013.

[27] L. Molyneaux, "Development of an Underground Mine Scout Robot, 2016".

[28] W. Lei, Z. Hua, T. Ming, and etal., "Research Progress of Elastomer Bulletproof Materials," *Functional Materials for Informatics*, vol. 6, pp. 8–12, 2014.

[29] IEC 60079-0-2007, "Electrical apparatus for explosive gas atmospheres - Part 0: General requirements".

[30] W. Xiao-Wei, H. Jin-Ying, and Z. Xu-Dong, "Antidetonation Properties on Composite Sandwich Structure with Polyurea Elastomer," *Engineering Plastics Application*, vol. 5, pp. 63–68, 2017.

[31] H. U. Songxia, L. I. U. Peili, and X. U. Chunying, "Recent Advance of Spray Polyurea Technology," *Chemical Propellants & Polymeric Materials*, vol. 3, 2010.

[32] J. Campbell, *Complete casting handbook: metal casting processes, metallurgy, techniques and design*, Butterworth-Heinemann, 2015.

[33] H. Stieger, "Magnesium Alloy Housing for Notebook Casing," *Special Casting and Non-ferrous Alloys*, vol. Z1, 73 pages, 2000.

[34] C. Yuan-Sheng, Z. Yanying, and D. Zhiming, "Liquid Forging Integral Forming Technology for Aluminum Matrix Composites with Local Reinforcement," *Special Casting & Nonferrous Alloys*, vol. 3, pp. 231–233, 2010.

[35] D. D. L. Chung and D. Chung, *Carbon fiber composites*, Butterworth-Heinemann, 2012.

[36] R. Remsburg, *Advanced thermal design of electronic equipment*, Springer Science & Business Media, 2011.

[37] J. Wadden, A. Lyashevsky, S. Gurumurthi, V. Sridharan, and K. Skadron, "Real-world design and evaluation of compiler-managed GPU redundant multithreading," *ACM SIGARCH Computer Architecture News*, vol. 42, no. 3, pp. 73–84, 2014.

[38] GB 3836.1-2010, "Explosive atmospheres-Part 1:Equipment-General requirements".

[39] J. Vertut, *Teleoperation and robotics: applications and technology*, Springer Science & Business Media, 2013.

[40] A. Kelly, N. Chan, H. Herman et al., "Real-time photorealistic virtualized reality interface for remote mobile robot control," *International Journal of Robotics Research*, vol. 30, no. 3, pp. 384–404, 2011.

[41] S. Tadokoro, "Rescue robotics challenge," in *Proceedings of the 2010 IEEE Workshop on Advanced Robotics and Its Social Impacts, ARSO 2010*, pp. 92–98, Republic of Korea, 2010.

[42] S. T. Shivappa, M. M. Trivedi, and B. D. Rao, "Audiovisual information fusion in human-computer interfaces and intelligent environments: a survey," *Proceedings of the IEEE*, vol. 98, no. 10, pp. 1692–1715, 2010.

[43] M. Bellone, A. Messina, and G. Reina, "A new approach for terrain analysis in mobile robot applications," in *Proceedings of the 2013 IEEE International Conference on Mechatronics, ICM 2013*, pp. 225–230, Italy, March 2013.

[44] R. Murphy, S. Tadokoro, and A. Kleiner, *A. Disaster robotics*, Springer, 2016.

[45] H. Kai and M. Xianmin, "Research on avoidance obstacle strategy of coal underground inspection robot based on binocular vision," in *Proceedings of the 29th Chinese Control and Decision Conference, CCDC 2017*, pp. 6732–6737, China, May 2017.

[46] G. Gong and H. Zhu, "A portable embedded explosion gas detection and identification device based on intelligent electronic nose system," *Sensor Review*, vol. 36, no. 1, pp. 57–63, 2016.

[47] Wikipedia, "Simulation[EB/OL]," https://en.wikipedia.org/wiki/Simulation.

[48] G. Tiantai, Z. Xiaojun, and Z. Genxing, *Various Meanings of Virtual Testing*, vol. 5, Machine Tool & Hydraulics, 3-5 edition, 2003.

[49] Wikipedia, "Hardware In-the-loop Simulation[EB/OL]," https://en.wikipedia.org/wiki/Hardware-in-the-loop_simulation.

[50] S. Jiayuan, M. Xiuyun, and D. Yan, *Hardware In-the-loop Simulation*, National Defence Industry Press, Beijing, China, 2008.

[51] L. Yan-Bin and J. Guang, "Development of Hardware in-the-loop Simulation," *Ome Information*, vol. 1, pp. 27–32, 2003.

[52] Y. Xinyu and H. Keli, "Techniques of Real -time Control in the Hardware in-the -loop Simulation Stystem," *Computer Simulation*, vol. 1, pp. 33–36, 2000.

[53] z. Xing and W. Lian-dong, "Application of Scaled Model Technique in Design of RF Simulation Anechoic Chamber," *Telecommunication Engineering*, vol. 5, pp. 32–35, 2008.

[54] N. B. Ruparelia, "Software development lifecycle models," *ACM SIGSOFT Software Engineering Notes*, vol. 35, no. 3, p. 8, 2010.

[55] Y. Leau B, W. Loo K, and W. Tham Y, "Software development life cycle AGILE vs traditional approaches," in *Proceedings of the International Conference on Information and Network Technology*, vol. 37, pp. 162–167, 2012.

[56] W. Wang, W. Dong, Y. Su, D. Wu, and Z. Du, "Development of search-and-rescue robots for underground coal mine applications," *Journal of Field Robotics*, vol. 31, no. 3, pp. 386–407, 2014.

[57] Y. Yao-hai, "Safety Requirements of Electrical Explosion Protected Equipments Used in Coal Industry," *Electric Explosion Protection*, vol. 2, 2008.

[58] X. Zhang, A. Wang, and R. Gu, "Discussion on Application of 10 kV Flame Proof and Positive Pressure Type Equipment to Underground Mine," *Coal Science and Technology*, vol. 1, 2012.

[59] W. B. WANG, "Reveal the national robot authentication authority! [EB/OL]," http://www.360doc.com/content/17/0227/12/18550343_632376961.shtml.

[60] S. Tadokoro, S. Seki, and H. Asama, "Priority issues of disaster robotics in Japan," in *Proceedings of the IEEE Region 10 Humanitarian Technology Conference, R10-HTC 2013*, pp. 41–46, Japan, 2013.

[61] J. Green, "Mine rescue robots requirements: Outcomes from an industry workshop," in *Proceedings of the 6th Robotics and Mechatronics Conference, (RobMech '13)*, pp. 111–116, South Africa, 2013.

[62] CCiD, "National Robot Testing and Evaluation Center [EB/OL]," 2016, http://www.ccidgroup.com/techdeve/7380.htm.

[63] China Standardization Working Group for Special Tasks Robots, "Profile of the Organizer [EB/OL]," http://cst-robot.com/ui/html/organzation.html.

[64] China Standardization Working Group for Special Tasks Robots, "(General technical conditions for rehabilitation training robot). The two national standards will be successfully held in shenzhen [EB/OL]," 2017, http://cst-robot.com/ui/html/news_detail.html?ID=2bbc4886-7742-40e4-b1db-7abdd61066c9.

[65] Chinese government net, "Notice on the issuance of management measures for the extraction and use of enterprise safe production costs [EB/OL]," 2012, http://www.gov.cn/zwgk/2012-02/29/content_2079240.htm.

[66] F. Gui, "The method for the extraction and use of enterprise safe production costs," *Traffic Accounting*, vol. 4, pp. 93–96, 2012.

[67] DB13T 1483-2011, "Underground mine detection robot," https://wenku.baidu.com/view/54312577e009581b6ad9eb11.html.

[68] ZB J28 002—90, "The General Technical Requirements for Spray Painting Robots".

[69] JB/T 9182-1999, "The General Technical Requirements for Spray Painting Robots," Tech. Rep., State Administration of Machinery Industry, Beijing, china, 1999.

[70] JB/T 9182-2014, "Spray-painting robot-General specifications".

[71] Y. Da-Ming, L. Wang, J. Chen, and W. Zhao-Hong, "Basic Safety Capability Requirement of Coal Mine Disaster-relief Robot," in *Safety in Coal Mines*, vol. 12, pp. 104–107, 2009.

[72] Y. Cun-Juan, "Experimental research on secondary blasting of coal dust and gas induced by gas explosion," *China Safety Science Journal*, vol. 12, pp. 29–32, 2014.

[73] L. M. Pejic, J. G. Torrent, E. Querol, and K. Lebecki, "A new simple methodology for evaluation of explosion risk in underground coal mines," *Journal of Loss Prevention in the Process Industries*, vol. 26, no. 6, pp. 1524–1529, 2013.

[74] X. Rong, R. Song, X. Song, and Y. Li, "Mechanism and explosion-proof design for a coal mine detection robot," in *Proceedings of the International Conference on Advanced in Control Engineering and Information Science, 9CEIS '11)*, pp. 100–104, China, 2011.

[75] Y. W. Li, S. R. Ge, and H. Zhu, "Explosion-proof design for coal mine rescue robots," *Advanced Materials Research*, vol. 211-212, pp. 1194–1198, 2011.

[76] J. Trevelyan, R. W. Hamel, and S. C. Kang, *Robotics in hazardous applications*, Springer, 2016.

[77] P. Novák, J. Babjak, T. Kot, P. Olivka, and W. Moczulski, "Exploration Mobile Robot for Coal Mines," in *Proceedings of the International Workshop on Modelling and Simulation for Autonomous Systems*, vol. 9055 of *Lecture Notes in Computer Science*, pp. 209–215, Springer International Publishing, 2015.

[78] D. N. Ray, R. Das, B. Sebastian, B. Roy, and S. Majumder, "Design and Analysis Towards Successful Development of a Tele-Operated Mobile Robot for Underground Coal Mines," in *Robotics and Factories of the Future*, Lecture Notes in Mechanical Engineering, pp. 589–602, Springer India, New Delhi, 2016.

8

A Family of Hyperbolic-Type Explicit Force Regulators with Active Velocity Damping for Robot Manipulators

Fernando Reyes-Cortés,[1] **César Chávez-Olivares**,[2] **and Emilio J. González-Galván**[3]

[1]*Facultad de Ciencias de la Electrónica, Benemérita Universidad Autónoma de Puebla, 72570 Puebla, PUE, Mexico*
[2]*Departamento de Ingeniería Robótica, Universidad Autónoma de Aguascalientes, Campus Sur, Av. Prolongación Mahatma Gandhi No. 6601, 20340 Aguascalientes, AGS, Mexico*
[3]*Facultad de Ingeniería, Universidad Autónoma de San Luis Potosí, Av. Manuel Nava 8, Zona Universitaria, 78290 San Luis Potosí, SLP, Mexico*

Correspondence should be addressed to Fernando Reyes-Cortés; fernando.reyes@correo.buap.mx

Academic Editor: L. Fortuna

This paper addresses the explicit force regulation problem for robot manipulators in interaction tasks. A new family of explicit force-control schemes is presented, which includes a term driven by a large class of saturated-type hyperbolic functions to handle the force error. Also, an active velocity damping term with the purpose of obtaining energy dissipation on the contact surface is incorporated plus compensation for gravity. In order to ensure asymptotic stability of the closed-loop system equilibrium point in Cartesian space, we propose a strict Lyapunov function. A force sensor placed at the end-effector of the robot manipulator is used in order to feed back the measure of the force error in the closed-loop, and an experimental comparison of the performance $\mathscr{L}_2$-norm between 5 explicit force control schemes, which are the classical proportional-derivative (PD), arctangent, and square-root controls and two members of the proposed control family, on a two-degree-of-freedom, direct-drive robot manipulator, is presented.

1. Introduction

The use of robotic systems has increased during the last four decades. Nowadays, several industrial applications continue to benefit from robotic systems in order to automate repetitive or hazardous maneuvers. It is well known that, in applications involving an automatic process in which the robot is moving freely within its workspace or executes tasks with no interaction with the environment, a control system based on position or motion control represents a viable alternative. However, in applications where the manipulator performs interaction tasks, the manipulator encounters environmental constraints and the interaction forces are not negligible [1–3].

In this situation, the use of motion-control strategies will produce an unsuccessful execution of the task; the adequate implementation of an interaction task requires, in addition to motion/position control strategies, force control and a detailed and accurate modeling of the system and the environment [4–9]. A particular type of interaction control is the impedance control scheme which can regulate environmental forces according to the specified mechanical impedance such as inertial, stiffness, and damping effects. Impedance control belongs to the category of indirect force control, which may not require measurements of contact forces; in addition, the user cannot specify a desired contact force [2, 4, 9–12].

In contrast, force sensing and force control are essential for the successful execution of tasks involving interaction between the manipulator and the environment. The explicit force control strategies are the most suitable option for controlling interaction forces between the robot manipulator and the environment. Such control schemes have a large number of applications in medical and industrial environments, as they are capable of increasing the speed and quality in applications such as deburring, grinding of surfaces, assembly processes, and other operations [3, 10, 13–16].

Some types of force control can be cited, such as [17] where two auxiliary, automatically controlled wings of a monohull ship used to attenuate the roll due to the waves are presented. These wings are variably dipped into the water in order to introduce a force that reduces the oscillations on the ship. Two different compensators for the roll effect have been designed, the first one using classical frequency domain techniques and the second being an adaptive, linear quadratic compensator with a multilayer perceptron neural network. Simulations and experimental tests performed on the ship are presented. In [18], an impedance function is proposed to specify a desired force directly in a robust robot force tracking impedance scheme, which employs a neural network as compensator to cancel out all uncertainties; simulation studies are presented. In [19], the robot dynamics are decomposed into force, position, and redundant joint subspaces; then, using that decomposition in a neural network-based adaptive control scheme for hybrid force/position in Lyapunov sense is proposed. Numerical simulation studies were performed in the model of a two-link, rigid robot manipulator. In [20], an approach based on hybrid force/position with fuzzy PID structure with unspecified robot dynamics in the presence of external disturbances, including experimental results on a PUMA-560 robot, is presented.

Unlike the impedance control schemes, the concept of explicit force control is to consider a desired force rather than the position or velocity references. Then the force error is defined as the difference between a desired force and the actual interaction force, measured with a force sensor mounted on the end-effector, enabling the possibility of ensuring force regulation. The explicit force control selects the directions in which the end-effector position should be controlled and those directions in which the forces controlled by the end-effector are applied on some environment, for a given task. This type of control strategies has two feedback loops for separated, position, and force control and is referred to as hybrid force/position control [1, 21–34].

The original hybrid force/position control concept was proposed in [1]; the basic idea is to divide the robot manipulator space into two orthogonal spaces using of a selection matrix. Position control and force control are applied on two spaces to achieve the capability of compliance control. Explicit force control involves the direct command and measurement of force values in order to control the contact force to a desired value.

Considering that explicit force control is an important topic as research area, it currently represents an open problem of interest for the scientific community and several research efforts have focused on this topic during the last four decades. As a result, different control schemes have been proposed in the literature; for example, we can cite the fundamental works [1, 5, 24, 26, 32]. Progressively this theme has been growing; for instance, [35] describes the design and implementation of a hybrid force/motion scheme to control the interaction force of a robot manipulator, using a PD controller with gain-scheduled linear parameter-varying controller; however, although the reference for this control scheme does not include a stability proof, it presents experimental results on a six-degree-of-freedom manipulator. The paper [36] proposes the design of a robust adaptive neural network-based hybrid position/force control with PD-type structure for manipulators. It includes a stability analysis in the sense of Lyapunov, but only simulation results with a model of a two-link robot manipulator are presented. In [37], the robustness and stability of disturbance observer based on explicit linear force control systems are analyzed. The validity of the proposals is verified with experimental results on a linear dc motor.

In [38], explicit force regulators for robot manipulators are presented. The control structure includes a nonlinear term formed by square-root-type functions to drive the force error, with an active velocity damping plus compensation of gravity forces. The stability analysis is performed in Lyapunov sense; in order to obtain asymptotic stability, LaSalle's theorem is used and experimental results on a two-degree-of-freedom robot are presented. In [39], an optimization-based approach for the regulation of excessive or insufficient forces at the end-effector level is introduced. Their approach minimizes the interaction force error at the robot end-effector and considers the linear impedance model and penalty functions in order to handle the constraints on the interaction forces; stability and convergence properties are analyzed taking into account force limits. The proposed scheme is validated via experimental results for a robotic grasping task. The work [40] presents the state of the art on compliant control algorithms applied to both stiff and soft joints based on electric motors for robot manipulators; there is a widely used control structure for all the works presented in this survey, which is the simple PD control.

Recently, in [16] a double active observer control is proposed to control the desired interaction and motion compensation under contact with the heart, relying on robot force control techniques. In [34], a position-based force explicit control strategy based on online trajectory prediction, with proportional-integral-type control structure, is presented. In this work, the experimental results show that the proposed method has a good control effect on the contact force. In [41], the regulation of the clamping force of a robot manipulator gripper is analyzed and discussed, when applied to rescue activities in the ruins after disasters; for this problem, a hybrid force/position control on a cascade PID-scheme and fuzzy sliding mode is presented; the control algorithms were validated with experimental results, whose force errors were acceptable. Force trajectories during isometric force-matching task involving contractions are described in [42]; a PID control with different gains and frequencies is introduced to examine the nature of the contractions in many individuals, which vary substantially. This variability can be explained by discrete PID with varying model parameters to adjust the muscle commands. The position and force control task presented in [43] has been defined considering the manipulator and environment models; asymptotic stability of the control systems is demonstrated, and experimental study of the issue is presented.

Most of the previous works use explicit force control strategies with PD or PID schemes; however, in general, there is a lack of proof on asymptotic stability analysis with different control structures; only a limited number of works have proposed strict Lyapunov functions. Therefore, our main motivation is in the theoretical and practical aspects of the problem. The objective of this paper is to present the design and analysis of a new explicit force control family with a term driven by a large class of saturated-type hyperbolic functions to handle the force error for the problem of force regulation in interaction tasks. This control structure includes also an active velocity damping term with the purpose of obtaining energy dissipation on the contact surface. To ensure global asymptotic stability of the closed-loop system equilibrium point in Cartesian space, we propose a strict Lyapunov function. Additionally, we present in this work another contribution, consisting of the mathematical development of a Cartesian dynamics model property (on the vector of centripetal and Coriolis forces in Cartesian space, see Appendix), which is useful for stability analysis.

The paper includes also an experimental comparison of the performance obtained between 5 explicit force control schemes; they are the classical proportional-derivative (PD), arctangent, square-root controls, and two members of the proposed control family on a two-degree-of-freedom, direct-drive robot manipulator. A force sensor placed at the end-effector of the robot manipulator is used in order to feed back the measure of the force error in the closed-loop, which is given as the difference between a desired and actual force. The Cartesian explicit force control is converted to joint torque signal using the transpose Jacobian introduced by Takegaki and Arimoto [44].

The paper is organized as follows. In Section 2, the joint-space dynamic model of robot manipulators and its relationship with the Cartesian dynamic model are presented; it also describes the most important properties used in the stability analysis of the control schemes. In Section 3, the new family of explicit force regulators is presented, as well as the design of the strict function required to carry out the asymptotic stability analysis in Lyapunov sense. In Section 4, the experimental results on a two-degree-of-freedom, direct-drive robot manipulator including a performance comparison of the control schemes under evaluation are presented. Finally, our conclusions are in Section 5.

2. Dynamic Model of the Robot Manipulator

It is a well-known fact that the dynamic model of a robot manipulator with rigid links is described by a second-order system formed by nonlinear differential equations. This model describes a relationship between the generalized joint torques/forces applied by the actuators and the motion of the robot, which incorporates the position, velocity, and acceleration joints. In joint space, the dynamic model of a robot manipulator with n degrees of freedom is defined by [32]

$$\boldsymbol{\tau} = \mathbf{M}(\mathbf{q})\ddot{\mathbf{q}} + \mathbf{C}(\mathbf{q},\dot{\mathbf{q}})\dot{\mathbf{q}} + \mathbf{g}(\mathbf{q}) + \mathbf{f}(\dot{\mathbf{q}}), \tag{1}$$

where $\mathbf{q}, \dot{\mathbf{q}}, \ddot{\mathbf{q}} \in \mathbb{R}^n$ denote the joint displacements, velocities, and acceleration values, respectively, $\boldsymbol{\tau} \in \mathbb{R}^{n\times 1}$ stands for the vector of generalized forces/torques applied, $\mathbf{M}(\mathbf{q}) \in \mathbb{R}^{n\times n}$ is the symmetric positive-definite inertia matrix, $\mathbf{C}(\mathbf{q},\dot{\mathbf{q}}) \in \mathbb{R}^{n\times n}$ is the matrix of centripetal and Coriolis torques, $\mathbf{g}(\mathbf{q}) \in \mathbb{R}^n$ is the vector of gravitational torques, and, finally, $\mathbf{f}(\dot{\mathbf{q}}) \in \mathbb{R}^n$ is the friction torque vector. For analysis purposes, only the viscous-friction model is considered in this paper; that is, $\mathbf{f}(\dot{\mathbf{q}}) = \mathbf{B}\dot{\mathbf{q}}$, where $\mathbf{B}$ is a diagonal matrix formed by the viscous-friction coefficients of each joint.

Moreover, the above model (1) can be expressed in terms of coordinates associated with a coordinate system attached to the tool by using a coordinate transformation [6, 25, 45]. For the formulation of the model in Cartesian space, it is assumed that the manipulator performs in a nonsingular configuration.

Let us consider the relationship between the joint velocities $\dot{\mathbf{q}}$ and the end-effector Cartesian velocities $\dot{\mathbf{x}}$, given by $\dot{\mathbf{x}} = \mathbf{J}(\mathbf{q})\dot{\mathbf{q}}$, where $\mathbf{J}(\mathbf{q})$ is referred to as the analytical Jacobian matrix; that is, $\mathbf{J}(\mathbf{q}) = \partial\mathbf{h}(\mathbf{q})/\partial\mathbf{q} \in \mathbb{R}^{n\times n}$, where $\mathbf{h}(\mathbf{q})$ is the direct kinematic. A well-known relationship exists between the applied forces $\mathbf{f}_\chi$ on the end-effector and the applied torques $\boldsymbol{\tau}$ in the joints, which is referred to as the Jacobian Transposed Controller [44]:

$$\boldsymbol{\tau} = \mathbf{J}^T(\mathbf{q})\,\mathbf{f}_\chi. \tag{2}$$

The Cartesian dynamic model of a robot manipulator is [45]

$$\mathbf{M}_\chi\ddot{\mathbf{x}} + \mathbf{C}_\chi\dot{\mathbf{x}} + \mathbf{g}_\chi + \mathbf{f}_{f\chi} = \mathbf{f}_\chi - \mathbf{f}_e, \tag{3}$$

where the term $\mathbf{f}_e$ represents the external forces applied to the end-effector, such as the forces appearing in an interaction task. The terms $\mathbf{M}_\chi$, $\mathbf{C}_\chi$, $\mathbf{g}_\chi$, and $\mathbf{f}_{f\chi}$ are described as

$$\begin{aligned}
\mathbf{M}_\chi &= \mathbf{J}^{-T}(\mathbf{q})\,\mathbf{M}(\mathbf{q})\,\mathbf{J}^{-1}(\mathbf{q}) \\
\mathbf{C}_\chi &= \mathbf{J}^{-T}(\mathbf{q})\,\mathbf{C}(\mathbf{q},\dot{\mathbf{q}})\,\mathbf{J}^{-1}(\mathbf{q}) - \mathbf{M}_\chi\dot{\mathbf{J}}(\mathbf{q})\,\mathbf{J}^{-1}(\mathbf{q}) \\
\mathbf{g}_\chi &= \mathbf{J}^{-T}(\mathbf{q})\,\mathbf{g}(\mathbf{q}) \\
\mathbf{f}_{f\chi} &= \mathbf{B}_\chi\dot{\mathbf{x}} = \underbrace{\mathbf{J}^{-T}(\mathbf{q})\,\mathbf{B}\mathbf{J}^{-1}(\mathbf{q})}_{\mathbf{B}_\chi}\dot{\mathbf{x}}.
\end{aligned} \tag{4}$$

We use the notation $\dot{\mathbf{J}}(\mathbf{q})$ to indicate the time derivative of the Jacobian matrix, $\dot{\mathbf{J}}(\mathbf{q}) = (d/dt)\mathbf{J}(\mathbf{q})$, and the inverse of transpose Jacobian matrix is denoted by $\mathbf{J}^{-T}(\mathbf{q}) = [\mathbf{J}^{-1}(\mathbf{q})]^T$.

Next, some well-known properties that are useful in the analysis and design of our controller scheme are presented.

Property 1. The inertia matrix $\mathbf{M}_\chi > 0$ is symmetric and definite positive (on a workspace without singularity).

$$\begin{aligned}
&\mathbf{M}_\chi > 0 \Longleftrightarrow \\
&\mathbf{M}_\chi^{-1} > 0.
\end{aligned} \tag{5}$$

Property 2. The matrix $\mathbf{M}_\chi$ is upper and lower bounded by

$$\lambda_{\mathrm{m}}\left\{\mathbf{M}_\chi\right\}\mathbf{I} \le \mathbf{M}_\chi \le \lambda_{\mathrm{M}}\left\{\mathbf{M}_\chi\right\}\mathbf{I}, \tag{6}$$

where $\lambda_{\mathrm{M}}\{\mathbf{M}_\chi\}$ and $\lambda_{\mathrm{m}}\{\mathbf{M}_\chi\}$ stand for the maximum and minimum eigenvalues of matrix $\mathbf{M}_\chi$, respectively.

Property 3. $\dot{\mathbf{M}}_\chi = \mathbf{C}_\chi + \mathbf{C}_\chi^T$; the matrix $\dot{\mathbf{M}}_\chi - 2\mathbf{C}_\chi$ is skew-symmetric and it satisfies

$$\frac{1}{2}\dot{\mathbf{x}}^T\left[\dot{\mathbf{M}}_\chi - 2\mathbf{C}_\chi\right]\dot{\mathbf{x}} \equiv 0. \tag{7}$$

Property 4. The norm of vector $\mathbf{C}_\chi\dot{\mathbf{x}}$ is bounded, as indicated below:

$$\left\|\mathbf{C}_\chi\dot{\mathbf{x}}\right\| \le k_{cx}\left\|\dot{\mathbf{x}}\right\|^2 \tag{8}$$

for some $k_{cx} \in \mathbb{R}_+$. As part of the contributions of this article, a brief demonstration of this property is included in the Appendix.

The following decoupled model is considered for modeling the compliant environment [3]:

$$\mathbf{f}_e = \mathbf{K}_e\left(\mathbf{x} - \mathbf{x}_e\right), \tag{9}$$

where $\mathbf{K}_e$ is a diagonal matrix referred to as the environment's stiffness matrix and the term $\mathbf{x}_e$ denotes the contact position with the environment, while $\mathbf{x}$ represents the actual position of the end-effector.

3. A New Family of Explicit Force Regulator Schemes

In this section, a new family of force control schemes with hyperbolic-type structure for robot manipulators is presented. Consider the following family with large class of explicit force control schemes plus gravity compensation, given by

$$\mathbf{f}_\chi = \mathbf{K}_p\begin{bmatrix}\dfrac{\cosh^{k-1}\left(\tilde{f}_1\right)\sinh\left(\tilde{f}_1\right)}{1+\cosh^k\left(\tilde{f}_1\right)}\\ \dfrac{\cosh^{k-1}\left(\tilde{f}_2\right)\sinh\left(\tilde{f}_2\right)}{1+\cosh^k\left(\tilde{f}_2\right)}\\ \vdots\\ \dfrac{\cosh^{k-1}\left(\tilde{f}_n\right)\sinh\left(\tilde{f}_n\right)}{1+\cosh^k\left(\tilde{f}_n\right)}\end{bmatrix} - \mathbf{K}_v\dot{\mathbf{x}} + \mathbf{g}_\chi + \mathbf{f}_e, \tag{10}$$

where $k \in \mathbb{N}_+$ is a positive integer, $\mathbf{K}_p, \mathbf{K}_v \in \mathbb{R}^{n\times n}$ are the proportional and derivative gains, which are positive-definite diagonal matrices, and $\tilde{\mathbf{f}}(t)$ is the force error vector, which is defined as the difference between a desired force $\mathbf{f}_d$ and the actual force $\mathbf{f}(t)$ on the end-effector, measured from the force sensor; $\tilde{\mathbf{f}}(t) = \mathbf{f}_d - \mathbf{f}(t)$. The applied torques on the robot joints are derived from (2); that is, $\boldsymbol{\tau} = \mathbf{J}^T(\mathbf{q})\mathbf{f}_\chi$.

The first term in the proposed control scheme (10) represents a large class of hyperbolic-type regulators to drive the force error; this one is derived from energy shaping of the artificial potential energy [44]. The second term acts as an active velocity damping; the next component includes the Cartesian gravity forces $\mathbf{g}_\chi$; for implementation purposes, partial knowledge of these gravity forces is required, and the last term $\mathbf{f}_e$ represents the interaction forces.

The closed-loop equation expressed in state variables $\left[\tilde{\mathbf{f}}\ \dot{\mathbf{x}}\right]^T \in \mathbb{R}^{2n}$ is obtained by combining the dynamics model (3) and control law (10):

$$\frac{d}{dt}\begin{bmatrix}\tilde{\mathbf{f}}\\ \dot{\mathbf{x}}\end{bmatrix} = \begin{bmatrix} -\mathbf{K}_e\dot{\mathbf{x}} \\ \mathbf{M}_\chi^{-1}\left[\mathbf{K}_p\begin{bmatrix}\dfrac{\cosh^{k-1}\left(\tilde{f}_1\right)\sinh\left(\tilde{f}_1\right)}{1+\cosh^k\left(\tilde{f}_1\right)}\\ \dfrac{\cosh^{k-1}\left(\tilde{f}_2\right)\sinh\left(\tilde{f}_2\right)}{1+\cosh^k\left(\tilde{f}_2\right)}\\ \vdots\\ \dfrac{\cosh^{k-1}\left(\tilde{f}_n\right)\sinh\left(\tilde{f}_n\right)}{1+\cosh^k\left(\tilde{f}_n\right)}\end{bmatrix} - \left[\mathbf{K}_v + \mathbf{B}_\chi\right]\dot{\mathbf{x}} - \mathbf{C}_\chi\dot{\mathbf{x}}\right]\end{bmatrix}. \tag{11}$$

In order to demonstrate that the state origin is an equilibrium point, it is necessary to consider the Jacobian matrix $\mathbf{J}(\mathbf{q})$ with full rank; for example, $\mathrm{rank}[\mathbf{J}(\mathbf{q})] = n$; in other words, the Cartesian dynamic model requires a singularity-free workspace. The following assumptions are taken into account:

(a) The Jacobian matrix $\mathbf{J}(\mathbf{q})$ is full rank and its spectral norm $\|\mathbf{J}(\mathbf{q})\| < k_J$ is bounded for all $\mathbf{q} \in \mathbb{R}^n$, where k_J is a positive constant, $k_J \in \mathbb{R}_+$.

(b) According to the environment's stiffness and friction models, the stiffness $\mathbf{K}_e$ and viscous-friction $\mathbf{B}_\chi$ matrices are diagonal and positive-definite matrices, respectively; then $\left[\mathbf{K}_e + \mathbf{B}_\chi\right]\dot{\mathbf{x}} = \mathbf{0} \Leftrightarrow \dot{\mathbf{x}} = \mathbf{0}$.

The first component of the closed-loop equation (11) satisfies $-\mathbf{K}_e\dot{\mathbf{x}} = \mathbf{0} \in \mathbb{R}^n \Leftrightarrow \dot{\mathbf{x}} = \mathbf{0} \in \mathbb{R}^n$, due to the fact that the matrix $\mathbf{K}_e$ is a diagonal definite positive matrix. For the second component in (11), we use Property 1 of the inertia matrix; then the first term inside the brackets composed of the proportional gain $\mathbf{K}_p$ and the vector of hyperbolic functions results in the vector $\mathbf{0} \in \mathbb{R}^n \Leftrightarrow \tilde{\mathbf{f}} = \mathbf{0} \in \mathbb{R}^n$, since $\mathbf{K}_p$ is a positive-definite matrix and each component $\cosh^{k-1}(\tilde{f}_i)\sinh(\tilde{f}_i)/(1+\cosh^k(\tilde{f}_i)) = 0 \Leftrightarrow \sinh(\tilde{f}_i) = 0 \Leftrightarrow \tilde{f}_i = 0$, $\forall k \in \mathbb{N}_+$, and $i = 1,\ldots,n$. We have also $\mathbf{C}_\chi\dot{\mathbf{x}} = 0$ if $\dot{\mathbf{x}} = \mathbf{0} \in \mathbb{R}^n$ and $\mathbf{C}_\chi \neq \mathbf{0} \in \mathbb{R}^{n\times n}$ in the singularity-free workspace. Therefore, the closed-loop system (11) is autonomous differential equation and the state space origin exists and it is a local equilibrium point.

3.1. Control Problem Statement. The problem of explicit force control consists of finding a Cartesian control scheme $\mathbf{f}_\chi$ such that the force error vector $\tilde{\mathbf{f}}(t)$ and Cartesian velocity $\dot{\mathbf{x}}(t)$ converge asymptotically stably to equilibrium point of the closed-loop equation (11) at least for sufficiently small initial conditions $\left[\tilde{\mathbf{f}}(0), \dot{\mathbf{x}}(0)\right]^T$; then the following objective is satisfied:

$$\lim_{t\to\infty}\begin{bmatrix}\tilde{\mathbf{f}}(t)\\ \dot{\mathbf{x}}(t)\end{bmatrix} = \mathbf{0} \in \mathbb{R}^{2n}. \tag{12}$$

3.2. Proposition. Consider the closed-loop system (11), under conditions sufficiently small; then the equilibrium point $\left[\tilde{\mathbf{f}}(t)\ \ \dot{\mathbf{x}}(t)\right] \in \mathbb{R}^{2n}$ is locally asymptotically stable and the control objective (12) is achieved.

Proof. To carry out the local stability analysis for the dynamic system (11), we use Lyapunov's direct method through the following strict Lyapunov function (a strict Lyapunov function is a positive-definite function, whose time derivative along the trajectories of closed-loop system yields a negative-definite function), which is composed by the sum of the robot manipulator kinetic and artificial potential energy, and a cross-term between force error $\tilde{\mathbf{f}}$ and velocity $\dot{\mathbf{x}}$:

$$V\left(\dot{\mathbf{x}},\tilde{\mathbf{f}}\right) = \frac{1}{2}\dot{\mathbf{x}}^T\mathbf{M}_\chi\dot{\mathbf{x}} + \mathcal{U}_a\left(\mathbf{K}_p,\mathbf{K}_e^{-1},\tilde{\mathbf{f}}\right) - \frac{\varepsilon_0}{1+\|\tilde{\mathbf{f}}\|}\tilde{\mathbf{f}}^T\mathbf{M}_\chi\dot{\mathbf{x}}, \tag{13}$$

where ε_0 is a positive constant number defined in the following interval:

$$\min\left\{\sqrt{\frac{\rho'\lambda_{\mathrm{m}}\{\mathbf{K}_e^{-1}\mathbf{K}_p\}}{k\lambda_{\mathrm{M}}\{\mathbf{M}_\chi\}}}, \frac{\lambda_{\mathrm{m}}\{\mathbf{K}_v\}+\lambda_{\mathrm{m}}\{\mathbf{B}_\chi\}+k_{cx}}{2\lambda_{\mathrm{M}}\{\mathbf{K}_e\}\lambda_{\mathrm{M}}\{\mathbf{M}_\chi\}},\right.$$
$$\left.\frac{\lambda_{\mathrm{m}}\{\mathbf{K}_p\}\left[\lambda_{\mathrm{m}}\{\mathbf{K}_v\}+\lambda_{\mathrm{m}}\{\mathbf{B}_\chi\}+k_{cx}\right]}{2\lambda_{\mathrm{m}}\{\mathbf{K}_p\}\lambda_{\mathrm{M}}\{\mathbf{K}_e\}\lambda_{\mathrm{M}}\{\mathbf{M}_\chi\}+(1/4)\left[\lambda_{\mathrm{m}}\{\mathbf{K}_v\}+\lambda_{\mathrm{m}}\{\mathbf{B}_\chi\}\right]^2}\right\} \tag{14}$$
$$> \varepsilon_0 > 0,$$

where ρ' is a positive constant, $\lambda_{\mathrm{m}}\{\mathbf{K}_e^{-1}\mathbf{K}_p\}$ is the minimum eigenvalue of the product $\mathbf{K}_e^{-1}\mathbf{K}_p$, the minimum eigenvalue of the environment's inverse stiffness matrix is $\lambda_{\mathrm{m}}\{\mathbf{K}_e^{-1}\}$, and $\lambda_{\mathrm{M}}\{\mathbf{K}_p\}$, $\lambda_{\mathrm{M}}\{\mathbf{K}_v\}$, $\lambda_{\mathrm{M}}\{\mathbf{B}_\chi\}$ are the maximum eigenvalues of the proportional and derivative gains and viscous-friction matrix, respectively. It is important to note that the constant ε_0 is only required for analysis purposes and its numerical value is not required for the control scheme. □

The artificial potential energy function $\mathcal{U}_a(\mathbf{K}_p,\mathbf{K}_e^{-1},\tilde{\mathbf{f}})$ is a continuously differentiable and positive-definite function, given by

$$\mathcal{U}_a\left(\mathbf{K}_p,\mathbf{K}_e^{-1},\tilde{\mathbf{f}}\right) = \frac{1}{k}\begin{bmatrix}\sqrt{\ln\left(1+\cosh^k\left(\tilde{f}_1\right)\right)-\ln(2)}\\ \sqrt{\ln\left(1+\cosh^k\left(\tilde{f}_2\right)\right)-\ln(2)}\\ \vdots\\ \sqrt{\ln\left(1+\cosh^k\left(\tilde{f}_n\right)\right)-\ln(2)}\end{bmatrix}^T \cdot \mathbf{K}_p\mathbf{K}_e^{-1}\begin{bmatrix}\sqrt{\ln\left(1+\cosh^k\left(\tilde{f}_1\right)\right)-\ln(2)}\\ \sqrt{\ln\left(1+\cosh^k\left(\tilde{f}_2\right)\right)-\ln(2)}\\ \vdots\\ \sqrt{\ln\left(1+\cosh^k\left(\tilde{f}_n\right)\right)-\ln(2)}\end{bmatrix}. \tag{15}$$

The candidate Lyapunov function (13) corresponds to the total energy of the closed-loop (11) and can be rewritten as

$$V\left(\dot{\mathbf{x}},\tilde{\mathbf{f}}\right) = \frac{1}{2}\left[\dot{\mathbf{x}} - \frac{\varepsilon_0}{1+\|\tilde{\mathbf{f}}\|}\tilde{\mathbf{f}}\right]^T\mathbf{M}_\chi\left[\dot{\mathbf{x}} - \frac{\varepsilon_0}{1+\|\tilde{\mathbf{f}}\|}\tilde{\mathbf{f}}\right] + \mathcal{U}_a\left(\mathbf{K}_p,\mathbf{K}_e^{-1},\tilde{\mathbf{f}}\right) - \frac{1}{2}\frac{\varepsilon_0^2}{\left[1+\|\tilde{\mathbf{f}}\|\right]^2}\tilde{\mathbf{f}}^T\mathbf{M}_\chi\tilde{\mathbf{f}}. \tag{16}$$

Now, we will give sufficient conditions to make $V(\dot{\mathbf{x}},\tilde{\mathbf{f}})$ a positive-definite function. The first term of the right side of (16) is a definite positive function with respect to $\tilde{\mathbf{f}}$ and $\dot{\mathbf{x}}$, because inertia matrix $\mathbf{M}_\chi$ is definite positive. Next, we provide upper and lower bounds on the second and third terms of the Lyapunov function (16).

The artificial energy function (15) has a unique and global minimum at $\tilde{\mathbf{f}} = \mathbf{0} \in \mathbb{R}^n$, and it is not a decreasing function; then there exists positive constant $\rho' \in \mathbb{R}_+$, such that $(1/k)\rho'\lambda_{\mathrm{m}}\{\mathbf{K}_e^{-1}\mathbf{K}_p\}\|\tilde{\mathbf{f}}\|^2 \le \mathcal{U}_a(\mathbf{K}_p, \mathbf{K}_e^{-1}, \tilde{\mathbf{f}})$. Therefore, the second and third terms of function (16) satisfy

$$
\begin{aligned}
&\left[\frac{1}{k}\rho'\lambda_{\mathrm{m}}\left\{\mathbf{K}_e^{-1}\mathbf{K}_p\right\} - \varepsilon_0^2\lambda_{\max}\left\{\mathbf{M}_\chi\right\}\right]\left\|\tilde{\mathbf{f}}\right\|^2 \\
&\le \frac{1}{k}\left\|\begin{bmatrix}\ln\left(1+\cosh^k\left(\tilde f_1\right)\right) - \ln(2) \\ \ln\left(1+\cosh^k\left(\tilde f_2\right)\right) - \ln(2) \\ \vdots \\ \ln\left(1+\cosh^k\left(\tilde f_n\right)\right) - \ln(2)\end{bmatrix}\right\|\lambda_{\min}\left\{\mathbf{K}_p\mathbf{K}_e^{-1}\right\} \\
&\quad - \varepsilon_0^2\frac{\left\|\tilde{\mathbf{f}}\right\|^2}{\left[1+\left\|\tilde{\mathbf{f}}\right\|\right]^2}\lambda_{\max}\left\{\mathbf{M}_\chi\right\} \\
&\le \mathcal{U}_a\left(\mathbf{K}_p, \mathbf{K}_e^{-1}, \tilde{\mathbf{f}}\right) - \frac{1}{2}\frac{\varepsilon_0^2}{\left[1+\left\|\tilde{\mathbf{f}}\right\|\right]^2}\tilde{\mathbf{f}}^T\mathbf{M}_\chi\tilde{\mathbf{f}}.
\end{aligned}
\tag{17}
$$

On the left-hand side of the inequality (17), to ensure that $\left[(1/k)\rho'\lambda_{\mathrm{m}}\{\mathbf{K}_e^{-1}\mathbf{K}_p\} - \varepsilon_0^2\lambda_{\max}\{\mathbf{M}_\chi\}\right]$ is a positive expression, ε_0 must be in the interval $0 < \varepsilon_0 < \sqrt{\rho'\lambda_{\mathrm{m}}\{\mathbf{K}_e^{-1}\mathbf{K}_p\}/k\lambda_{\max}\{\mathbf{M}_\chi\}}$, which is satisfied by the first term of (14). Hence, the Lyapunov function $V(\dot{\mathbf{x}}, \tilde{\mathbf{f}})$ given by (13) is a definite positive function in local form.

The total derivative of the candidate Lyapunov function (13) with respect to time is

$$
\begin{aligned}
\dot V\left(\dot{\mathbf{x}}, \tilde{\mathbf{f}}\right) &= \dot{\mathbf{x}}^T\mathbf{M}_\chi\ddot{\mathbf{x}} + \frac{1}{2}\dot{\mathbf{x}}^T\dot{\mathbf{M}}_\chi\dot{\mathbf{x}} \\
&\quad - \begin{bmatrix}\dfrac{\cosh^{k-1}\left(\tilde f_1\right)\sinh\left(\tilde f_1\right)}{1+\cosh^k\left(\tilde f_1\right)} \\ \dfrac{\cosh^{k-1}\left(\tilde f_2\right)\sinh\left(\tilde f_2\right)}{1+\cosh^k\left(\tilde f_2\right)} \\ \vdots \\ \dfrac{\cosh^{k-1}\left(\tilde f_n\right)\sinh\left(\tilde f_n\right)}{1+\cosh^k\left(\tilde f_n\right)}\end{bmatrix}^T \mathbf{K}_p\mathbf{K}_e^{-1}\underbrace{\mathbf{K}_e\dot{\mathbf{x}}}_{\dot{\mathbf{f}}} \\
&\quad - \frac{\varepsilon_0}{1+\left\|\tilde{\mathbf{f}}\right\|}\tilde{\mathbf{f}}^T\mathbf{M}_\chi\ddot{\mathbf{x}} - \frac{\varepsilon_0}{1+\left\|\tilde{\mathbf{f}}\right\|}\tilde{\mathbf{f}}^T\underbrace{\dot{\mathbf{M}}_\chi}_{\mathbf{C}_\chi^T+\mathbf{C}_\chi}\dot{\mathbf{x}} \\
&\quad + \frac{\varepsilon_0}{1+\left\|\tilde{\mathbf{f}}\right\|}\underbrace{\dot{\mathbf{x}}^T\mathbf{K}_e}_{\dot{\mathbf{f}}^T}\mathbf{M}_\chi\dot{\mathbf{x}} \\
&\quad - \varepsilon_0\frac{\overbrace{\dot{\mathbf{x}}^T\mathbf{K}_e}^{\dot{\mathbf{f}}^T}\tilde{\mathbf{f}}}{\left\|\tilde{\mathbf{f}}\right\|\left[1+\left\|\tilde{\mathbf{f}}\right\|\right]^2}\tilde{\mathbf{f}}^T\mathbf{M}_\chi\dot{\mathbf{x}}.
\end{aligned}
\tag{18}
$$

Substituting the closed-loop equation (11), we also use Property 3; then it yields,

$$
\begin{aligned}
\dot V\left(\dot{\mathbf{x}}, \tilde{\mathbf{f}}\right) &= \cancel{\dot{\mathbf{x}}^T\mathbf{K}_p\begin{bmatrix}\dfrac{\cosh^{k-1}\left(\tilde f_1\right)\sinh\left(\tilde f_1\right)}{1+\cosh^k\left(\tilde f_1\right)} \\ \dfrac{\cosh^{k-1}\left(\tilde f_2\right)\sinh\left(\tilde f_2\right)}{1+\cosh^k\left(\tilde f_2\right)} \\ \vdots \\ \dfrac{\cosh^{k-1}\left(\tilde f_n\right)\sinh\left(\tilde f_n\right)}{1+\cosh^k\left(\tilde f_n\right)}\end{bmatrix}} \\
&\quad - \dot{\mathbf{x}}^T\left[\mathbf{K}_v + \mathbf{B}_\chi\right]\dot{\mathbf{x}}^T \cancelto{0}{- \dot{\mathbf{x}}^T\mathbf{C}_\chi\dot{\mathbf{x}}^T + \frac{1}{2}\dot{\mathbf{x}}^T\dot{\mathbf{M}}_\chi\dot{\mathbf{x}}} \\
&\quad \cancel{- \begin{bmatrix}\dfrac{\cosh^{k-1}\left(\tilde f_1\right)\sinh\left(\tilde f_1\right)}{1+\cosh^k\left(\tilde f_1\right)} \\ \dfrac{\cosh^{k-1}\left(\tilde f_2\right)\sinh\left(\tilde f_2\right)}{1+\cosh^k\left(\tilde f_2\right)} \\ \vdots \\ \dfrac{\cosh^{k-1}\left(\tilde f_n\right)\sinh\left(\tilde f_n\right)}{1+\cosh^k\left(\tilde f_n\right)}\end{bmatrix}^T \mathbf{K}_p\dot{\mathbf{x}}} \\
&\quad - \frac{\varepsilon_0}{1+\left\|\tilde{\mathbf{f}}\right\|}\tilde{\mathbf{f}}^T\mathbf{K}_p\begin{bmatrix}\dfrac{\cosh^{k-1}\left(\tilde f_1\right)\sinh\left(\tilde f_1\right)}{1+\cosh^k\left(\tilde f_1\right)} \\ \dfrac{\cosh^{k-1}\left(\tilde f_2\right)\sinh\left(\tilde f_2\right)}{1+\cosh^k\left(\tilde f_2\right)} \\ \vdots \\ \dfrac{\cosh^{k-1}\left(\tilde f_n\right)\sinh\left(\tilde f_n\right)}{1+\cosh^k\left(\tilde f_n\right)}\end{bmatrix} \\
&\quad + \frac{\varepsilon_0}{1+\left\|\tilde{\mathbf{f}}\right\|}\tilde{\mathbf{f}}^T\left[\mathbf{K}_v + \mathbf{B}_\chi\right]\dot{\mathbf{x}} + \frac{\varepsilon_0}{1+\left\|\tilde{\mathbf{f}}\right\|}\tilde{\mathbf{f}}^T\cancel{\mathbf{C}_\chi}\dot{\mathbf{x}} \\
&\quad - \frac{\varepsilon_0}{1+\left\|\tilde{\mathbf{f}}\right\|}\tilde{\mathbf{f}}^T\left[\mathbf{C}_\chi^T + \cancel{\mathbf{C}_\chi}\right]\dot{\mathbf{x}} + \underbrace{\frac{\varepsilon_0}{1+\left\|\tilde{\mathbf{f}}\right\|}}_{\varepsilon_0/(1+\|\tilde{\mathbf{f}}\|)\le\varepsilon_0}\dot{\mathbf{x}}^T\mathbf{K}_e\mathbf{M}_\chi\dot{\mathbf{x}} \\
&\quad - \varepsilon_0\frac{\dot{\mathbf{x}}^T\mathbf{K}_e\tilde{\mathbf{f}}}{\left\|\tilde{\mathbf{f}}\right\|\left[1+\left\|\tilde{\mathbf{f}}\right\|\right]^2}\tilde{\mathbf{f}}^T\mathbf{M}_\chi\dot{\mathbf{x}},
\end{aligned}
\tag{19}
$$

where, after cancelling some terms, we obtain

$$
\dot V\left(\dot{\mathbf{x}}, \tilde{\mathbf{f}}\right) = -\dot{\mathbf{x}}^T\left[\mathbf{K}_v + \mathbf{B}_\chi - \varepsilon_0\mathbf{K}_e\mathbf{M}_\chi\right]\dot{\mathbf{x}}^T
$$

$$
\begin{aligned}
&-\frac{\varepsilon_0}{1+\|\tilde{\mathbf{f}}\|}\tilde{\mathbf{f}}^T\mathbf{K}_p\begin{bmatrix}\dfrac{\cosh^{k-1}(\tilde{f}_1)\sinh(\tilde{f}_1)}{1+\cosh^k(\tilde{f}_1)}\\ \dfrac{\cosh^{k-1}(\tilde{f}_2)\sinh(\tilde{f}_2)}{1+\cosh^k(\tilde{f}_2)}\\ \vdots\\ \dfrac{\cosh^{k-1}(\tilde{f}_n)\sinh(\tilde{f}_n)}{1+\cosh^k(\tilde{f}_n)}\end{bmatrix}\\
&+\frac{\varepsilon_0}{1+\|\tilde{\mathbf{f}}\|}\tilde{\mathbf{f}}^T\left[\mathbf{K}_v+\mathbf{B}_\chi\right]\dot{\mathbf{x}}\\
&-\underbrace{\frac{\varepsilon_0}{1+\|\tilde{\mathbf{f}}\|}\tilde{\mathbf{f}}^T\mathbf{C}_\chi^T\dot{\mathbf{x}}}_{(\varepsilon_0/(1+\|\tilde{\mathbf{f}}\|))\|\tilde{\mathbf{f}}\|\|\mathbf{C}_\chi\dot{\mathbf{x}}\|\le\varepsilon_0 k_{cx}\|\dot{\mathbf{x}}\|^2}\\
&-\varepsilon_0\frac{\dot{\mathbf{x}}^T\mathbf{K}_e\tilde{\mathbf{f}}}{\|\tilde{\mathbf{f}}\|\left[1+\|\tilde{\mathbf{f}}\|\right]^2}\tilde{\mathbf{f}}^T\mathbf{M}_\chi\dot{\mathbf{x}}.
\end{aligned}
\tag{20}
$$

Note that the second term of the right side in (20) corresponds to the gradient of the artificial potential energy $(\partial/\partial\tilde{\mathbf{f}})\mathcal{U}_a(\mathbf{K}_p,\mathbf{K}_e^{-1},\tilde{\mathbf{f}})$; it is a vector composed of saturated functions with hyperbolic-type structure, for each ith component $\cosh^{k-1}(\tilde{f}_i)\sinh(\tilde{f}_i)/(1+\cosh^k(\tilde{f}_i)) = (\cosh^k(\tilde{f}_i)/(1+\cosh^k(\tilde{f}_i)))(\sinh(\tilde{f}_i)/\cosh(\tilde{f}_i)) = (\cosh^k(\tilde{f}_i)/(1+\cosh^k(\tilde{f}_i)))\tanh(\tilde{f}_i)$; note also that $1+\cosh^k(\tilde{f}_i) > \cosh^k(\tilde{f}_i) \Leftrightarrow (\cosh^k(\tilde{f}_i)/(1+\cosh^k(\tilde{f}_i))) < 1$; then $-1 \le \cosh^{k-1}(\tilde{f}_i)\sinh(\tilde{f}_i)/(1+\cosh^k(\tilde{f}_i)) \le 1$; there exist positive constants $\rho_2,\rho_2' \in \mathbb{R}_+$, such that $\rho_2'|\tilde{f}_i| \le (\cosh^{k-1}(\tilde{f}_i)\sinh(\tilde{f}_i)/(1+\cosh^k(\tilde{f}_i)))\tilde{f}_i \le \rho_2|\tilde{f}_i|$; therefore it satisfies

$$
\rho_2'\|\tilde{\mathbf{f}}\|^2 < \tilde{\mathbf{f}}^T\frac{\partial}{\partial\tilde{\mathbf{f}}}\mathcal{U}_a\left(\mathbf{K}_p,\mathbf{K}_e^{-1},\tilde{\mathbf{f}}\right) < \rho_2\|\tilde{\mathbf{f}}\|^2,\quad \forall\tilde{\mathbf{f}}\neq\mathbf{0}\in\mathbb{R}^n.
\tag{21}
$$

In other words, the second term of function (20) satisfies the following condition:

$$
\begin{aligned}
&\frac{\varepsilon_0}{1+\|\tilde{\mathbf{f}}\|}\|\tilde{\mathbf{f}}\|^2\lambda_{\mathrm{m}}\{\mathbf{K}_p\}\\
&\le\frac{\varepsilon_0}{1+\|\tilde{\mathbf{f}}\|}\tilde{\mathbf{f}}^T\mathbf{K}_p\begin{bmatrix}\dfrac{\cosh^{k-1}(\tilde{f}_1)\sinh(\tilde{f}_1)}{1+\cosh^k(\tilde{f}_1)}\\ \dfrac{\cosh^{k-1}(\tilde{f}_2)\sinh(\tilde{f}_2)}{1+\cosh^k(\tilde{f}_2)}\\ \vdots\\ \dfrac{\cosh^{k-1}(\tilde{f}_n)\sinh(\tilde{f}_n)}{1+\cosh^k(\tilde{f}_n)}\end{bmatrix}\\
&\le\frac{\varepsilon_0}{1+\|\tilde{\mathbf{f}}\|}\|\tilde{\mathbf{f}}\|^2\lambda_{\mathrm{M}}\{\mathbf{K}_p\},
\end{aligned}
\tag{22}
$$

where $\lambda_{\mathrm{m}}\{\mathbf{K}_p\}$ and $\lambda_{\mathrm{M}}\{\mathbf{K}_p\}$ stand for the maximum and minimum eigenvalues of matrix $\mathbf{K}_p$, respectively. Using Property 4, the fourth term from (23) satisfies $-(\varepsilon_0/(1+\|\tilde{\mathbf{f}}\|))\tilde{\mathbf{f}}^T\mathbf{C}_\chi^T\dot{\mathbf{x}} \le (\varepsilon_0/(1+\|\tilde{\mathbf{f}}\|))\|\tilde{\mathbf{f}}\|\|\mathbf{C}_\chi\dot{\mathbf{x}}\| \le \varepsilon_0 k_{cx}\|\dot{\mathbf{x}}\|^2$. Next, by taking (22) and Cartesian dynamic model property (2), the expression for $\dot{V}(\dot{\mathbf{x}},\tilde{\mathbf{f}})$ (20) takes the form

$$
\begin{aligned}
&\dot{V}\left(\dot{\mathbf{x}},\tilde{\mathbf{f}}\right)\\
&\le-\left[\lambda_{\min}\{\mathbf{K}_v+\mathbf{B}_\chi\}+k_{cx}-\varepsilon_0\lambda_{\mathrm{M}}\{\mathbf{K}_e\mathbf{M}_\chi\}\right]\|\dot{\mathbf{x}}\|^2\\
&\quad-\frac{\varepsilon_0}{1+\|\tilde{\mathbf{f}}\|}\lambda_{\mathrm{m}}\{\mathbf{K}_p\}\|\tilde{\mathbf{f}}\|^2+\underbrace{\frac{\varepsilon_0}{1+\|\tilde{\mathbf{f}}\|}}_{\varepsilon_0/(1+\|\tilde{\mathbf{f}}\|)\le\varepsilon_0}\left[\lambda_{\mathrm{m}}\{\mathbf{K}_v+\mathbf{B}_\chi\}\right]\|\tilde{\mathbf{f}}\|\|\dot{\mathbf{x}}\|\\
&\quad+\underbrace{\varepsilon_0\frac{\|\mathbf{x}\|\lambda_{\mathrm{M}}\{\mathbf{K}_e\}\cancel{\|\tilde{\mathbf{f}}\|}}{\cancel{\|\tilde{\mathbf{f}}\|}\left[1+\|\tilde{\mathbf{f}}\|\right]^2}\|\tilde{\mathbf{f}}\|\lambda_{\mathrm{M}}\{\mathbf{M}_\chi\}\|\mathbf{x}\|}_{\le\varepsilon_0\lambda_{\mathrm{M}}\{\mathbf{K}_e\}\lambda_{\mathrm{M}}\{\mathbf{M}_\chi\}\|\dot{\mathbf{x}}\|^2}\\
&\le-\left[\lambda_{\min}\{\mathbf{K}_v\}+\lambda_{\min}\{\mathbf{B}_\chi\}+k_{cx}-2\varepsilon_0\lambda_{\mathrm{M}}\{\mathbf{K}_e\}\lambda_{\mathrm{M}}\{\mathbf{M}_\chi\}\right]\|\dot{\mathbf{x}}\|^2\\
&\quad-\frac{\varepsilon_0}{1+\|\tilde{\mathbf{f}}\|}\lambda_{\mathrm{m}}\{\mathbf{K}_p\}\|\tilde{\mathbf{f}}\|^2+\varepsilon_0\left[\lambda_{\mathrm{m}}\{\mathbf{K}_v\}+\lambda_{\mathrm{m}}\{\mathbf{B}_\chi\}\right]\|\tilde{\mathbf{f}}\|\|\dot{\mathbf{x}}\|.
\end{aligned}
\tag{23}
$$

The time derivative of the Lyapunov function becomes

$$
\dot{V}\left(\dot{\mathbf{x}},\tilde{\mathbf{f}}\right)\le-\begin{bmatrix}\|\dot{\mathbf{x}}\|\\ \|\tilde{\mathbf{f}}\|\end{bmatrix}^T\underbrace{\begin{bmatrix}\underbrace{\left[\lambda_{\min}\{\mathbf{K}_v\}+\lambda_{\min}\{\mathbf{B}_\chi\}+k_{cx}-2\varepsilon_0\lambda_{\mathrm{M}}\{\mathbf{K}_e\}\lambda_{\mathrm{M}}\{\mathbf{M}_\chi\}\right]}_{Q_{11}} & -\dfrac{\varepsilon_0}{2}\left[\lambda_{\mathrm{m}}\{\mathbf{K}_v\}+\lambda_{\mathrm{m}}\{\mathbf{B}_\chi\}\right]\\ -\dfrac{\varepsilon_0}{2}\left[\lambda_{\mathrm{m}}\{\mathbf{K}_v\}+\lambda_{\mathrm{m}}\{\mathbf{B}_\chi\}\right] & \dfrac{\varepsilon_0}{1+\|\tilde{\mathbf{f}}\|}\lambda_{\mathrm{m}}\{\mathbf{K}_p\}\end{bmatrix}}_{\mathbf{Q}}\begin{bmatrix}\|\dot{\mathbf{x}}\|\\ \|\tilde{\mathbf{f}}\|\end{bmatrix}<0.
\tag{24}
$$

In order to ensure that matrix $\mathbf{Q}$ is positive definite, the number ε_0 must satisfy the second condition of the left side of (14); then the element $Q_{11} > 0$; also the third condition of the left side in the expression (14) must be satisfied; then the determinant of matrix $\mathbf{Q}$ is positive. Considering that ε_0 satisfies those considerations, we can conclude that $\dot{V}(\dot{\mathbf{x}},\tilde{\mathbf{f}})$ is a negative-definite function; note that, due to the fact that both conditions $V(\dot{\mathbf{x}},\tilde{\mathbf{f}}) > 0$ and $\dot{V}(\dot{\mathbf{x}},\tilde{\mathbf{f}}) < 0$ are satisfied, the proposed Lyapunov function is a strict Lyapunov function. Therefore, according to Lyapunov's direct method, we conclude local asymptotic stability of the origin $\left[\tilde{\mathbf{f}}^T\ \dot{\mathbf{x}}^T\right]^T = \mathbf{0}\in\mathbb{R}^{2n}$ of the closed-loop equation (11), which means that both state variables $\tilde{\mathbf{f}}(t)$ and $\dot{\mathbf{x}}(t)$ asymptotically converge to zero, as $t\to\infty$.

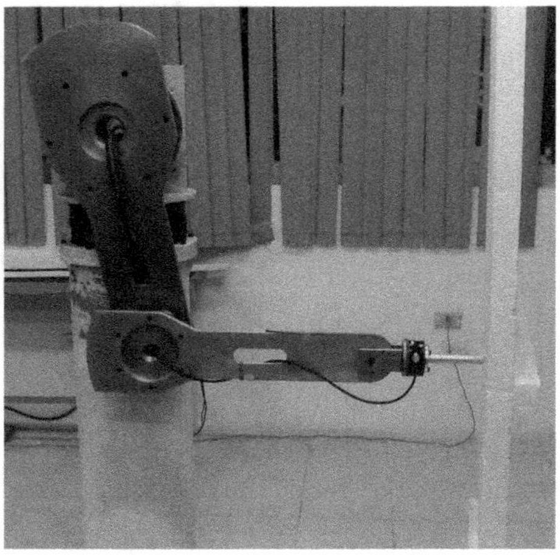

FIGURE 1: Experimental robot "Rotradi."

4. Experimental Results

The experimental results were obtained by using an experimental platform known as "Rotradi" [46], shown in Figure 1. This platform, designed and built at the Robotics Laboratory of Benemérita Universidad Autónoma de Puebla (BUAP) in Mexico, is a two-degree-of-freedom, direct-drive robot. Rotradi consists of two 6061 aluminum links, actuated by brushless, direct-drive servo actuators, in order to drive the joints without a gear reduction. The motors used in Rotradi are DM-1150A and DM-1015B models from Parker Compumotor, for the shoulder and elbow joints, respectively. Since the servomotors are operated in torque mode, they act as a torque source; in this mode, an analog voltage is used as a reference of the torque signal, in proportional scale. The features of the servomotors are shown in Table 1. A motion-control board of Precision MicroDynamics Inc. is included in the robot system; this device is designed for reading the encoders and generating reference voltages. The control algorithms, written in C code, run in real time with a 2.5 ms sample period, on a Pentium computer.

With the purpose of performing force sensing, a six-axis force/torque sensor model ATI FT Gamma SI-130-10 is mounted at the end-effector of the robot manipulator. The sensor is capable of measuring force in the range of ±130 N and torque within the range of ±10 Nm. The sensor is connected to a 3.6 GHz Pentium IV computer, in which the signals are processed using a Visual C++ application. The communication between computers (for robot control and force-sensor readings) was enabled by using a communication protocol based on interruption signals, sent via parallel port.

The complete dynamic model for our experimental robot is reported in [46]. However, the gravitational torques vector and the Jacobian matrix are written in an explicit form in order to explain the experiments performed in this paper. The gravitational torques vector is given by

$$\begin{aligned} g_1(\mathbf{q}) &= 54.813\sin(q_1) - 1.182\cos(q_1) \\ &\quad + 2.145\sin(q_1+q_2) \\ &\quad + 0.089\cos(q_1+q_2), \\ g_2(\mathbf{q}) &= 2.145\sin(q_1+q_2) + 0.089\cos(q_1+q_2), \end{aligned} \tag{25}$$

while the Jacobian matrix is given as follows:

$$\mathbf{J}(\mathbf{q}) = \begin{bmatrix} 0.45\cos(q_1) + 0.69\cos(q_1+q_2) & 0.69\cos(q_1+q_2) \\ 0.45\sin(q_1) + 0.69\sin(q_1+q_2) & 0.69\sin(q_1+q_2) \end{bmatrix}. \tag{26}$$

We present an experimental comparison of five explicit force regulators on the two-degree-of-freedom, direct-drive robot manipulator. The evaluated schemes are the simple proportional-derivative (PD), denoted as $\mathbf{f}_{\chi_{\mathrm{PD}}}$, (arctangent) Atan and (square-root) Sqrt control schemes [38], represented as $\mathbf{f}_{\chi_{\mathrm{Atan}}}$ and $\mathbf{f}_{\chi_{\mathrm{Sqrt}}}$, respectively, and two hyperbolic-type members of the proposed control family, denoted by $\mathbf{f}_{\chi_1}$ and $\mathbf{f}_{\chi_2}$, for exponents $k = 1$ and $k = 2$, respectively.

The structure of these regulators, evaluated on the two-degree-of-freedom robot manipulator, is given by

$$\mathbf{f}_{\chi_{\mathrm{PD}}} = \mathbf{K}_p \begin{bmatrix} \tilde{f}_x \\ \tilde{f}_y \end{bmatrix} - \mathbf{K}_v \begin{bmatrix} \dot{x}_x \\ \dot{x}_y \end{bmatrix} + \mathbf{g}_\chi + \mathbf{f}_e \tag{27}$$

$$\mathbf{f}_{\chi_1} = \mathbf{K}_p \begin{bmatrix} \dfrac{\sinh(\tilde{f}_x)}{1+\cosh(\tilde{f}_x)} \\ \dfrac{\sinh(\tilde{f}_y)}{1+\cosh(\tilde{f}_y)} \end{bmatrix} - \mathbf{K}_v \begin{bmatrix} \dot{x}_x \\ \dot{x}_y \end{bmatrix} + \mathbf{g}_\chi + \mathbf{f}_e \tag{28}$$

$$\begin{aligned} \mathbf{f}_{\chi_2} &= \mathbf{K}_p \begin{bmatrix} \dfrac{\cosh(\tilde{f}_x)\sinh(\tilde{f}_x)}{1+\cosh^2(\tilde{f}_x)} \\ \dfrac{\cosh(\tilde{f}_y)\sinh(\tilde{f}_y)}{1+\cosh^2(\tilde{f}_y)} \end{bmatrix} - \mathbf{K}_v \begin{bmatrix} \dot{x}_x \\ \dot{x}_y \end{bmatrix} + \mathbf{g}_\chi \\ &\quad + \mathbf{f}_e \end{aligned} \tag{29}$$

$$\mathbf{f}_{\chi_{\mathrm{Atan}}} = \mathbf{K}_p \begin{bmatrix} \operatorname{atan}(\tilde{f}_x) \\ \operatorname{atan}(\tilde{f}_y) \end{bmatrix} - \mathbf{K}_v \begin{bmatrix} \operatorname{atan}(\dot{x}_x) \\ \operatorname{atan}(\dot{x}_y) \end{bmatrix} + \mathbf{g}_\chi + \mathbf{f}_e \tag{30}$$

$$\mathbf{f}_{\chi_{\mathrm{Sqrt}}} = \mathbf{K}_p \begin{bmatrix} \dfrac{\tilde{f}_x}{\sqrt{1+\tilde{f}_x^2}} \\ \dfrac{\tilde{f}_y}{\sqrt{1+\tilde{f}_y^2}} \end{bmatrix} - \mathbf{K}_v \begin{bmatrix} \dfrac{\dot{x}_x}{\sqrt{1+\dot{x}_x^2}} \\ \dfrac{\dot{x}_y}{\sqrt{1+\dot{x}_y^2}} \end{bmatrix} + \mathbf{g}_\chi + \mathbf{f}_e. \tag{31}$$

All evaluated control schemes (27)–(31) have the following desired force references $\mathbf{f}_d = [f_{xd} \ f_{yd}]^T = [5.0 \ 0.1]^T$ [N]; and the force error signals are given as $\tilde{\mathbf{f}} = [\tilde{f}_x \ \tilde{f}_y]^T = [f_{xd} - f_x \ f_{yd} - f_y]^T$, while the gain matrices $\mathbf{K}_p$ and $\mathbf{K}_v$ are tuned in order to prevent saturation of the actuators, which is a complicated and time-consuming process where several

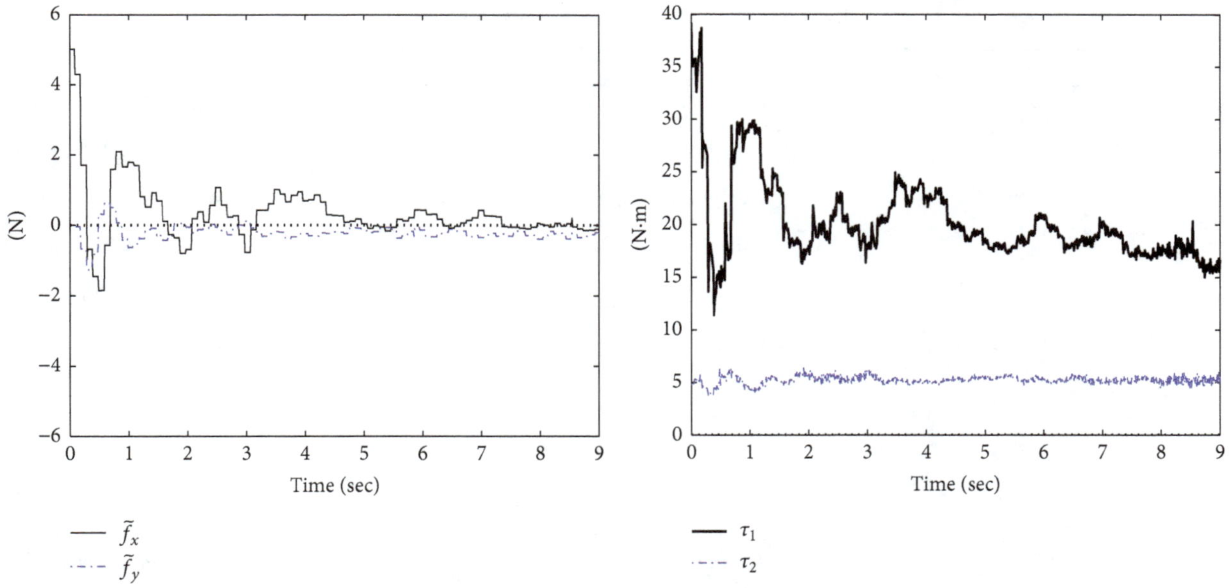

(a) Force error for the PD control $\mathbf{f}_{\chi \mathrm{PD}}$

(b) Applied torques of the PD scheme $\boldsymbol{\tau}_{\mathrm{PD}} = \mathbf{J}(\mathbf{q})^T \mathbf{f}_{\chi \mathrm{PD}}$

FIGURE 2: Experimental results for the PD control $\mathbf{f}_{\chi \mathrm{PD}}$.

TABLE 1: Robot arm servo actuators.

Joint	Model	Max. torque [Nm]	Resolution [cpr]
(1) Shoulder	DM-1150A	150	1,024,000
(2) Elbow	DM-1015B	15	1,024,000

trials were necessary in order to ensure a small transient state and overshoot and smaller steady-state error. The tuning used for each evaluated force control scheme is presented in Table 2. The hyperbolic structure in (28) and (29) has two arguments: one of them is the force error $\tilde{\mathbf{f}}$, measured in Newtons; the second one is a constant with unit value (for simplicity, it does not appear in these equations), measured in Newton^{-1}, so that the whole argument is dimensionless.

In order to test the performance of these control structures, an interaction experiment is designed; the experiment consists of placing the robot at an initial position (the force sensor attached to the tool at the end of the robot manipulator is placed at the contact position with the environment through a tangent-hyperbolic position control [47]), and the second link is normal to the plane of the wall. Once positioned, the tool attached to the force sensor is located close to the wall, allowing a slight touch on the surface. For this experiment, it is not necessary to control the end-effector orientation. The initial conditions of the force-regulation experiment are depicted in Figure 1.

Now, the experimental evaluation of the three controllers is described. The force error and the applied torques from PD control $\mathbf{f}_{\chi \mathrm{PD}}$ are presented in Figure 2. The force error signals for the PD control corresponding to two components f_x and f_y are depicted in Figure 2(a), which presents convergence properties. It can be noted that each component has a short transient (4.5 sec), and both directions tend asymptotically to small neighborhood of zero; the steady-state force errors at time $t = 9$ sec have small values, specifically $[\tilde{f}_x(t), \tilde{f}_y(t)]^T = [0.2, 0.25]^T$ N. Figure 2(b) describes the applied control torques $\boldsymbol{\tau}_{\mathrm{PD}} = \mathbf{J}(\mathbf{q})^T \mathbf{f}_{\chi \mathrm{PD}}$ at the shoulder and elbow joints for the PD scheme. Note that the two torque signals are within the range of operation (see Table 1).

In Figure 3, the results for force error and applied torque of the Atan control are presented. Note that the experimental curves for both force error components $[\tilde{f}_x, \tilde{f}_y]^T$ converge to zero; the transitory response has small oscillations; for this control, its transient phase ($t = 4.5$ sec) is evolved for a shorter time than transient response of the PD control; the steady-state force errors at time $t = 9$ sec have small values, $[\tilde{f}_1(t), \tilde{f}_2(t)]^T = [0.18, 0.29]^T$ N, and the control algorithm $\mathbf{f}_{\chi \mathrm{Atan}}$ achieves the force-regulation objective. The torque signals for shoulder and elbow joints avoid saturation of the actuators.

Figure 4 describes the experimental results for the Sqrt control scheme; the transitory response is approximately 4.5 sec; the force error signals reach the steady-state error $[\tilde{f}_x(t), \tilde{f}_y(t)]^T = [-0.35, -0.31]^T$ N, and the applied torques $\mathbf{J}(\mathbf{q})^T \mathbf{f}_{\chi \mathrm{Sqrt}}$ clearly evolve inside the prescribed limits shown in Table 1.

The force error and applied torque signals for the hyperbolic controller $\mathbf{f}_{\chi 1}$ with $k = 1$ are shown in Figure 5. The experimental results for the force errors $\tilde{\mathbf{f}}$ are depicted in Figure 5(a), which evolve until reaching steady-state errors

TABLE 2: Gain matrices $\mathbf{K}_p$ and $\mathbf{K}_v$ for the force regulators.

Regulator	$\mathbf{K}_p$	$\mathbf{K}_v$
Proportional	diag(16.0, 5.0) dimensionless	diag(450.0, 150.0) N-sec/m
Hyperbolic, $k = 1$	diag(65.0, 5.0) N	diag(650.0, 150.0) N-sec/m
Hyperbolic, $k = 2$	diag(55.0, 5.0) N	diag(700.0, 150.0) N-sec/m
Atan	diag(30.0, 5.0) N	diag(400.0, 150.0) N
Sqrt	diag(39.0, 5.0) N	diag(400.0, 150.0) N

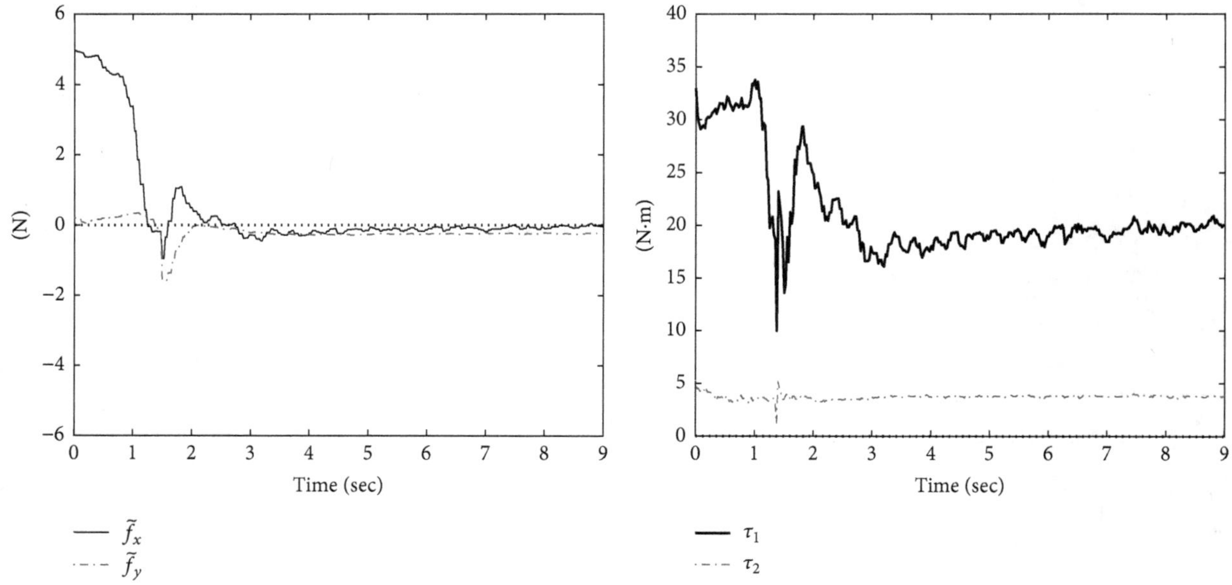

(a) Force error for the Atan controller

(b) Applied Atan control torques

FIGURE 3: Experimental results for the Atan control $\boldsymbol{\tau}_{\text{Atan}} = \mathbf{J}(\mathbf{q})^T \mathbf{f}_{\chi_{\text{Atan}}}$.

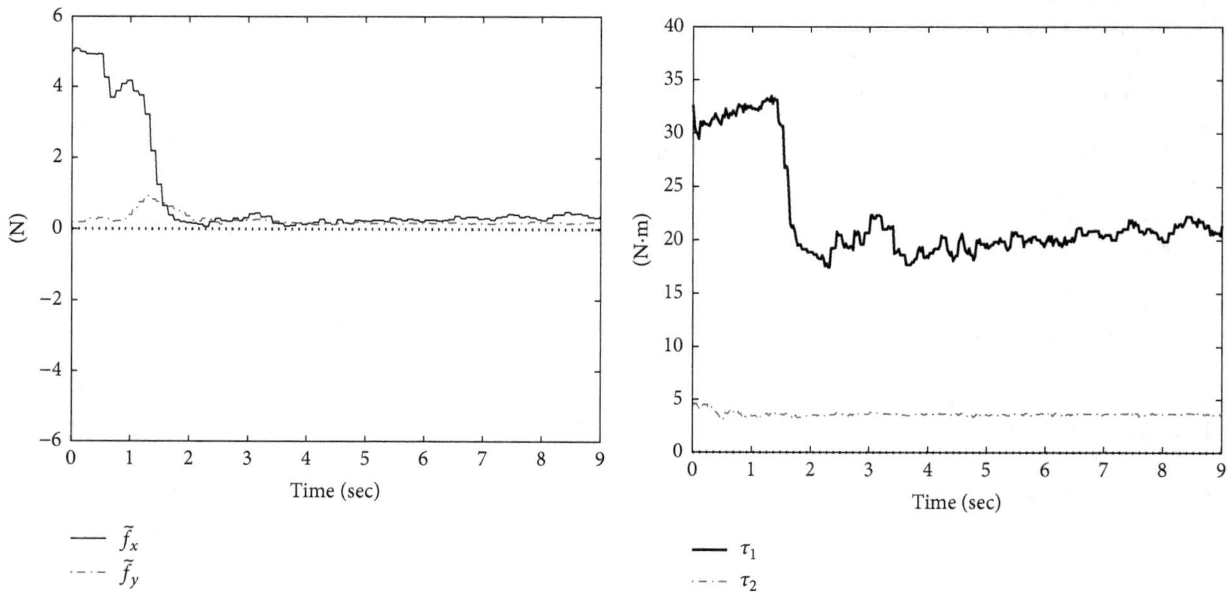

(a) Force error for the Sqrt controller

(b) Applied Sqrt control torques

FIGURE 4: Experimental results for the Sqrt control $\boldsymbol{\tau}_{\text{Sqrt}} = \mathbf{J}(\mathbf{q})^T \mathbf{f}_{\chi_{\text{Sqrt}}}$.

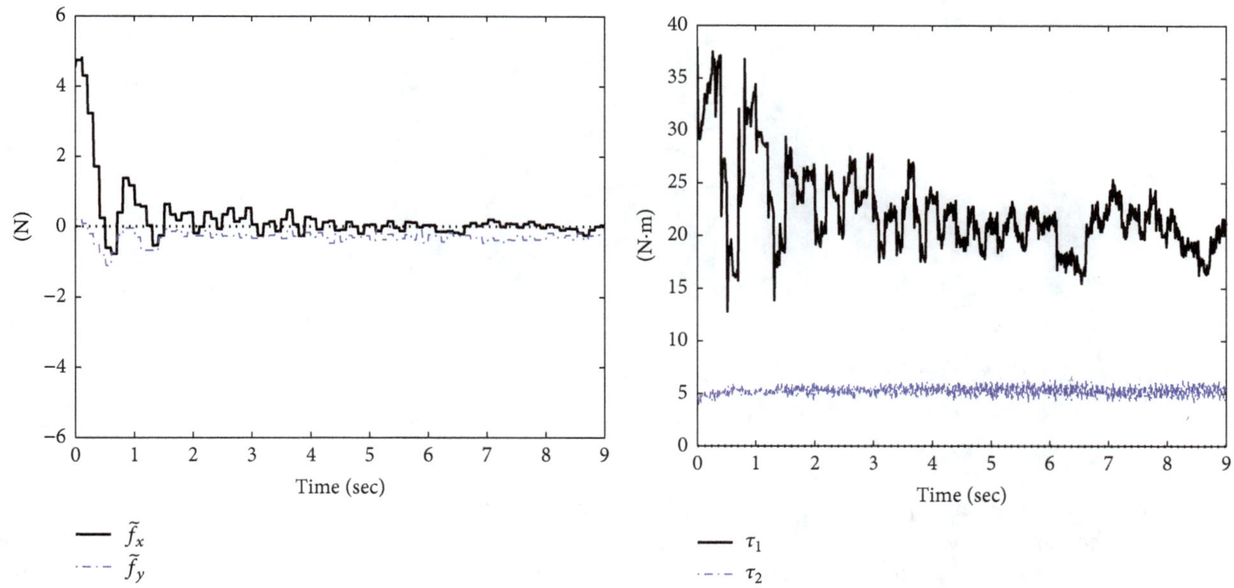

(a) Force error for the hyperbolic control scheme with $k = 1$

(b) Applied hyperbolic control torques with $k = 1$

FIGURE 5: Experimental results of the hyperbolic control $\boldsymbol{\tau}_1 = \mathbf{J}(\mathbf{q})^T \mathbf{f}_{\chi_1}$ for $k = 1$.

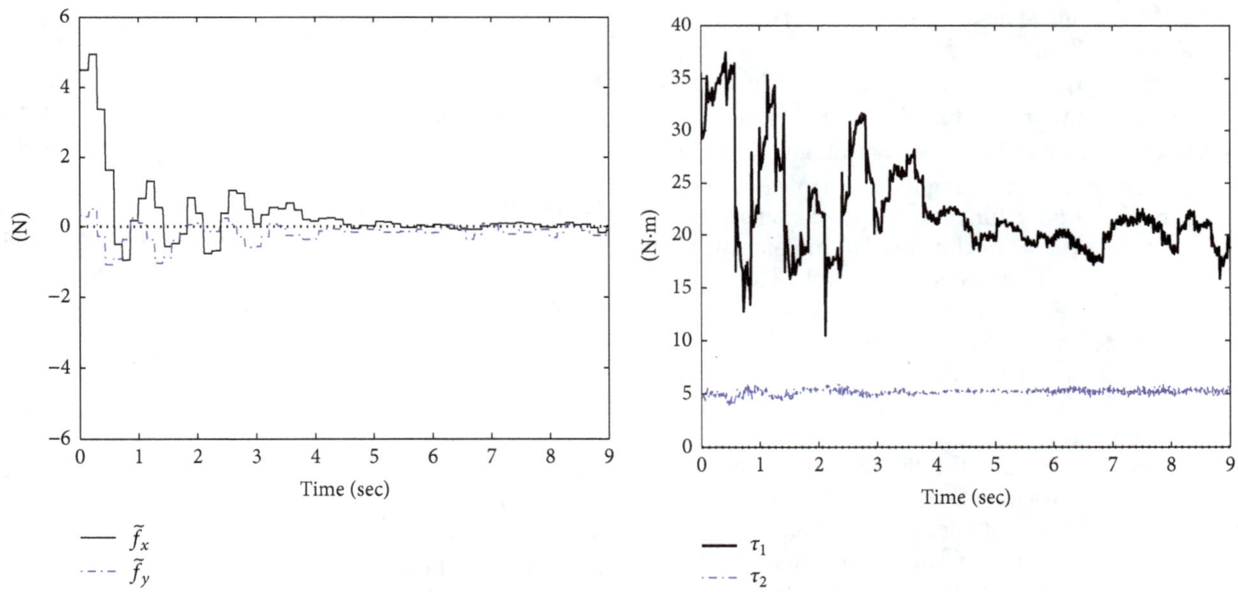

(a) Force error of the hyperbolic controller, $k = 2$

(b) Applied hyperbolic control torques, $k = 2$

FIGURE 6: Experimental results for the hyperbolic control $\boldsymbol{\tau}_2 = \mathbf{J}(\mathbf{q})^T \mathbf{f}_{\chi_2}$ with $k = 2$.

(t = 9 sec) $[-0.1, 0.15]^T$ N; the transient state (4.5 sec) is similar to the PD control. However, it is smoother than the transient state of the PD control. The applied torques of the control $\boldsymbol{\tau}_1 = \mathbf{J}(\mathbf{q})^T \mathbf{f}_{\chi_1}$ are presented in Figure 5(b); these signals are inside the prescribed limits.

The experimental results for the hyperbolic control scheme for $\mathbf{f}_{\chi_2}$ are depicted in Figure 6; the force errors $\tilde{\mathbf{f}}$ have short transient (approximately 4.5 sec) and a smaller steady-state force error (t = 9 sec) $[0.08, 0.09]^T$ N than PD, Atan, and Sqrt schemes as can be seen in Figure 6(a).

Observe that the transient phase is brief for all control schemes; the amplitude of the oscillation peaks is lower for the Sqrt, $\mathbf{f}_{\chi_1}$, and $\mathbf{f}_{\chi_2}$ controls compared to Atan and PD controls; and the steady-state force errors of the control

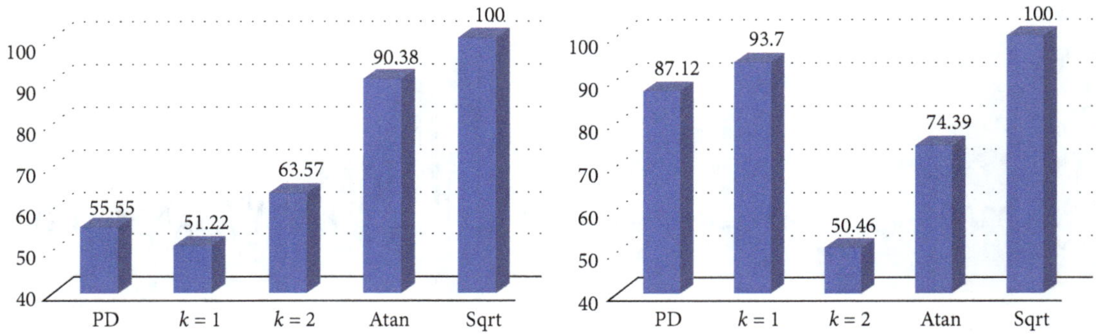

(a) Normalized $\mathscr{L}_2$-norm of the transitory state

(b) Normalized $\mathscr{L}_2$-norm of the steady state

FIGURE 7: Performance indices $\mathscr{L}_2$ of evaluated control schemes.

$\mathbf{f}_{\chi 2}$ are smaller than PD, Atan, Sqrt, and $\mathbf{f}_{\chi 1}$ controls. As a consequence, the novel hyperbolic controllers yield an increased control performance for the explicit force problem of robot manipulators. Steady-state force errors are present due to the presence of static friction at joints and due to the lack of a model for this type of friction in the environment. It is important to note that, despite the presence of nonmodeled friction phenomena, the force errors are acceptably small. The results for the applied torques $\boldsymbol{\tau}_2 = \mathbf{J}(\mathbf{q})^T \mathbf{f}_{\chi 2}$ are shown in Figure 6(b). The profile of these torque signals clearly evolves within the physical limits.

The hyperbolic-type control schemes work as follows: for large errors $\tilde{f}_i$, the hyperbolic function $\cosh^{m-1}(\tilde{f}_i)\sinh(\tilde{f}_i)/(1+\cosh^m(\tilde{f}_i))$ returns a bounded control output avoiding input saturation in the servo amplifier; when the errors $\tilde{f}_i$ are small, that hyperbolic function is approximately linear; then it works like proportional controller. In contrast, with PD control, $\mathbf{f}_{\chi \mathrm{PD}}$ can demand a large amount of torque and it is not saturated. The derivative action of the Atan and Sqrt schemes is bounded, while that for the proposed hyperbolic-type control is linear, demanding the effect of damping injection as needed. It is important to emphasize that, in the experiments presented herein, the applied control torques are guaranteed to be within the linear zone of the servo amplifier. Therefore, problems associated with linear control, related to vibration, thermal effects, mechanical noise, and nonmodeled dynamics, are avoided. Linear control is highly demanding in terms of gain tuning, as compared to the scheme proposed herein. In accordance with the analytical results, the proposed family of control schemes provides good performance.

In order to evaluate and compare the performances of the controllers in a scalar form, the norm $\mathscr{L}_2$ of the force error is used; it is an objective numerical measure of the force error $\tilde{\mathbf{f}}$ [20]; a smaller norm $\mathscr{L}_2\{\tilde{\mathbf{f}}\}$ represents smaller force error and it means the best performance of the evaluated control:

$$\mathscr{L}_2\{\tilde{\mathbf{f}}\} = \sqrt{\frac{1}{T}\int_0^T \left\|\tilde{\mathbf{f}}(\sigma)\right\|^2 d\sigma}, \tag{32}$$

where T is the time of the experiment. The norm $\mathscr{L}_2\{\tilde{\mathbf{f}}\}$ measures the root-mean-square "average" of the force error $\tilde{\mathbf{f}}$.

The performance comparison $\mathscr{L}_2\{\tilde{\mathbf{f}}\}$ of the evaluated control schemes took into account the transitory and steady states; that is, two requirements were considered on five evaluated controllers. The first one consisted of getting the performance of the transient response and the second aimed at achieving steady-state response. Once the performance was evaluated for each controller through the $\mathscr{L}_2\{\tilde{\mathbf{f}}\}$-norm, then the results are normalized with respect to the largest value and presented in percent scale for the transitory phase in Figure 7(a). The time interval to measure the $\mathscr{L}_2\{\tilde{\mathbf{f}}\}$-norm in the transitory phase is considered at $t \in [0, 4.5]$ sec. The best performance for the transient state corresponds to the hyperbolic-type control with exponent $k = 1$, corresponding to the smallest index, 51.22%, with respect to the largest value of the Sqrt control. Note that, for the transitory response shown in Figure 4(a), the Euclidean norm square of the force error $\|\tilde{\mathbf{f}}\|^2$ for the control $\mathbf{f}_{\chi \mathrm{Sqrt}}$ shows a larger area than other evaluated schemes; therefore a poor performance $\mathscr{L}_2\{\tilde{\mathbf{f}}\}$ is obtained for the Sqrt control. The PD control was the second best performance, followed by hyperbolic-type control (exponent $k = 1$) with a good performance.

The comparative results of performance indexes for the steady state are shown in Figure 7(b). The smaller $\mathscr{L}_2\{\tilde{\mathbf{f}}\}$ shows that the $k = 2$ hyperbolic-type control with 50.46% is the one with the best performance, because the steady-force errors presented less error than the rest of the evaluated control algorithms; the Sqrt control obtained the largest performance index and therefore a poor performance.

The total performance index for each evaluated control scheme would be the average of the sum of indexes of the transitory and steady states; then $\mathscr{L}_{2_{f_{\chi \mathrm{PD}}}} = 71.335$, $\mathscr{L}_{2_{f_{\chi \mathrm{Atan}}}} = 82.385$, $\mathscr{L}_{2_{f_{\chi \mathrm{Sqrt}}}} = 100$, $\mathscr{L}_{2_{f_{\chi 1}}} = 72.46$, and $\mathscr{L}_{2_{f_{\chi 2}}} = 57.015$, which correspond to PD, Atan, Sqrt, $k = 1$, and $k = 2$ hyperbolic-type controls, respectively. Therefore, the best performance corresponds to the hyperbolic-type control scheme $\mathbf{f}_{\chi 2}$.

5. Conclusion

In this paper, a new family of explicit force control schemes is presented. The proposed control family includes a large class of saturated-type hyperbolic function terms; the hyperbolic structure depends on k as a power function. It includes also an active velocity damping term plus gravity compensation. Our proposal of force regulators is supported by a stability analysis in Lyapunov sense using a strict Lyapunov function. The asymptotic stability of the closed-loop system equilibrium point in Cartesian space is guaranteed. The mathematical development of Property 4 for the Cartesian dynamics model has been completed; this property is very useful for the stability analysis.

In order to show the performance of the controllers, experimental results have been obtained. An experimental comparison of the performances obtained with the classical proportional-derivative (PD), Atan, and Sqrt controls and two members of the proposed control family on a two-degree-of-freedom, direct-drive robot manipulator is presented. A force sensor (a force/torque sensor ATI FT Gamma SI-130-10) is mounted at the end-effector of the robot manipulator with the purpose of measuring the contact force in order to feed back the force error in the closed-loop. The experimental evidence shows that the controller in transitory state with the best performance is the hyperbolic structure with $k = 1$, followed by the hyperbolic controller with $k = 2$, while in steady state the best performance corresponds to the hyperbolic control with exponent $k = 2$. The performance of the proposed control is improved compared with traditional PD control scheme and it is a good candidate for industrial applications. The effectiveness of the proposed control scheme can be concluded.

Appendix

The Cartesian dynamic model has some useful properties for stability analysis of robot controllers. Among these properties, Property 4 is widely cited in the literature of Cartesian robot control; however, we did not find in the literature the proof that the norm of vector $\mathbf{C}_\chi \dot{\mathbf{x}}$ is bounded. Therefore, the mathematical development for Property 4 is presented in this section. For the validity of the Cartesian dynamic model, it is assumed that the Jacobian matrix is full rank.

Proof of Property 4. First, it is necessary to prove that $\|\dot{\mathbf{J}}(\mathbf{q})\| \leq k_{j(q)}\|\dot{\mathbf{q}}\|$, for some $k_{j(q)} \in \mathbb{R}_+$, is true. Since $\sqrt{\dot{q}_i^2} \leq \sqrt{\dot{q}_1^2 + \cdots + \dot{q}_i^2 + \cdots + \dot{q}_n^2}$, then $\dot{q}_i \leq \|\dot{\mathbf{q}}\|$. Also,

$$\begin{aligned} \left\|\dot{\mathbf{J}}(\mathbf{q})\right\| &= \left\|\frac{\partial \mathbf{J}(\mathbf{q})}{\partial q_1}\dot{q}_1 + \cdots + \frac{\partial \mathbf{J}(\mathbf{q})}{\partial q_n}\dot{q}_n\right\| \\ &\leq \left\|\frac{\partial \mathbf{J}(\mathbf{q})}{\partial q_1}\dot{q}_1\right\| + \cdots + \left\|\frac{\partial \mathbf{J}(\mathbf{q})}{\partial q_n}\dot{q}_n\right\| \\ &\leq \left[\sum_{i=1}^{n}\left\|\frac{\partial \mathbf{J}(\mathbf{q})}{\partial q_i}\right\|\right]\|\dot{\mathbf{q}}\| = k_{j(q)}\|\dot{\mathbf{q}}\|. \end{aligned} \tag{A.1}$$

For the next step, consider that $\|\mathbf{C}(\mathbf{q},\dot{\mathbf{q}})\dot{\mathbf{q}}\| \leq k_{c1}\|\dot{\mathbf{q}}\|^2$. Also, since $\dot{\mathbf{x}} = \mathbf{J}(\mathbf{q})\dot{\mathbf{q}}$, then $\|\dot{\mathbf{q}}\| \leq \|\mathbf{J}^{-1}(\mathbf{q})\|\|\dot{\mathbf{x}}\|$; hence,

$$\begin{aligned} \left\|\mathbf{C}_\chi \dot{\mathbf{x}}\right\| &= \left\|\mathbf{J}^{-T}(\mathbf{q})\,\mathbf{C}(\mathbf{q},\dot{\mathbf{q}})\,\dot{\mathbf{q}} - \mathbf{M}_\chi \dot{\mathbf{J}}(\mathbf{q})\,\dot{\mathbf{q}}\right\| \\ &\leq \left\|\mathbf{J}^{-1}(\mathbf{q})\right\| k_{c1}\|\dot{\mathbf{q}}\|^2 + \left\|\mathbf{M}_\chi\right\| k_{j(q)}\|\dot{\mathbf{q}}\|^2 \\ &\leq \left\|\mathbf{J}^{-1}(\mathbf{q})\right\|^2 \left(\left\|\mathbf{J}^{-1}(\mathbf{q})\right\| k_{c1} + \left\|\mathbf{M}_\chi\right\| k_{j(q)}\right)\|\dot{\mathbf{x}}\|^2 \\ &= k_{cx}\|\dot{\mathbf{x}}\|^2. \end{aligned} \tag{A.2}$$

Therefore, $\|\mathbf{C}_\chi \dot{\mathbf{x}}\| \leq k_{cx}\|\dot{\mathbf{x}}\|^2$. □

Conflicts of Interest

The authors declare that there are no conflicts of interest regarding this paper.

Acknowledgments

This work is partially supported by BUAP, Mexico (Project VIEP–00110).

References

[1] M. H. Raibert and J. J. Craig, "Hybrid posiition/force control of manipulators," *Journal of Dynamic Systems, Measurement, and Control*, vol. 103, no. 2, pp. 126–133, 1981.

[2] N. Hogan, "Impedance control - An approach to manipulation. I - Theory. II - Implementation. III - Applications," *ASME Transactions Journal of Dynamic Systems and Measurement Control B*, vol. 107, pp. 1–24, mar 1985.

[3] B. Siciliano and K. Oussama, *Handbook of Robotics*, Springer, 2nd edition, 2016.

[4] H. Kazerooni, T. B. Sheridan, and P. K. Houpt, "Robust Compliant Motion for Manipulators, Part I: The Fundamental Concepts of Compliant Motion," *IEEE Journal on Robotics and Automation*, vol. 2, no. 2, pp. 83–92, 1986.

[5] D. E. Whitney, "Historical Perspective and State of the Art in Robot Force Control," *International Journal of Robotics Research*, vol. 6, no. 1, pp. 3–14, 1987.

[6] O. Khatib, "A unified approach for motion and force control of robot manipulators: the operational space formulation," *IEEE Transactions on Robotics and Automation*, vol. 3, no. 1, pp. 43–53, 1987.

[7] R. Carelli and R. Kelly, "An adaptive impedance/force controller for robot manipulators," *Institute of Electrical and Electronics Engineers Transactions on Automatic Control*, vol. 36, no. 8, pp. 967–971, 1991.

[8] S. P. Patarinski and R. G. Botev, "Robot force control: A review," *Mechatronics*, vol. 3, no. 4, pp. 377–398, 1993.

[9] M. Mendoza-Gutiérrez, F. Reyes, I. Bonilla-Gutiérrez, and E. González-Galván, "Proportional–derivative impedance control of robot manipulators for interaction tasks," *Proceedings of the Institution of Mechanical Engineers, Part I: Journal of Systems and Control Engineering*, vol. 225, no. 3, pp. 315–329, 2011.

[10] B. Siciliano and L. Villani, *Robot Force Control*, Kluwer Academic Publishers, Boston, MA, USA, 1999.

[11] I. Bonilla-Guti, F. Reyes, M. Mendoza-Guti, and E. J. Gonz, "A Dynamic-compensation Approach to Impedance Control of Robot Manipulators," *Journal of Intelligent Robotic Systems*, vol. 63, pp. 51–73, 2011.

[12] M. Mendoza, I. Bonilla, F. Reyes, and E. Gonzalez-Galvan, "A Lyapunov-based design tool of impedance controllers for robot manipulators," *Kybernetika*, vol. 48, no. 6, pp. 1136–1155, 2012.

[13] S. Chiaverini, B. Siciliano, and L. Villani, "A survey of robot interaction control schemes with experimental comparison," *IEEE/ASME Transactions on Mechatronics*, vol. 4, no. 3, pp. 273–285, 1999.

[14] A. Denève, S. Moughamir, L. Afilal, and J. Zaytoon, "Control system design of a 3-DOF upper limbs rehabilitation robot," *Computer Methods and Programs in Biomedicine*, vol. 89, no. 2, pp. 202–214, 2008.

[15] H. Mehdi and O. Boubaker, "Stiffness and Impedance Control Using Lyapunov Theory for Robot-Aided Rehabilitation," *International Journal of Social Robotics*, vol. 4, no. 1, pp. 107–119, 2012.

[16] R. Cortesão and M. Dominici, "Robot Force Control on a Beating Heart," *IEEE/ASME Transactions on Mechatronics*, vol. 22, no. 4, pp. 1736–1743, 2017.

[17] L. Fortuna and G. Muscato, "A roll stabilization system for a monohull ship: modeling, identification, and adaptive control," *IEEE Transactions on Control Systems Technology*, vol. 4, no. 1, pp. 18–28, 1996.

[18] S. Jung and T. C. Hsia, "Robust neural force control scheme under uncertainties in robot dynamics and unknown environment," *IEEE Transactions on Industrial Electronics*, vol. 47, no. 2, pp. 403–412, 2000.

[19] N. Kumar, V. Panwar, N. Sukavanam, S. P. Sharma, and J.-H. Borm, "Neural network based hybrid force/position control for robot manipulators," *International Journal of Precision Engineering and Manufacturing*, vol. 12, no. 3, pp. 419–426, 2011.

[20] H. Chaudhary, V. Panwar, R. Prasad, and N. Sukavanam, "Adaptive neuro fuzzy based hybrid force/position control for an industrial robot manipulator," *Journal of Intelligent Manufacturing*, vol. 27, no. 6, pp. 1299–1308, 2016.

[21] T. Yoshikawa, T. Sugie, and M. Tanaka, "Dynamic Hybrid Position/Force Control of Robot Manipulators—Controller Design and Experiment," *IEEE Journal on Robotics and Automation*, vol. 4, no. 6, pp. 699–705, 1988.

[22] R. Lozano and B. Brogliato, "Adaptive hybrid force-position control for redundant manipulators," *Institute of Electrical and Electronics Engineers Transactions on Automatic Control*, vol. 37, no. 10, pp. 1501–1505, 1992.

[23] D. Jeon and M. Tomizuka, "Learning Hybrid Force and Position Control of Robot Manipulators," *IEEE Transactions on Robotics and Automation*, vol. 9, no. 4, pp. 423–431, 1993.

[24] R. Volpe and P. Khosla, "A theoretical and experimental investigation of explicit force control strategies for manipulators," *Institute of Electrical and Electronics Engineers Transactions on Automatic Control*, vol. 38, no. 11, pp. 1634–1650, 1993.

[25] S. Chiaverini and L. Sciavicco, "Parallel approach to force/position control of robotic manipulators," *IEEE Transactions on Robotics and Automation*, vol. 9, no. 4, pp. 361–373, 1993.

[26] S. K. Singh and D. O. Popa, "An Analysis of Some Fundamental Problems in Adaptive Control of Force and Impedance Behavior: Theory and Experiments," *IEEE Transactions on Robotics and Automation*, vol. 11, no. 6, pp. 912–921, 1995.

[27] G. Ferretti, G. Magnani, and P. Rocco, "Toward the implementation of hybrid position/force control in industrial robots," *IEEE Transactions on Robotics and Automation*, vol. 13, no. 6, pp. 838–845, 1997.

[28] G. Zeng and A. Hemami, "An overview of robot force control," *Robotica*, vol. 15, no. 5, pp. 473–482, 1997.

[29] C. Natale, *Interaction Control of Robot Manipulators: Six-degrees-of-freedom Tasks*, Springer Tracts in Advanced Robotics, Springer, 2003.

[30] Q. H. Xia, S. Y. Lim, M. H. Ang Jr., and T. M. Lim, "Parallel force and motion control using adaptive observer-controller," in *Proceedings of the 2008 IEEE International Conference on Systems, Man and Cybernetics, SMC 2008*, pp. 3143–3149, Singapore, October 2008.

[31] Y. Karayiannidis and Z. Doulgeri, "Robot force/position tracking on a surface of unknown orientation," in *European Robotics Symposium 2008*, vol. 44 of *Springer Tracts Adv. Robot.*, pp. 253–262, Springer, Berlin, 2008.

[32] B. Siciliano, L. Sciavicco, L. Villani, and G. Oriolo, *Robotics: Modelling, Planning and Control*, Springer Publishing Company, Incorporated, 2010.

[33] S. A. M. Dehghan, M. Danesh, and F. Sheikholeslam, "Adaptive hybrid force/position control of robot manipulators using an adaptive force estimator in the presence of parametric uncertainty," *Advanced Robotics*, vol. 29, no. 4, pp. 209–223, 2015.

[34] X. Sheng, L. Xu, and Z. Wang, "A position-based explicit force control strategy based on online trajectory prediction," *International Journal of Robotics and Automation*, vol. 32, no. 1, pp. 93–100, 2017.

[35] S. M. Hashemi, U. Gürcüoğlu, and H. Werner, "Interaction control of an industrial manipulator using LPV techniques," *Mechatronics*, vol. 23, no. 6, pp. 689–699, 2013.

[36] H. P. Singh and N. Sukavanam, "Stability analysis of robust adaptive hybrid position/force controller for robot manipulators using neural network with uncertainties," *Neural Computing and Applications*, vol. 22, no. 7-8, pp. 1745–1755, 2013.

[37] E. Sariyildiz and K. Ohnishi, "On the explicit robust force control via disturbance observer," *IEEE Transactions on Industrial Electronics*, vol. 62, no. 3, pp. 1581–1589, 2015.

[38] C. A. Chavez-Olivares, F. Reyes-Cortes, and E. J. Gonzalez-Galvan, "On Explicit Force Regulation with Active Velocity Damping for Robot Manipulators," *Automatika*, vol. 56, no. 4, pp. 478–490, 2015.

[39] R. D. J. Portillo-Vélez, A. Rodriguez-Angeles, and C. A. Cruz-Villar, "An optimization-based impedance approach for robot force regulation with prescribed force limits," *Mathematical Problems in Engineering*, vol. 2015, Article ID 918301, 13 pages, 2015.

[40] A. Calanca, R. Muradore, and P. Fiorini, "A review of algorithms for compliant control of stiff and fixed-compliance robots," *IEEE/ASME Transactions on Mechatronics*, vol. 21, no. 2, pp. 613–624, 2016.

[41] Q. Wu, X. Wang, B. Chen, H. Wu, and Z. Shao, "Development and hybrid force/position control of a compliant rescue manipulator," *Mechatronics*, vol. 46, pp. 143–153, 2017.

[42] J. L. Dideriksen, D. F. Feeney, A. M. Almuklass, and R. M. Enoka, "Control of force during rapid visuomotor force-matching tasks can be described by discrete time PID control algorithms," *Experimental Brain Research*, vol. 235, no. 8, pp. 2561–2573, 2017.

[43] P. Gierlak and M. Szuster, "Adaptive position/force control for robot manipulator in contact with a flexible environment," *Robotics and Autonomous Systems*, vol. 95, pp. 80–101, 2017.

[44] M. Takegaki and S. Arimoto, "New feedback method for dynamic control of manipulators," *Journal of Dynamic Systems, Measurement and Control*, vol. 103, no. 2, pp. 119–125, 1981.

[45] C. Ott, *Cartesian Impedance Control of Redundant and Flexible Joint Robots*, Springer, 2008.

[46] C. A. Chávez-Olivares, F. Reyes-Cortés, E. J. González-Galván, M. O. Mendoza-Gutiérrez, and I. Bonilla-Gutiérrez, "Experimental evaluation of parameter identification schemes on a direct-drive robot," *Proceedings of the Institution of Mechanical Engineers, Part I: Journal of Systems and Control Engineering*, vol. 226, no. 10, pp. 1419–1431, 2012.

[47] R. Kelly, V. Santibanez, and F. Reyes, "On saturated-proportional derivative feedback with adaptive gravity compensation of robot manipulators," *International Journal of Adaptive Control and Signal Processing*, vol. 10, no. 4-5, pp. 465–479, 1996.

9

Tracking a Subset of Skeleton Joints: An Effective Approach towards Complex Human Activity Recognition

Muhammad Latif Anjum,[1] Stefano Rosa,[2] and Basilio Bona[2]

[1] *School of Electrical Engineering and Computer Science (SEECS), National University of Sciences and Technology (NUST), H-12, Islamabad, Pakistan*
[2] *Department of Control and Computer Engineering (DAUIN), Politecnico di Torino, Corso Duca degli Abruzzi 24, 10129 Torino, Italy*

Correspondence should be addressed to Muhammad Latif Anjum; latif.anjum@seecs.edu.pk

Academic Editor: Yinlai Jiang

We present a robust algorithm for complex human activity recognition for natural human-robot interaction. The algorithm is based on tracking the position of selected joints in human skeleton. For any given activity, only a few skeleton joints are involved in performing the activity, so a subset of joints contributing the most towards the activity is selected. Our approach of tracking a subset of skeleton joints (instead of tracking the whole skeleton) is computationally efficient and provides better recognition accuracy. We have developed both manual and automatic approaches for the selection of these joints. The position of the selected joints is tracked for the duration of the activity and is used to construct feature vectors for each activity. Once the feature vectors have been constructed, we use a Support Vector Machines (SVM) multiclass classifier for training and testing the algorithm. The algorithm has been tested on a purposely built dataset of depth videos recorded using Kinect camera. The dataset consists of 250 videos of 10 different activities being performed by different users. Experimental results show classification accuracy of 83% when tracking all skeleton joints, 95% when using manual selection of subset joints, and 89% when using automatic selection of subset joints.

1. Introduction

After a successful revolution in industrial robotics, the focus of the robotics community has now been shifted to humanoids and social robotics. For robots to invade social sphere of humans, the most important task for them is to understand what is going on around them or more specifically to understand what activities the humans around them are performing. A successful recognition of the activities being performed by humans will enable the robots to respond appropriately and is, therefore, an important step towards natural human robot interaction. This paper makes an attempt to contribute to the field and presents an algorithm to help robots classify activities being performed in front of them. The key advantage of our approach to activity recognition based on skeleton tracking is the selection of joints most relevant to the activity being performed. This not only reduces the computational complexity of the algorithm but also enhances the recognition accuracy as shown in experimental results. In the previous work ([1, 2]), we presented a method for skeleton tracking based activity classification based on the manual selection of subset joints depending upon the activity. With this paper, we extend that approach to automatic selection of subset joints most relevant to the activity. We also improve our dataset to include more activities and tested our algorithm on an activity involving two individuals. Our set of activities include *waving, checking wrist watch, doing a sit-stand exercise, sitting in a chair and drinking water, picking something from ground, pointing in a direction*, and *handshake between two individuals*. Additionally, we test our algorithm on three signals from umpire in a Cricket match.

We use the Robot Operating System (ROS) (http://www.ros.org/) to develop our algorithm. We use a modified version of the OpenNI skeleton tracker (http://wiki.ros.org/openni_tracker) to track the position of selected joints. As opposed to other approaches [3], we are not using the position of all joints to construct feature vectors. Since only a few joints are undergoing a change in position for any given activity (considered as the joints contributing towards the activity), we select only those joints contributing the most towards an activity to construct feature vectors, thereby increasing computational efficiency and recognition performance of the algorithm. We present manual as well as automatic approaches for the selection of these joints and the experimental results for both approaches have been separately presented. The experimental results have been obtained on a purposely built dataset of activities. The dataset has been made public and constructed specifically to suit the requirements of skeleton tracking using ROS based OpenNI skeleton tracker. The contributions of this paper are (1) presenting a computationally efficient activity recognition algorithm based on tracking the position of a subset of skeleton joints, (2) provision of ROS based OpenNI skeleton tracker friendly RGB-D dataset of human activities, and (3) classification of activities involving multiple individuals.

The dataset along with all the ROS packages have been made public and can be accessed through the website of our research group, LabRob (http://www.polito.it/Labrob). The dataset contains video files of all activities in OpenNI file format (.oni).

2. Related Work

The research in activity recognition took a great step forward after the advent of cameras that can capture depth images and videos [4]. Microsoft Kinect is a low cost and readily available sensor and is being used extensively in many computer vision and robotics applications [5]. Zhu and Fujimura demonstrated the advantage of using depth images for human body pose tracking [6]. Their approach used key-point method where they detected and tracked anatomical landmarks (key-points) of human body to reconstruct the human pose. Activity recognition using depth videos has also been presented by Koppula et al. [7]. They have combined skeleton tracking and object affordances to construct feature vectors and have then used structural support vector machines (SSVM) for training and testing. Their results were obtained after testing the algorithm on a purposely built dataset of their own which included only the activities involving one individual. Droeschel et al. presented a person awareness and gesture recognition approach for joint attention in a domestic environment [8]. They used time-of-flight cameras to classify between *showing* and *pointing* gestures and presented their results with high accuracy.

The fact that all skeleton joints do not contribute equally towards an activity was used by Jiang et al. [9]. They analyzed mean contribution of each skeleton joint for various action classes and utilized the contribution ratio of the joints to classify the actions. Their approach is efficient as shown by the experimental results but they are still tracking the positions of all skeleton joints. The idea of using selected joints for activity recognition has recently been utilized by Wu et al. [10]. They only focus on informative body parts such as head and hands to construct Histograms of Located Displacements (HOLD) and Local Depth Motion Maps (L-DMM) based Gabor representation. The discriminative nature of Gabor representations help classify human activities. Our approach is an extension of the idea of using selected skeleton joints for activity recognition.

FIGURE 1: A user makes a surrender pose in front of a camera.

The remainder of this paper is organized as follows. Section 3 describes method for tracking the skeleton joints, followed by Section 4, where the construction scheme for feature vectors is discussed. Both manual and automatic approaches to the selection of subset joints are discussed in this section. Section 5 describes the dataset we created to experimentally test our algorithm along with other available RGB-D datasets. The SVM training and testing procedure is described in Section 6, followed by the description of experimental results in Section 7. Section 8 concludes the paper.

3. Tracking Joints' Positions

OpenNI ROS package for skeleton tracking is available to all ROS users. The package works with a Kinect camera connected to a PC and can track fifteen joints and body parts in human skeleton. These joints include both hands, both elbows, head, neck, both shoulders, torso, left and right hips, both knees, and both feet. The 3D positions and rotation quaternions of all these joints in space are published with reference to the center of camera frame, */openni_depth_frame*, as a set of ROS transforms (*/tf*). These joints are shown in Figure 1 where an individual makes a surrender pose in front of the camera.

We have modified the OpenNI skeleton tracker package to work with off-line recorded videos in our dataset. The videos have been recorded in OpenNI file format with the extension (.oni) and contain both the RGB and depth information needed for OpenNI skeleton tracker.

4. Construction of Feature Vectors for SVM

In order to use SVM, we need to extract certain features from activity videos and construct feature vectors corresponding to every activity video in our dataset. We use the position of selected joints in every activity as features. Using position of all joints to construct feature vectors is inefficient because,

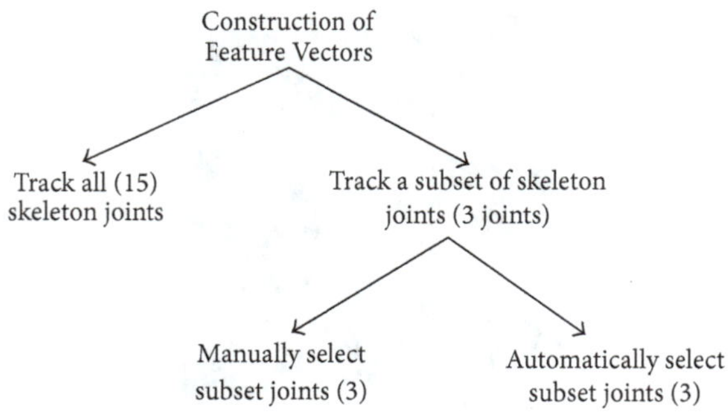

FIGURE 2: Methods developed for selecting feature vectors.

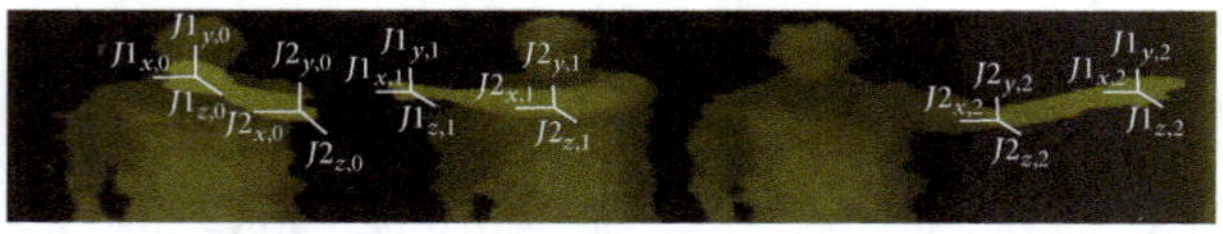

FIGURE 3: Position of two joints across three frames.

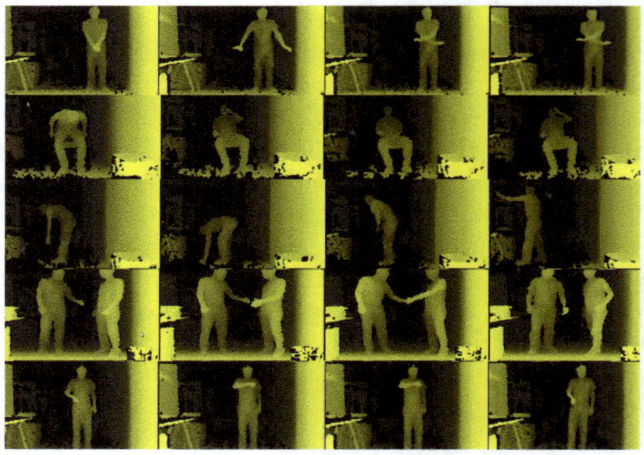

FIGURE 4: Screenshot images at four different positions during the dead ball, sit and drink, pick something from ground, handshake, and check watch activities,respectively.

for any given activity, there is only a subset of joints contributing towards the activity. For example, for *waving hello* activity, only the position of right hand and right elbow is changing while the position of all other joints remains the same. Using the position of all joints to construct feature vectors will increase the similarity index of feature vectors that will decrease recognition efficiency. We have, however, experimented with both approaches and tabulated results for both. Figure 2 shows possible methods to select feature vectors. For the selection of subset joints, we have developed both manual and automatic approaches explained in Sections 3 and 4.

4.1. Tracking All Skeleton Joints. This approach is the simplest where we use the position of all fifteen (15) joints with reference to *torso* to construct feature vectors for each activity. Let these joints be $J1, J2, J3, \dots, J15$. We have the 3D position of each joint in successive frames during the whole activity (shown in Figure 3 with three successive frames). If we represent the x component of the position of joint $J1$ in frame number 0 as $J1_{x,0}$, y and z components as $J1_{y,0}$ and $J1_{z,0}$, respectively, we can formulate feature vectors (FV) as in (1). The subscripts $\{0, 1, \dots, n\}$ are the frame number

$$
\begin{aligned}
\mathrm{FV} = \Big\{&A1, \big\{\left(J1_{x,0}, J2_{x,0}, \dots, J15_{x,0}\right), \\
&\left(J1_{y,0}, J2_{y,0}, \dots, J15_{y,0}\right), \left(J1_{z,0}, J2_{z,0}, \dots, J15_{z,0}\right)\big\}, \\
&\big\{\left(J1_{x,1}, J2_{x,1}, \dots, J15_{x,1}\right), \left(J1_{y,1}, J2_{y,1}, \dots, J15_{y,1}\right), \\
&\left(J1_{z,1}, J2_{z,1}, \dots, J15_{z,1}\right)\big\}, \dots, \\
&\big\{\left(J1_{x,n}, J2_{x,n}, \dots, J15_{x,n}\right), \left(J1_{y,n}, J2_{y,n}, \dots, J15_{y,n}\right), \\
&\left(J1_{z,n}, J2_{z,n}, \dots, J15_{z,n}\right)\big\}\Big\},
\end{aligned}
\tag{1}
$$

where n is the number of frames in a given activity and its value depends on the length of each activity. The first element, $A1$, of each feature vector is the label of the activity required for SVM based training program. Each feature vector FV is, therefore, a row vector and has $(45 \times n) + 1$ elements if we use all skeleton joints as given in (1).

A word is in order regarding the selection of *torso* as the reference point. If we consider camera position as the reference, joints' position will depend on the position of user in front of the camera. This will result in different recognition results for the same activity being performed at different positions in front of the camera. To avoid this, reference point is selected on the user's body to ensure that the position of joints is not dependent on position of the user in front of the camera; that is, the user can stand near the camera, far from it, or walk in front of the camera and the algorithm works in all the cases as long as the whole skeleton remains visible to the camera. *Torso* provides a good reference point because it is located in the middle of human skeleton and remains stationary during most of common human activities. It does change its position during our *sit-stand* activity but experimental results show it works well even for activities involving motion of torso.

4.2. Tracking a Subset of Skeleton Joints. As has been discussed in Section 4.1, using all skeleton joints to construct feature vectors adds to the similarity index of feature vectors resulting in poor efficiency. Our approach is to select only those joints contributing significantly towards the activity to construct feature vectors. We have to decide how many joints are required to constitute the subset for feature vector construction. Repeated experiments with all activities have shown that three joints are sufficient for the construction of feature vectors for the activities in our dataset. Selecting the number of joints could be tricky: tracking fewer joints involves the risk of losing important information while tracking more joints (specifically the ones not undergoing any motion) involves the risk of increasing similarity index between activities. Three joints in each activity have been selected to form the subset of joints for feature vector construction. Once the joints in the subset have been selected, a modified version of (1) is used to construct feature vectors as shown in

$$\begin{aligned} \text{FV} = \{&A1, \{(J1_{x,0}, J2_{x,0}, J3_{x,0}), (J1_{y,0}, J2_{y,0}, J3_{y,0}), \\ &(J1_{z,0}, J2_{z,0}, J3_{z,0})\}, \{(J1_{x,1}, J2_{x,1}, J3_{x,1}), \\ &(J1_{y,1}, J2_{y,1}, J3_{y,1}), (J1_{z,1}, J2_{z,1}, J3_{z,1})\}, \ldots, \\ &\{(J1_{x,n}, J2_{x,n}, J3_{x,n}), (J1_{y,n}, J2_{y,n}, J3_{y,n}), \\ &(J1_{z,n}, J2_{z,n}, J3_{z,n})\}\}. \end{aligned} \tag{2}$$

Since we have reduced the number of joints to only three, each feature vector FV here will be a row vector with $(9 \times n) + 1$ elements as given in (2). The number of elements in each feature vector has been reduced five times with our approach of tracking a subset of skeleton joints, thereby reducing the computational cost. For the selection of subset joints, we have developed both manual and automatic selection approaches described below.

4.2.1. Manual Selection of Subset Joints. A manual approach for selecting a subset of skeleton joints has been developed to test the accuracy of our algorithm. Only those joints that are contributing the most towards the activity are selected and the selection is made based on human intuition and the knowledge of the activity. For example, *waving* and *checking wrist watch* activities include a distinct motion of right arm, so the position of right hand and right elbow can be tracked for these two activities. Similarly, the position of right hand and right knees can be tracked for a *leg bye signal* activity. Since we have an activity that involves two users *(handshake)*, we have tracked the position of right hand and right elbow of the second user in addition to the right hand of first user for a *handshake* activity. Some complex and long activities like *sitting on a chair and drinking* and *sit-stand* involve the motion of many joints while other activities like *waving hello* and *checking wrist watch* involve motion of only two joints. Table 1 lists the joints tracked during manual selection for each activity along with the point of reference.

Table 1: The summary of the joints and their reference points used during manual selection of subset joints where the /right_hand_2 shows the right hand of second user during handshake activity.

Activity	Joints tracked	Reference
Pointing in a direction	/left_hand	/torso
	/left_elbow	/torso
	/left_hand	/torso
Sitting on a chair and drinking from bottle	/right_elbow	/torso
	/right_hand	/torso
	/left_hand	/torso
Picking an object from ground	/right_hand	/torso
	/head	/torso
	/right_shoulder	/torso
Sit-stand exercise	/right_foot	/torso
	/left_foot	/torso
	/left_knee	/torso
Checking wrist watch	/left_hand	/torso
	/left_elbow	/torso
	/left_hand	/torso
Waving hello	/right_hand	/torso
	/right_elbow	/torso
	/right_elbow	/torso
Handshake between two individuals	/right_hand	/torso
	/right_hand_2	/torso_2
	/right_elbow	/torso
Dead ball signal	/right_hand	/torso
	/left_hand	/torso
	/head	/torso
Four signal	/right_hand	/torso
	/right_elbow	/torso
	/right_elbow	/torso
Leg bye signal	/right_hand	/torso
	/right_knee	/torso
	/right_knee	/torso

4.2.2. Automatic Selection of Subset Joints. Robots do not have intuition and prior knowledge of the activity and therefore the manual joint selection strategy cannot be used in practice. We need a system to automatically select joints from the skeleton contributing the most towards any given activity. For this purpose, we track the position of all joints with respect to a single reference, (*/torso*) and constitute position vectors, v_i, of each joint as given in

$$\begin{aligned} v_i = \{&(Ji_{x,0}, Ji_{y,0}, Ji_{z,0}), (Ji_{x,1}, Ji_{y,1}, Ji_{z,1}), \ldots, \\ &(Ji_{x,n}, Ji_{y,n}, Ji_{z,n})\}, \end{aligned} \tag{3}$$

where v_i is the position vector of ith joint.

Once the position vectors of all joints have been obtained with respect to one single reference, we compute the variance of all position vectors in the activity. The joints with highest variance of position vectors are the ones undergoing highest motion (change in position) and are therefore contributing the most towards the activity. Three joints with highest variances are considered for the construction of feature vectors. Once joints are selected, (2) (only for three selected joints) is used to construct feature vectors.

Some activities such as *sit-stand* activity may involve motion of some joints that do not define that activity. For example one person performing *sit-stand* activity might be moving his/her hands along the way, while another might not. The movement of hands therefore does not define the *sit-stand* activity. Our automatic joint selection procedure, however, selects three joints undergoing highest motion irrespective of the nature of joints. Therefore, joints selected could be different for the same activity being performed by different individuals.

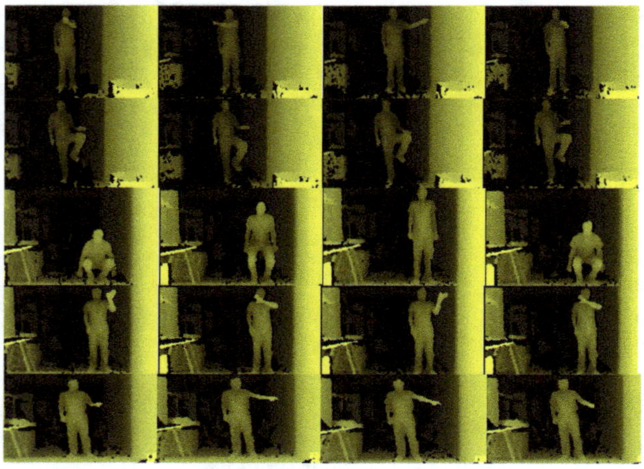

Figure 5: Screenshot images at four different positions during the four signal, leg bye signal, sit stand, wave, and point in a direction activities, respectively.

5. Creating the Dataset

Since most of the work in gesture and activity recognition has been previously done using RGB videos and images, there are only a few activity datasets available involving RGB-D videos. An RGB-D dataset, CAD-120, was presented by Koppula et al. [7]. Their dataset is provided in two formats: as correlated RGB and depth images and as RGB-D text format. Both of these formats require further processing to convert the correlated RGB and depth images into an OpenNI video format to run ROS OpenNI skeleton tracker. Another such dataset, presented by [11], is not suitable for OpenNI skeleton tracker because their videos do not start with the surrender pose (Figure 1). The surrender pose is necessary for ROS based OpenNI tracker to calibrate the user and start tracking.

We have therefore constructed a dataset of our own and made it public through the website of our group, http://www.polito.it/Labrob. Our dataset is purposely built to be used with ROS based OpenNI skeleton tracker and should serve as a reference for future research in Kinect based activity recognition. The dataset was initially presented in [1] and we have now added few more activities to the dataset. The dataset now consists of a total of 10 activities being performed by different users. The activities include 6 daily life activities: *waving hello, checking wrist watch, picking something from ground and placing it over a cupboard, sit-stand, sitting on a chair and drinking from bottle*, and *pointing in a direction*. We have additionally included a *handshake between the two individuals* activity to test our algorithm for activities involving multiple individuals. Finally our dataset includes three distinct signals from the umpire in a Cricket match. These signals include *four signal, leg bye signal*, and *dead ball signal*. Screenshot images from these activities are shown in Figures 4 and 5. We have recorded 25 videos of each activity in the dataset so the dataset contains a total of 250 videos. The videos are available in OpenNI depth video format (.oni) with the resolution of 640 × 480.

Selection of a particular activity to be included in our dataset required a bit of discussion. There are countless daily life human activities which can be included in our dataset. Since our approach to activity classification is based on tracking position of skeleton joints, our first guiding principle was to select activities that involve movement of skeleton joints. The selection of signals from umpire in a Cricket match was influenced by the fact that those signals involved extensive movement of joints (both hands and legs). There is no end of the possible human daily life activities, but 10 activities were considered sufficient enough to verify our approach. Other activities can be added to the dataset in future.

6. Activity Classification Using SVM

Once we have constructed the feature vectors of each activity using (1) or (2) (whichever is applicable), we use Support Vector Machines (SVM) to classify the activities. The features data set needs to be split into training and testing data for SVM. Out of 25 videos of each activity, we use 15 videos of each activity (a total of 150 videos) for training the SVM and the algorithm is tested on 10 remaining videos of each activity (a total of 100 videos). The SVM training data matrix is constructed using

$$\text{trainData} = \begin{bmatrix} (\text{FV1}_1, \text{FV1}_2, \text{FV1}_3, \dots, \text{FV1}_{15})^T \\ (\text{FV2}_1, \text{FV2}_2, \text{FV2}_3, \dots, \text{FV2}_{15})^T \\ \vdots \\ (\text{FV10}_1, \text{FV10}_2, \text{FV10}_3, \dots, \text{FV10}_{15})^T \end{bmatrix}, \quad (4)$$

where FV1, FV2, ..., FV10 represent the feature vectors of corresponding activities constructed individually using (1) or (2), while the subscript represents the number of video of each activity. Since we have used first 15 videos in our dataset for training, the subscript goes from 1 to 15. As a result, the training data matrix will have an order of $150 \times ((9 \times n) + 1)$, if (2) is used, or an order of $150 \times ((45 \times n) + 1)$, if (1) is used. Since n is the number of frames in a given activity and its value depends upon the length of each activity, it will be different for each activity resulting in various rows of training data matrix (4) having different number of elements. To counter this, the number of columns in training data matrix is based on maximum value of n (for the longest activity) and remaining matrices are padded with zeros. The testing data matrix for SVM is similarly constituted using

$$\text{testData} = \begin{bmatrix} (\text{FV1}_{16}, \text{FV1}_{17}, \text{FV1}_{18}, \dots, \text{FV1}_{25})^T \\ (\text{FV2}_{16}, \text{FV2}_{17}, \text{FV2}_{18}, \dots, \text{FV2}_{25})^T \\ \vdots \\ (\text{FV10}_{16}, \text{FV10}_{17}, \text{FV10}_{18}, \dots, \text{FV10}_{25})^T \end{bmatrix}, \quad (5)$$

where the testing data matrix is of the order of $100 \times ((9 \times n) + 1)$, if (2) is used, or of the order of $100 \times ((45 \times n) + 1)$ if (1) is used. This matrix will classify all 100 test videos in one go. The number of rows of test data matrix can be changed based on the number of videos to be classified. Clearly, none of the video used in training is used for testing the algorithm.

Support vector machines are machine learning tools that are used to classify data consisting of high dimensional feature space. The theory is based on selecting optimal hyperplanes that separate various classes of data. In general, the hyperplanes can be represented by equation $f(x) = \beta_0 + \beta^T x$, where β and β_0 are weight vectors and bias, respectively, and x consists of training data points closest to the hyperplane. The SVM algorithm selects β_0 and β to maximize the distance between hyperplanes and x, called the margin. The margin can be expressed as $M = 2/\|\beta\|$ where $\|\beta\|$ is the norm of β. Hence the problem of maximizing M is essentially equivalent to problem of minimizing a β dependent function $F(\beta)$ subject to some constraints. This can be expressed by optimization equation:

$$\begin{aligned} &\underset{\beta,\beta_0}{\text{minimize}} \quad F(\beta) = \frac{2}{\|\beta\|^2} \\ &\text{subject to} \quad A_i\left(\beta_0 + \beta^T x_i\right) \geq 1 \quad \forall i, \end{aligned} \tag{6}$$

where A_i are the labels introduced in the training set. This optimization problem is solved using Lagrange multipliers to find the optimal values of β and β_0. For further readings on SVM, [12] can be consulted.

Since we have multiple categories for classification, we have used SVM multiclass classifier available in OpenCV library, LibSVM. Gaussian kernel has been used for training the algorithm. The algorithm has been implemented using ROS environment and a final ROS package for SVM based training and testing has been made available. The steps required to classify any incoming activity are summarized in flow diagram shown in Figure 6.

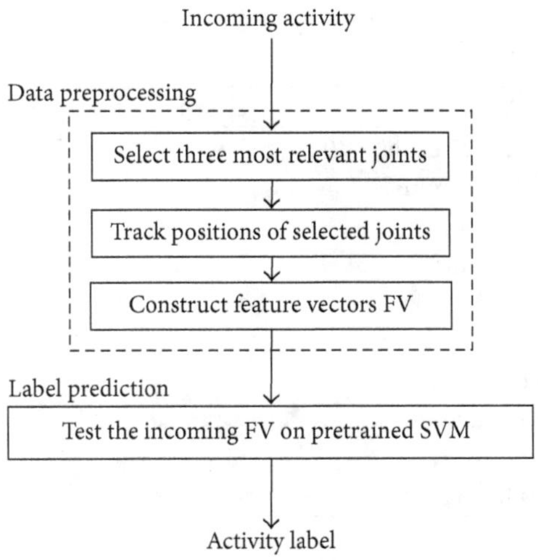

FIGURE 6: Flow diagram detailing the steps required to classify an incoming activity.

TABLE 2: Confusion matrix for activity recognition when using all skeleton joints.

No.	Activity	1	2	3	4	5	6	7	8	9	10
1	Point in a direction	8								2	
2	Pick from ground		7		2						1
3	Check watch			9			1				
4	Sit-stand				9	1					
5	Sit and drink				2	8					
6	Wave				1		8			1	
7	Handshake							9	1		
8	Dead ball signal		1						9		
9	Four signal						2			8	
10	Leg bye signal							1		1	8

7. Experimental Results

Experimental results were obtained on a subset of activity videos that were not used for SVM based training. Experimental results are separately presented for three cases: (a) when tracking all skeleton joints, (b) when manually selecting a subset of skeleton joints to be tracked, and (c) when automatically selecting a subset of joints to be tracked.

7.1. Recognition Results When Using All Skeleton Joints. We selected 10 videos for each activity for testing the algorithm that were not previously used for training and therefore a total of 100 videos of 10 activities. The SVM based testing algorithm successfully classified 83 videos out of 100, resulting in an accuracy of 83%. As has been emphasized earlier, using all skeleton joints for feature vectors construction only adds to the similarity index of the feature vectors resulting in lower accuracy. Table 2 shows the confusion matrix for this case.

The experimental results indicate the problem with tracking all skeleton joints. The activities are being confused with each other, especially the activities which are similar or involve motion of same joints. This is because most of the joints being tracked are stationary in all activities and are adding to the similarity index of the activities resulting in poor recognition accuracy.

7.2. Recognition Results When Using a Subset of Skeleton Joints. We have two strategies for selecting subset joints. Experimental results are presented separately for both of them.

7.2.1. Recognition Results for Manual Selection of Subset Joints. The same videos were used to test the algorithm for manual selection of subset joints. The SVM based testing algorithm successfully classified 95 videos out of 100, resulting in an accuracy of 95%. Table 3 shows the confusion matrix for this case. As has been explained in Section 4, the manual selection of joints is made based on the intuition and the knowledge of the activity, so only those joints having distinct motion pattern in any activity were selected resulting in a higher accuracy.

The activities that are remarkably different based on the motion of joints have been classified with 100% accuracy. These activities include *pointing in a direction*, *picking from*

Table 3: Confusion matrix for activity recognition when using manual selection of joints.

No.	Activity	1	2	3	4	5	6	7	8	9	10
1	Point in a direction	10									
2	Pick from ground		10								
3	Check watch			10							
4	Sit-stand				10						
5	Sit and drink				1	9					
6	Wave						9			1	
7	Handshake							10			
8	Dead ball signal		1						9		
9	Four signal						2			8	
10	Leg bye signal										10

Table 4: Confusion matrix for activity recognition when using automatic selection of joints.

No.	Activity	1	2	3	4	5	6	7	8	9	10
1	Point in a direction	10									
2	Pick from ground		10								
3	Check watch			8			2				
4	Sit-stand				10						
5	Sit and drink				1	9					
6	Wave						8		1	1	
7	Handshake			1				9			
8	Dead ball signal			2					8		
9	Four signal						1		1	8	
10	Leg bye signal									1	9

ground, checking watch, sit stand, leg bye signal, and *handshake.* Notable confusions are *waving hello* and *four signal* because both of these activities include the motion of right hand in a not so distinct trajectory.

7.2.2. Recognition Results for Automatic Selection of Subset Joints. Automatic selection of joints is required for the robots to use the system in real time. As has been explained in Section 4, all joints are tracked with respect to a single reference and three joints with highest variances in position vectors are selected. Experimental results show that our algorithm successfully classifies 89 out of 100 test videos resulting in an accuracy of 89%. Table 4 shows the confusion matrix when using automatic joint selection.

7.3. Results Discussion. The experimental results show that tracking a subset of skeleton joints easily outperforms the approach of tracking whole skeleton. The selection of these subset joints is however tricky and we have presented both manual and automatic approaches to this. The experimental results show that the algorithm is more accurate when using manual selection of joints as compared to automatic selection of joints. This is understandable given the fact that joints are manually selected to make their motion trajectories distinct as possible for each activity. The results obtained with automatic selection of joints are also very encouraging and we are confusing only with activities that are similar to each other based on the motion of the joints involved. The algorithm has not only been tested on simple and small gestures but also on fairly long and complex activities. For example, the *sit and drink* activity involves a person sitting down on a chair, picking up a bottle of water, and then drinking from it. The experimental results are very encouraging and show the potential of the use of skeleton tracking for activity recognition. The skeleton tracking using depth images and videos is strikingly simple that eliminates many conventional headaches one encounters while using RGB images and videos such as background scene, colors, and lighting conditions. The approach is very useful in restricted social environments where robots have to identify a fixed number of activities and gestures and respond accordingly. For example, a coffee serving robot in a coffee bar can be trained with a specific set of activities and gestures the customers usually perform.

The question of what will happen if an unknown activity (an activity on which the algorithm has not been previously trained) is given as input to the classifier needs a bit of discussion. The SVM, as any other machine learning algorithm, requires training on all possible outcomes. If an activity from outside the training set comes in, the classifier will still classify it to one of the activities it was trained with. The unknown activity will be labeled as the one it resembles the most from the training set. Our approach, as has been written already, is useful in a restricted environment where robots have to classify a fixed number of activities. One possible way out is to train the algorithm for the negative world that includes all activities not included in the training set. That will require an extensive dataset of negative world activities and extensive research.

8. Conclusion

All human gestures and activities somewhat involve the motion of some of their joints and body parts. Any approach to activity recognition based on skeleton tracking, therefore, has the potential to classify any human gesture or activity. We have presented our approach to activity recognition for natural human-robot interaction where we use a subset of human skeleton joints to differentiate between the activities being performed. This subset is constituted using the skeleton joints contributing the most towards an activity. The selection of these joints can be intuitively made by humans while training the robot. We have also presented an automatic approach to the selection of these subset joints during the training phase. Experimental results for both approaches have been presented. Once the subset of joints is constituted, we use the 3D position of these joints for the period of the activity to construct feature vectors which are then used for SVM based training and testing of the approach. We have been able to obtain an accuracy of 95% when using manual selection of joints and an accuracy of 89% when using automatic selection of joints. The recognition accuracy when tracking all skeleton joints is also evaluated for comparison and is found to be lower than our algorithm.

Disclosure

This research was conducted during the PhD study of one of the authors at Politecnico di Torino, Italy, and was funded by Higher Education Commission (HEC), Government of Pakistan through its Faculty Development program, HRD-UESTPs/UETs, 2012–2015.

Competing Interests

The authors declare that there is no conflict of interests regarding the publication of this paper.

References

[1] M. L. Anjum, O. Ahmad, S. Rosa, J. Yin, and B. Bona, "Skeleton tracking based complex human activity recognition using kinect camera," in *Proceedings of the 6th International Conference on Social Robotics (ICSR '14)*, pp. 23–33, Sydney, Australia, 2014.

[2] M. L. Anjum, *Vision tracking and activity recognition: towards natural human-robot interaction [Ph.D. thesis]*, Department of Mechatronics, Politecnico di Torino, Torino, Italy, 2015.

[3] L. A. Schwarz, A. Mkhitaryan, D. Mateus, and N. Navab, "Human skeleton tracking from depth data using geodesic distances and optical flow," *Image and Vision Computing*, vol. 30, no. 3, pp. 217–226, 2012.

[4] P. Henry, M. Krainin, E. Herbst, X. Ren, and D. Fox, "RGB-D mapping: using kinect-style depth cameras for dense 3D modeling of indoor environments," *The International Journal of Robotics Research*, vol. 31, no. 5, pp. 647–663, 2012.

[5] R. A. El-Laithy, J. Huang, and M. Yeh, "Study on the use of Microsoft Kinect for robotics applications," in *Proceedings of the IEEE/ION Position, Location and Navigation Symposium (PLANS '12)*, pp. 1280–1288, Myrtle Beach, SC, USA, April 2012.

[6] Y. Zhu and K. Fujimura, "A bayesian framework for human body pose tracking from depth image sequences," *Sensors*, vol. 10, no. 5, pp. 5280–5293, 2010.

[7] H. S. Koppula, R. Gupta, and A. Saxena, "Learning human activities and object affordances from RGB-D videos," *International Journal of Robotics Research*, vol. 32, no. 8, pp. 951–970, 2013.

[8] D. Droeschel, J. Stückler, D. Holz, and S. Behnke, "Towards joint attention for a domestic service robot—person awareness and gesture recognition using time-of-flight cameras," in *Proceedings of the IEEE International Conference on Robotics and Automation (ICRA '11)*, pp. 1205–1210, May 2011.

[9] M. Jiang, J. Kong, G. Bebis, and H. Huo, "Informative joints based human action recognition using skeleton contexts," *Signal Processing: Image Communication*, vol. 33, pp. 29–40, 2015.

[10] M.-Y. Wu, T.-Y. Chen, K.-Y. Chen, and L.-C. Fu, "Daily activity recognition using the informative features from skeletal and depth data," in *Proceedings of the IEEE International Conference on Robotics and Automation (ICRA '16)*, pp. 1628–1633, May 2016.

[11] B. Ni, G. Wang, and P. Moulin, "RGBD-HuDaAct: a color-depth video database for human daily activity recognition," in *Proceedings of the IEEE International Conference on Computer Vision Workshops (ICCV Workshops '11)*, Barcelona, Spain, November 2011.

[12] N. Cristianini and J. Shawe-Taylor, *An Introduction to Support Vector Machines and Other Kernel-based Learning Methods*, Cambridge University Press, Cambridge, UK, 2000.

10

A Computing Model of Selective Attention for Service Robot based on Spatial Data Fusion

Huanzhao Chen and Guohui Tian

School of Control Science and Engineering, Shandong University, Jinan, Shandong 250061, China

Correspondence should be addressed to Guohui Tian; g.h.tian@sdu.edu.cn

Academic Editor: L. Fortuna

Robots and humans are facing the same problem: they all need to face a lot of perceptual information and choose valuable information. Before the robots provide services, they need to complete a robust real-time selective attention process in the domestic environment. Visual attention mechanism is an important part of human perception, which enables humans to select the visual focus on the most potential interesting information. It also could dominate the allocation of computing resource. It also could focus human's attention on valuable objects in the home environment. Therefore we are trying to transfer visual attention selection mechanism to the scene analysis of service robots. This will greatly improve the robot's efficiency in perception and processing information. We proposed a computing model of selective attention which is biologically inspired by visual attention mechanism, which aims at predicting focus of attention (FOA) in a domestic environment. Both static features and dynamic features are composed in attention selection computing process. Information from sensor networks is transformed and incorporated into the model. FOA is selected based on a winner-take-all (WTA) network and rotated by inhibition of return (IOR) principle. The experimental results showed that this approach is robust to the partial occlusions, scale-change illumination, and variations. The result demonstrates the effectiveness of this approach with available literature on biological evidence. Some specific domestic service tasks are also tailored to this model.

1. Introduction

Humans are able to choose the salient part of the visual scene and focus on this area. Building a computing model to detect such salient has become a popular approach. Its advantage lies in the allocation of computing resources in the following cognition and analysis operations. The most popular definition for selective attention has been proposed by the psychologist J. William. The potential of an area or a location in an image to appear distinct from the rest in the scene is roughly referred to as its visual salience. It is believed that the saliency is affected by top-down knowledge and bottom-up stimulus-driven processes [1]. Saliency computing and selective attention mechanism are widely used in many applications including object of interest image segmentation, visual scene analysis, and object recognition [2].

Here a question is raised about how to choose a specific focus of attention based on multiple formations that describe the visual scene. Based on this theory, most popular approaches of bottom-up attention, Itti model, hypothesized that many kinds of feature maps on image feed into a unique representation called saliency map. The saliency map is a two-dimensional scalar map representation while it topographically represents activity saliency of image and is used to choose the salient location. Specifically when this location is salient, it is referred to as an active location of a saliency map, no matter the facts that it corresponds to a thing of an obvious colour in a dark background or to a moving object in a still background. Switching attention based on this scalar topographical representation to focus on the most salient location could draw attention on the area of the highest activity in the saliency map[3].

The concept of "salient" is usually mentioned in the process of bottom-up computations. The term "saliency" characterizes specific parts in a scene that might be users or objects that are assumed to be relative to the neighbouring area [4].

FIGURE 1: Illustration of intelligent space system.

The representation of saliency in a specific scene is computationally introduced in the process of choosing spatial area. For selection and convergence process, when one wins the spatial competition in one or more feature maps at different spatial scales, the related location is defined as salient. After that, detailed measurement of saliency integrating multiple kinds of feature maps is encoded by saliency map. In that way it provides potential heuristic information for choosing attention on salient locations, without consideration of the detailed feature responses that made those locations salient. The saliency map is where the attention spotlight should be placed as well as the most salient location in the scene [5]. A proper neural architecture to find the area needing user's attention from the generated saliency map is winner-take-all network, which implements a neural distributed maximum detector [4]. Using this mechanism, however, brings another computing problem: how to lay attention on the different saliency location or winning area in a saliency map. One plausible computing strategy has already got biological experimental support. It contains transiently inhibiting neurons for current focus of attention in presenting saliency map. The currently attended location is then suppressed. This progress will be repeated and generates attention recording. This kind of inhibitory strategy aims at transferring the FOA. This phenomenon has been named as "inhibition of return" (IOR) and became a popular way of simulating selective attention mechanism [5]. By using IOR process, we are able to rapidly find and shift the focus of attention on different salient locations with decreasing saliency, rather than being bound to attend only to the location of maximal saliency at a given time.

We also use intelligent space to obtain information about the users in a comprehensive way. Intelligent space is capable of improving both perception and executive skills of service robots. Figure 1 illustrates the overview of intelligent space. Users are capable of communicating with both robots and machines, with the help of intelligent space. Many useful services could be provided by actuators discretely installed in the space. The main suggestion of the intelligent space is to install sensors and actuators separately in domestic space, so that service robot could provide timely, stably, and intelligent service with limited configuration [6]. The sensor network can capture human motion information as the user moves in everyday life scenes. In the meantime, intelligent space can offer some service to users by its actuators. It is important to fuse the data from different resources as the sensor network of intelligent space is discretely distributed in target environment. Cellular neural network is an effective way to deal with spatiotemporal temporal processing [7]. Here we choose spatial transformation as a mathematical way in data fusion operation. This mathematical approach is relatively simple and straightforward comparing with neural network. In conclusion, intelligent space is utilized in our research aiming to develop both perception skills and executing skills for robots.

In this paper, we proposed a computing model of visual attention based on data fusion using intelligent space. The model is turned out to select visual attention accurately and robustly by a set of experiments.

2. Related Work

The popular attention selection model is inspired by the biological structure and the behaviour of human visual system demonstrations [1, 8]. Usually, saliency map is proposed based on topographical combination with multiscale features in colour images. Gabor filtering and center-surround operator are used to process the features. Center-surround mechanism is inspired by photoreceptor interactions and visual receptive field organization. It consists of the center and the surround as the feature values at the corresponding image pixels derived from responses to stimuli [9]. A dynamical neural network based is built on WTA network and IOR mechanism, to choose focus of attention and to balance the saliency map. A few other saliency computing models based on Itti model for salience computation are presented [10–13].

In recent years, some models are proposed by new approaches other than Itti's approach. Zhao proposed a multi-context deep learning framework for salient object detection where local context and global context are all concerned [14]. Wang presented a deep neural network based saliency detection mechanism by integrating global search and local estimation [15, 16]. Liu presented saliency tree as a novel saliency detection framework which could provide a hierarchical representation of saliency for generating high-quality regional and pixel-wise saliency maps [17]. Zhang proposed a saliency-guided unsupervised feature learning framework for scene analysis [18]. Wei proposed an algorithm for simultaneously localizing objects and discovering object classes based on bottom-up multiple class learning [19]. Lin proposed a novel salient object detection algorithm by integrating multi-level features including local contrast, global contrast, and background priors [20]. Chen proposed a protoobjects computational model by predicting fixations using a combination of saliency and mid-level representations of shape [21].

Most models of fixation prediction operation at the feature level are like the classical Itti model. Others emphasize that salient object is more important while defining it needs researchers annotation. Most of the existing models consider low-level visual features and locations of objects. The major difference among them is the method for dealing with incoming sensor information, feature types, and computing saliency. For domestic service robot, there are two main shortages. At first, considering the complex environment of domestic environment, blocking is a big trouble. Then human behaviour is a key component of robot attention for service

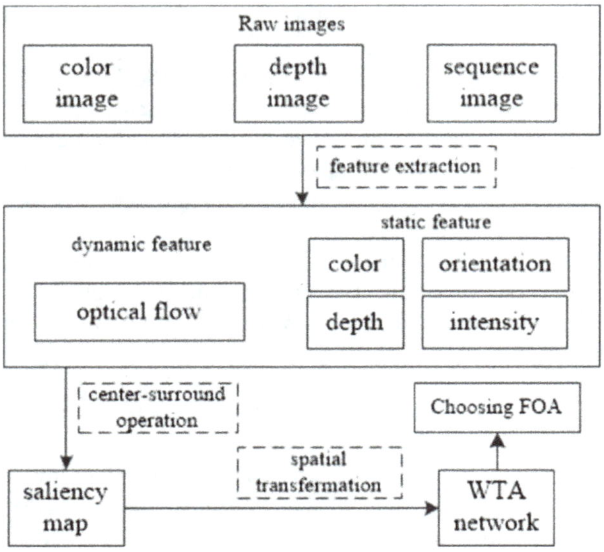

FIGURE 2: Illustration of selective attention model procedure.

task, so that human behaviour deserves more interests in the saliency computing process. Unlike other studies which have emphasize elaborately designed low-level feature descriptors, our approach proposed a saliency model with the help of intelligent space platform. So more information from different cameras is mixed and computed. This has the potential to improve the accuracy and robustness of the saliency computing.

3. A Saliency Computing Model of Selective Attention

The general procedure of saliency computing model for selective attention is illustrated in Figure 2. Raw images (colour image, time ordered image sequences, and depth image) are obtained from cameras distribution from intelligent space. At first, features are extracted using Gabor filter and difference of Gaussian filter. Besides, dynamic features are calculated by computing of optical flow. Then the homogeneous transformation matrix is applied for representing geometric relations between cameras from different coordinate system. After that a saliency map is proposed for nonlinearly combining the features and feeds them into a WTA network. By integrating with inhibition of return method, the FOA is finally selected and rotated.

3.1. Feature Extraction and Processing. For this computing model, the input is captured as colour images and depth images. Colour images are obtained, being digitized at a specific resolution. Based on dyadic Gaussian pyramids and being created in few spatial scales, this is considered as progressively low-pass filter. The colour images are subsampled to produce horizontal and vertical image-reduction factors. Depth images are obtained by infrared sensor of Microsoft Kinect and are capable of detecting user movement in key areas.

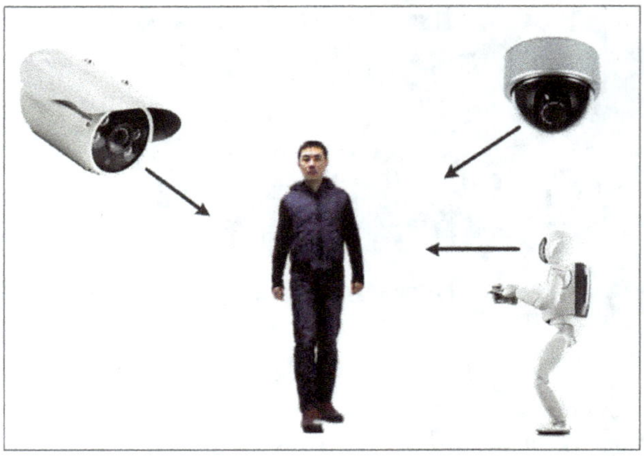

FIGURE 3: Illustration of multiview situation from sensor network in intelligent space.

Static features could be extracted from one single image and represent some static character (like outline and chromatic aberration). Colour images are obtained from ceiling camera from intelligent space platform, as shown in Figure 3. As b, g, and r refer to the blue, green, and red from the colour images, then we build intensity feature map$I = (r + g + b)/3$. This measures the intensity value of all colour channels of an image. Four broadly tuned colour channels are created as follows:

$$B = b - \frac{(g + r)}{2} \quad (1)$$

$$R = r - \frac{(g + b)}{2} \quad (2)$$

$$G = g - \frac{(b + r)}{2} \quad (3)$$

$$Y = \frac{(b + r)}{2} - \frac{|r - g|}{2} - b \quad (4)$$

for red, green, red, and yellow. Based on these colour channels, four Gaussian pyramids are created as $R(\sigma)$, $G(\sigma)$, $B(\sigma)$, and $Y(\sigma)$, while the scale is $\sigma \in [0, 1, 2, 3, \ldots, 8]$. By using Gaussian pyramid in image processing to extract multiscale features and downsampling, we defined the difference of cross scale as $\ominus$ between two feature maps based on center-surround operation. This operation is generated based on point-by-point subtraction and interpolation. This computing process is highly relative to the sensitive features which is shown in a series of maps $I(c, s)$, with $s = c + \delta$, $c \in [2, 3, 4]$ and $s \in [3, 4]$.

$$I(c, s) = |I(c) \ominus I(s)| \quad (5)$$

By simulating the colour-double-opponent system in parts of human brain, chromatic and spatial opponency is constructed for the red/green, green/red, blue/yellow, and yellow/blue colour pairs in our model.

$$RG(c, s) = |(R(c) - G(c)) \ominus (G(s) - R(s))| \quad (6)$$

$$BY(c, s) = |(B(c) - Y(c)) \ominus (B(s) - Y(s))| \quad (7)$$

Gabor filters are the product of a cosine grating and a 2D Gaussian envelope, approximating the receptive field sensitivity profile (impulse response) of orientation selective neurons in primary visual cortex. Next, we extracted information of local orientation from based on Gabor pyramids $O(\sigma, \theta)$, where $\sigma \in [0, 1, 2, ..., 8]$ represents the scale and $\theta \in [0^\circ, 45^\circ, 90^\circ, 135^\circ]$ is the preferred orientation.

$$G(x, y, \theta) = \exp\left(-\frac{x'^2 + \gamma^2 y'^2}{2\delta^2}\right)\cos\left(2\pi\frac{x'^2}{\lambda} + \psi\right) \quad (8)$$

Orientation feature maps $O(c, s, \theta)$ are encoded as a group local orientation contrast between the center and surround scales:

$$O(c, s, \theta) = |O(c, \theta) \ominus (s, \theta)| \quad (9)$$

Here we obtain a depth image based on infrared camera from Microsoft Kinect. At first, Kinect captures Speckle Pattern in different scales and it is saved as Primary or Reference Pattern. Then, the infrared camera detects the objects and gets its corresponding Speckle Pattern. The depth information is obtained by matching Primary or Reference Pattern and triangulation method of ground resistance test. Finally it generates a depth image to describe the spatial information by using 3D Reconstruction and Local offset. After a normalization operation, we can get depth map m_d to describe the spatial relationship of the environment.

Next the dynamic features from image sequences are extracted by computing optical flow. The mobile objects (normally users and robots) are salient part and deserve more attention from service robots. Suppose the image intensity function $I(x, t)$ is

$$I(x, t) \approx I(x + \delta x, t + \delta t) \quad (10)$$

where δx is the displacement of the local image region at (x, t) after time δt. The left side of the equation is expanded based on a Taylor series:

$$I(x, t) = I(x, t) + \nabla I \cdot \delta x + \delta t \cdot I_t + O^2 \quad (11)$$

where $\nabla I = (I_x, I_y)$ and I_t are the first-order partial derivatives of $I(x, t)$ and O^2 is the second- and higher-order terms, which are assumed negligible. Subtract $I(x, t)$ on both sides while ignoring O^2 and dividing by δt:

$$\nabla I \cdot (v) + I_t = 0 \quad (12)$$

This equation is called optical flow constraint equation. Suppose (u, v) is the image velocity, while $\nabla I = (I_x, I_y)$ is the spatial intensity gradient. Optical flow is defined as the distribution of the velocity in given part of the image. The optical flow at given time is described as feature map m_{of}.

3.2. Space Transformation Using Coordination Transform. In this chapter, we proposed a systematic and generalized method about three-dimensional coordinate transform. That is because the service robot and visual devices under real circumstances are from different area, with respect to different reference coordinates. We use a 3D Euclidean space-homogeneous coordinate representation to develop matrix transformations. This operation includes scaling, rotating, translating, and perspective transformation. So features can be flexibly selected to generate a saliency map and focus the attention accurately. In this way, some common problems like pattern noise can be successfully solved.

The homogeneous coordinate representation implies that the n-dimensional vector could be represented by a $n+1$ component vector. Thus in a 3D space a position vector and an augmented vector $\mathbf{w}$ can be used to represent that a position vector $\mathbf{p}$ is in homogeneous coordinate representation.

$$\mathbf{w} \times \mathbf{p} = (wp_x, wp_y, wp_z, w)^t \quad (13)$$

In physical world the coordinates corresponded to homogeneous coordinate as follows:

$$p_x = \frac{wp_x}{w}, \quad p_y = \frac{wp_y}{w}, \quad p_z = \frac{wp_z}{w} \quad (14)$$

The component w represents the scale factor of the homogeneous coordinate. The homogeneous transformation matrix is a 4×4 matrix. This matrix is used to map a homogenous coordinate position vector from one coordinate system to the other coordinate system. The homogeneous transformation matrix T is as follows:

$$T = \begin{bmatrix} \mathbf{R}_{3\times3} & \mathbf{P}_{3\times1} \\ \mathbf{f}_{1\times3} & w_{1\times1} \end{bmatrix} = \begin{bmatrix} RotationMatrix & PositionVector \\ PerspectiveTransf & ScalingFactor \end{bmatrix} \quad (15)$$

The two components, respectively, are the 3×3 rotation matrix and the 3×1 position vector. The vector is the rotated coordinate system origin with respect to the reference system. For the second row, one component is a 4×4 homogeneous transformation matrix and the other component of T represents *ScalingFactor*. The matrix maps a vector expressed in homogeneous coordinates with respect of the *OUVW* coordinate system to the reference coordinate *OXYZ* system. That is,

$$\hat{p}_{xyz} = T\hat{p}_{uvw} \quad (16)$$

$$T = \begin{bmatrix} n_x & s_x & a_x & p_x \\ n_y & s_y & a_y & p_y \\ n_z & s_z & a_z & p_z \\ 0 & 0 & 0 & 1 \end{bmatrix} = \begin{bmatrix} \mathbf{n} & \mathbf{s} & \mathbf{a} & \mathbf{p} \\ 0 & 0 & 0 & 1 \end{bmatrix} \quad (17)$$

As is shown, (17) represents a homogeneous transformation matrix in a 3D space. The space transfer operation is used to composite an overall saliency map based on the saliency maps from different cameras. In conclusion, the homogeneous transformation matrix geometrically represents the location of a fixed coordinate system with respect to a reference coordinate system from other visual resources.

3.3. Building Saliency Map and Finding FOA. In this chapter, we compute saliency by building a saliency map referring to biological principles to describe the value of saliency in the given image. The most salient location in the specific coordinate is referred to as FOA. It represents the maximum area of the space that the robot should focus on at given time. Next we need to propose a FOA selection and inhibition mechanism to switch FOA among saliency maps. There are two major steps which are biologically plausible. At first, winner-take-all network is used to model the saliency score all over the environment and identify FOA. Then Argmax function is used to identify of the saliency map against the highest contributing score of the saliency map activity at the winning location. The neurons of WTA network include a single capacitance, a leakage conductance, and a voltage threshold. A prototypical spike will be produced as the charge delivered by synaptic input reaches the threshold of the neuron. Finally the saliency became the neurons feed into a winner-take-all (WTA) neural network. There is one and only one neuron that could be activated by the WTA network at one time. Other neurons are inhibited at the same time to ensure that the only activated exclusively neuron represents the FOA in space.

We can build four "feature maps" corresponding to the features extracted by methods proposed in previous chapter. Intensity map m_i is obtained in (1). Colour map m_c describes colour features obtained by (2) and (3). Orientation feature map is $m_c = O(c, s, \theta)$. Depth map m_d is extracted from sensors in Kinect. Moreover, the dynamic feature map m_{of} obtained from the computing of optical flow in (8). Here we use entropy method to choose the weigh for every part of saliency amount. The entropy can be obtained by the following equation:

$$e_s = \sum_{k=1}^{4} e_k \tag{18}$$

where e_k, $k \in [1, 2, 3, 4]$, means the entropy for m_c,m_i,m_o and m_d. e_s is the summation of e_k.The final saliency map S is generated by accumulation static and the former feature maps as follows:

$$S = \frac{e_1}{e_s} m_c + \frac{e_2}{e_s} m_i + \frac{e_3}{e_s} m_o + \frac{e_4}{e_s} m_d + m_{of} \tag{19}$$

In this way, features are nonlinearly accumulated by their entropy contribution in normalization coefficient so that the result will be more precise and comprehensive. Then the final saliency map S from one specific resource is generated.

The neurons in the saliency map S could receive activation signals from previous process and remain independent of each other. Every neuron of the saliency should activate the corresponding WTA network neuron. Based on biological mechanism, the neuron could be fast activated according to saliency map on salient place of space. Here some relative process of choosing FOA is included as follows. First, the FOA in original state is not fixed. Another winner neuron with salient place will be excited. Then, all other neurons are suppressed to make the FOA exclusively and wait for next signal. At last IOR mechanism is activated in saliency computing in order to switch the FOA dynamically based on saliency map [22]. The FOA could be unconstrained transferred from the previous FOA to the next FOA while other areas are suppressed. In order to locate the FOA, the winning position is

$$FOA = \operatorname{argmax} S \tag{20}$$

In this way, the FOA which is highly relative to service task can be picked out. This mechanism can be verified by experiments in the following chapter.

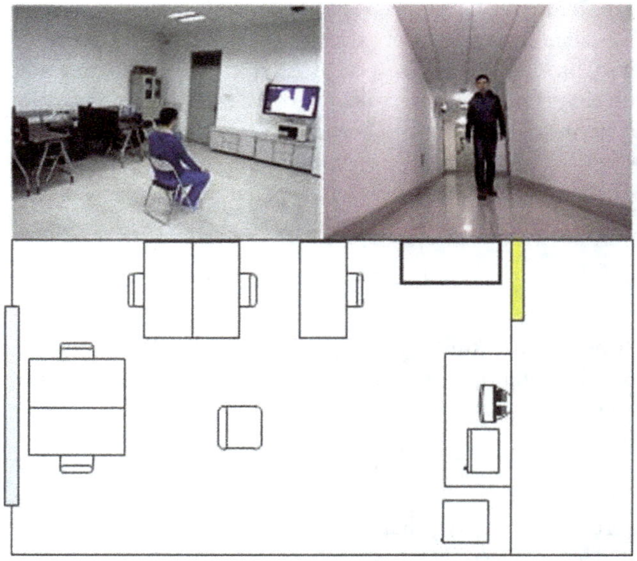

FIGURE 4: Illustration of intelligent space system.

4. Experiments

In this chapter we present a few experiments in real images to demonstrate the computing model of selective attention model. In order to stimulate indoor environment for service robot, we build a room and interconnected corridor experiment environment based on intelligent space, as is shown in Figure 4.

It is a simple environment by contrast as there are fewer obstacles and a simple background in the corridor. However the room is a sophisticated environment as there are various kinds of furniture and electronics. All these objects will make it hard for attention selection. We built an intelligent space platform. There are a few cameras, including ceiling camera and robot sensors, as is shown in Figure 5. This sensor network could be able to observe the target environment in a more comprehensive way. We can obtain colour images, depth images, and time sequence images based on our

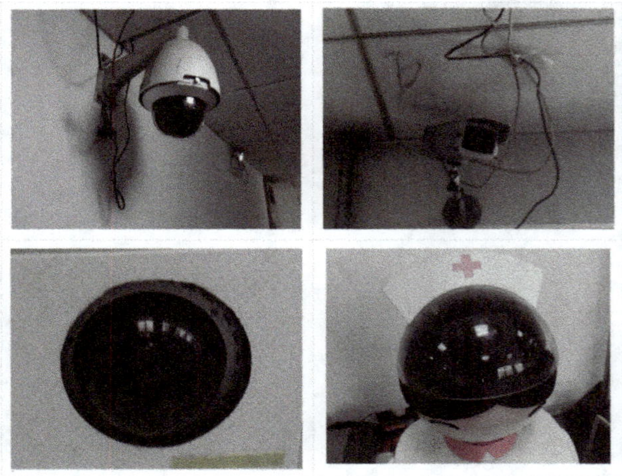

FIGURE 5: Illustration of cameras in intelligent space system.

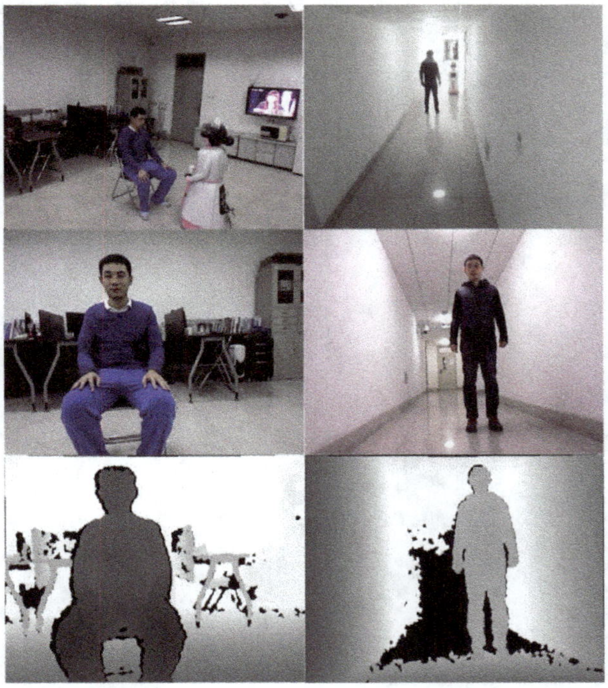

FIGURE 6: Original colour image and depth image in a corridor.

approach. Based on this configuration, a series of experiments are performed in this real scenario.

General Performance. The first experiment is deployed in the corridor. Figure 6 illustrates the original images (colour images and depth images) both from a room and a corridor. There are one user and a service robot in the experiment. This is in accordance with the scenes of daily life. The first two rows are colour images from ceiling camera and robot. The bottom image is depth image from robot. We also capture a series of time ordered image sequences. It implies that colour image contains the global colour features from users and objects. Depth image emphasizes the saliency about users located in specific area. Then we calculated the optical flow

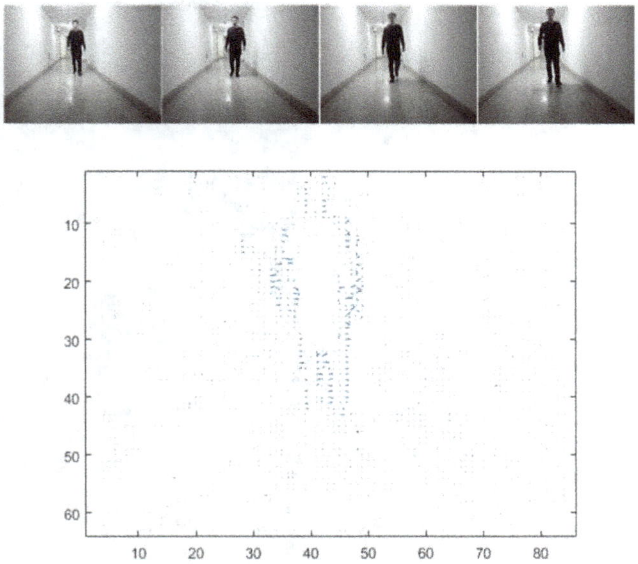

FIGURE 7: Illustration of cameras in intelligent space system.

using the image sequence. The result of saliency computing experiment based on our approach is illustrated in Figure 7. We can see that as user is walking in the corridor, the outline of the user is clearly shown in the chart. As depth image is static and captured in fixed area, the user's movement as dynamic feature is more accurate to reveal the saliency of the target environment. It turns out to be a good supplementary method.

After that, the final result of the saliency computing result is shown in Figure 8. The first row is saliency maps while the bottom picture visualizes the FOA in the environment. From Figure 8 we can see the model can find the user from the corridor accurately. In the common sense, a human waiter should also focus on the user and prepare to offer some service. So our result is in agreement with biological sense. As for the indoor environment, despite the fact that the background is complex, we can see that our attention is in the user. By common sense, the model chooses the FOA accurately. Due to the existing of depth information, the user can be more attractive than the background.

Comparison Experiment with Existing Model. Then we make two comparison experiments with traditional computing approach as is shown in Figure 9. The first row is the original colour images while one is from a corridor and others are from a room. The second row is result of traditional model, while the last row is result of our model. By comparison, we can see that the result of Itti model is coarse and inaccurate. Some other parts which are less salient are chosen (like desks and chairs). That is because traditional approaches are mainly focused on the saliency parts in image. Normally the human user is of great importance and should be the focus of the service tasks. On the contrary, our model is able to choose FOA more precisely and more reasonably, due to the process of depth image and event item which emphasize

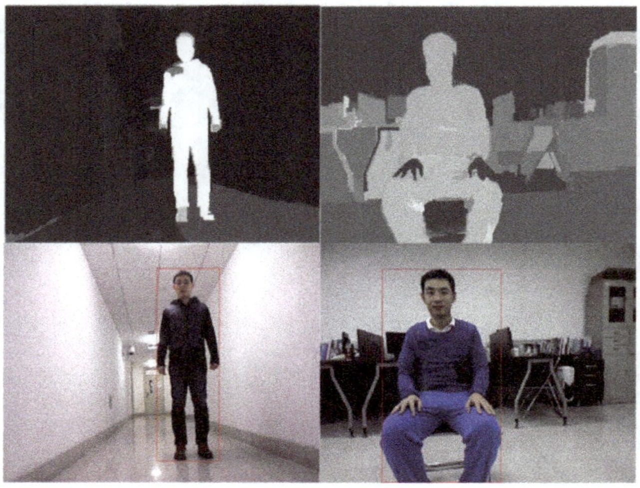

FIGURE 8: General performance experiment.

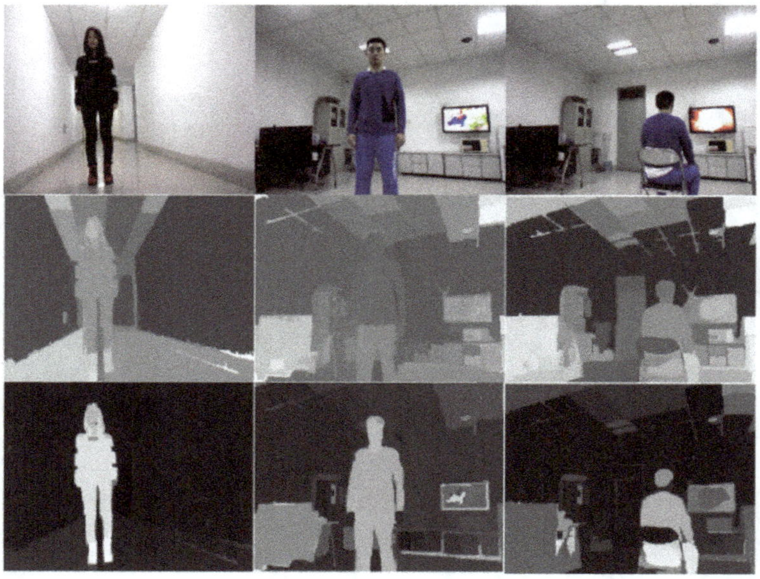

FIGURE 9: Comparison experiment with existing model.

discrete information about intelligent space device or human behaviour.

Influence of Noise. Figure 10 is an example of influence of noise. The top-left is an image with salt and pepper noise while the top-right image is a depth image. The bottom images are the result of saliency map and corresponding FOA. As is shown in the figure, the user and service robot are chosen as saliency part. The influence of noise has been eliminated and the model chosen is right FOA by common sense. In such images, our model was also in better performance as our subjective perception of saliency. Due to the depth information, noise in colour images can be excluded from the result. This operation improves the robustness of the saliency computing model.

5. Discussion

Selective visual attention is capable of finding attention timely on salient objects or users in the domestic environment. This kind of skill enables service robots to locate salient things in a sophisticated visual scene. As it makes human able to find possible food, predator, or mate in the outer world. This paper focuses on developing computational models for service robots that describe the detailed process on attention selection in a specific scene. This has been an important challenge for both robotic and computational neuroscience.

We proposed a computing model for visual attention based on spatial data fusion. Using images spatially independently from the sensor network is applied for visualizing static occlusion culling. We also captured dynamic features

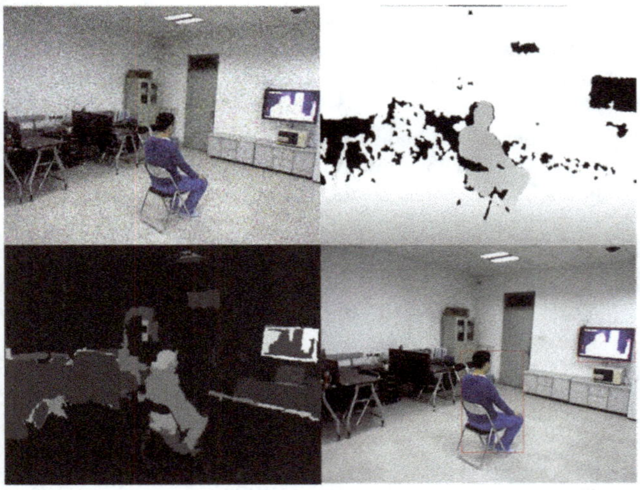

Figure 10: Illustration of cameras in intelligent space system.

in order to reflect human motion states. This is a key component of potentially interesting information for a service robot. When performing in a real environment both quantitatively and qualitatively, the application of our approach has been found to correspond more accurately to locations and saliency maps of eye fixations in visual environment than traditional approaches. Our computing method approach is inspired by the framework of Itti model and has also been found to be more useful in performing better performance than any other same state-of-the-art existing same saliency computing approaches. The better performance of the proposed model has been confirmed through a series of experiments. Most traditional approaches on saliency computing merely are based on low-level features from colour image. This paper covers spatial information and dynamic features which are important to be concerned in scene analysis for service robot. Our model shows great improvement compared with traditional methods both in task-related accuracy and in noise resistance capability. What is more, compared with traditional approaches, it produces better results for attention selection. To some extent, it is observed that the result using our saliency computing approach is consistent manual annotations.

At present, there are a few shortages in our research. Some low-level visual features in multiscales are considered in the research such as depth, hues, orientation, and intensities. While biological principle (like the comprehensive shape features) derived is extensible and could be used to more feature dimension. In the future, new approaches with less computing cost, incurred through multiplicative top-down weight on bottom-up saliency maps, which will combine both goal-driven and stimulus-driven attention should be concerned. Also it should optimize the efficiency of guiding on potential target locations, simultaneously being productive to change visual stimulus situations. Furthermore, when robot operates in unconstrained environments where unexpected events such as accidents may occur, the attention system based on saliency computing needs to bridge the attention selection to further application in further research. This has great potentials to improve its guiding capabilities for more visual computing tasks.

6. Conclusions

In this paper, we proposed a novel saliency computing model for selective attention based on human visual system and traditional bottom-up saliency computing approach. This new computing model with feature fusion is built inspired by the framework of Itti model. It performs salience computing based on multiple information fusions. The main contribution of this paper is to build a computing model for visual attention selection for service robots based on spatial data fusion. By introducing colour images and depth information from different locations, our model solved the problem of image matching under complex condition like noise corruption. The result of our model is consistent with the biological evidence of study of visual attention in human psychophysics and optimized target detection speed, tailored to a specific domestic service task.

Visual attention mechanism plays an important role in the human visual system and for robots and other machines [1]. There are two main goals for developing the selective attention model. At first, selective attention model can be used to cope with massive complexity perception information. This model can prioritize saliency areas in visual images. More potential operations exist, especially for large image information, that need to be processed or if real-time processing is needed for service robots. The other goal is the ability to support task plans. This is always necessary especially for service task performs in a domestic and sophisticated unknown environment. As service robot usually works in the same environment with human. It is reasonable to transfer the human attention mechanism to service robots in order to perform service tasks.

The novel computing model of selective attention proposed in the paper is a biological plausible approach. Furthermore, it could extend to other applications not only in saliency computing related approaches. A large bunch of potential opportunity stays in modifying and updating the saliency computing framework of Itti model and referring to recent research in biology. In the future, it would be interesting in exploiting performance of the saliency computing model built on the application of the proposed computing model. Besides service task inducing in our research, more application of the extracted saliency computing model can be present to segmentation, bionic cognition, and so on.

Conflicts of Interest

The authors declare that they have no conflicts of interest.

References

[1] F. Baluch and L. Itti, "Mechanisms of top-down attention," *Trends in Neurosciences*, vol. 34, no. 4, pp. 210–224, 2011.

[2] M. Cheng, N. J. Mitra, X. Huang, P. H. Torr, and S. M. Hu, "Global contrast based salient region detection," *IEEE Transactions on Pattern Analysis and Machine Intelligence*, vol. 37, no. 3, pp. 569–582, 2015.

[3] L. Itti and C. Koch, "Computational modelling of visual attention," *Nature Reviews Neuroscience*, vol. 2, no. 3, pp. 194–203, 2001.

[4] C. Koch and S. Ullman, "Shifts in selective visual attention: towards the underlying neural circuitry," *Human Neurobiology*, vol. 4, no. 4, pp. 219–227, 1985.

[5] H.-W. Kwak and H. Egeth, "Consequences of allocating attention to locations and to other attributes," *Perception & Psychophysics*, vol. 51, no. 5, pp. 455–464, 1992.

[6] H. Hashimoto, "Intelligent space - how to make spaces intelligent by using DIND?" in *Proceedings of the SMC2002: IEEE International Conference on Systems, Man and Cybernetics*, pp. 14–19, Yasmine Hammamet, Tunisia, 2002.

[7] L. Fortuna, P. Arena, D. Balya, and A. Zarandy, "Cellular neural networks: a paradigm for nonlinear spatio-temporal processing," *IEEE Circuits Systems Magazine*, vol. 1, no. 4, pp. 6–21, 2001.

[8] L. Itti, C. Koch, and E. Niebur, "A model of saliency-based visual attention for rapid scene analysis," *IEEE Transactions on Pattern Analysis and Machine Intelligence*, vol. 20, no. 11, pp. 1254–1259, 1998.

[9] D. Sen and M. Kankanhalli, "A bio-inspired center-surround model for salience computation in images," *Journal of Visual Communication and Image Representation*, vol. 30, pp. 277–288, 2015.

[10] X. Hou, J. Harel, and C. Koch, "Image signature: highlighting sparse salient regions," *IEEE Transactions on Pattern Analysis and Machine Intelligence*, vol. 34, no. 1, pp. 194–201, 2012.

[11] R. Achanta and S. Süsstrunk, "Saliency detection using maximum symmetric surround," in *Proceedings of the 17th IEEE International Conference on Image Processing (ICIP '10)*, pp. 2653–2656, September 2010.

[12] T. Liu, Z. Yuan, J. Sun et al., "Learning to detect a salient object," *IEEE Transactions on Pattern Analysis and Machine Intelligence*, vol. 33, no. 2, pp. 353–367, 2011.

[13] C. Yang, L. H. Zhang, H. C. Lu, X. Ruan, and M.-H. Yang, "Saliency detection via graph-based manifold ranking," in *Proceedings of the 26th IEEE Conference on Computer Vision and Pattern Recognition (CVPR '13)*, pp. 3166–3173, Portland, Ore, USA, June 2013.

[14] R. Zhao, W. Ouyang, H. Li, and X. Wang, "Saliency detection by multi-context deep learning," in *Proceedings of the IEEE Conference on Computer Vision and Pattern Recognition, CVPR 2015*, pp. 1265–1274, IEEE, Massachusetts, Mass, USA, June 2015.

[15] T. Chen, L. Lin, L. Liu, X. Luo, and X. Li, "DISC: deep image saliency computing via progressive representation learning," *IEEE Transactions on Neural Networks and Learning Systems*, vol. 27, no. 6, pp. 1135–1149, 2016.

[16] J. Pan, E. Sayrol, X. Giro-I-Nieto, K. McGuinness, and N. E. O'Connor, "Shallow and deep convolutional networks for saliency prediction," in *Proceedings of the 2016 IEEE Conference on Computer Vision and Pattern Recognition, CVPR 2016*, pp. 598–606, usa, July 2016.

[17] Z. Liu, W. Zou, and O. Le Meur, "Saliency tree: a novel saliency detection framework," *IEEE Transactions on Image Processing*, vol. 23, no. 5, pp. 1937–1952, 2014.

[18] F. Zhang, B. Du, and L. Zhang, "Saliency-guided unsupervised feature learning for scene classification," *IEEE Transactions on Geoscience and Remote Sensing*, vol. 53, no. 4, pp. 2175–2184, 2015.

[19] X. H. Shen, Y. Wu, and Evanston, "A unified approach to salient object detection via low rank matrix recovery," in *Proceedings of the IEEE Conference on Computer Vision and Pattern Recognition (CVPR '12)*, pp. 853–860, Providence, RI, USA, June 2012.

[20] M. Lin, C. Zhang, and Z. Chen, "Global feature integration based salient region detection," *Neurocomputing*, vol. 159, no. 1, pp. 1–8, 2015.

[21] Y. Chen and G. Zelinsky, "Computing Saliency over Proto-Objects Predicts Fixations During Scene Viewing," *Journal of Vision*, vol. 17, no. 10, p. 209, 2017.

[22] M. I. Posner, "Components of visual orienting," *Attention Performance*, vol. 32, no. 4, pp. 531–556, 1984.

11

Modeling and Kinematics Simulation of a Mecanum Wheel Platform in RecurDyn

Yunwang Li,[1] Sumei Dai,[2] Yuwei Zheng,[1] Feng Tian,[1] and Xucong Yan[1]

[1] *School of Mechatronic Engineering, China University of Mining and Technology, Xuzhou 221116, China*
[2] *School of Mechanical and Electrical Engineering, Xuzhou Institute of Technology, Xuzhou 221111, China*

Correspondence should be addressed to Sumei Dai; sumei-dai@hotmail.com

Academic Editor: Long Cheng

The innovative method of modeling and kinematics simulation in RecurDyn are proposed, taking a Mecanum wheel platform(MWP) for omnidirectional wheelchair as research object. In order to study the motion characteristics and mobile performance of the MWP, the virtual prototype simulation model is established in SolidWorks, and virtual prototype simulation is carried out in RecurDyn. The experience of simulation for the MWP in RecurDyn is introduced, and the simulation steps and points for attention are described detailedly. The working states of the mobile system in real environment have been simulated through virtual simulation experiments. Four typical motion models including moving forward, moving laterally, moving laterally in the direction of 45°, and rotation have been simulated in RecurDyn. The simulation results exactly reflect the motion of the MWP. By comparing the simulation results with the theoretical results, there are acceptable errors that are relatively less overall in the simulation results. The simulation results can be used to predict the performance of the platform and evaluate the design rationality, and design quality can be improved according to the exposed problem. This paper can provide reference for the simulation of mobile platform by using RecurDyn.

1. Introduction

With the development of computer technology and the expansion requirements of engineering application, virtual prototyping technology (VPT) [1–4] has gradually come into people's vision. The virtual prototyping technology, of which the cores are mechanical system kinematics, dynamics, and control theory, is a new engineering technology and has developed rapidly through mature three-dimensional computer graphics technology, graph-based user interface technology, and mature market.

Recursive Dynamic (RecurDyn) [5] is a latest generation of multibody system simulation optimization software developed by FunctionBay company based on recursive algorithm. The relative coordinate system motion equation theory and the complete recursive algorithm are used to solve problems, which make it ideally suitable for solving the multibody dynamics problem of large-scale and complicated contact.

The Mecanum wheel, a kind of omnidirectional wheel, has got a wider range of applications in the field of mobile robots in recent years [6–9]. The movements in any direction on flat surface can be achieved by a combination of three Mecanum wheels or more than three ones. Mecanum wheel platform (MWP) has a high application value in the industrial field. The kinematic characteristic of MWP is the basis of the research on motion control research and application of MWP. The motion simulation of virtual prototyping is an effective research method.

In order to study the motion characteristics and mobile performance of a MWP for an omnidirectional wheelchair, the virtual prototype simulation model was established using SolidWorks software, and virtual prototype simulation was carried out in RecurDyn, before its physical prototype of MWP is being manufactured. The simulation results have promoted the improvement of the prototype design. In this paper, the experience of simulation for the MWP by RecurDyn is introduced, and the simulation steps and points for attention are described detailedly. In the simulation environment of RecurDyn, the simulation analysis of contact collision problem can be carried out quickly and efficiently.

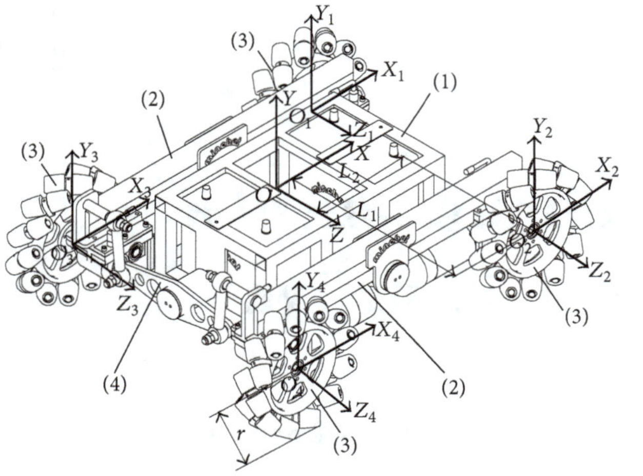

FIGURE 1: 3D model of the MWP. (1) Main body. (2) Rocker arm. (3) Mecanum wheel. (4) Differential mechanism.

The working state of the mobile system in real environment can be simulated through a variety of virtual simulation experiments. The simulation results can be used to predict the performance of the platform and evaluate the rationality of the design, and design quality can be improved according to the exposed problem.

2. Virtual Prototype Simulation Model of the Platform

The 3D modeling function of RecurDyn is relatively weak, so the 3D virtual prototype model for simulating in RecurDyn is usually developed in other CAD software in order to improve the modeling efficiency. SolidWorks, an interactive CAD/CAE/CAM software, is powerful in the function of design and modeling and widely applied in the design field. In this paper, designing, modeling and assembling of the MWP were completed by SolidWorks. Before being imported into RecurDyn, the model should be simplified to reduce the amount of data processing and improve the efficiency of simulation calculation. Some parts, such as bearings, bolts, and electrical components, that have less impact on the simulation can be ignored, and the main subject required for study must be retained.

2.1. Three-Dimensional Model of Virtual Prototype. The MWP in this paper is mainly composed of four parts: main body, two rocker arms, four Mecanum wheels, and differential mechanism, as shown in Figure 1 [10]. The components such as battery and electrical components are placed in the main body. The rocker arms and differential mechanism are used to combine the main body and the four Mecanum wheels that are driven independently. With the differential mechanism, the platform can adapt to terrain passively, the four wheels can keep contact with the ground simultaneously, and the load of the four wheels is approximately equal. The platform can move omnidirectionally with the Mecanum wheels. In this paper, the virtual prototype simulation model of the MWP is developed by SolidWorks.

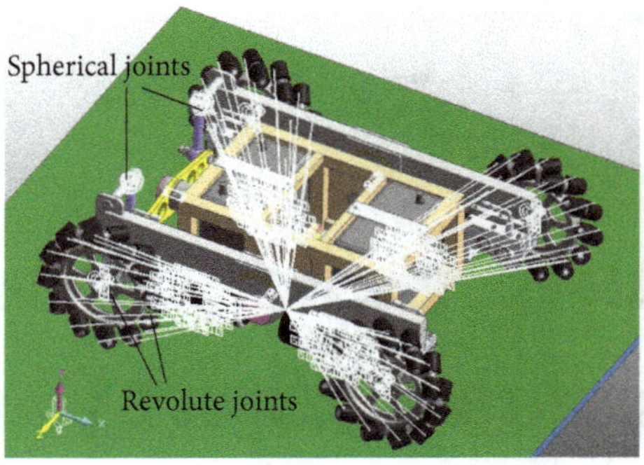

FIGURE 2: Simulation model of the MWP.

2.2. Creating of Joints and Motions. The simplified assembly model can be imported into RecurDyn in Parasolid format. In RecurDyn system, the material and name of the parts and components should be changed, and the Joints between the parts should be defined.

The accuracy of the simulation analysis depends on the accuracy of the above Joints creating. The Joints between the components are created under the premise of confirming the constructional and kinematic relation of the parts. In this simulation model, 103 revolute Joints and 4 spherical Joints are created. 96 of these revolute Joints are between rollers and roller axes of the Mecanum wheel. The 4 spherical Joints are in differential mechanism.

By defining relevant motion functions of the revolute Joints between Mecanum wheels and the rocker arms, the virtual prototype can be enabled to simulate motion according to the predetermined requirements. The STEP functions are used to add motion to system in order to prevent the system simulation abnormity caused by abrupt speed variation in the simulation process. The STEP function, a built-in 3-order polynomial approximation of the step function software, usually used to define a relatively smooth STEP function, is suitable for loading the velocity drive smoothly for the platform.

2.3. Contacts Creating and Parameter Setting. Although the Joints between the parts of the simulation model have been created and the above settings have been completed, the whole model will penetrate the ground under the action of gravity. In order to simulate the motion of the platform, the Contacts between the rollers of the wheels and the ground need to be added. 96 Contacts are generated at one time by using secondary development tool "ProcessNet" to define the above Contacts. This method of Contacts-generating can save operation time and improve modeling efficiency. When the Joints, the Motions, and the Contacts have been defined, the simulation model of MWP has been set up, as shown in Figure 2. According to the Hertz contact theory, the Contact parameters are defined as follows: spring coefficient is 1400 N/mm, damping coefficient is 10, dynamic friction coefficient is 0.5, stiffness exponent is 2.2, the outer layer material of the roller is set as rubber, and the rest parameters are kept the system default value, as shown in Figure 3.

TABLE 1: Driving parameters setting of the platform in four typical motion states.

Motion mode	$\omega_1/\text{rad}*\text{s}^{-1}$	$\omega_2/\text{rad}*\text{s}^{-1}$	$\omega_3/\text{rad}*\text{s}^{-1}$	$\omega_4/\text{rad}*\text{s}^{-1}$
Straight-line motion	ω	ω	ω	ω
Lateral motion	ω	$-\omega$	$-\omega$	ω
Oblique motion at 45°	0	ω	ω	0
Original-place rotation	ω	$-\omega$	ω	$-\omega$

FIGURE 3: Contact parameters setting.

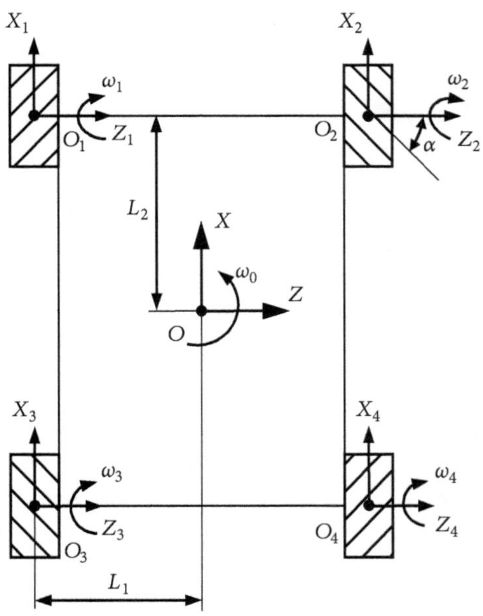

FIGURE 4: Structural schematic diagram of the MWP.

3. Kinematic Model and Simulation

3.1. Kinematic Model. The MWP in this paper is a typical four-wheel longitudinal symmetrical layout structure [11], as shown in Figure 4. The coordinate system is set to match the RecurDyn. The bias angles of all the rollers are α and the diagonal lines in Figure 4 indicate the layout of the rollers in contact with the ground. On the assumption that the roller does not skid against the ground and the platform moves on the horizontal plane, the inverse kinematics equation of the system can be established by kinematics analysis [12]:

$$V_w = J(\alpha) \cdot V_o, \tag{1}$$

where $V_w = [\omega_1 \ \omega_2 \ \omega_3 \ \omega_4]^T$ is the velocity of the wheel. $V_o = [v_x \ v_z \ \omega_o]^T$ is the generalized velocity of the center point of the platform in the coordinate system XOZ. $J(\alpha)$ is the Jacobi matrix of the inverse equations of motion.

$$J(\alpha) = \frac{1}{r}\begin{bmatrix} 1 & \dfrac{1}{\tan\alpha} & -\dfrac{L_1\tan\alpha + L_2}{\tan\alpha} \\ 1 & -\dfrac{1}{\tan\alpha} & \dfrac{L_1\tan\alpha + L_2}{\tan\alpha} \\ 1 & -\dfrac{1}{\tan\alpha} & -\dfrac{L_1\tan\alpha + L_2}{\tan\alpha} \\ 1 & \dfrac{1}{\tan\alpha} & \dfrac{L_1\tan\alpha + L_2}{\tan\alpha} \end{bmatrix}. \tag{2}$$

The expression of its generalized inverse matrix $J(\alpha)^+$ is

$$J(\alpha)^+ = \left(J(\alpha)^T J(\alpha)\right)^{-1} J(\alpha)^T, \quad J(\alpha)^+ J(\alpha) = I_3. \tag{3}$$

In this model, $\alpha = 45°$, and the forward kinematics equation of the system is obtained:

$$\begin{bmatrix} v_x \\ v_z \\ \omega_0 \end{bmatrix} = \frac{r}{4} \cdot \begin{bmatrix} 1 & 1 & 1 & 1 \\ 1 & -1 & -1 & 1 \\ -\dfrac{1}{L_1+L_2} & \dfrac{1}{L_1+L_2} & -\dfrac{1}{L_1+L_2} & \dfrac{1}{L_1+L_2} \end{bmatrix} \begin{bmatrix} \omega_1 \\ \omega_2 \\ \omega_3 \\ \omega_4 \end{bmatrix}. \tag{4}$$

3.2. Simulation Results and Analysis. According to the inverse kinematics equation of formula (1), the platform moves along the X-axis, when $v_x \neq 0$, $v_z = 0$, and $\omega_0 = 0$, and the rotation speed of each Mecanum wheel is equal to $\omega_1 = \omega_2 = \omega_3 = \omega_4 = \omega$.

A constant angular velocity drive function is added to the revolute pair between each wheel and the rocker arm: $\omega(t) =$ STEP(TIME, 0, 0, 0.2, PI), then the movement of the platform along X-axis can be simulated. Four typical motion states of Mecanum wheel are set to analyze the kinetic characteristics, as shown in Table 1.

TABLE 2: Comparison between simulation results and theoretical results of the platform motion velocity.

Movement type	Simulation results	Theoretical results	Relative error
Rectilinear movement $\lvert v_x \rvert/\mathrm{mm}*\mathrm{s}^{-1}$	438.98	439.82	0.19%
Lateral movement $\lvert v_z \rvert/\mathrm{mm}*\mathrm{s}^{-1}$	433.57	439.82	1.42%
Oblique movement at 45°			
$\lvert v_x \rvert/\mathrm{mm}*\mathrm{s}^{-1}$	217.41	219.91	1.14%
$\lvert v_z \rvert/\mathrm{mm}*\mathrm{s}^{-1}$	217.24	219.91	1.21%
Original-place rotation $\lvert \omega_0 \rvert/\mathrm{rad}*\mathrm{s}^{-1}$	0.685	0.691	0.87%

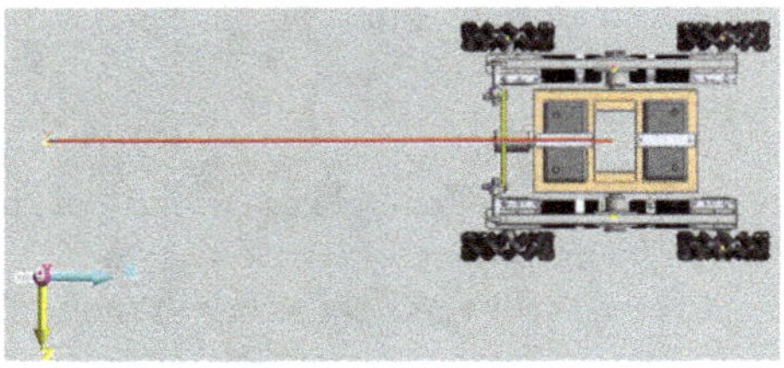

FIGURE 5: Trajectory of the MWP in straight-line motion.

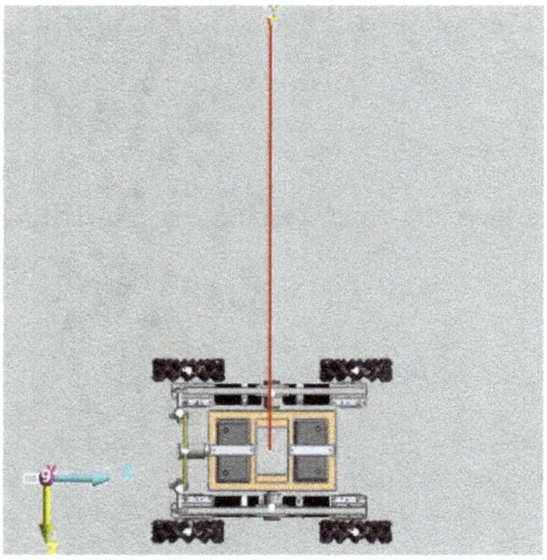

FIGURE 6: Trajectory of the MWP in lateral motion.

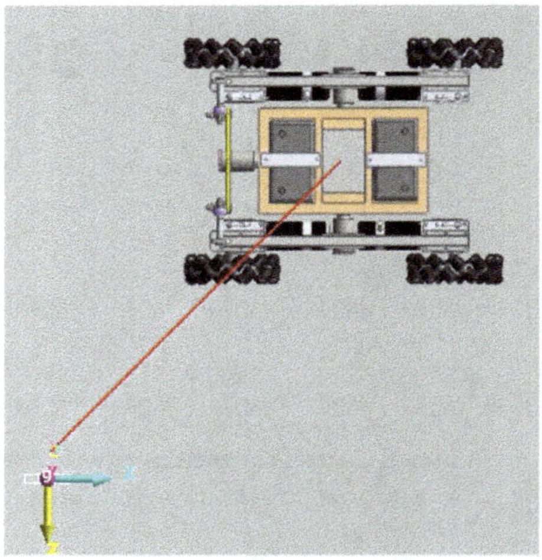

FIGURE 7: Trajectory of the MWP in oblique motion at 45°.

In order to better show the motion state of the platform in the simulation process, the trajectory of related Marker Point can be displayed using the Marker Trace command during simulation animation. In the simulation, a Marker is set at the center of the mobile platform. The trajectories of the platform's center in the above four motion modes are shown in Figures 5–8.

Figures 5, 6, and 7 show that the trajectory of the platform's center is a nearly straight line and the orientation of the platform stays the same during its movement. Figure 8 shows the simulation process of the original rotation of the mobile platform. The trajectory of the center of the MWP is a complete circle in the course of the simulation.

The displacement-time curve of the Marker in vertical direction (Y-axis direction) in straight-line motion is shown in Figure 9. It can show that the MWP falls to the ground freely under gravity at first stage of the simulation; the displacement in vertical direction changes rapidly. After the wheels touch the ground, the platform approximately vibrates up and down periodically. The amplitude of vibration is within 1 mm, which can be ignored comparing with the size of the platform. The curve also shows that the Mecanum wheels bring vibration to the mobile platform during the motion.

The velocity-time curves in X-axis and Z-axis direction in oblique motion at 45° are drawn according to the simulation results, as shown in Figure 10. From Figure 10, the velocity of the platform increases rapidly from 0 to expected value in the X, Z direction within 0.2 s, consistent with the variation trend of the defined STEP function. Then the velocity values in the X, Z direction fluctuate within a certain range. The velocities in the two directions are basically the same that show the platform move laterally in the direction of 45° with horizontal line commendably.

According to formula (4), if $r = 140$ mm, $L_1 = 313.0$ mm, and $L_2 = 323.8$ mm, theoretical results of the platform motion velocity can be obtained. The simulation results are compared with the theoretical results in Table 2. It can be known from Table 2 that there exist varying degrees of errors in simulation results, and the simulation error is relatively less overall. The error is caused by many factors, such as model error, assembly errors, software errors, roller deformation, and skidding. The impact of each factor on the errors still needs to be further studied and is not discussed in detail in this paper. By comparing all simulation results, the rectilinear

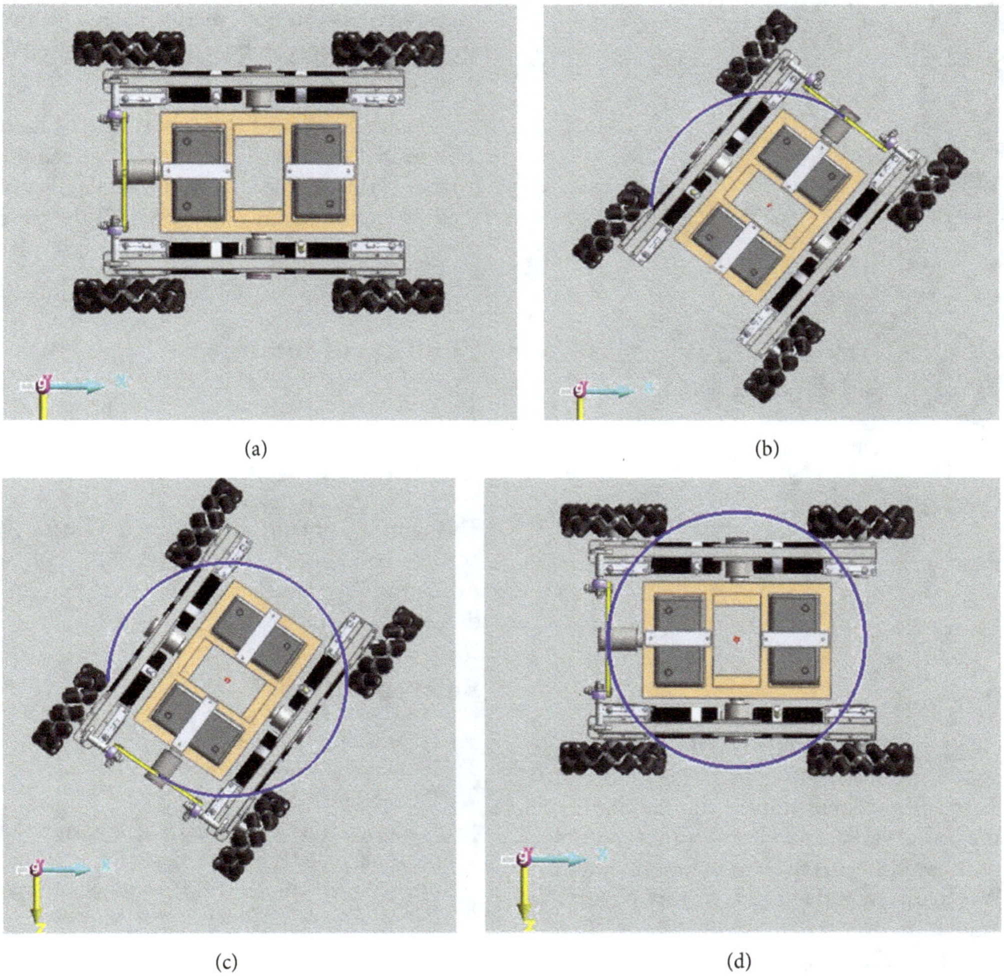

(a) (b) (c) (d)

FIGURE 8: Trajectory of the MWP during the original-place rotation.

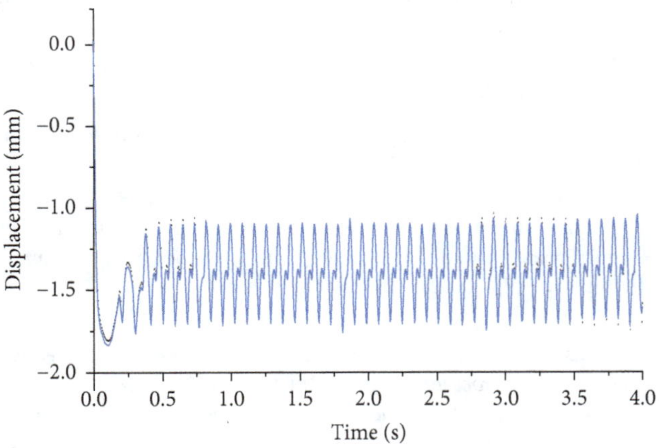

FIGURE 9: Displacement-time curve in the Y direction in straight-line motion.

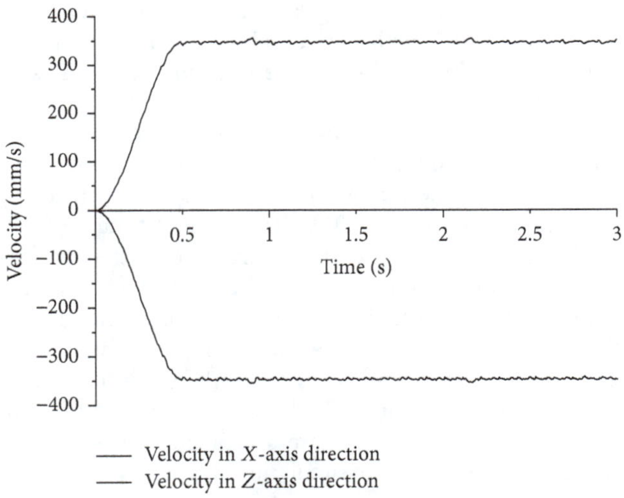

FIGURE 10: Velocity-time curves in X-axis and Z-axis direction in oblique motion at 45°.

movement's relative error of velocity is minimum, and the error in lateral motion is greater, because the Mecanum wheel is equivalent to the ordinary wheel when the MWP moves forward straightly. The structure of the Mecanum wheel has little influence on the movement of the mobile platform during straight motion but great during lateral motion.

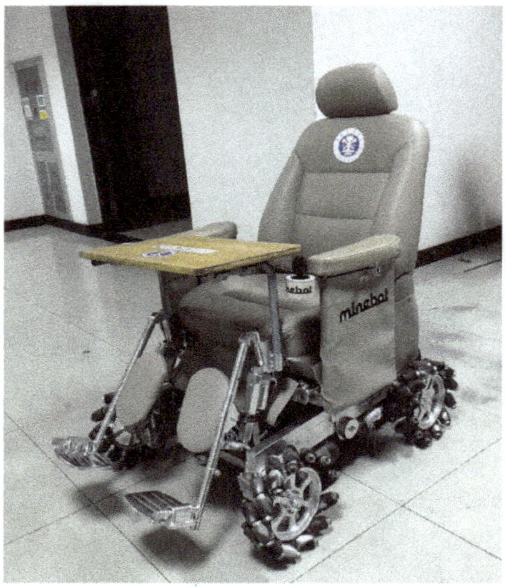

FIGURE 11: The photo of physical prototype of omnidirectional wheelchair.

4. Conclusion

In order to study the motion characteristics and mobile performance of a Mecanum wheel platform for an omnidirectional wheelchair, the virtual prototype simulation model is established using SolidWorks, and virtual prototype simulation is carried out based on RecurDyn, before its physical prototype of MWP being manufactured. In this paper, the experience of simulation for the MWP by RecurDyn is introduced, and the simulation steps and points for attention are described detailedly.

The working states of the mobile system in real environment have been simulated through a variety of virtual simulation experiments. Four typical motion modes including moving forward, moving laterally, moving laterally in the direction of 45° with horizontal line, and rotation have been simulated by RecurDyn. The trajectory of the related Marker that is set at the center of the mobile platform has been displayed using the Marker Trace command during simulation animation. The trajectories in the above four motion modes show the motion state of the platform clearly in the simulation process. The displacement-time curve of the Marker set on the MWP in vertical direction (Y-axis direction) in straight-line motion is drawn. The velocity-time curves of the platform in X-axis and Z-axis direction in oblique motion at 45° are drawn according to the simulation results. The trajectories and curves exactly reflect the motion of the MWP. By comparing the simulation results with the theoretical results, there are acceptable errors that are relatively less overall in the simulation results.

The simulation results can be used to predict the performance of the platform and evaluate the design rationality, and design quality can be improved according to the exposed problem. On the basis of RecurDyn simulation, the design of Mecanum wheel mobile platform has been further optimized, and the physical prototype of omnidirectional wheelchair was designed and manufactured, as shown in Figure 11.

This paper shares the experience of modeling and simulation in RecurDyn. The simulation example of the MWP further verifies the feasibility and scalability of the simulation method involved in this paper. This paper can provide reference for the simulation of mobile platform by using RecurDyn.

Conflicts of Interest

The authors declare that they have no conflicts of interest.

Acknowledgments

This work was financially supported by the National Natural Science Foundation of China (no. 51675518) and the Priority Academic Program Development of Jiangsu Higher Education Institutions.

References

[1] G.-L. Xiong, B.-H. Li, and X.-D. Chai, "Virtual prototyping technology," *Journal of System Simulation*, vol. 13, no. 1, pp. 106–109, 2001.

[2] F. Zorriassatine, C. Wykes, R. Parkin, and N. Gindy, "A survey of virtual prototyping techniques for mechanical product development," *Proceedings of the Institution of Mechanical Engineers, Part B: Journal of Engineering Manufacture*, vol. 217, no. 4, pp. 513–530, 2003.

[3] G. Lu, G. Xue, and Z. Chen, "Design and implementation of virtual interactive scene based on unity 3D," *Advanced Materials Research*, vol. 317-319, pp. 2162–2167, 2011.

[4] J. Wang, K. Bennett, and E. Guinness, "Virtual Astronaut for Scientific Visualization—A Prototype for Santa Maria Crater on Mars," *Future Internet*, vol. 4, no. 4, pp. 1049–1068, 2012.

[5] *RecurDynTM/Solver Theoretical Manual: 5th Revision. FunctionBay Inc., 2005.*

[6] A. Gfrerrer, "Geometry and kinematics of the Mecanum wheel," *Computer Aided Geometric Design*, vol. 25, no. 9, pp. 784–791, 2008.

[7] L. Lin C and Y. Shih H, "Modeling and adaptive control of an omni-mecanum-wheeled robot," *Intelligent Control & Automation*, vol. 2013, no. 2, pp. 166–179, 2013.

[8] A. Shimada, S. Yajima, P. Viboonchaicheep, and K. Samura, "Mecanum-wheel vehicle systems based on position corrective control," in *Proceedings of the IECON 2005: 31st Annual Conference of IEEE Industrial Electronics Society*, pp. 2077–2082, November 2005.

[9] J. A. Cooney, W. L. Xu, and G. Bright, "Visual dead-reckoning for motion control of a mecanum-wheeled mobile robot," *Mechatronics*, vol. 14, no. 6, pp. 623–637, 2004.

[10] L. Yunwang, T. Feng, and D. Sumei, "Rocker type omnidirectional mobile platform based on Mecanum wheel," GBCN204250202U.

[11] Y. Wang and D. Chang, "Motion performance analysis and layout selection for motion system with four Mecanum wheels," *Journal of Mechanical Engineering*, vol. 45, no. 5, pp. 307–316, 2009.

[12] M. O. Tatar, C. Popovici, D. Mandru, I. Ardelean, and A. Plesa, "Design and development of an autonomous omni-directional mobile robot with Mecanum wheels," in *Proceedings of the 2014 19th IEEE International Conference on Automation, Quality and Testing, Robotics, AQTR 2014*, May 2014.

12

Fuzzy-Sliding Mode Force Control Research on Robotic Machining

Shou-yan Chen, Tie Zhang, and Yan-biao Zou

School of Mechanical and Automotive Engineering, South China University of Technology, Guangzhou, Guangdong 510640, China

Correspondence should be addressed to Tie Zhang; merobot@scut.edu.cn

Academic Editor: Keigo Watanabe

The low stiffness has limited the applications of robot to machining process. In this paper, a fuzzy-sliding mode control scheme is proposed to manage the oscillation and chatter appearing in machining operation by adjusting the feed rate. The robotic machining dynamics is first analyzed to identify the parameters with focus on the system stiffness and the behavior during machining process. A controller consisting of a fuzzy estimation enginery which can determine the control gain coefficients according to system status and a sliding mode controller which is used to guarantee convergence and global stability of the system is then proposed. Simulations and experiments results show that, in comparison with open loop and fuzzy-PID control scheme, the fuzzy-sliding mode control scheme can reduce the amplitude and period of oscillation.

1. Introduction

Industrial robot is now worldwide applied in many fields such as material transfer, machining, and assembling [1–3]. As for machining, many studies have been reported and the results indicated that some critical issues, including trajectory error, material removal rate, and contacting force, are needed to be addressed.

To identify the relationship between removal rate and contacting force, Domroes and Krewet [4] compared the performance of force dependent feed rate control and orthogonal force control and reached a conclusion that the removal rate at constant force varies around a mean value. This statement is also agreed upon by Karayiannidis and Doulgeri [5], who proposed an adaptive leaning controller to identify the surface condition with the use of force and joint position/velocity measurements. The results show that the controller can reduce the distortion and tracking error caused by frictional contacts and uncertainties on the contacted surface. For unknown object and environment, Kiguchi and Fukuda [6] proposed an intelligent controller which was constructed on the basis of adaptive fuzzy neural position/force strategy, while Žarko and Vlastimir [7] proposed adaptive neuro-fuzzy-genetic control schemes for explicit force robot control. However, the negative effects on machining performance, caused by the low stiffness of robot, have not been discussed in these studies.

The influence of the low stiffness of robot on machining has been noticed by Matsuoka et al. [8] since 1996, who studied the characteristics of robotic milling operation and proposed a method to reduce the effect caused by low stiffness. To further explore the influence of low stiffness on machining error, Abele et al. [9] identified the robot stiffness by two methods and conducted machining experiments whose results indicated that the path displacement, caused by robot low stiffness, plays a significant role in robotic machining error. With a view of this, many researchers have proposed some measures to compensate for the deformation. Zhang et al. [10] analyzed the coupling relationship between the joint rigidity and Cartesian rigidity and established an online compensation model. Wang et al. [11] present a compensation scheme based on joint stiffness model to reduce the machining deformation caused by machining force.

This paper presents a fuzzy-sliding mode control scheme to compensate for the deformation and oscillation caused by machining force. The robotic machining dynamics is first

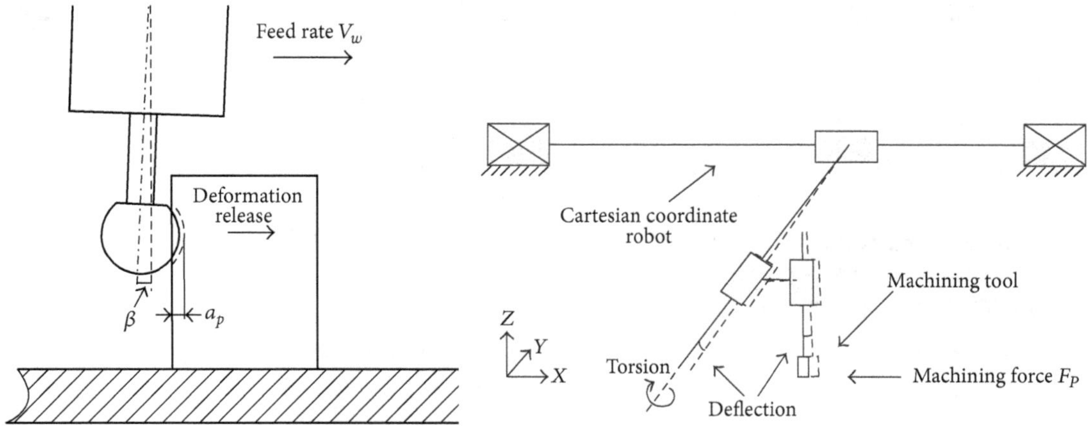

FIGURE 1: Generation of cutting deformation.

analyzed to identify the parameters with focus on the system stiffness and its behavior during machining process. With the knowledge of machining behavior, a sliding mode controller is then constructed based on fuzzy rules to manage the complexity of robotic machining. Simulations and experiments are subsequently conducted to evaluate the performance of the control scheme.

2. Cutting Dynamics

Compared with CNC, the stiffness of robotic is lower, which may result in larger path deviation and oscillation during machining process. Before establishing the dynamics model of robotic machining, the formation of deformation is illustrated.

2.1. Robotic Deformation. As shown in Figure 1, the rotating tool moves towards the workpiece at a feed rate V_w and extrudes the workpiece to perform machining. When machining force is applied to the machining tool, deflections and torsion on robotic system occur and result in a position offset on end-effector. To facilitate the following discussion, this position offset is regarded as a deflection on end-effector and the deflection results in an unmachined area $\sigma_t = [\sigma_x, \sigma_y, \sigma_z]^{\mathrm{T}}$ (labeled as deformation in the following discussion). The deflection is gradually released as the machining continues.

The generation and release process of deflection can be observed in Figure 2. The cutting force reaches the peak value at 10 N when the encoder displays that the robot reaches the required depth. However, the machining force does not go back to 0 N immediately and performs a sustained oscillation, decreasing from around 3 N to 1 N, which indicates that the machining operation is still conducting. Therefore, it can be deduced that a deformation σ_t occurs in machining process which can lead to a deviation between actual path and planned path as shown in Figure 3. The deformation σ_t can be considered as the integration of speed difference

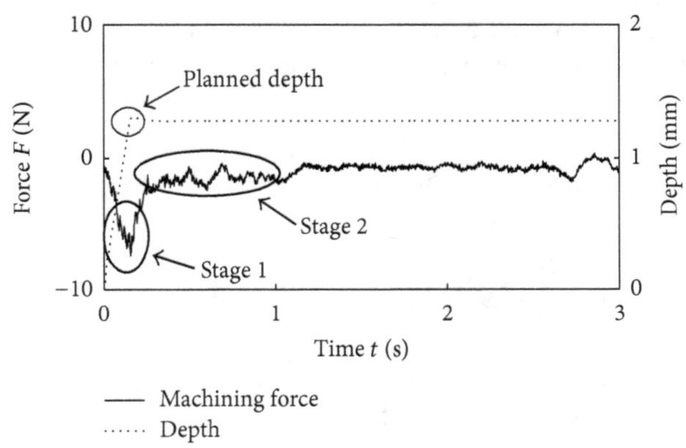

FIGURE 2: Deformation release process.

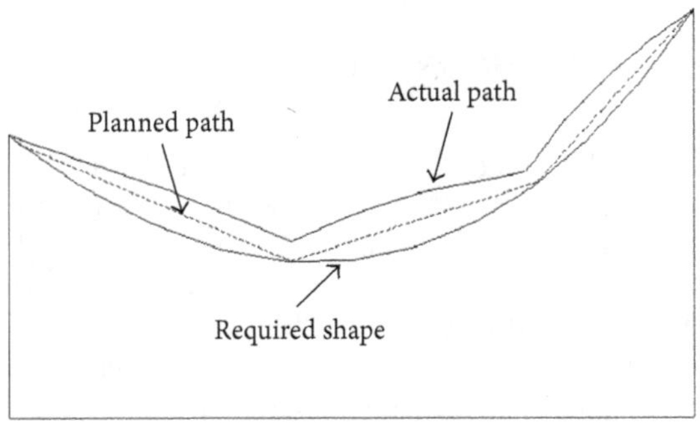

FIGURE 3: Deviation between planned path and actual path.

$V_c(t)$ between feed rate V_w and removal rate V_e, which can be expressed as

$$\sigma_t = \sum V_c(t_i)\,t = \sum \left(V_w(t_i) - V_e(t_i)\right) t. \qquad (1)$$

As shown in Figure 2, an oscillation of force appears at stage 2. This is due to the release of deformation which occurs when $V_w(t_i) \leq V_e(t_i)$. The oscillation force can be expressed as

$$\begin{aligned} F_o(t_i) &= K_s\sigma_t(t_{i-1}) * \sin(wt+\varphi) \\ \sigma_t(t_{i-1}) &= \sigma_t(t_i) - (V_e(t_i) - V_w(t_i))\,t, \end{aligned} \tag{2}$$

where $F_o(t_i)$ is the force caused by the deformation release; w and φ are the robotic system coefficients. Therefore, the force caused by deformation can be expressed as

$$F_s(t_i) = \begin{cases} K_s\sigma_t(t_{i-1}) & V_w(t_i) \geq V_e(t_i) \\ K_s\sigma_t(t_{i-1}) * \sin(wt+\varphi) & V_w(t_i) \leq V_e(t_i). \end{cases} \tag{3}$$

Compared with stage 1, the force at stage 2 is much smaller, while the feed rate decreases from V_w to 0. Based on this, an assumption can be made that the values of feed rate and deformation σ_t have significant influences on the force acting on cutter.

2.2. Robotic Cutting Dynamics. General cutting dynamic model can be expressed as

$$F_p(t_i) = M\ddot{X}_p(t_i) + C\dot{X}_p(t_i) + K_pX_p(t_i). \tag{4}$$

Considering the influence of deformation, the dynamic model in Cartesian frame can be written as

$$\begin{aligned} F(t_i) &= F_p(t_i) + F_s(t_i) \\ F(t_i) &= M\ddot{X}_p(t_i) + C\dot{X}_p(t_i) + K_pX_p(t_i) \\ &\quad + K_s\sigma_t(t_{i-1}), \end{aligned} \tag{5}$$

where $F = [F_x, F_y, F_z]$ represents the machining force, f_x is the force perpendicular to the chamfer surface, and f_y is the force tangential to the chamfer surface. Machining force consists of contact force in normal direction and friction force in tangent direction. Considering the fact that the machining quality in normal direction of workpiece is more important, this paper focuses on the control of machining deformation in normal direction. F_p represents the cutting force; F_s represent the force caused by deformation; M is the mass coefficient; C is the damping coefficient; $\sigma_t(t_{i-1})$ is the deformation at time t_{i-1}; K_p is the process stiffness; K_s is the static stiffness. According to preliminary experiment result, a preconclusion can be made that the value of K_s is much larger than the one of K_p. $X_p(t_i)$, $\dot{X}_p(t_i)$, and $\ddot{X}_p(t_i)$ are the position of machining tool and its first-order derivative and second-order derivative, where $X_p = [x_x, x_y, x_z]^T$ and x_x is the cutting position perpendicular to the chamfer surface which is regarded as machining depth $x_x = a_p$; x_y is the cutting position tangential to the chamfer surface. Considering the fact that the desired machining depth is a_p^*, (5) can be rewritten as

$$M\ddot{a}_p^*(t_i) + C\dot{a}_p^*(t_i) + K_pa_p^*(t_i) + K_s\sigma_x(t_{i-1}) = F_x \tag{6}$$

$$\begin{aligned} a_p^*(t_i) &= a_{sp}^*(t_i) - \sigma_x(t_{i-1}) \\ &= a_{sp}^*(t_i) - \sum_{1}^{i-1}\left(V_{wx}(t_j) - V_{ex}(t_j)\right)t, \end{aligned} \tag{7}$$

where V_{wx} and V_{ex} are the feed rate and removal rate component perpendicular to the chamfer surface and can be calculated by

$$\begin{aligned} V_{wx} &= |V_w|\sin\theta \\ V_{ex} &= |V_e|\sin\theta \end{aligned} \tag{8}$$

and θ is the machining angle. Define the machining force error perpendicular to the chamfer surface as $e_x = F_x - F_x'$, while the machining depth error caused by robotic deformation is $e_p = \sigma_x(t) = a_p(t) - a_p^*(t)$. According to the coupling relationship illustrated in (7), the desired machining depth $a_p^*(t)$ and machining force F_x' can be achieved by the modifications of θ, $|V_w|$, and $a_{sp}^*(t)$. The values of θ and $a_{sp}^*(t)$ in this research are constant sets and the focus of this paper is to adjust $|V_w|$ to reduce the force deviation e_x. An alternative solution to reduce deformation is to slow down feed rate V_w. The decrease of approaching velocity can lead to a decrease of machining efficiency. Since machining efficiency is one of the key elements needed to be considered, this paper aims to reduce initial overshoot and machining deformation on the premise that the machining efficiency is guaranteed.

3. Fuzzy-Sliding Mode Control Scheme

Define V_w as the control output; then

$$\dot{F} = f(F_x) + g(F_x)V_w, \tag{9}$$

where $g(F_x)$ is the gain coefficient of the control law; $f(F_x)$ is the uncertainties of robotic system. According to the analysis above, the machining force and deformation are affected by machining angle and feed rate. To deal with the complexity of machining force caused by the shift of machining angle, a sliding mode control scheme based on fuzzy rules is proposed in this part.

3.1. Sliding Mode Control Scheme for Machining Process. The advantage of sliding mode is to transform the nth order tracking problem into first-order stability problem. Define the sliding surface as

$$S = \left(\frac{d}{dt} + \lambda\right)e_x(t_i) = 0, \tag{10}$$

where λ is a positive constant, and the control law is

$$V_w = u = \dot{S} = -H_kS - H_c\,\mathrm{sgn}(S). \tag{11}$$

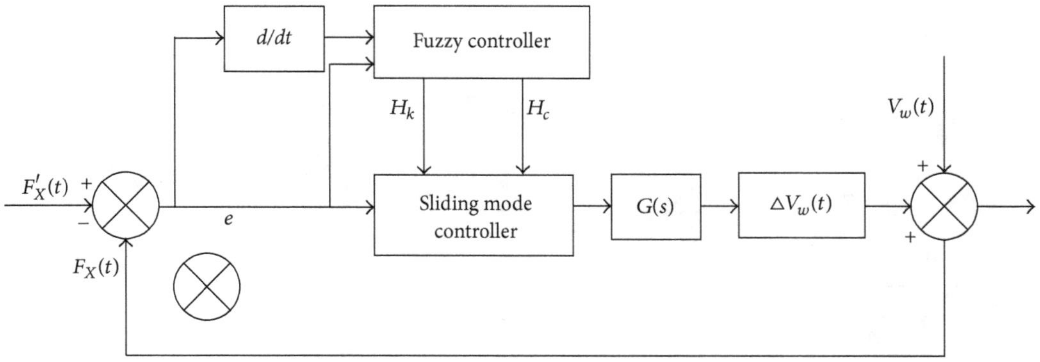

FIGURE 4: Fuzzy-sliding mode control scheme.

H_k and H_c are the gain coefficient of sliding mode control. When the system status achieves desired status within given precision ε at time t_i, there are

$$\begin{gathered} \left|e_x\left(t_i\right)\right| \leq \varepsilon, \\ S=0. \end{gathered} \tag{12}$$

Equation (10) can be rewritten as

$$\begin{gathered} \lambda e_x\left(t_i\right)+\dot{e}_x\left(t_i\right)=0 \\ \lambda e_x\left(t_i\right)+e_x\left(t_i\right)-e_x\left(t_{i-1}\right)=0 \\ \left|e_x\left(t_{i-1}\right)\right| \leq(\lambda+1) \varepsilon . \end{gathered} \tag{13}$$

Deduce

$$\left|e_x\left(t_{i-n}\right)\right| \leq(\lambda+1)^n \varepsilon . \tag{14}$$

In the same way, we can get

$$\begin{aligned} & \left|-g\left(F_x\right) H_c\left(t_i\right) \operatorname{sgn}\left(S\left(t_i\right)\right)\right| \leq(\lambda+1) \varepsilon \\ & \left(-g\left(F_x\right) H_k\left(t_{i-n}\right) S\left(t_{i-n}\right)\right. \\ & \quad \left.-g\left(F_x\right) H_c\left(t_{i-n}\right) \operatorname{sgn}\left(S\left(t_{i-n}\right)\right)\right)^n \leq(\lambda+1)^{n+1} \varepsilon . \end{aligned} \tag{15}$$

Lyapunov is used to ensure the global stability:

$$\begin{gathered} L_p=\frac{1}{2} S^2 \\ \dot{L}_p=\left(-H_k S-H_c \operatorname{sgn}(S)\right) S \leq 0 . \end{gathered} \tag{16}$$

Therefore, the gain coefficients H_c and H_k can be modified online during machining process when they are subject to (15) and (16).

3.2. Fuzzy-Sliding Mode Control Scheme. Fuzzy rules are used to determine the real-time value of H_c and H_k by evaluating the status of system. The inputs fuzzy variables are force deviation e_x and its first-order derivative $\dot{e}_x$, and their domains are $e_x=\{-10,-5,0,5,10\}$ and $\dot{e}_x=\{-10,-5,0,5,10\}$. The domains of H_c and H_k are $H_c=\{-2,-1,0,1,2\}$ and $H_k=\{-0.2,-0.1,0,0.1,0.2\}$. With use of Gaussian membership function and center of gravity method [12],

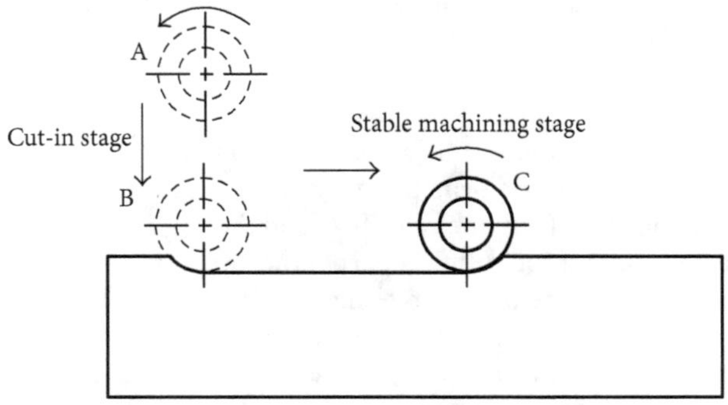

FIGURE 5: Machining path.

$$\begin{aligned} H_c\left(t_i\right) & =\frac{\sum_1^l H_{ci} \mu_i}{\sum_1^l \mu_i} \\ H_k\left(t_i\right) & =\frac{\sum_1^l H_{ki} \mu_i}{\sum_1^l \mu_i}, \end{aligned} \tag{17}$$

where $i=[1,2,\ldots,l]$; l is the number of fuzzy sets; μ_i is grade of membership. The control sketch is shown in Figure 4.

4. Robotic Machining Simulation and Experiments

4.1. Simulation. Before conducting machining experiments, simulations including open loop control, fuzzy-PID control, and fuzzy-sliding mode control are carried out to get a preview of the control scheme performance. The machining path for the simulation is shown in Figure 5, which can be classified as cut-into stage and stable machining stage. At cut-into stage, the cutter starts its path at position A and moves towards workpiece at the feed rate V_w perpendicular to the workpiece. After reaching position B, the cutter shifts its machining path which is parallel to the workpiece surface. The simulation parameters are original feed rate V_w = 2 mm/s; machining depth a_p = 2 mm; abrasive radius R_s = 6 mm; force target is 5 N.

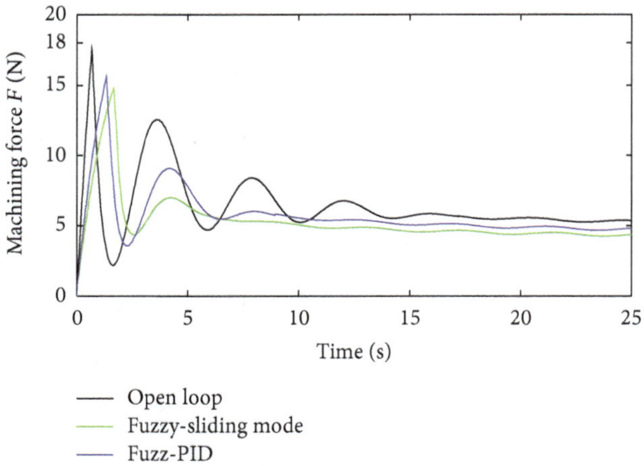

FIGURE 6: Control performance comparisons (fuzzy-PID, fuzzy-sliding mode, and open loop).

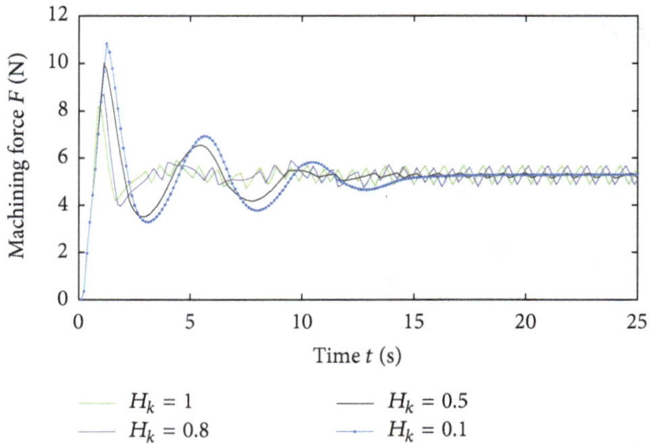

FIGURE 7: Impact of sliding mode control gain on cutting force.

The simulation results are shown in Figure 6. According to (7) and (8), the deformation value in x direction, named as σ_x, is impact by feed rate value $|V_w|$ and the angle θ. When the machining path of cutter is perpendicular to the workpiece surface, the value of feed rate has a significant influence on the robotic deformation in x direction. As shown in Figure 6, the peak value of machining force appears when the cutter reaches position B. When no control is conducted on feed rate, the force peak value of open loop control is around 19 N. The force peak values of fuzzy-PID and fuzzy-sliding mode control are around 15 N, 4 N lower than the one of open loop control, which is in compliance with (7). The machining force then decreases and oscillates for a while. The oscillation period of open loop control is around 14 seconds which is 5 seconds more than the one of fuzzy-PID control and almost 8 seconds more than the one of fuzzy-sliding mode control. Besides, the oscillation amplitude of fuzzy-sliding mode control is around 3 N, 2 N less than the one of fuzzy-PID control. Therefore, compared with the other two control schemes, fuzzy-sliding mode control has the advantage in reducing the period and amplitude of oscillation and the peak value of machining force. The machining depth of the simulation is 2 mm, which is similar to machining depth of rough machining process. And semifinishing and finishing machining process will be conducted after rough machining process. Considering the fact that the aim of proposed control scheme is to reduce the scale and time period of vibration and deformation, and the vibration and deformation could not be thoroughly eliminated, the time period of initial oscillation is acceptable.

Another simulation is performed to evaluate the impact of the gain coefficients H_c and H_k on the performance of control scheme. According to (9) and (11), the change of feed rate is influenced by H_c and H_k, which can eventually affect the change of machining force. As shown in Figure 7, the peak value of machining force decreases from 11 N to 8 N as the value of H_k increases from 0.1 to 1. A reduction appears in the period and amplitude of oscillation when H_k increases. However, while H_k is larger than 0.5, an obvious chatter shows up in stable machining stage, and the amplitude of it increases as the value of H_k increases, which can be illustrated by (14). Similar performance appears in the evaluation of H_c.

Therefore, a preconclusion can be made that fuzzy-sliding mode control has the advantage in reducing the peak value of machining force, oscillation period, and amplitude. The modification of gain coefficients H_c and H_k can modify the rate of convergence. However, due to the influence of H_c and H_k on the chattering amplitude, the increase of both gain coefficients should be bounded.

4.2. Machining Experiments Based on Fuzzy-Sliding Mode Control

4.2.1. Robotic Machining System. The robotic machining system used for experiment is composed of Cartesian coordinate robot, expended rotating platform, force sensor, machining tool, and workpiece. As shown in Figure 8, the robot, consisting of Cartesian coordinate robot and expended rotating platform, has an open loop and serial structure, which is similar to general industrial robots. Therefore, it can completely illustrate the rigidity behavior of general industrial robots. The force sensor and machining tool are installed on the end of axis Z, while the workpiece is installed on the rotating platform.

4.2.2. Experiments. The experiments conducted on plane part of the workpiece and the parameters are shown in Figure 9. The machining path is the same as the one conducted in the simulations. The machining parameters are original feed rate $V_w = 2$ mm/s; machining depth $a_p = 0.8$ mm; abrasive radius $R_s = 6$ mm; abrasive width $L_s = 5.5$ mm; rotational speed of cutter is 4000 r/min. The aim of the machining force is 4 N. Three kinds of measurement methods are used in this paper, including force sensor, three-coordinate measuring system, and roughness tester.

According to (6) and (7) and simulation (1), the machining force error is significantly impacted by the feed rate value $|V_w|$ while the machining path is perpendicular to the workpiece surface. As shown in Figure 10, a considerable increase in machining force appears in cut-in stage and

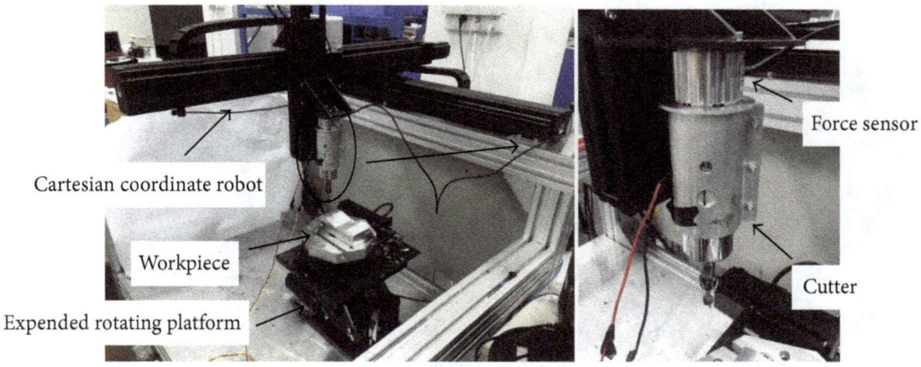

FIGURE 8: Architecture of robotic cutting system ((1) *XYZ* Cartesian coordinate robot; (2) rotating platform; (3) six-axis force sensor; (4) cutting tool; (5) workpiece).

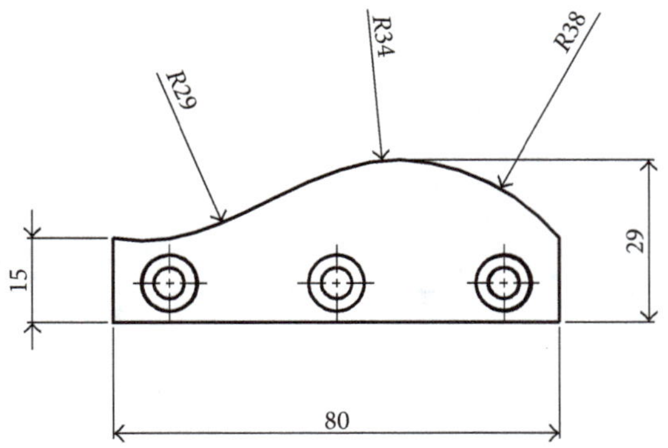

FIGURE 9: Workpiece (aluminum alloy).

reaches the peak value of 17 N at the 3rd second. After that the machining force decreases and oscillates with amplitude of 7 N which is caused by the release of deformation. The stable machining stage start at 9th second and the fluctuation of chatter is around 5 N, while the average force is 4.5 N. The machining force value at cut-in stage is higher than the one at stable machining stage which implies that the change of machining angle has a significant influence on the change of machining force in x direction. This performance is in compliance with (5) and (6). The dimensional inspections are conducted on a three-coordinate measuring system (Global advantage 05.07.05). The actual machining depth is around 0.75 mm, and the amplitude of fluctuation is around 0.06 mm. The roughness of machining surface is Ra 1.821 μm.

The experiment result under fuzzy-PID control is shown in Figure 11. According to (5) and (6), the modification of feed rate $|V_w|$ can change the machining force value. Therefore the peak value is around 14 N, 3 N less than the one of open loop control. The machining force oscillates with amplitude of 6 N and achieves stable machining at 7th second, 2 seconds less than one of open loop control. The fluctuation of chatter is around 3 N which is more smooth than the one of open loop control, and the average force is 4 N. The actual machining depth is around 0.75 mm and the amplitude of fluctuation is around 0.04 mm. The roughness of machining surface is Ra 1.432 μm.

According to (5) and (6), the modification of feed rate $|V_w|$ can change the machining force value. Therefore, the experiment uses fuzzy-sliding mode control to modify the feed rate according to the real-time status of system in order to smooth the fluctuation. The experiment result under fuzzy-sliding mode control is shown in Figure 12. The peak value is around 14 N, which is almost the same as the one of fuzzy-PID control. The machining force oscillates with amplitude of 2 N, less than the ones of other two control schemes, and achieves stable machining at 5th second. Compared with fuzzy-PID control, fuzzy-sliding mode control achieves a reduction of 2 seconds in oscillation period and a reduction of 4 N in oscillation amplitude. The chatter amplitude decreases from 3 N to 2 N. The reductions in the amplitude and period of oscillation result in a smoother force curve which implies verifying the validity of (8) and (10), indicating that a smooth force fluctuation can be achieved with the use of fuzzy-sliding control scheme. The actual machining depth is around 0.75 mm and the amplitude of fluctuation is around 0.03 mm. The roughness of machining surface is Ra 1.292 μm.

5. Conclusion

With an open loop and serial structure, the stiffness of robot is low and easily results in deformation and vibration during machining process. The generation of deformation and the robotic machining mechanism are explored in this study to construct the dynamic model. Due to the complexity of robotic machining, simple control scheme is not sufficient enough to manage the force fluctuation that occurs during machining. With the view of this, a sliding mode control scheme based on fuzzy rule is proposed to reduce the peak value of machining force, the oscillation amplitude and period, and the amplitude of chatter. The simulations and relevant experiments are then conducted to evaluate the performance of fuzzy-sliding mode control. The results show that fuzzy-sliding mode control has the advantages in reducing the oscillation period and amplitude compared with fuzzy-PID control, making the reductions of 4 N and 2 seconds. The force peak values of fuzzy-PID control and fuzzy-sliding mode control are lower than the one of open loop control, achieving a reduction of 4 N.

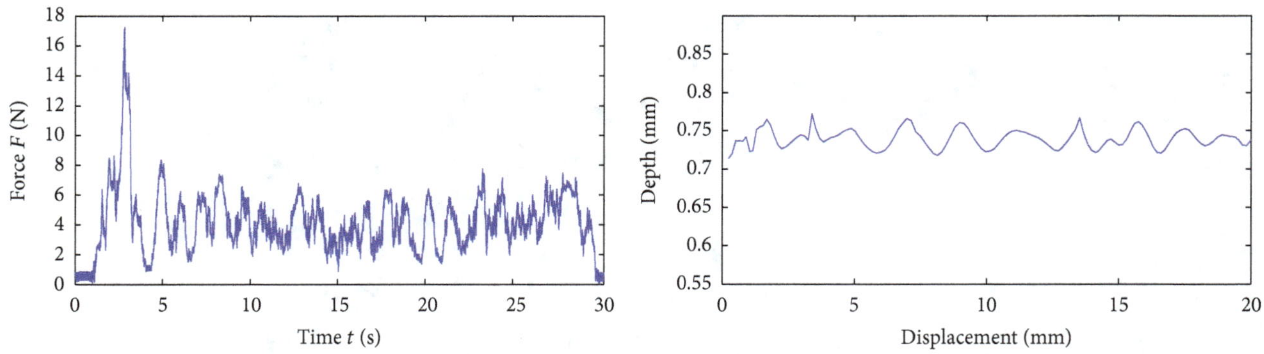

Figure 10: Experiment result of robotic machining under open loop control.

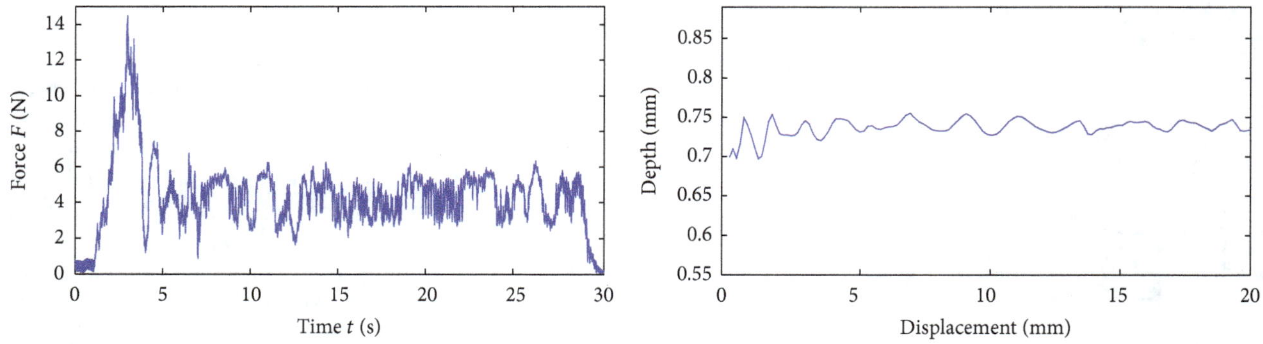

Figure 11: Experiment result of robotic machining under fuzzy-PID control.

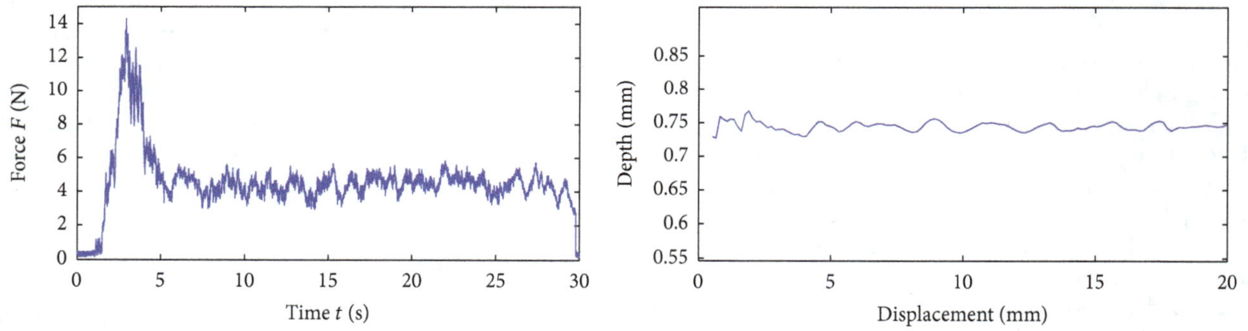

Figure 12: Experiment result of robotic machining under fuzzy-sliding mode control.

Current studies are mostly focused on system integration and force control, yet few of them concentrated on feed rate control. This study fills the gap in this field and provides a reference to later research.

Conflicts of Interest

The authors declare that they have no conflicts of interest.

Acknowledgments

This project is sponsored by National Science and Technology Major Project of China (no. 20152X04005006), Science and Technology Planning Project of Guangdong Province, China (nos. 2014B090921004, 2014B090920001, and 2015B010918002), and Science and Technology Major Project of Guangzhou, China (no. 201505031617224).

References

[1] Y. Chen and F. Dong, "Robot machining: recent development and future research issues," *International Journal of Advanced Manufacturing Technology*, vol. 66, no. 9–12, pp. 1489–1497, 2013.

[2] L. Ding, H. Wu, Y. Yao, and Y. Yang, "Dynamic model identification for 6-DOF industrial robots," *Journal of Robotics*, vol. 2015, Article ID 471478, pp. 1–9, 2015.

[3] A. Olabi, R. Béarée, O. Gibaru, and M. Damak, "Feedrate planning for machining with industrial six-axis robots," *Control Engineering Practice*, vol. 18, no. 5, pp. 471–482, 2010.

[4] F. Domroes and C. Krewet, "Application and analysis of force control strategies to deburring and grinding," *Modern Mechanical Engineering*, vol. 3, no. 2, pp. 11–18, 2013.

[5] Y. Karayiannidis and Z. Doulgeri, "Adaptive control of robot contact tasks with on-line learning of planar surfaces," *Automatica*, vol. 45, no. 10, pp. 2374–2382, 2009.

[6] K. Kiguchi and T. Fukuda, "Intelligent position/force controller for industrial robot manipulators—application of fuzzy neural networks," *IEEE Transactions on Industrial Electronics*, vol. 44, no. 6, pp. 753–761, 1997.

[7] M. Žarko and D. Vlastimir, "Hybrid industrial robot compliant motion control," *Facta Universitatis. Series Automatic Control and Robotics*, vol. 7, no. 1, pp. 99–110, 2008.

[8] S.-I. Matsuoka, K. Shimizu, N. Yamazaki, and Y. Oki, "High-speed end milling of an articulated robot and its characteristics," *Journal of Materials Processing Technology*, vol. 95, no. 1–3, pp. 83–89, 1999.

[9] E. Abele, M. Weigold, and S. Rothenbücher, "Modeling and identification of an industrial robot for machining applications," *CIRP Annals - Manufacturing Technology*, vol. 56, no. 1, pp. 387–390, 2007.

[10] H. Zhang, J. Wang, G. Zhang et al., "Machining with flexible manipulator: toward improving robotic machining performance," in *Proceeding of the International Conference on Advanced Intelligent Mechatronics*, pp. 1127–1132, Calif, USA, 2005.

[11] J. Wang, H. Zhang, and T. Fuhlbrigge, "Improving machining accuracy with robot deformation compensation," in *Proceedings of the IEEE/RSJ International Conference on Intelligent Robots and Systems (IROS '09)*, pp. 3826–3831, October 2009.

[12] Y. Bai, N. Guo, and G. Agbegha, "Fuzzy interpolation and other interpolation methods used in robot calibrations," *Journal of Robotics*, vol. 2012, no. 2, Article ID 376293, pp. 69–74, 2012.

Gas Concentration Prediction based on the Measured Data of a Coal Mine Rescue Robot

Xiliang Ma and Hua Zhu

School of Mechanical and Electrical Engineering, China University of Mining and Technology, Xuzhou, Jiangsu 221116, China

Correspondence should be addressed to Xiliang Ma; xlma818@sina.com

Academic Editor: Giovanni Muscato

The coal mine environment is complex and dangerous after gas accident; then a timely and effective rescue and relief work is necessary. Hence prediction of gas concentration in front of coal mine rescue robot is an important significance to ensure that the coal mine rescue robot carries out the exploration and search and rescue mission. In this paper, a gray neural network is proposed to predict the gas concentration 10 meters in front of the coal mine rescue robot based on the gas concentration, temperature, and wind speed of the current position and 1 meter in front. Subsequently the quantum genetic algorithm optimization gray neural network parameters of the gas concentration prediction method are proposed to get more accurate prediction of the gas concentration in the roadway. Experimental results show that a gray neural network optimized by the quantum genetic algorithm is more accurate for predicting the gas concentration. The overall prediction error is 9.12%, and the largest forecasting error is 11.36%; compared with gray neural network, the gas concentration prediction error increases by 55.23%. This means that the proposed method can better allow the coal mine rescue robot to accurately predict the gas concentration in the coal mine roadway.

1. Introduction

Coal mine accidents have occurred frequently in our country. Coal mine accidents caused by a gas outburst, gas explosion, and gas suffocation accounted for 70%~80% of all coal mine accidents [1], which not only cause great losses to the coal mining enterprises but also bring huge economic losses and psychological harm to the enterprise employees, family members, and the country. After a coal mine accident, the timeliness of an emergency rescue is very important. Rescue and relief workers have a small window of opportunity to enter the underground scene of the accident for rescue purposes. However, a complex and dangerous coal mine environment is a great security threat for rescue workers. With the development of science and technology, the mine rescue robot can be used for coal mine disaster relief to detect the situation within the disaster areas, to find personnel in distress, and to provide decision-making support for the rescue.

Gas disaster accidents often destroy roadway ventilation facilities, damage ventilation systems, and cause methane to flow from the coal wall and the goaf, most likely forming gas accumulation in the disaster areas. Furthermore, there is the risk of elevated temperature, coal dust, and other flammable and explosive gases produced from coal dry distillation, further increasing the risk of gas explosion. Not extinguishing the fire or new flames produced by secondary factors may cause a secondary gas explosion at any time. To ensure that the coal mine rescue robot can perform the detection and search and rescue mission, measures must be taken to avoid secondary disasters such as gas explosion caused by concentration. Therefore, predicting gas concentrations early has very important significance.

A comprehensive analysis of domestic and foreign research on gas concentration forecast shows that, at present, the research mainly focuses on the prediction of gas emissions, gas concentration prediction algorithms or methods, and so on. Özgen Karacan [2] proposes a principle component analysis and artificial neural network-based approach to predict the ventilation methane emission rates of US longwall mines. Runqiua et al. [3] put forward a prediction gas emission method based on self-organizing data mining, and Hu et al. [4] established prediction models of the coalbed methane content using quantification theory based on the limited data obtained from the exploratory bores. The prediction method and algorithm of the gas concentration are compared with the prediction model [5] or an improved prediction model [6] to improve the accuracy of prediction. By using an effective prediction method or algorithm [7–9] to validate the practicality of the prediction [10] and using the combination forecasting method or algorithm [11–14] to accurately predict the concentration range, the trend of the gas concentration can be tracked effectively. However, there are some problems in these studies. Predicting the emission quantity is mostly based on data statistics or data mining. Searching for a gas change law and predicting the amount of gas emission in a mine or a range often require a large amount of data. Sometimes, data from the past few decades are needed, resulting in a large amount of forecast work. Furthermore, among the research in the gas concentration prediction method or algorithm, the accuracy and effectiveness of the method or algorithm are of greater concern, and the practicability of the algorithm is seldom considered. If the gray system and neural network are combined to form the gray neural network, it can take full advantage of the gray system's data modeling and the neural network's nonlinear mapping features. With the help of the gray mode, the gray neural network uses new data after the accumulation of new data to model. To a certain extent, this process weakens the randomness of the original data, and it is easy to find the transformation of the data. This approach is suitable when partial information is known, some information is unknown, there is a small sample, there is poor information, or there is uncertainty in the system. The establishment of a gray neural network can solve the problem of insufficient data quantity for gas concentration, but it also has its own inherent defects, such as randomness in the weight and threshold in the initialization process, falling into the local optimum during network evolution, and the inability to further modify the relevant parameters. Therefore, to improve the accuracy and applicability of gray neural network forecasts, the gray neural network is combined with the optimization algorithm. Currently, more common optimization algorithms include the particle swarm algorithm, ant colony algorithm, and genetic algorithm. The particle swarm optimization algorithm's main problem is that it is easy to produce premature convergence (especially when dealing with complex multimodal search problems) and poor local search optimization capabilities. The PSO algorithm is at the local minimum, which is mainly due to the loss of diversity in the search space. The ant colony algorithm is a typical probabilistic algorithm. The parameter settings in the ant colony algorithm are usually determined by experiments, which are closely related to human experience, and it is difficult to optimize the performance of the algorithm. The genetic algorithm can find the optimal solution of the problem using probability, but it does not take advantage of the excellent evolution immature subset of the information provided in the optimization process. In addition, there are slow convergence and premature convergence defects. The quantum genetic algorithm is introduced into the genetic optimization algorithm, which makes the genetic operation more effective. A quantum bit was introduced into the chromosome coding in the quantum state vector based on the narration. Thus, a chromosome can express the superposition of multiple states, reduce the size of the population, and increase the diversity of the population. At the same time, the application of the concept of a quantum gate can promote the convergence of the algorithm, and the quantum gate mutation operation can be used.

The phase characteristics of quantum entanglement can break the traditional genetic algorithm to produce several individual concept hybrid offspring. Using the quantum entanglement of quantum hybrid design can exchange information in the entire population, making it easy to find great individual populations. Hybrid quantum entanglement with quantum information can be communicated throughout the population, making it easy to find great individual populations.

This paper combines the actual work requirements of the coal mine rescue robots with a small amount of measurement data of the robot, and the factors affecting the gas concentration (such as temperature, wind speed, and distance) are considered. The gas concentration 10 meters in front of the coal mine rescue robot is predicted using the gray neural network variation for eigenvalues in uncertain systems. The prediction accuracy of the gray neural network is improved by optimizing the quantum genetic algorithm parameters of the gray neural network. The safety of the coal mine rescue robot's task execution is ensured by avoiding gas explosion accidents because of either the coal mine rescue robot itself or the ground friction sparks. This provides an effective basis for the rescue decision.

2. Model Establishment

2.1. Gray Neural Network. Grey system theory is founded by Professor Deng Julong in 1982, an expansion theory from the method of white-black-gray box, which is a new research method for the issue with a few new original data and uncertain features. A gray neural network is the combination of gray theory and a neural network, which can predict the development and change of the characteristic value of the uncertain system. The original sequence of the system $x_t^{(0)}$ $(t = 1, 2, \ldots, N - 1)$, after a cumulative sequence gives $x_t^{(1)}$, and its exponential growth can be used as a function or differential equation for data fitting and prediction. To facilitate representation, redefine the original sequence $x(t)$ after sequence x_t^1 is represented as a cumulative sequence, and the result is expressed as a forecast.

The differential predictive model of the parameters in a gray neural network is

$$\frac{dy}{dt} + ay_1 = b_1y_2 + b_2y_3 + \cdots + b_{n-1}y_n, \tag{1}$$

where $y_1, y_2, \ldots, y_n$ are the input parameters for the system; y_1 is the system output parameter; and $a, b_1, b_2, \ldots, b_n$ are the

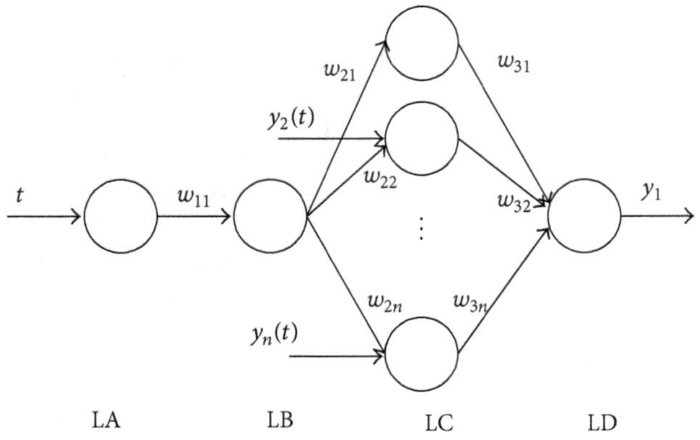

FIGURE 1: Gray neural network structure topology.

coefficient differential equations. The time response formula of (1) is

$$\begin{aligned} z(t) &= \left(y_1(0) - \frac{b_1}{a}y_2(t) - \frac{b_2}{a}y_3(t) - \cdots - \frac{b_{n-1}}{a}y_n(t)\right) \\ &\quad \cdot e^{-at} + \frac{b_1}{a}y_2(t) + \frac{b_2}{a}y_3(t) + \cdots + \frac{b_{n-1}}{a}y_n(t). \end{aligned} \tag{2}$$

Let $d = (b_1/a)y_2(t) - (b_2/a)y_3(t) - \cdots - (b_{n-1}/a)y_n(t)$. Equation (2) can be simplified to

$$\begin{aligned} z(t) &= \left((y_1(0) - d)\cdot\frac{e^{-at}}{e^{-at}+1} + d\cdot\frac{1}{e^{-at}+1}\right) \\ &\quad \cdot\left(e^{-at}+1\right) \\ &= \left((y_1(0) - d)\cdot\left(1 - \frac{e^{-at}}{e^{-at}+1}\right) + d\cdot\frac{1}{e^{-at}+1}\right) \\ &\quad \cdot\left(e^{-at}+1\right) \\ &= \left((y_1(0) - d) - y_1(0)\cdot\frac{e^{-at}}{e^{-at}+1} + 2d\cdot\frac{1}{e^{-at}+1}\right) \\ &\quad \cdot\left(e^{-at}+1\right). \end{aligned} \tag{3}$$

Mapping (3) to the BP neural network gives the gray neural network with N input parameters and 1 output parameter, and the network topology is shown in Figure 1.

In the graph, t is the sequence number, and $y_2(t), y_3(t), \ldots, y_n(t)$ are the network input parameters, $w_{21}, w_{22}, \ldots, w_{2n}$ and $w_{31}, w_{32}, \ldots, w_{3n}$ are the network weight values, and y_1 is the network forecast value. LA, LB, LC, and LD represent the four layers in the gray neural network.

Let $2b_1/a = u_1, 2b_2/a = u_2, \ldots, 2b_{n-1}/a = u_{n-1}$, so the network weights can be expressed as follows:

$$\begin{aligned} w_{11} &= a; \\ w_{21} &= -y_1(0); \\ w_{22} &= u_1, \ldots, w_{2n} = u_{n-1}; \\ w_{31} &= w_{32} = \cdots = w_{3n} = 1 + e^{-at}. \end{aligned} \tag{4}$$

The LD layer output node threshold is

$$\theta = \left(1 + e^{-at}\right)\left(d - y_1(0)\right). \tag{5}$$

The gray neural network training process is as follows:

(1) Initialize the gray neural network and the parameters.

(2) Each output is calculated for each input sample:

LA layer: $a = w_{11}t$.

LB layer: $b = f(w_{11}t) = 1/(1 + e^{-w_{11}t})$.

LC layer: $c_1 = bw_{21}$, $c_2 = y_2(t)bw_{22}, \ldots, c_n = y_n(t)bw_{2n}$.

LD layer: $d = w_{31}c_1 + w_{32}c_2 + \cdots + w_{3n}c_n - \theta_{y1}$.

(3) Calculate the error of the predicted output and the actual value, and the error formula of each layer is as follows:

LD layer: $\delta = d - y_1(t)$.

LC layer: $\delta_1 = \delta(1+e^{-w_{11}t})$, $\delta_2 = \delta(1+e^{-w_{11}t}), \ldots, \delta_n = \delta(1+e^{-w_{11}t})$.

LB layer: $\delta_{n+1} = (1/(1 + e^{-w_{11}t}))(1 - 1/(1 + e^{-w_{11}t}))(w_{21}\delta_1 + w_{22}\delta_2 + \cdots + w_{2n}\delta_n)$.

(4) Adjust the threshold formula in accordance with the following error:

LB to LC layer: $w_{21} = -y_1(0), w_{22} = w_{22} - u_1\delta_2 b, \ldots, w_{2n} = w_{2n}u_{n-1}\delta_n$.

LA to LB layer: $w_{11} = w_{11} + at\delta_{n+1}$;

Adjust the threshold: $\theta = (1 + e^{-w_{11}t})((w_{22}/2)y_2(t) + (w_{23}/2)y_3(t) + \cdots + (w_{2n}/2)y_n(t) - y_1(0))$.

(5) Determine whether the set error is reached. If it is reached, the training is over, and if not, return to step (2).

2.2. Quantum Genetic Algorithm

2.2.1. Quantum Genetic Algorithm Coding. The common coding methods of the quantum genetic evolutionary algorithm are binary coding, decimal coding, and symbol coding. A novel coding scheme based on quantum bit is used in the quantum evolutionary strategy. Its most essential feature is the full use of superposition and the coherence of quantum states, as well as the entangled states of the quantum bits. The smallest unit of information is a quantum bit, or a qubit. In the quantum genetic algorithm, a qubit may be in $|1\rangle$ or $|0\rangle$, in an intermediate state of $|1\rangle$ and $|0\rangle$, and in different superpositions of $|1\rangle$ and $|0\rangle$. Therefore, a qubit state is expressed as follows:

$$|\psi\rangle = \alpha|0\rangle + \beta|1\rangle . \quad (6)$$

In the formula, α and β can be complex, and they represent the probability amplitude of the corresponding state and satisfy the condition $\alpha^2 + \beta^2 = 1$.

A system with m quantum bits can be expressed as 2^m states, namely,

$$\begin{bmatrix} \alpha_1, \alpha_2, \ldots, \alpha_m \\ \beta_1, \beta_2, \ldots, \beta_m \end{bmatrix}. \quad (7)$$

In the formula, $\alpha_i^2 + \beta_i^2 = 1, i = 1, 2, \ldots, m$.

2.2.2. Quantum Variation. By the basic assumptions of quantum mechanics, n-bit quantum registers can be in 2^n ground state of coherent superposition state $|\psi\rangle$; that is to say, it can simultaneously represent the 2^n number. The relationship between superposition state and ground state can be expressed as $|\psi\rangle = \sum_i C_i|\psi_i\rangle$, where C_i is the probability amplitude of the state ψ_i, and $|C_i|^2$ express the probability of ψ collapse to the ground state ψ_i, when subjected to quantum computing systems and measuring instruments observation entangled.

Quantum coherence [15] is an important mechanism for quantum computing, using a special link between the quantum states; other quantum states can be launched from one or more quantum states. One of the main principles of quantum computing is to make each ground state of superposition state change the relative phase between them through quantum gate interference effect. For example, superposition state and quantum gate are given as follows:

$$|\psi'\rangle = \frac{2}{\sqrt{5}}|0\rangle + \frac{1}{\sqrt{5}}|1\rangle = \frac{1}{\sqrt{5}}\begin{pmatrix} 2 \\ 1 \end{pmatrix},$$
$$U = \frac{1}{\sqrt{2}}\begin{pmatrix} 1 & 1 \\ 1 & -1 \end{pmatrix}. \quad (8)$$

Quantum gate action in superposition state as a result is as follows:

$$|\psi'\rangle = \frac{3}{\sqrt{10}}|0\rangle + \frac{1}{\sqrt{10}}|1\rangle . \quad (9)$$

From (9) it can be seen that the probability amplitude of ground state $|1\rangle$ decreases, while the probability amplitude of $|0\rangle$ increases. The quantum system is coherent when its ground state is in linear superposition state. The linear superposition of coherent system will disappear when interacting with the surrounding environment, the process called decoherence.

Another important mechanism for quantum computing is quantum entanglement. The so-called entanglement refers to some states existing in two interaction subsystems; it cannot be represented as tensor product of two subsystems' state but rather as some entanglement form of the subsystem state. Some certain state of an entangled quantum system does not correspond to the state of the subsystem; certain correlation exists between the various subsystems. For example, $|00\rangle + |11\rangle$ cannot be expressed as a tensor product of two subsystems' state. Otherwise, there must be four numbers a_1, a_2, b_1, and b_2, satisfying

$$(a_1|0\rangle + b_1|1\rangle) \otimes (a_2|0\rangle + b_2|1\rangle) = |00\rangle + |11\rangle . \quad (10)$$

Furthermore,

$$(a_1|0\rangle + b_1|1\rangle) \otimes (a_2|0\rangle + b_2|1\rangle)$$
$$= a_1a_2|00\rangle + a_1b_2|01\rangle + a_2b_1|10\rangle + b_1b_2|11\rangle . \quad (11)$$

Suppose that (11) holds; then get $a_1b_2 = 0$ and $a_2b_1 = 0$, which could indicate $a_1a_2 = 0$ or $b_1b_2 = 0$. Obviously, (10) and (11) should not be established at the same time because of its self-contradiction. Therefore, in the quantum system, the entire system cannot use quantum state of each qubit but is described by entangled quantum states. The operation of a few of entangled qubits will not only change the state of the qubits but also change the other qubits in which they are entangled.

The most essential feature of the quantum genetic algorithm is making full use of the superposition and coherence of quantum states, as well as the entangled states of the quantum bits. The quantum unitary transformation process applies the information processing process to the classical genetic algorithm. Using the quantum bit coding chromosome, the evolution of the population is realized through quantum mutation. A rotating quantum gate is a type of quantum gate that evolves by the rotation of the quantum gates. The information of the best individual can be added to the mutation to guide the evolution and speed up the convergence of the algorithm. The adjustment operation of the quantum rotation gate is shown in Table 1.

TABLE 1: Adjustment strategy of the quantum rotation gate.

x_i	b_i	$f(x) < f(b)$	θ	$S(\alpha_i, \beta_i)$			
				$g > 0$	$g < 0$	$\alpha_i = 0$	$\beta_i = 0$
0	0	F	0	0	0	0	0
0	0	T	0	0	0	0	0
0	1	F	0	0	0	0	0
0	1	T	0.05π	−1	+1	±1	0
1	0	F	0.01π	−1	+1	±1	0
1	0	T	0.025π	+1	−1	0	±1
1	1	F	0.005π	+1	−1	0	±1
1	1	T	0.025π	+1	−1	0	±1

x_i and b_i, respectively, refer to solution x and binary bits corresponding to the ith quantum bit of the current best individual B. $f(x)$ is the fitness function. The variation of the rotation angle controls the convergence speed of the algorithm for the direction of the rotation angle to ensure the algorithm's convergence.

The rotational quantum gate ensures that the algorithm quickly converges into a chromosome with higher fitness. When $x_i = 0$, $b_i = 0$, $f(x) \geq f(b)$, to make the

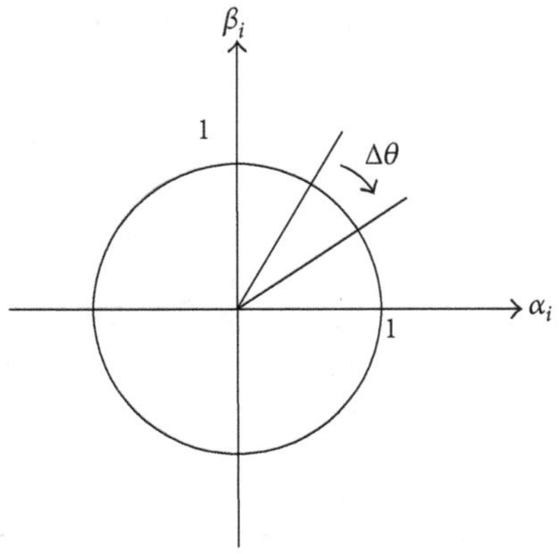

FIGURE 2: Quantum spin in the revolving door program.

TABLE 2: Chromosomes before cross-interference.

Chromosome	Genetic factor				
1	A(1)	A(2)	A(3)	A(4)	A(5)
2	B(1)	B(2)	B(3)	B(4)	B(5)
3	C(1)	C(2)	C(3)	C(4)	C(5)
4	D(1)	D(2)	D(3)	D(4)	D(5)
5	E(1)	E(2)	E(3)	E(4)	E(5)

current converge to a higher degree of adaptation of the chromosome, $|\alpha_i^t|^2$ should be larger. Therefore, if (α_i, β_i) are in the first and fourth quadrants, θ should rotate clockwise, and if (α_i, β_i) are in the second and third quadrants, θ should rotate counterclockwise, as shown in Figure 2.

2.2.3. Quantum Cross. The optimal value is often not precisely known in the actual data optimization process. The optimal value established in the optimization process often already counts the extreme value out, thus causing the system to fall into the local optimum. Quantum crossover is an effective method to prevent the system from the local optimal solution. The quantum crossover evolution strategy is an effective means to prevent the system from falling into local optima. In the genetic algorithm, the crossover is a means of searching the optimal solution, and its role is to achieve structural information exchange between individuals. According to the characteristics of the prediction of the gas concentration, the method of cross-interference is a full cross-interference in this paper [16], which uses quantum coherence properties and constructed a new crossover for ordinary chromosome. All the chromosomes in the population are involved in the crossover operation. Suppose that there are initially five chromosomes and that each chromosome has five genes, as shown in Table 2. Take chromosome 1 from the first genetic factor, chromosome 2 from the second genetic factor, chromosome 3 from the third genetic factor, chromosome 4 from the fourth genetic factor, and chromosome 5 from the fifth genetic factor. After the whole interference crossover, the first chromosome forms a new chromosome: A(1)-B(2)-C(3)-D(4)-E(5), as shown in Table 3.

TABLE 3: Chromosomes after cross-interference.

Chromosome	Genetic factor				
1	A(1)	B(2)	C(3)	D(4)	E(5)
2	B(1)	C(2)	D(3)	E(4)	A(5)
3	C(1)	D(2)	A(3)	A(4)	B(5)
4	D(1)	E(2)	B(3)	B(4)	C(5)
5	E(1)	A(2)	C(3)	C(4)	D(5)

The purpose of this cross-interference is to fully exploit the information of all chromosomes in the population to improve the locality and one-sidedness. This method can also be applied to the evolution of the common chromosome. When there is premature population evolution, it can process new impetus into evolutionary processes by producing a new individual.

2.2.4. Fitness Function. A fitness function is the evaluation criterion of the algorithm to the individual's "good and bad," and its value is positive.

This paper intends to use the quantum genetic algorithm optimization gray neural network parameters to predict the roadway gas concentration. Therefore, the error of the gray neural network's prediction value of the training data and its actual value are used as the fitness function as follows:

$$Q = \text{sum}\left(\left|O_{Pi} - O_{Ri}\right|\right). \tag{12}$$

O_{Pi} is used to predict the gas concentration, and O_{Ri} is used for the actual gas concentration value.

3. Experimental Analysis

3.1. Experimental Process. The coal mine rescue robot and the related sensor position are shown in Figure 3. The mine rescue robot is the dual arm robot. Position 1 is connected internally with the photoelectric encoder to detect the robot's walking distance and moving speed. Position 2 is the robot's power supply and drive unit. Position 3 is the control unit of the robot, where the instruction of the host computer is processed and the robot's state and the environmental information are collected and processed. Position 4 is a multiparameter detector, which can simultaneously detect methane, carbon dioxide, carbon monoxide, oxygen, temperature, humidity, and other parameters. Position 5 is the wind speed and direction sensors.

The gas concentration prediction experiments based on the coal mine rescue robot-measured data are carried out in the coal mine tunnel. The coal mine rescue robot's walking speed is 1 m/s, and after every 0.5 meters, all sensor data are uploaded (but also according to instructions to the upload data). The core of the robot control unit is the embedded industrial computer, which is powerful at processing data.

The main factors influencing the gas flow are air flow motion, diffusion motion caused by concentration gradient, and floating motion caused by density difference, and so on, while in the tunnel it is mainly affected by the air flow motion and diffusion motion. Gas diffusion is the migration form

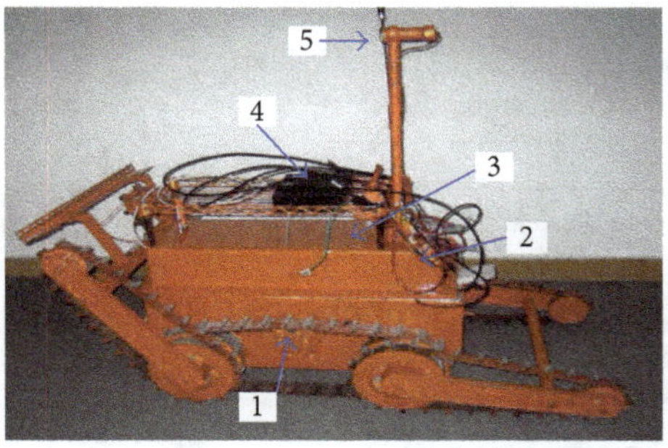

FIGURE 3: Coal mine rescue robot and the related sensor location map. 1, the displacement sensor; 2, power supply and drive unit; 3, the control unit; 4, multiparameter detector; and 5, the wind speed and direction sensors.

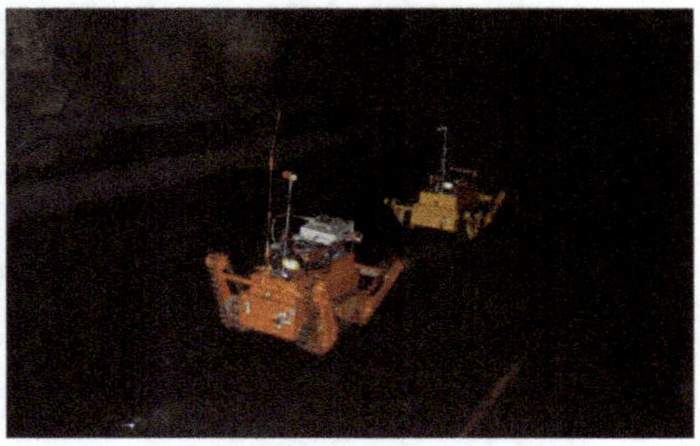

FIGURE 4: Coal mine rescue robot walking on the roadway.

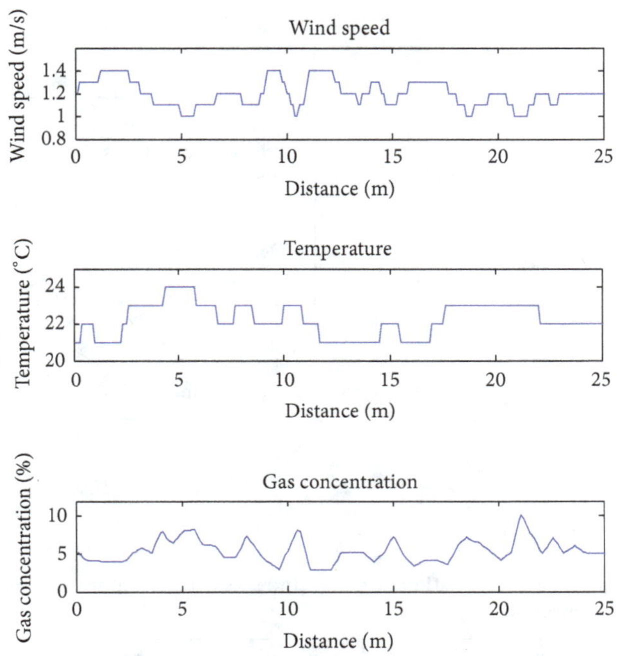

FIGURE 5: Wind speed, temperature, and gas concentration data from the coal mine roadway.

due to thermal motion of gas molecules, which on the one hand is due to the gas concentration being higher than the concentration of gas in airflow and on the other hand is also affected by temperature difference and ventilation eddy current.

The diffusion motion obeys Fick's law of diffusion; Fick's first law expression is

$$J = -D\frac{dC}{dx}, \tag{13}$$

where D is gas diffusion coefficient, C is the concentration of gas molecules, dC/dx is concentration gradient, sign "–" indicates a direction opposite to the concentration gradient diffusion direction, indicating that diffusion component is made of high concentration to low concentration diffusion region, and J is diffusion flux. The diffusion coefficient depends on the pressure, temperature, and composition of gas mixture, usually determined by experiment. Document [17] gives the diffusion coefficient by experimental determination as follows:

$$D = 1.103 \times 10^{-2} T^{1.75} \frac{(M_A + M_B) / (M_A M_B)}{P (V_A + V_B)^2}. \tag{14}$$

Among them, V_A and V_B are the volume of corresponding gas component; P is the gas pressure; T is temperature; M_A and M_B are the molecular weight of corresponding gas component.

The air in airflow produces the friction and impact on gas, which force the gas moving with the airflow. Assuming the gas concentration c, the gas flow of the surface can be calculated by the convection transport principle, namely, amount of gas flux: $\vec{J} = C\vec{\mu}$, where $\vec{J}$ is the gas diffusion flux and $\vec{\mu}$ is the absolute wind speed along each coordinate axis.

Figure 4 shows the coal mine rescue robots walking on the roadway. Joint search and rescue tasks require multiple robots, and each robot needs to probe different regions. The sample data are the gas and related environmental data in the first 20 meters, collected by the coal mine rescue robot; the wind speed, temperature, and gas concentration data of the mine rescue robot on the coal mine roadway are shown in Figure 5. The coal mine rescue robot predicts the gas concentration in the first 10 meters of the coal mine.

3.2. Experimental Data Processing and Analysis. In this paper, the quantum genetic algorithm is used to optimize the gray neural network to construct the prediction model of the gas concentration in the coal mine underground roadway based on the gas concentration, temperature, and wind speed data collected by the coal mine rescue robots. The construction flow of the forecasting model is shown in Figure 6.

When the coal mine rescue robots predict the gas concentration of the roadway in the coal mine, the main influencing factors are the current position's gas concentration, the ambient temperature, and the wind speed. At the same time, the more the input features, the better the prediction effect of the forecasting model. In general, the greater the wind speed, the more dispersed the gas concentration distribution, and the higher the temperature, the more gathered the gas concentration distribution. The correlations among wind speed, temperature, and gas concentration were analyzed, the correlation coefficients of wind speed with gas concentration and temperature with gas concentration are calculated, and they are shown in Table 4.

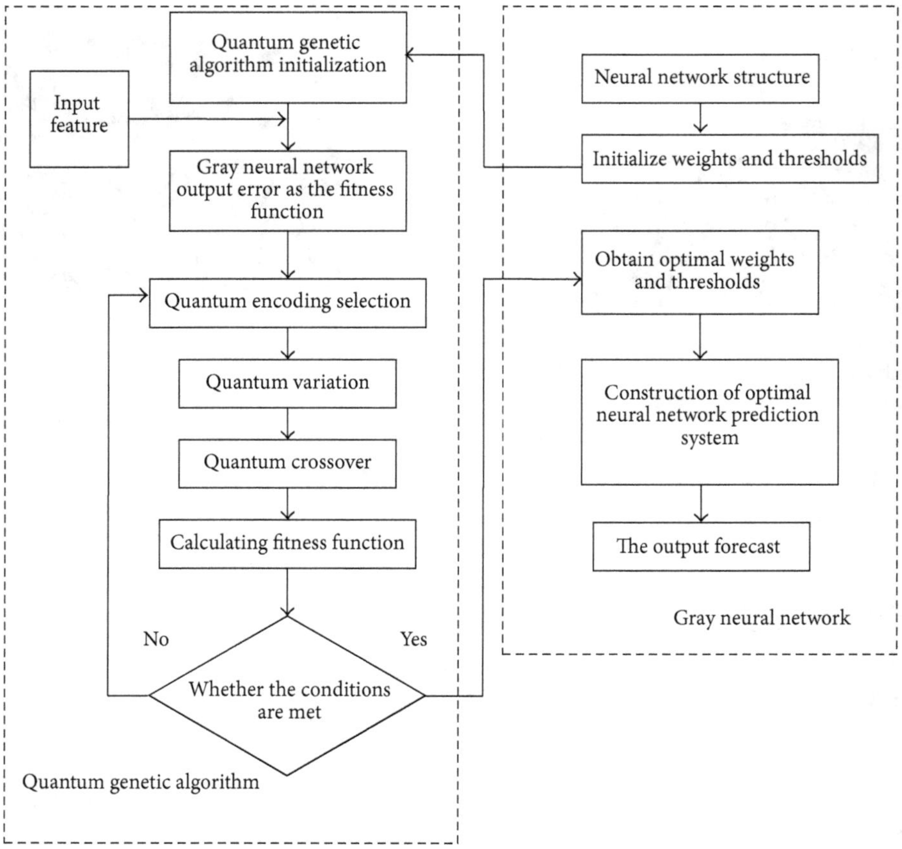

FIGURE 6: Construction flow of the forecasting model.

TABLE 4: The correlations among wind speed, temperature, and gas concentration.

The correlation among wind speed and gas concentration	0.5701
The correlation among temperature and gas concentration	0.6774

It can be seen from the calculation results of correlation that the correlation among wind speed and gas concentration belongs to the moderate degree, and the correlation among temperature and gas concentration belongs to strongly correlated degree. Wind speed and temperature are the major factors that affect the distribution of gas concentration.

Therefore, in this paper, the gas concentration of 10 meters is forecasted using the sensor information of the current position and the first meter. The gray neural network is suitable for the small sample and the uncertain system, which has partial information and unknown information and can predict the behavior and laws of the system by the extraction and exploitation of the known information.

A gray neural network was designed in this paper. The LA layer has one neuron, and the LB layer has one neuron. Because of six input vectors, the LC layer has seven neurons, and the output layer LD layer has one neuron. The initial number of training steps of the gray neural network is 100. The change in the training error of the gray neural network is shown in Figure 7, which shows that, after 6 steps of basic training, the gray neural network converges. The convergence rate is fast, but the network is in the local optimum. The weights of the gray neural network cannot be modified. After the gray neural network training is completed, the test data are used for testing. The test data are inputted into the trained gray neural network to predict the gas concentration 10 meters outside the current position of the coal mine rescue

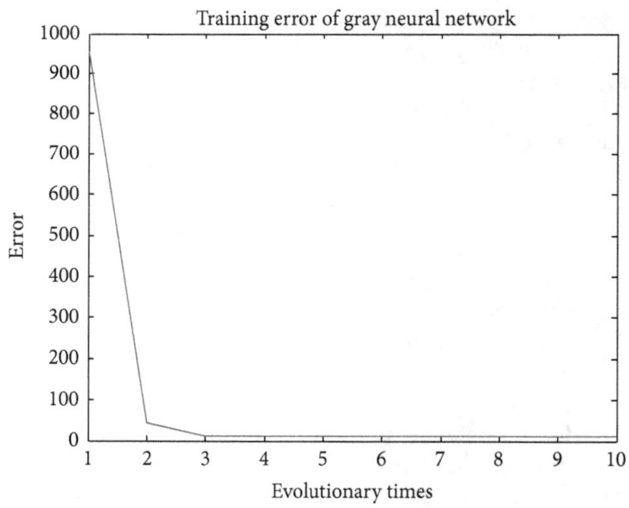

FIGURE 7: The change in the training error of the gray neural network.

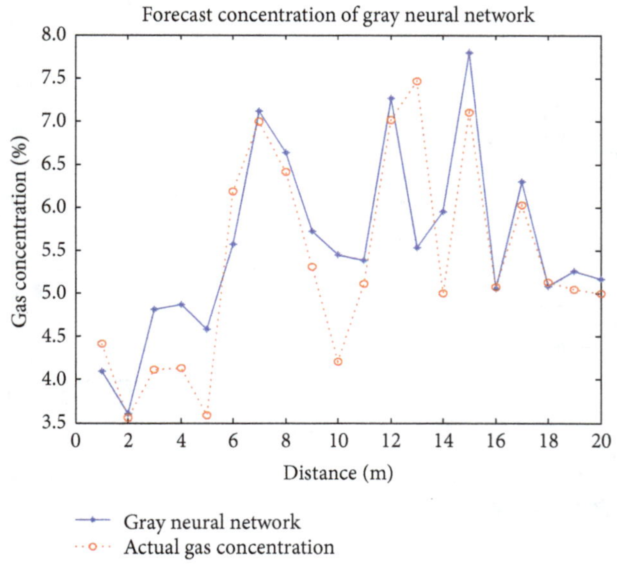

FIGURE 8: Comparison of the gray neural network forecast value and the actual value.

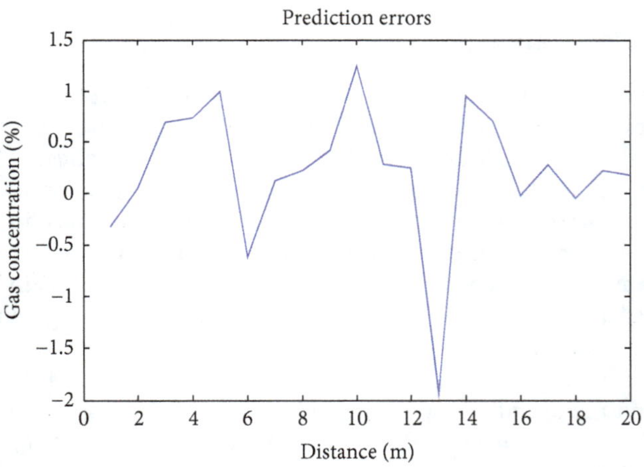

FIGURE 9: The prediction errors of the gray neural network.

robots. Figure 8 compares the gray neural network forecast and the actual gas concentration. Figure 9 is the prediction errors of the gray neural network.

In Figure 8, blue represents the gray neural network forecast and red the actual gas concentration. The gray neural network forecast values are more accurate at a few points, such as 7-8 m and 16–18 m. For example, the predictive value of gas concentration at 7 m is 7.137, while the real value of gas concentration at 7 m is 7, so the prediction deviation is 0.137, and the forecast error is $E_7 = |(x^{(1)}(7) - \hat{x}^{(1)}(7))/\hat{x}^{(1)}(7)| = 0.137/7.137 = 1.92\%$, but at some points, the difference between the predicted value and the actual value is still relatively large, such as 2–5 m and 9–12 m. For example, the predictive value of gas concentration at 10 m is 5.5, while the real value of gas concentration at 10 m is 4.25, so the prediction deviation is 1.25, and the forecast error is $E_{10} = |(x^{(1)}(10) - \hat{x}^{(1)}(10))/\hat{x}^{(1)}(10)| = 1.25/5.5 = 22.73\%$, which is significantly higher than the acceptable deviation range.

TABLE 5: Neural network parameter results of the optimization of the quantum genetic algorithm.

Parameter	a	b_1	b_2	b_3	b_4	b_5	b_6
Parameter values	0.1476	0.3073	0.1287	0.1485	0.1688	0.3297	0.2620

The overall average error of the gray neural network forecast is $E = (1/20)\sum_{i=1}^{20}|x^{(1)}(i) - \hat{x}^{(1)}(i)|/x^{(1)}(i) = 20.37\%$. The cause of prediction error is various; the main reason has four aspects: First, the prediction error is caused by the measurement error of the gas concentration and sensor used in this paper had already been certified and licensed; second, the prediction error is caused by the selection of predictive factor and the gas concentration is easily affected by gas emission, wind speed, temperature, and concentration of CO_2. Due to the small effect of CO_2 concentrations, it can be removed by using principal component analysis; therefore, the gas concentration, temperature, and wind speed were been selected as a predictor of gas concentration prediction model; third, adjusting the parameters in the prediction model of grey neural network can improve the convergence speed and stability training error, but how to accurately select the error threshold needs repeat experiments according to the system; fourth, the number of sample datasets will also affect the prediction error; the more the sample data, the better the prediction effect; however, more samples' data means that walking speed of mine rescue robot must be slowed down; this is bound to delay the rescue time and reduce the role of mine rescue robot.

However, because the gray neural network threshold weights are randomly initialized, the network training evolution process can easily fall into the local optimum, and every prediction result is not the same. Being easily trapped in the local minima can be avoided by combining a genetic algorithm with a neural network. The neural network is prone to overfitting and has poor generalization ability; to solve this problem, the quantum genetic algorithm optimizes the parameters a, b_1, b_2, b_3, b_4, b_5, and b_6 of the neural network. The quantum genetic algorithm uses entity coding and the quantum interference method to train the data of the gray neural network. The error of the gray neural network's prediction value of the training data is used as the fitness function. The population size of the quantum genetic algorithm is 30, the genetic factor is 7, and the number of iterations is 70. The optimal individual fitness value of the quantum genetic algorithm changes with the number of iterations, as shown in Figure 10.

Figure 10 shows that the optimal fitness value of the individual fitness gradually tends to be steady after 70 steps of training, and the optimal individual fitness value is 4.29. The gray neural network parameters after training, a, b_1, b_2, b_3, b_4, b_5, and b_6, are shown in Table 5.

The optimal gray neural network parameters are brought into the gray neural network; the test data were inputted into the gray neural network optimized by the quantum genetic algorithm to predict the gas concentration 10 meters outside the current position of the coal mine rescue robots. Figure 11

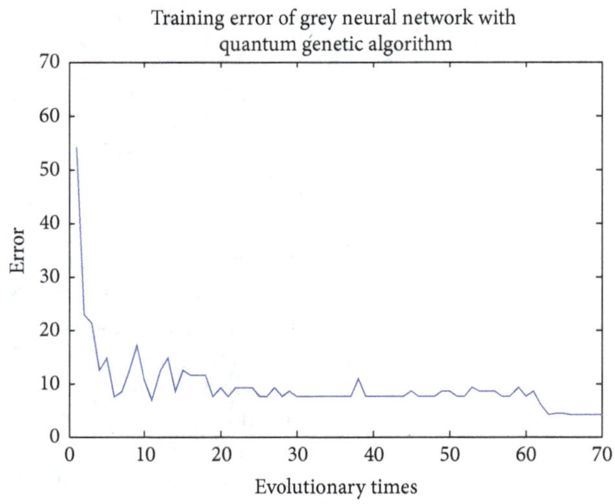

FIGURE 10: The change in the training error of the gray neural network with quantum genetic algorithm.

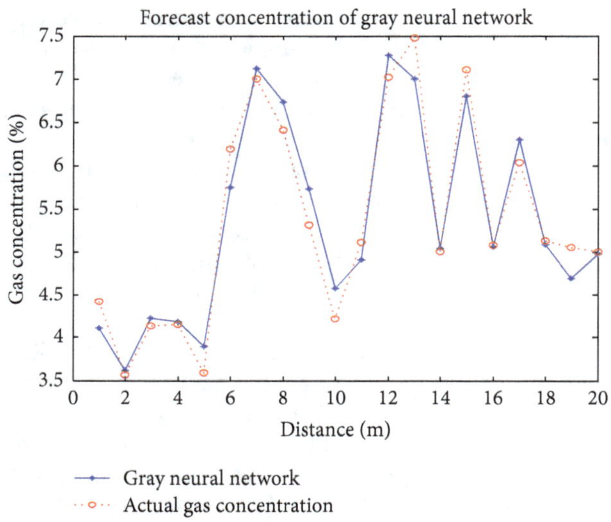

FIGURE 11: Forecast output of the gray neural network optimized by the quantum genetic algorithm.

compares the gray neural network forecast and the actual gas concentration. Figure 12 shows the prediction errors of the gray neural network optimized by the quantum genetic algorithm.

The graph shows that the gas concentration of the gray neural network after the optimization of the quantum genetic algorithm is more accurate. For 5 m, 13 m, 15 m, and 19 m, the predictive value of gas concentration at 5 m is 3.96, while the real value of gas concentration at 5 m is 3.51, so the prediction deviation is 0.45, and the forecast error is $E_5 = |(x^{(1)}(5) - \hat{x}^{(1)}(5))/\hat{x}^{(1)}(5)| = 0.45/3.96 = 11.36\%$, the optimized gray neural network forecast value is close to the actual value, and the overall forecast error is $E = (1/20)\sum_{i=1}^{20} |x^{(1)}(i) - \hat{x}^{(1)}(i)|/x^{(1)}(i) = 9.12\%$. Compared with the ordinary gray neural network, the overall accuracy gas concentration prediction increases by (20.37% − 9.12%)/20.37% = 55.23%. Therefore, the optimized gray neural network can better help coal mine rescue robots to accurately predict the gas concentration.

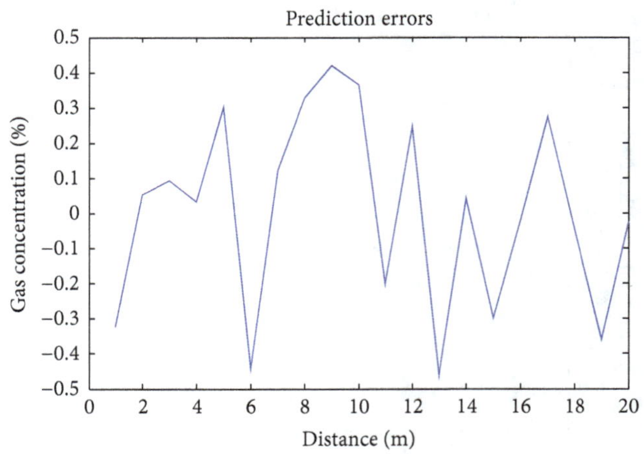

FIGURE 12: Prediction errors of the gray neural network optimized by the quantum genetic algorithm.

4. Conclusions

After a coal mine accident, in order for the coal mine rescue robot to successfully execute detection and search and rescue mission, this paper addresses the problem of predicting the roadway gas concentration in front of the coal mine rescue robot based on the real-time data collection of the temperature, wind speed, and gas concentration. The gray neural network cannot fully execute its forecasting capabilities because the gray neural network training can easily fall into the local optimum. The difference between the predicted gas concentration and the actual value is large, and the overall average forecast error is 22.73%. Combining the quantum genetic algorithm and the gray neural network together can solve the shortcomings of the gray neural network and obtain a more accurate prediction of the gas concentration. The experiment shows that the gray neural network optimized by the quantum genetic algorithm is more accurate for predicting the gas concentration. The overall prediction error is 9.12%, and the largest forecasting error is 11.36%. The gas concentration's overall prediction accuracy is improved by 55.23% compared with the ordinary gray neural network, and predicting the gas concentration of the gray neural network optimized by the quantum genetic algorithm is an effective gas concentration prediction method for the coal mine rescue robot in an underground tunnel.

Competing Interests

The authors declare that they have no competing interests.

Acknowledgments

This work is supported by grant of the National 863 Program of China (no. 2012AA041504) and the Policy Guidance Class Plan of Jiangsu Province (no. BY2015024-2).

References

[1] A.-R. Zhao, "Prediction method of gas concentration based on wavelet analysis and BP neural network," *Coal Mine Machinery*, vol. 35, pp. 258–260, 2014.

[2] C. Özgen Karacan, "Modeling and prediction of ventilation methane emissions of U.S. longwall mines using supervised artificial neural networks," *International Journal of Coal Geology*, vol. 73, no. 3-4, pp. 371–387, 2008.

[3] L. Runqiua, S. Shilianga, W. Aiyoua, L. Wenkea, and Z. Hongpingc, "Research on prediction of gas emission based on self-organizing data mining in coal mines," *Procedia Engineering*, vol. 84, pp. 779–785, 2014.

[4] X. Hu, S. Yang, X. Zhou, G. Zhang, and B. Xie, "A quantification prediction model of coalbed methane content and its application in Pannan coalfield, Southwest China," *Journal of Natural Gas Science and Engineering*, vol. 21, pp. 900–906, 2014.

[5] G.-X. Jing, S.-M. Xu, X.-W. Heng, and C.-Q. Li, "Research on the prediction of gas emission quantity in coal mine based on grey system and linear regression for one element," *Procedia Engineering*, vol. 26, supplement, pp. 1585–1590, 2011.

[6] T. Han, S. Wu, and P. Wang, "Prediction of gas concentration based on residual correction of Markov Chain," *Industry and Mine Automation*, vol. 40, pp. 28–31, 2014.

[7] S.-G. Cao, Y.-B. Liu, and Y.-P. Wang, "A forecasting and forewarning model for methane hazard in working face of coal mine based on LS-SVM," *Journal of China University of Mining and Technology*, vol. 18, no. 2, pp. 172–176, 2008.

[8] S. Wei-Hua and Z. Hong-Wei, "Regional prediction of coal and gas outburst hazard based on multi-factor pattern recognition," *Procedia Earth and Planetary Science*, vol. 1, no. 1, pp. 347–353, 2009.

[9] X.-H. Zhao, G. Wang, K.-K. Zhao, and D.-J. Tan, "On-line least squares support vector machine algorithm in gas prediction," *Mining Science and Technology*, vol. 19, no. 2, pp. 194–198, 2009.

[10] X. Sun, "Coalmine gas prediction based on fuzzy neural network," *Journal of Anhui University of Technology and Science*, vol. 29, no. 3, pp. 229–232, 2012.

[11] Y. Liu, Q. Zhao, and W. Hao, "Study of gas concentration prediction based on genetic algorithm and optimizing BP neural network," *Mining Safety & Environmental Protection*, vol. 42, pp. 56–60, 2015.

[12] D. Gan and Z. Zhao, "The optimal combination prediction method and its application in gas concentration prediction," *Journal of Shandong University of Science and Technology*, vol. 31, pp. 55–61, 2012.

[13] R. Guo and G. Xu, "Research on coal mine gas concentration multi-sensor prediction model based on information fusion and GA-SVM," *China Safety Science Journal*, vol. 23, pp. 33–38, 2013.

[14] J.-Y. Zhang, H. Xu, J. Chen, and X.-D. Cao, "Application research of support vector machine based on particle swarm optimization in gas concentration forecasting," *Industry and Mine Automation*, vol. 10, pp. 32–35, 2010.

[15] A. O. Caldeira and A. J. Leggett, "Quantum tunnelling in a dissipative system," *Annals of Physics*, vol. 149, no. 2, pp. 374–456, 1983.

[16] S.-Y. Yang, F. Liu, and L.-C. Jiao, "The quantum evolutionary strategies," *Acta Electronica Sinica*, vol. 29, no. 12, pp. 1873–1877, 1873.

[17] E. N. Fuller, P. D. Schettler, and J. C. Giddings, "New method for prediction of binary gas-phase diffusion coefficients," *Industrial & Engineering Chemistry*, vol. 58, no. 5, pp. 18–27, 1966.

14

Mathematical Modeling and Modal Switching Control of a Novel Tiltrotor UAV

Zhiwei Kong and Qiang Lu

School of Automation, Hangzhou Dianzi University, Hangzhou, China

Correspondence should be addressed to Qiang Lu; lvqiang@hdu.edu.cn

Academic Editor: L. Fortuna

This paper concentrates on the flight control of a novel tiltrotor aircraft with fixed wings. This kind of aircraft has two flight modes and a transition mode. In the phase of vertical take off and landing (VTOL), the aircraft can operate as a quadrotor helicopter. And in the phase of horizontal flight, the aircraft is in the normal airplane mode. The transition mode is between these two flight modes. In this part of work, a novel tiltrotor aircraft was presented since transition mode is achieved by tilting the front dual tiltrotor (DTR) and the mathematical model was established. The classical PID method was used during the phase of VTOL and the numerical results were given and the simulation shows good control effect. A nonlinear control law based on backstepping was proposed to achieve a stable transition from vertical flight to horizontal flight. And the numerical results show that the flight mode could transit stably which shows the effectiveness of the control approach. Finally, the vertical flight experiment has been carried out on DTR aircraft and the attitude was stable.

1. Introduction

Tiltrotor aircraft attracted much researchers' attention in the field of aeronautics and astronautics since 1980s [1, 2]. The reason that tiltrotors have received such great attention is mainly because they combine the advantages of helicopters and fixed-wing aircraft [3, 4]. Helicopters can take off and land vertically, but they cannot fly horizontally at a high speed. Fixed-wing aircrafts can fly horizontally at a high speed, but they need a runway for takeoff and landing [4, 5]. Tiltrotor aircraft has the advantages of both helicopter and fixed-wing aircraft, which makes it have a wide application scene.

Bell Helicopter company developed the V22 Osprey tiltrotor aircraft in the 1990s, which became the world's first tiltrotor aircraft. Osprey immediately received widespread attention because of its excellent performance. At present, tiltrotor aircrafts have achieved great success in military and civil fields [6]. In military aspects, tiltrotor aircrafts can be used as carrier-based aircrafts and can complete the transportation task in harsh environment, such as deserts and islands. In civil aspects, they are often used in the tasks of pesticide spraying, power patrol inspection, terrain exploration, and disaster relief.

In previous studies on multimodal aircraft, many researchers have focused on quad tilt-wing (QTW) unmanned aerial vehicle (UAV) [3, 7–11]. This kind of aircraft transfers from quadrotor helicopter to airplane by turning wings. Oner [2] developed the QTW vehicle and the dynamic model is derived by using Newton-Euler formulation. And a linear quadratic regulator (LQR) controller was proposed to stabilize vertical flight. Benkhoud [3] utilized Model Predictive Control (MPC) approach to stabilize vehicle attitude and track its trajectory. And the simulation results showed the effectiveness of the approach. In engineering practice, the most widely used regulator control law is proportional, integral, and derivative control, referred to as PID control. Hancer [8] estimated the disturbance with a disturbance observer and utilized PID controller for robust hovering control. And the effectiveness of the controller was verified through experiments and simulations. Another kind of aircraft with similar performance called quad tiltrotor (QTR) UAV also attracted widespread attention [5, 6, 12–15]. This kind of aircrafts switches flight modal by turning rotors.

In [12], a convertible QTR was designed and a nonlinear controller based on dynamic inversion was given. Moreover, the trajectory of the aircraft was tracked. Flores [4, 16] have done some work on this subject. In [4], a QTR model with fixed wings was presented and a nonlineared controller was also given which is based on saturations and Lyapunov. And from the numerical simulations the control strategy obtained satisfactory results.

Some studies on the vertical takeoff and landing of quadrotor helicopter have achieved good results [1, 17–21]. Sliding model controller was used in quadrotor helicopter altitude and attitude control [1, 22]. Xu [22] divided quadrotor helicopter model into actuated subsystem and underactuated subsystem. Furthermore, sliding mode control laws were designed for the two subsystems and the vehicle's altitude and attitude were stabilized. In [18], a feedback linearizing controller and backstepping-like control law was proposed to control vehicle's motion which is restricted to yaw and vertical motions.

In recent years, there have been a large number of studies on QTW and QTR. Most of them studied the horizontal and vertical flight of this type of aircraft. However, there are few studies on the transition process. Therefore, solving the problem of switching the flight modal stably for QTR or QTW has great practical significance.

In this paper, a novel tiltrotor aircraft model with dual tiltable rotors was proposed as you can see in Figures 1, 3, and 4. It can take off vertically with four rotors and after transition the vehicle can fly horizontally with the front dual rotors. And the back two rotors will stop when the transition mode is completed. The modal switching control strategy is showed in Figure 2. When the DTR is in the phase of VTOL, DTR's kinetics model is the same as the quadrotor helicopter and the attitude controller is the quadrotor helicopter attitude controller. After the DTR reaches the reference value, the decision-making module will switch the kinetics model and controller into transition modal at the same time. The remainder of this paper is arranged as follows. In Section 2, the DTR kinetics model will be presented. In Section 3, PID method was used to control the position and attitude of the vertical flight phase. In Section 4, a nonlinear controller will be designed to implement the transition. In Section 5, the takeoff experiment was carried out on DTR aircraft. In Section 6, we will give a conclusion about this part of work and point out deficiencies and future improvements.

2. The Mathematical Model of the Tiltrotor UAV

2.1. Coordinate System and Force Analysis. In order to establish the mathematical model of the aircraft and analyze the force, the inertial coordinate and body coordinate are introduced. As you can see in Figure 4, $C_w(x_w, y_w, z_w)$ is the inertial coordinate that the origin of the coordinate is any point in space and x_w axis points north, y_w axis points east, and the z_w axis points vertically downwards to the center of the earth. $C_b(x_b, y_b, z_b)$ is the body coordinate that the origin of the coordinate is the center of the aircraft mass. x_b points to the aircraft head, the z_b axis is perpendicular to the x_b axis and points to the earth, and the y_b axis is determined by the right hand rule. The body coordinate is fixed relative to the fuselage and rotates as the aircraft rotates. The attitude angles ψ, θ, ϕ represent yaw, pitch, and roll, respectively. The transformation matrix Γ_{bw} from body coordinate to inertial coordinate is represented as

$$\Gamma_{bw} = \begin{bmatrix} c_\psi c_\theta & s_\phi s_\theta c_\psi - c_\phi s_\psi & c_\phi s_\theta c_\psi + s_\phi s_\psi \\ c_\theta s_\psi & s_\phi s_\theta s_\psi + c_\phi c_\psi & c_\phi s_\theta s_\psi - s_\phi c_\psi \\ -s_\theta & s_\phi c_\theta & c_\phi c_\theta \end{bmatrix} \quad (1)$$

where c_θ and s_θ represent $\cos\theta$ and $\sin\theta$, respectively.

The aircraft is mainly subjected to the pull of four rotors $T_i (i = 1, 2, 3, 4)$, gravity mg, wings lift T_L, and air drag T_D. The rotor tilting angle is expressed by α, and then the pull along the x_b axis can be expressed as

$$T_{px} = (T_1 + T_2)\sin(\alpha) \quad (2)$$

where $T_i = k\omega_i^2$ and ω_i is the rotors' angular velocity. And the pull along the z_b axis can be expressed as

$$T_{pz} = (T_1 + T_2)\cos(\alpha) + T_3 + T_4 \quad (3)$$

The speed along the x_b axis will keep increasing when the rotors are tilting. At this time, the lift of the wings and the air drag to the fuselage cannot be ignored. And the wings lift T_L and the air drag T_D can be calculated as follows [23]:

$$T_L = \frac{1}{2}\varepsilon\rho A V_{xz}^2 \quad (4)$$

$$T_D = p_1 V_{xz}^2 + p_2 V_{xz} \quad (5)$$

where ε is the lift coefficient, ρ is air density, A represents the area of the wing, and p_1 and p_2 are drag coefficient. V_{xz}^2 represents the vector sum of x_b axis and z_b axis velocity, i.e.,

$$V_{xz} = \sqrt{v_{bx}^2 + v_{bz}^2} \quad (6)$$

Suppose $L(\pm L_x, \pm L_y, \pm L_z)$ represents the coordinates of the rotor in the body frame; then the resultant force moment of the rotors $M_t = [M_x, M_y, M_z]^T$ can be expressed as

$$\begin{aligned} M_t &= \begin{bmatrix} T_1 \sin\alpha \\ 0 \\ -T_1\cos\alpha \end{bmatrix} \times \begin{bmatrix} L_x \\ -L_y \\ -L_z \end{bmatrix} + \begin{bmatrix} T_2 \sin\alpha \\ 0 \\ -T_2\cos\alpha \end{bmatrix} \times \begin{bmatrix} L_x \\ L_y \\ -L_z \end{bmatrix} \\ &+ \begin{bmatrix} 0 \\ 0 \\ -T_3 \end{bmatrix} \times \begin{bmatrix} -L_x \\ -L_y \\ -L_z \end{bmatrix} + \begin{bmatrix} 0 \\ 0 \\ -T_4 \end{bmatrix} \times \begin{bmatrix} -L_x \\ L_y \\ -L_z \end{bmatrix} \\ &= \begin{bmatrix} (-T_1\cos\alpha + T_2\cos\alpha - T_3 + T_4) L_y \\ -(T_1 + T_2)(L_x\cos\alpha - L_z\sin\alpha) + (T_3 + T_4) L_x \\ (-T_1 + T_2) L_y \sin\alpha \end{bmatrix} \end{aligned} \quad (7)$$

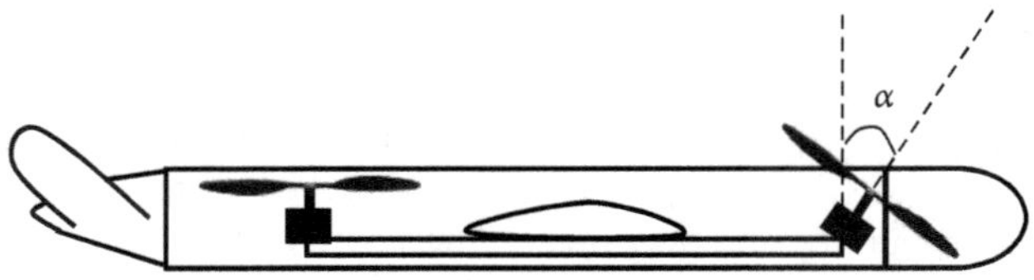

FIGURE 1: Tilting angle.

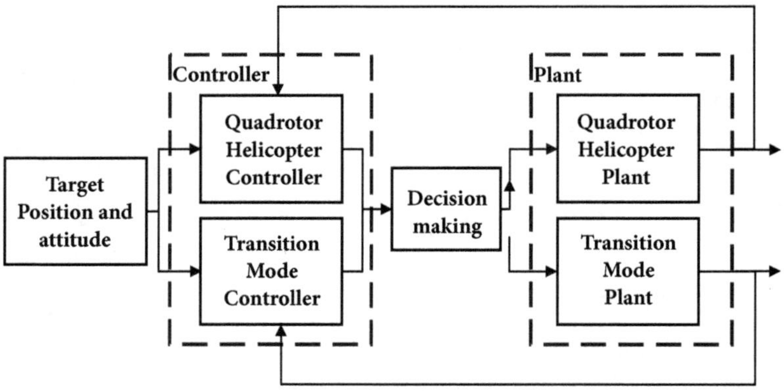

FIGURE 2: Modal switching control strategy.

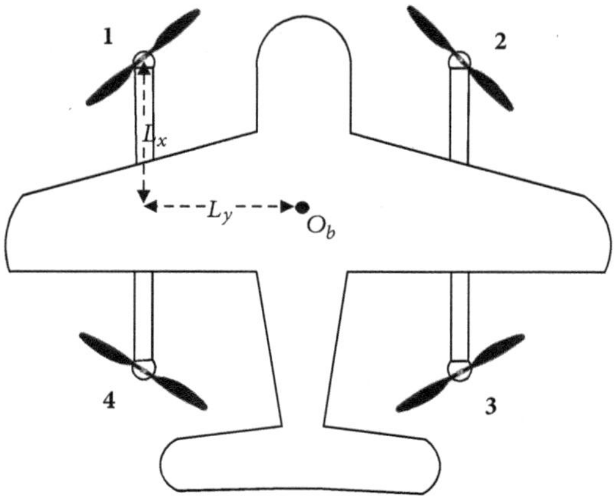

FIGURE 3: The top view of DTR.

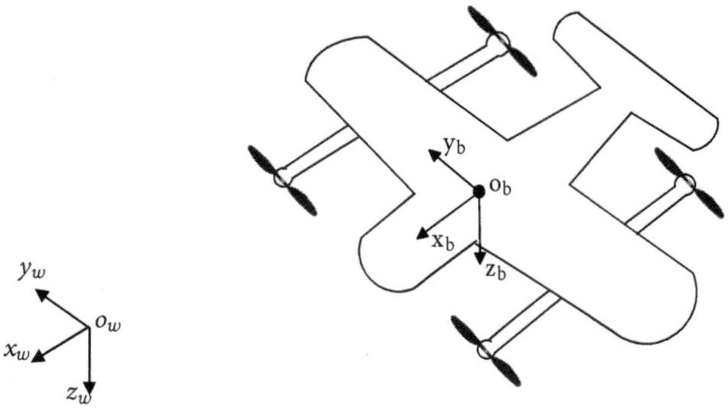

FIGURE 4: The inertial coordinate and body coordinate.

Assuming that air drag and lift act on the geometric center of the two wings, then the moment of air drag and lift can be expressed as

$$M_{ld} = \begin{bmatrix} -T_D \\ 0 \\ -T_L \end{bmatrix} \times \begin{bmatrix} 0 \\ -L_y \\ 0 \end{bmatrix} + \begin{bmatrix} -T_D \\ 0 \\ -T_L \end{bmatrix} \times \begin{bmatrix} 0 \\ L_y \\ 0 \end{bmatrix} = 0 \quad (8)$$

Due to the rotation of the rotors, the gyro effect and counteractive moment are produced. But when the tilting rotors have the same angular velocity as well as the fixed rotors, the gyro effect and counteractive moment will be eliminated. And in the following control strategy, the input $T_1 = T_2$ and $T_3 = T_4$ will be given. Therefore, the gyros effect and counteractive moment are not considered here. And gravity does not produce a moment due to it acts on the center of mass.

2.2. Kinetics Model. If $v(p, q, r)$ represents the aircraft roll, pitch, and yaw rates in body coordinates, $\gamma(I_x, I_y, I_z)$ represents the rotary inertia of body coordinates axis. From the Newton-Euler equation, the relationship of angular acceleration and moment can be expressed as

$$\begin{aligned} I_x\dot{p} + \left(I_z - I_y\right) qr &= \Sigma M_x \\ I_y\dot{q} + \left(I_x - I_z\right) rp &= \Sigma M_y \\ I_z\dot{r} + \left(I_y - I_x\right) pq &= \Sigma M_z \end{aligned} \quad (9)$$

substituting (7) into (9),

$$\begin{aligned} \dot{p} &= \frac{I_y - I_z}{I_x} qr - \frac{L_y \cos\alpha}{I_x} \left(-T_1 + T_2\right) + \frac{L_y}{I_x} \left(-T_3 + T_4\right) \\ \dot{q} &= \frac{I_z - I_x}{I_y} rp - \frac{L_x \cos\alpha - L_z \sin\alpha}{I_y} \left(T_1 + T_2\right) + \frac{L_x}{I_y} \left(T_3 + T_4\right) \\ \dot{r} &= \frac{I_x - I_y}{I_z} pq + \frac{L_y \sin\alpha}{I_z} \left(-T_1 + T_2\right) \end{aligned} \quad (10)$$

When the Euler angle is relatively small, it can be considered approximately as $p = \dot{\phi}$, $q = \dot{\theta}$, and $r = \dot{\psi}$.

According to Newton's second law $\vec{F} = m\vec{v}$ and transforming the resultant force from body coordinate to inertial coordinate, the dynamic model of the aircraft can be expressed as

$$\begin{aligned} m\ddot{x} &= c_\psi c_\theta T_{px} - \left(s_\phi s_\psi + c_\psi s_\theta c_\phi\right)\left(T_{pz} + 2T_L\right) - c_\psi c_\theta T_D \\ m\ddot{y} &= c_\theta s_\psi T_{px} - \left(c_\phi s_\theta s_\psi - s_\phi c_\psi\right)\left(T_{pz} + 2T_L\right) - c_\theta s_\psi T_D \\ m\ddot{z} &= -s_\theta T_{px} - c_\phi c_\theta \left(T_{pz} + 2T_L\right) + mg + s_\theta T_D \\ \ddot{\phi} &= c_1\dot{\theta}\dot{\psi} - c_2 \cos\alpha \left(-T_1 + T_2\right) + c_3 \left(-T_3 + T_4\right) \\ \ddot{\theta} &= c_4\dot{\psi}\dot{\phi} - \left(c_5 \cos\alpha - c_6 \sin\alpha\right)\left(T_1 + T_2\right) + c_7 \left(T_3 + T_4\right) \\ \ddot{\psi} &= c_8\dot{\phi}\dot{\theta} + c_9 \sin\alpha \left(-T_1 + T_2\right) \end{aligned} \quad (11)$$

where $c_1 = (I_y - I_z)/I_x$, $c_2 = c_3 = L_y/I_x$, $c_4 = (I_z - I_x)/I_y$, $c_5 = c_7 = L_x/I_y$, $c_6 = L_z/I_y$, $c_8 = (I_x - I_y)/I_z$, $c_9 = L_y/I_z$.

3. Vertical Takeoff Control Approach

In the phase of vertical takeoff, the classic quadrotor PID control method was used. The tilting angle $\alpha = 0$ when the DTR takes off at a low speed. Therefore, air drag T_D and wings lift T_L will be ignored. The aircraft makes a small angle movement in the hovering state, and then the dynamic model can be expressed as

$$\begin{aligned} m\ddot{x} &= -\left(s_\phi s_\psi + c_\psi s_\theta c_\phi\right)\left(T_1 + T_2 + T_3 + T_4\right) \\ m\ddot{y} &= -\left(c_\phi s_\theta s_\psi - s_\phi c_\psi\right)\left(T_1 + T_2 + T_3 + T_4\right) \\ m\ddot{z} &= -c_\phi c_\theta \left(T_1 + T_2 + T_3 + T_4\right) + mg \\ \ddot{\phi} &= c_1\dot{\theta}\dot{\psi} + c_2 \left(T_1 - T_2 - T_3 + T_4\right) \\ \ddot{\theta} &= c_4\dot{\psi}\dot{\phi} + c_5 \left(-T_1 - T_2 + T_3 + T_4\right) \\ \ddot{\psi} &= c_8\dot{\phi}\dot{\theta} + c_9 \left(-T_1 + T_2 - T_3 + T_4\right) \end{aligned} \quad (12)$$

The PID controller is a common feedback loop component in industrial control applications and consists of a proportional element P, an integral element I, and a derivative element D. The basis of PID control is proportional control, integral control can eliminate steady-state error but may increase overshoot, and derivative control can speed up large inertial system response speed and weaken overshoot trend.

The resultant force of the four rotors is used as the control input, and the dynamic model can be writen as

$$\begin{aligned} \ddot{x} &= \left(s_\phi s_\psi + c_\psi s_\theta c_\phi\right) u_1 \\ \ddot{y} &= \left(c_\phi s_\theta s_\psi - s_\phi c_\psi\right) u_1 \\ \ddot{z} &= c_\phi c_\theta u_1 + g \\ \ddot{\phi} &= c_1\dot{\theta}\dot{\psi} + u_2 \\ \ddot{\theta} &= c_4\dot{\psi}\dot{\phi} + u_3 \\ \ddot{\psi} &= c_8\dot{\phi}\dot{\theta} + u_4 \end{aligned} \quad (13)$$

The trajectory of the aircraft (x, y, z) and the yaw ϕ were tracked only since this is an underactuated system. Furthermore, to track the desire attitude θ_d, ϕ_d, replace the virtual inputs u_1 in (13) as follows:

$$\begin{aligned} u_{x1} &= \left(s_\phi s_\psi + c_\psi s_\theta c_\phi\right) u_1 \\ u_{y1} &= \left(c_\phi s_\theta s_\psi - s_\phi c_\psi\right) u_1 \\ u_{z1} &= c_\phi c_\theta u_1 \end{aligned} \tag{14}$$

And the control law is designed as

$$\begin{aligned} u_{x1} &= -r_{xp} x - r_{xi} \int x dt - r_{xd} \dot{x} \\ u_{y1} &= -r_{yp} y - r_{yi} \int y dt - r_{yd} \dot{y} \\ u_{z1} &= -r_{zp} z_e - r_{zi} \int z_e dt - r_{zd} \dot{z}_e - g + \ddot{z}_d \end{aligned} \tag{15}$$

where $z_e = z - z_d$ and z_d is the desire altitude. To satisfy the control law (15) the desired attitude angle θ_d, ϕ_d should be tracked. The attitude angle error is defined as

$$\begin{aligned} \theta_e &= \theta - \theta_d, \\ \dot{\theta}_e &= \dot{\theta} - \dot{\theta}_d \\ \phi_e &= \phi - \phi_d, \\ \dot{\phi}_e &= \dot{\phi} - \dot{\phi}_d \end{aligned} \tag{16}$$

where θ_d, ϕ_d can be calculated as

$$\begin{aligned} \theta_d &= \arcsin\left(\frac{c_\psi^2 u_{x1} + c_\psi s_\psi u_{y1}}{u_{z1}}\right) \\ \phi_d &= \arctan\left(\frac{c_\psi s_\psi u_{x1} - c_\psi^2 u_{y1}}{u_{z1}}\right) \end{aligned} \tag{17}$$

The attitude system control law is designed as

$$\begin{aligned} u_2 &= -r_{p2} \phi_e - r_{i2} \int \phi_e dt - r_{d2} \dot{\phi}_e + \ddot{\phi}_d \\ u_3 &= -r_{p3} \theta_e - r_{i3} \int \theta_e dt - r_{d3} \dot{\theta}_e + \ddot{\theta}_d \\ u_4 &= -r_{p4} \psi_e - r_{i4} \int \psi_e dt - r_{d4} \dot{\psi}_e + \ddot{\psi}_d \end{aligned} \tag{18}$$

The numerical results are showed in Figures 5, 6, 7, and 8. Figure 5 shows the aircraft's takeoff trajectory from (3,3) to (0,0) and the target altitude $z_d = 15$. Figure 6 shows the control inputs changes over time. In Figure 7, θ and ϕ converge to 0 and ψ reaches the target value $\pi/6$. Figure 8 shows the position changes along the axis of the inertia coordinate. From the simulation results the PID control approach shows good control effect on attitude and position in vertical flight phase.

4. Transition Flight Control Approach

In the transition mode, rotorl and rotor2 are always given the same input $T_1 = T_2$ as well as rotor3 and rotor4 $T_3 = T_4$. This input allocation method has a greater influence on θ than ϕ and ψ. Moreover, define the input $\tau_1 = T_1 = T_2$ and $\tau_2 = T_3 = T_4$. Therefore, the dynamic model during transition can be expressed as

$$\begin{aligned} \ddot{x} &= 2\tau_1 s_\alpha \frac{c_\psi c_\theta}{m} - \frac{2\left(s_\phi s_\psi + c_\psi s_\theta c_\phi\right)}{m} \left(\tau_1 c_\alpha + \tau_2 + T_L\right) \\ &\quad - \frac{c_\psi c_\theta}{m} T_D \\ \ddot{y} &= 2\tau_1 s_\alpha \frac{c_\theta s_\psi}{m} - \frac{2\left(c_\phi s_\theta s_\psi - s_\phi c_\psi\right)}{m} \left(\tau_1 c_\alpha + \tau_2 + T_L\right) \\ &\quad - \frac{c_\theta s_\psi}{m} T_D \\ \ddot{z} &= -2\tau_1 s_\alpha \frac{s_\theta}{m} - \frac{2 c_\phi c_\theta}{m} \left(\tau_1 c_\alpha + \tau_2 + T_L\right) + g + \frac{s_\theta}{m} T_D \\ \ddot{\phi} &= c_1 \dot{\theta} \dot{\psi} \\ \ddot{\theta} &= c_4 \dot{\psi} \dot{\phi} - 2\tau_1 \left(c_5 c_\alpha - c_6 s_\alpha\right) + 2 c_7 \tau_2 \\ \ddot{\psi} &= c_8 \dot{\phi} \dot{\theta} \end{aligned} \tag{19}$$

When the rotors tilting angle change from 0 to $\pi/2$, the key point to design the control law is to keep the height of the aircraft from falling and the pitch angle θ should be close to 0. That means the resultant force on the z_w axis should be in the negative direction, i.e.,

$$s_\theta T_{px} + c_\phi c_\theta \left(T_{pz} + 2T_L\right) + s_\theta T_D \geqq mg \tag{20}$$

Therefore, to realize the tilting process stably the main task under this input strategy is to control the following subsystem:

$$\begin{aligned} \ddot{z} &= -2\tau_1 s_\alpha \frac{s_\theta}{m} - \frac{2 c_\phi c_\theta}{m} \left(\tau_1 c_\alpha + \tau_2 + T_L\right) + g + \frac{s_\theta}{m} T_D \\ \ddot{\theta} &= c_4 \dot{\psi} \dot{\phi} - 2\tau_1 \left(c_5 c_\alpha - c_6 s_\alpha\right) + 2 c_7 \tau_2 \end{aligned} \tag{21}$$

In addition, at the equilibrium point of the system $\sin\theta \approx \theta$, $\cos\theta \approx 1$. Then, the nonlinear system is partially linearized as follows:

$$\begin{aligned} \ddot{z} &= -2\tau_1 s_\alpha \frac{\theta}{m} - \frac{2 c_\phi}{m} T + g + \frac{\theta}{m} T_D \\ \ddot{\theta} &= c_4 \dot{\psi} \dot{\phi} - 2\tau_1 K + 2 c_7 \tau_2 \end{aligned} \tag{22}$$

where $K = c_5 c_\alpha - c_6 s_\alpha$, $T = \tau_1 c_\alpha + \tau_2 + T_L$.

To stabilize the transition mode system (22), the backstepping approach is used to design the control laws τ_1 and τ_2. The basic idea of the backstepping approach is to decompose a complex nonlinear system into several subsystems that do not exceed the system order. Then, design partial Lyapunov

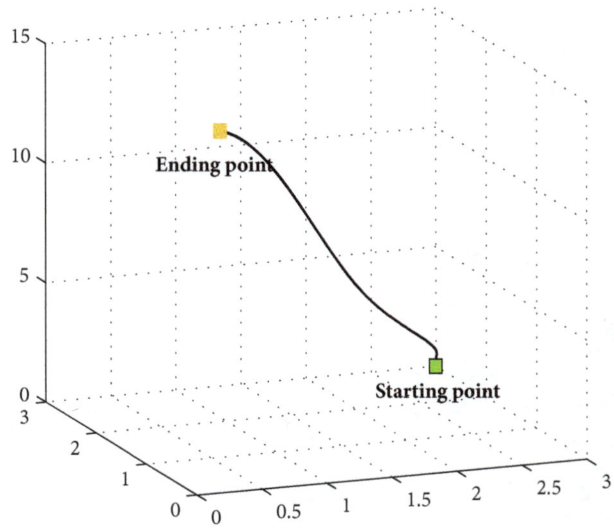

FIGURE 5: DTR takeoff trajectory.

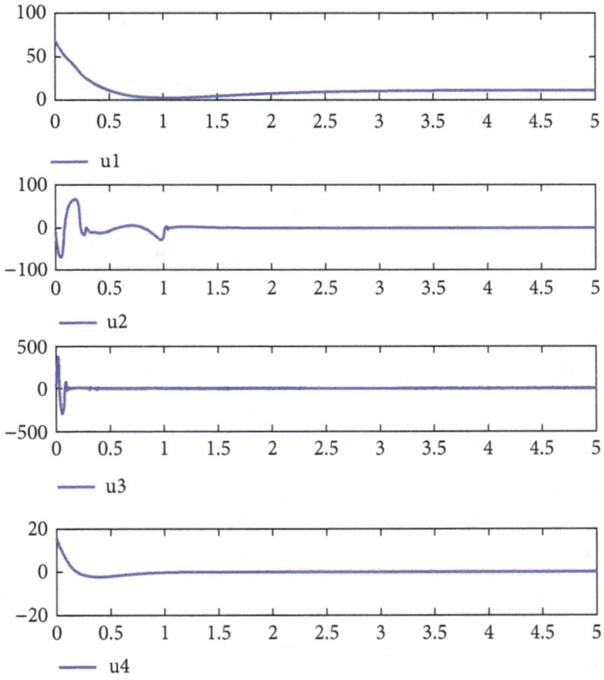

FIGURE 6: The control inputs of takeoff.

functions and intermediate virtual control for each subsystem until back to the entire system. Finally, they are integrated to complete the design of the entire control law.

Theorem 1. *Consider that the transition mode system (22) is a multi-input and multioutput coupling system. The equations of* $\ddot{\theta}$ *and* $\ddot{z}$ *could be separated to design the control law, respectively. If* $a = m(-z_3 - k_3\dot{z}_3 + g - k_4 z_4) + T_D\theta - 2c_\phi T_L - c_\phi K/c_7$, $b = 2s_\alpha\theta + 2c_\phi c_\alpha + 2c_\phi K/c_7$. *The control law can be given as* $\tau_1 = a/b$ *and* $\tau_2 = (k_1\dot{z}_1 + z_1 + k_2 z_2 + 2aK/b)/2c_7$.

Proof. The process of proof will be divided into three parts. In the first part, the Lyapunov function will be constructed and the stability of $\ddot{\theta}$ subsystem will be proved. In the second part, the stability of the $\ddot{z}$ subsystem will be proved. In the third part, the control input of the stabilization system will be given.

Part I: We Will Prove the Stability of $\ddot{\theta}$ *Subsystem.* The pitch subsystem is designed as follows:

$$\begin{aligned} \dot{\omega}_1 &= \omega_2 \\ \dot{\omega}_2 &= c_4\dot{\psi}\dot{\phi} - 2\tau_1 K + 2c_7\tau_2 \end{aligned} \tag{23}$$

where $\omega_1 = \theta$. The virtual input u_θ is introduced which is equal to the pitch rate $\dot{\omega}_1$. The error of pitch is defined as

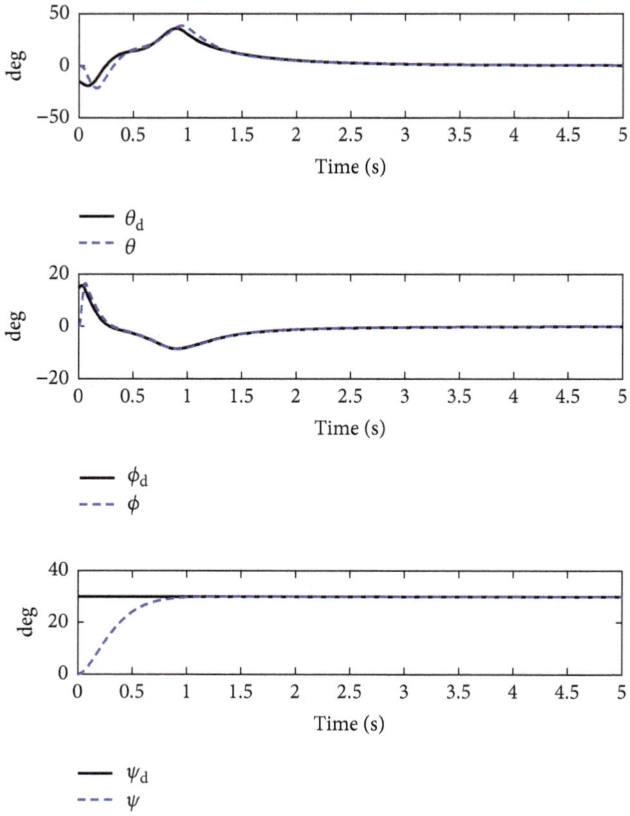

FIGURE 7: DTR attitude tracking effect.

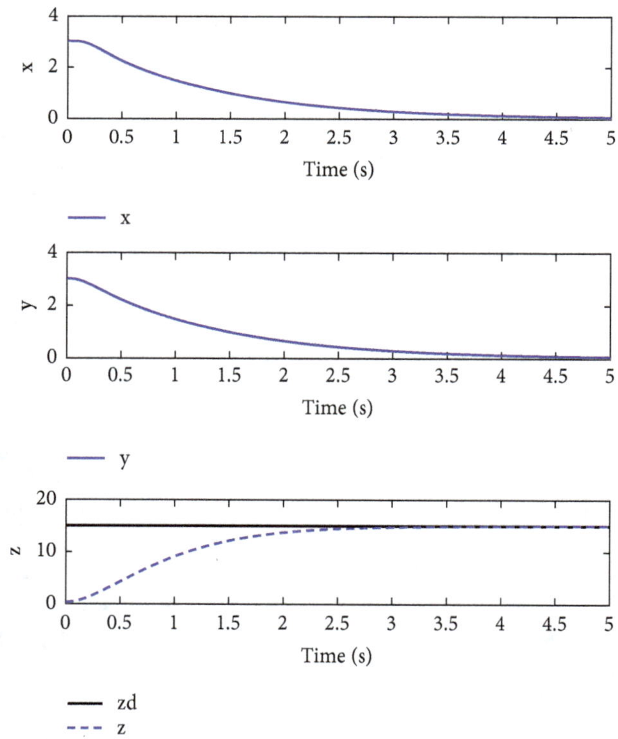

FIGURE 8: DTR position tracking effect.

$$z_1 = \theta_d - \omega_1 \tag{24}$$

Construct the Lyapunov function,

$$V(z_1) = \frac{1}{2} z_1^2 \tag{25}$$

The derivative of the Lyapunov function is

$$\dot{V}(z_1) = z_1 \dot{z}_1 = -z_1 u_\theta \tag{26}$$

To guarantee $\dot{V}(z_1) < 0$, suppose that

$$u_\theta = k_1 z_1 \tag{27}$$

where $k_1 > 0$. The error of the pitch rate is defined as

$$z_2 = u_\theta - \omega_2 \tag{28}$$

The derivative of z_2 is

$$\dot{z}_2 = k_1 \dot{z}_1 - \dot{\omega}_2 \tag{29}$$

And $\dot{z}_1$ can be obtained from (24) and (28),

$$\dot{z}_1 = -k_1 z_1 + z_2 \tag{30}$$

Construct the Lyapunov function,

$$V(z_1, z_2) = \frac{1}{2} z_1^2 + \frac{1}{2} z_2^2 \tag{31}$$

And the derivative of $V(z_1, z_2)$ is

$$\begin{aligned} \dot{V}(z_1, z_2) &= z_1 \dot{z}_1 + z_2 \dot{z}_2 \\ &= z_1(-k_1 z_1 + z_2) + z_2(k_1 \dot{z}_1 - \dot{\omega}_2) \end{aligned} \tag{32}$$

To make $\dot{V}(z_1, z_2) < 0$, suppose that

$$k_1 \dot{z}_1 - \dot{\omega}_2 = -k_2 z_2 - z_1 \tag{33}$$

where $k_2 > 0$. Thus,

$$\dot{V}(z_1, z_2) = -k_1 z_1^2 - k_2 z_2^2 \tag{34}$$

And $\dot{V}(z_1, z_2) < 0$ satisfies the condition for system stabilization.

Part II: We Will Prove the Stability of $\ddot{z}$ Subsystem. The altitude subsystem is designed as

$$\begin{aligned} \dot{\xi}_1 &= \xi_2 \\ \dot{\xi}_2 &= -2\tau_1 s_\alpha \frac{\theta}{m} - \frac{2c_\phi}{m} T + g + \frac{\theta T_D}{m} \end{aligned} \tag{35}$$

where $\xi_1 = z$. The virtual input u_z is introduced which is equal to the altitude rate $\dot{z}$. The altitude error is defined as

$$z_3 = z_d - \xi_1 \tag{36}$$

Construct the Lyapunov function,

$$V(z_3) = \frac{1}{2} z_3^2 \tag{37}$$

The derivative of the $V(z_3)$ is

$$\dot{V}(z_3) = z_3 \dot{z}_3 = -z_3 u_z \tag{38}$$

Suppose that $u_z = k_3 z_3$. Thus,

$$\dot{V}(z_3) = -k_3 z_3^2 \tag{39}$$

where $k_3 > 0$. The altitude rate error is defined as

$$z_4 = u_z - \xi_2 \tag{40}$$

From (37) and (41) z_3 can be obtained as

$$z_3 = -k_3 z_3 + z_4 \tag{41}$$

Construct the Lyapunov function $V(z_3, z_4)$ as

$$V(z_3, z_4) = \frac{1}{2} z_3^2 + \frac{1}{2} z_4^2 \tag{42}$$

The derivative of the $V(z_3, z_4)$ is

$$\begin{aligned} \dot{V}(z_3, z_4) &= z_3 \dot{z}_3 + z_4 \dot{z}_4 \\ &= z_3(-k_3 z_3 + z_4) + z_4(k_3 \dot{z}_3 - \dot{\xi}_2) \end{aligned} \tag{43}$$

To make $\dot{V}(z_3, z_4) < 0$, suppose that

$$k_3 \dot{z}_3 - \dot{\xi}_2 = -k_4 z_4 - z_3 \tag{44}$$

where $k_4 > 0$. Thus,

$$\dot{V}(z_3, z_4) = -k_3 z_3^2 - k_4 z_4^2 \tag{45}$$

And $\dot{V}(z_3, z_4) < 0$ satisfies the condition for system stabilization.

Part III: The Control Inputs Will Be Given. The first equation of τ_1 and τ_2 can be obtained from (23) and (33),

$$k_1 \dot{z}_1 - c_4 \dot{\psi} \dot{\phi} + 2\tau_1 K - 2c_7 \tau_2 = -k_2 z_2 - z_1 \tag{46}$$

The second equation of τ_1 and τ_2 can be obtained from (35) and (44),

$$k_3 \dot{z}_3 + \frac{2s_\alpha \theta \tau_1}{m} + \frac{2c_\phi T}{m} - g - \frac{T_D \theta}{m} = -k_4 z_4 - z_3 \tag{47}$$

Thus, the control law τ_1 and τ_2 can be calculated by (46) and (47),

$$\begin{aligned} \tau_1 &= \frac{a}{b} \\ \tau_2 &= \frac{k_1 \dot{z}_1 + z_1 + k_2 z_2 + 2aK/b}{2c_7} \end{aligned} \tag{48}$$

This proof is completed. □

Table 1: Simulation parameters of the DTR vehicle.

Parameters	Unit	Value
m	kg	4.50
I_x	$kg.m^2$	0.039
I_y	$kg.m^2$	0.042
I_z	$kg.m^2$	0.076
L_x	m	0.32
L_y	m	0.30
L_z	m	0.11
ρ	kg/m^3	1.29
A	m_2	0.20
k_1	1	1.20
k_2	1	1.20
k_3	1	2.20
k_4	1	2.20

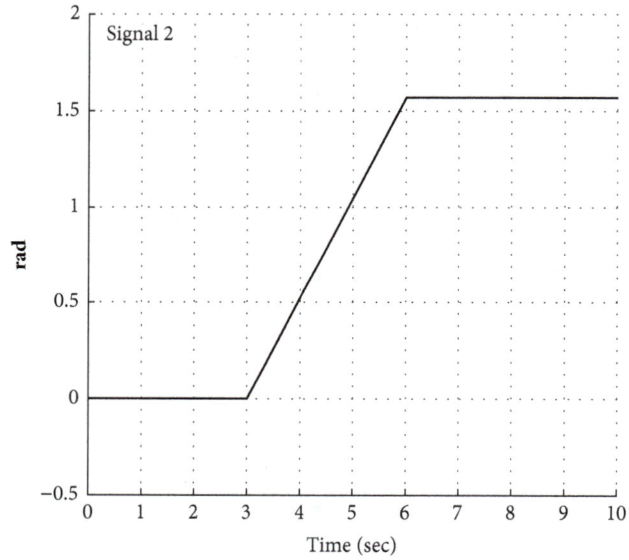

Figure 9: Tilting angle signal source.

The numerical results for the DTR UAV have been showed in Figures 10, 11, and 12. The Figure 9 represents the tilting angle signal source which changes from 0 to $\pi/2$ in 3 to 6 seconds. The system parameters are listed in Table 1.

In Figure 10, the DTR UAV takes off vertically until it reaches the target height $d1 = 15m$. The rotors start to tilt in the 3rd second and end in the 6th second and the height also reaches the reference height $zd2 = 40m$. After transition the vehicle flies horizontally at the reference height. In Figure 11, θ changes within 10 degrees and eventually converges to 0 from the beginning of the transition to the end. ϕ shows a slight changes and ψ slightly deviates from the reference value of the vertical takeoff. However, these tiny changes will not affect the balance of the aircraft. The Figure 12 shows the complete trajectory from takeoff to horizontal flight.

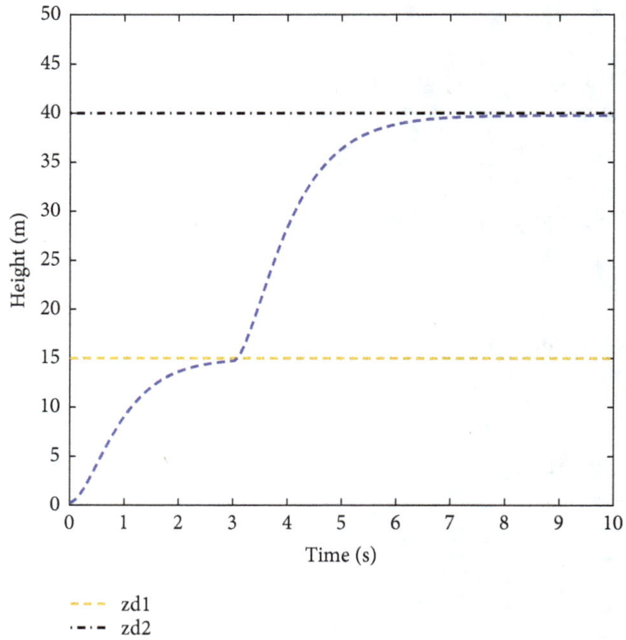

Figure 10: Altitude control during transition mode.

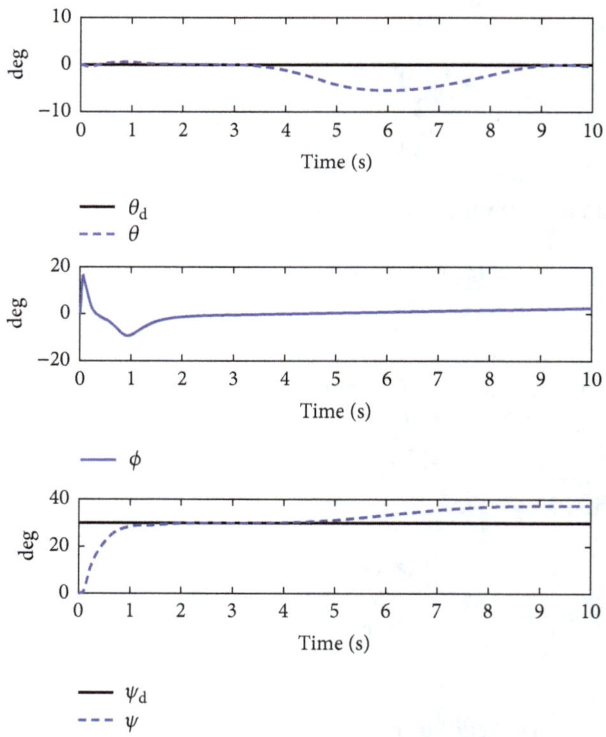

Figure 11: Attitude control during transition mode.

5. Experimental Results

The vertical flight experiment was performed on the DTR aircraft platform which is developed at laboratory. The fuselage framework is built by elastomeric polyurethane (EPU) and carbon fiber. The tilting angle is controlled by two actuators. In addition, the DTR aircraft is equipped with GPS, airspeed

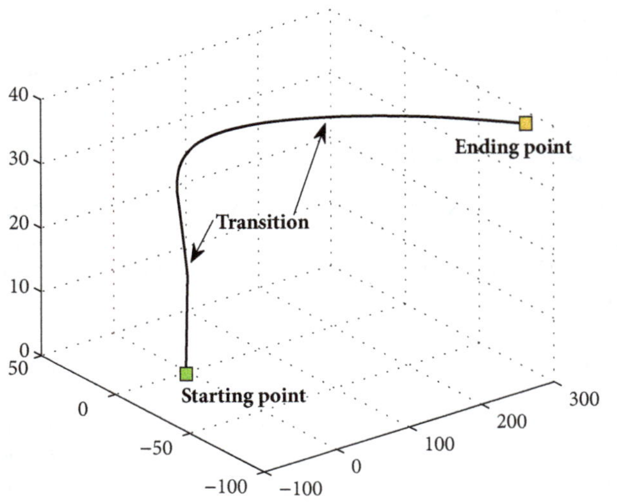

FIGURE 12: DTR trajectory from takeoff to horizontal flight.

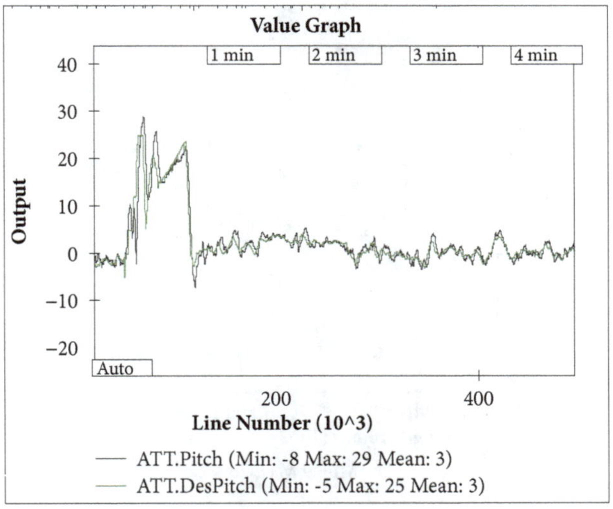

FIGURE 13: The pitch and desired pitch in experiment.

meter, and antenna for communication with the ground station software. The 10000mha high-rate lithium polymer battery is chosen and connected with electronic governor to supply the rotors. Moreover, the control system integrates 3-axis accelerometer and gyroscope to provide aircraft's attitude. The aircraft is equipped with the high precision digital air pressure sensor MS-5611 used to obtain height and connected with the compass through the I2C interface to calibrate the flight direction. The flight control system is based on PIXHAWK Autopilot. PIXHAWK is a dual-processor flight controller consisting of a 32 bit STM32F427 Cortex M4 core processor which contains 256KB RAM and 2 MB flash and a 32 bit STM32F103 coprocessor which is used to ensure safety when the core processor crashes.

The experiment was conducted in an outdoor environment without wind. The experiment results can be obtained by establishing a connection between the ground station software Mission Planner and the aircraft. The experiment results are showed in Figures 13–16 and Figure 18. Figures 13–16 show the DTR aircraft attitude and flight height. And the DTR aircraft is showed in Figure 17. Figure 18 shows the DTR aircraft's flight in the real experiment and the aircraft flight attitude was stable.

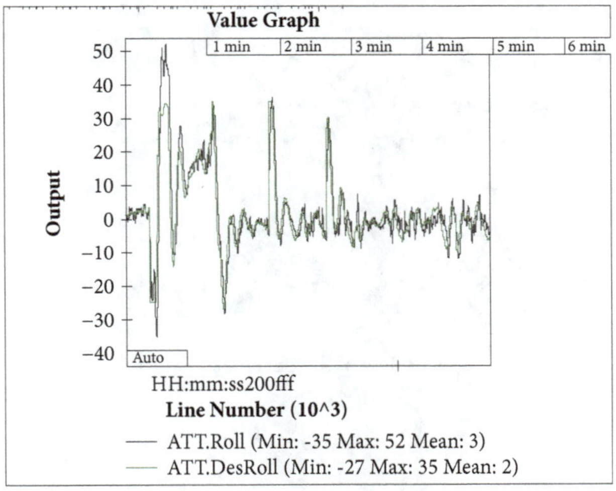

FIGURE 14: The roll and desired roll in experiment.

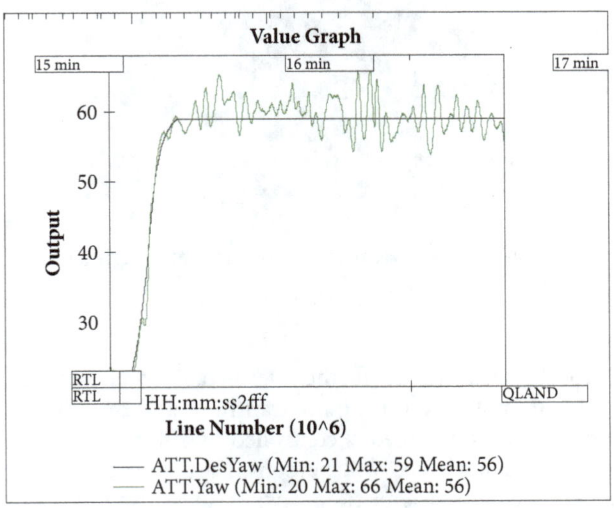

FIGURE 15: The yaw and desired yaw in experiment.

6. Conclusion

A novel tiltrotor aircraft has been presented with two tiltable rotors and two fixed rotors. The kinetics model of the DTR aircraft has been established. And control laws for vertical flight and transition are designed, respectively. The PID controller was designed for the vertical flight and a nonlinear controller based on backstepping approach was designed for the transition. The numerical results have shown the effectiveness of the control approaches for vertical flight and transition. Moreover, the vertical flight experiment was performed on the DTR aircraft platform and the flight attitude was stable. In this part of work, wind disturbance is not considered. Therefore, the robustness of the proposed

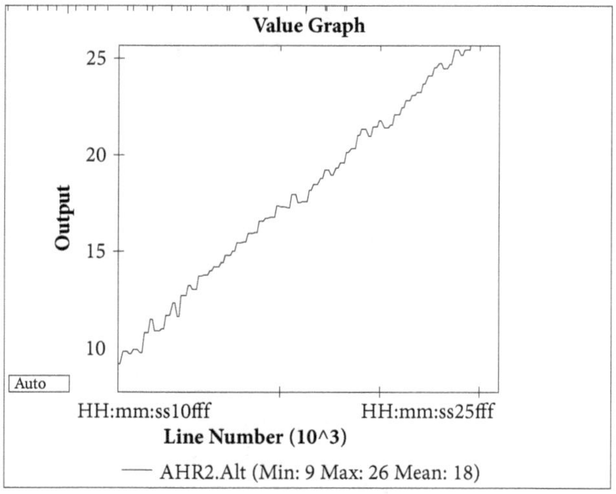

FIGURE 16: The flight height of the DTR aircraft.

FIGURE 17: Overview of the developed DTR aircraft.

FIGURE 18: The DTR aircraft vertical flight experiment.

approach needs to be further verified. The elevator and rudder are not used during the transition and the DTR aircraft attitude is completely controlled by rotors. In the next work, the disturbance will be considered into the kinetics model. The elevator and rudder will also be used to control the attitude of the DTR aircraft.

Conflicts of Interest

The authors declare that they have no conflicts of interest.

Acknowledgments

This work was supported in part by the Zhejiang Provincial Natural Science Foundation of China under Grant LY18F030008 and the National Natural Science Foundation of China under Grant 61375104.

References

[1] C.-C. Yih, "Flight control of a tilt-rotor quadcopter via sliding mode," in *Proceedings of the 2016 International Automatic Control Conference, CACS 2016*, pp. 65–70, Taichung, Taiwan, November 2016.

[2] K. T. Oner, E. Cetinsoy, E. Sirimoglu, C. Hancer, T. Ayken, and M. Unel, "LQR and SMC stabilization of a new unmanned aerial vehicle," *World Academy of Science, Engineering and Technology*, vol. 58, pp. 373–378, 2009.

[3] K. Benkhoud and S. Bouallègue, "Model predictive control design for a convertible Quad Tilt-Wing UAV," in *Proceedings of the 2016 4th International Conference on Control Engineering & Information Technology (CEIT)*, pp. 1–6, Hammamet, Tunisia, December 2016.

[4] G. Flores and R. Lozano, "Transition flight control of the quad-tilting rotor convertible MAV," in *Proceedings of the 2013 International Conference on Unmanned Aircraft Systems, ICUAS 2013*, pp. 789–794, Atlanta, Ga, USA, May 2013.

[5] G. R. Flores-Colunga and R. Lozano-Leal, "A nonlinear control law for hover to level flight for the Quad Tilt-rotor UAV," *IFAC Proceedings Volumes*, vol. 47, no. 3, pp. 11055–11059, 2014.

[6] F. Kendoul, I. Fantoni, and R. Lozano, "Modeling and control of a small autonomous aircraft having two tilting rotors," in *Proceedings of the 44th IEEE Conference on Decision and Control, and the European Control Conference, CDC-ECC '05*, pp. 8144–8149, Seville, Spain, December 2005.

[7] K. Benkhoud, S. Bouallègue, and M. Ayadi, "Rapid control prototyping of a quad-tilt-wing unmanned aerial vehicle," in *Proceedings of the 2017 International Conference on Control, Automation and Diagnosis, ICCAD 2017*, pp. 423–428, Hammamet, Tunisia, January 2017.

[8] C. Hancer, K. T. Oner, E. Sirimoglu, E. Cetinsoy, and M. Unel, "Robust hovering control of a quad tilt-wing UAV," in *Proceedings of the 36th Annual Conference of the IEEE Industrial Electronics Society, IECON 2010*, pp. 1615–1620, Glendale, Ariz, USA, November 2010.

[9] M. Sato and K. Muraoka, "Control performance improvement for QTW UAV by using feedforward gains," in *Proceedings of the 54th Annual Conference of the Society of Instrument and Control Engineers of Japan, SICE 2015*, pp. 824–829, Hangzhou, China, July 2015.

[10] S. Suzuki, R. Zhijia, Y. Horita et al., "Attitude control of quad rotors QTW-UAV with tilt wing mechanism," *Journal of System Design & Dynamics*, vol. 4, no. 3, pp. 416–428, 2007.

[11] T. Mikami and K. Uchiyama, "Design of flight control system for quad tilt-wing UAV," in *Proceedings of the 2015 International Conference on Unmanned Aircraft Systems, ICUAS 2015*, pp. 801–805, Denver, Colo, USA, June 2015.

[12] Q. Lin, Z. Cai, J. Yang, Y. Sang, and Y. Wang, "Trajectory tracking control for hovering and acceleration maneuver of Quad Tilt Rotor UAV," in *Proceedings of the 33rd Chinese Control Conference, CCC 2014*, pp. 2052–2057, Nanjing, China, July 2014.

[13] J. Lee, C. Yoo, Y.-S. Park et al., "An experimental study on time delay control of actuation system of tilt rotor unmanned aerial vehicle," *Mechatronics*, vol. 22, no. 2, pp. 184–194, 2012.

[14] X. Wang and L. Cai, "Mathematical modeling and control of a tilt-rotor aircraft," *Aerospace Science and Technology*, vol. 47, no. 12, pp. 473–492, 2015.

[15] S. Park, J. Bae, Y. Kim, and S. Kim, "Fault tolerant flight control system for the tilt-rotor UAV," *Journal of The Franklin Institute*, vol. 350, no. 9, pp. 2535–2559, 2013.

[16] G. Flores, I. Lugo, and R. Lozano, "6-DOF hovering controller design of the Quad Tiltrotor aircraft: Simulations and experiments," in *Proceedings of the 2014 53rd IEEE Annual Conference on Decision and Control, CDC 2014*, pp. 6123–6128, Los Angeles, Calif, USA, December 2014.

[17] A. Nemati and M. Kumar, "Modeling and control of a single axis tilting quadcopter," in *Proceedings of the 2014 American Control Conference, ACC 2014*, pp. 3077–3082, Portland, Ore, USA, June 2014.

[18] E. Altug, J. Ostrowski, and R. Mahony, "Control of a quadrotor helicopter using visual feedback," in *Proceedings of the International Conference on Robotics & Automation*, vol. 1, pp. 72–77, Washington, Wash, USA, 2002.

[19] H. Liu, Y. Bai, G. Lu, Z. Shi, and Y. Zhong, "Robust tracking control of a quadrotor helicopter," *Journal of Intelligent & Robotic Systems*, vol. 75, no. 3-4, pp. 595–608, 2014.

[20] T. Hamel, R. Mahony, and A. Chriette, "Visual servo trajectory tracking for a four rotor VTOL aerial vehicle," in *Proceedings of the EEE International Conference on Robotics & Automation*, vol. 3, pp. 2781–2786, Washington, Wash, USA.

[21] E. Kayacan and R. Maslim, "Type-2 fuzzy logic trajectory tracking control of quadrotor VTOL aircraft with elliptic membership functions," *IEEE/ASME Transactions on Mechatronics*, vol. 22, no. 1, pp. 339–348, 2017.

[22] R. Xu and Ü. Özgüner, "Sliding mode control of a class of underactuated systems," *Automatica*, vol. 44, no. 1, pp. 233–241, 2008.

[23] C. Papachristos, K. Alexis, and A. Tzes, "Hybrid model predictive flight mode conversion control of unmanned Quad-TiltRotors," in *Proceedings of the 2013 12th European Control Conference, ECC 2013*, pp. 1793–1798, Zurich, Switzerland, July 2013.

15

Toward Dynamic Monitoring and Suppressing Uncertainty in Wildfire by Multiple Unmanned Air Vehicle System

Sharon Rabinovich, Renwick E. Curry, and Gabriel H. Elkaim

Computer Engineering, University of California Santa Cruz, Santa Cruz, California, USA

Correspondence should be addressed to Sharon Rabinovich; srabinov@ucsc.edu

Academic Editor: L. Fortuna

Containing a wildfire requires an efficient response and persistent monitoring. A crucial aspect is the ability to search for the boundaries of the wildfire by exploring a wide area. However, even as wildfires are increasing today, the number of available monitoring systems that can provide support is decreasing, creating an operational gap and slow response in such urgent situations. The objective of this work is to estimate a propagating boundary and create an autonomous system that works in real time. It proposes a coordination strategy with a new methodology for estimating the periphery of a propagating phenomenon using limited observations. The complete system design, tested on the high-fidelity simulation, demonstrates that steering the vehicles toward the highest perpendicular uncertainty generates the effective predictions. The results indicate that the new coordination scheme has a large beneficial impact on uncertainty suppression. This study thus suggests that an efficient solution for suppressing uncertainty in monitoring a wildfire is to use a fleet of low-cost unmanned aerial vehicles that can be deployed quickly. Further research is needed on other deployment schemes that work in different natural disaster case studies.

1. Introduction

In any region undergoing some form of environmental distress, it is very important to detect changes occurring on the ground. In some cases, the environmental incident has spatial changes, and the incident spreads steadily. In other cases, it becomes difficult to follow the incident without knowing how it is evolving. Having a system that follows the event helps rescue human lives, monitor the incident, and allow the human responders to take better actions (as well as deploy assets in an optimal manner to mitigate the incident).

It is of great importance to monitor and respond to natural phenomena (e.g., fires) and national security disasters (e.g., emitting source). One needs to be able to explore a wide area and search for the source of hazardous substance emissions or the expansion of a fire front. In 2016 alone, Federal agencies reported 67,595 fires and an estimated cost of fire suppression of approximately 2 billion U.S. dollars [1]. In addition to financial loss and significant damage to the environment, wildfires threaten the lives of firefighters and civilians during these fire extinguishing operations [2].

There are two main reasons why a solution to fire tracking needs to be found. The first relates to modeling; it is very challenging to predict the fire frontier as a stochastic phenomenon dependent on weather conditions, terrain, and fuel (flammable materials) [3].

Secondly, operational aspects are exposed to severe limitations and constraints. The resources to respond to and monitor disasters are still quite limited. In the aviation section of the National Interagency Fire Center's annual summary of wildfire activity in 2012, there were many requests for large air tankers, which were Unable To be Filled (UTF). The number of cases of firefighters needing air tankers that were unavailable reached a high of 48 percent in 2012 [4] (This means that, in 2012, almost half of all requests for tankers to bomb fires were unanswered due to limited resources).

An incident with a dangerous spread area requires immediate exploration. Some examples are distributed fire spots and chemical threats; however, there are many others. This type of scenario requires surveillance to search for threats, but human observers are difficult to deploy because the task is dull, dirty, and/or dangerous. Wildfire monitoring missions

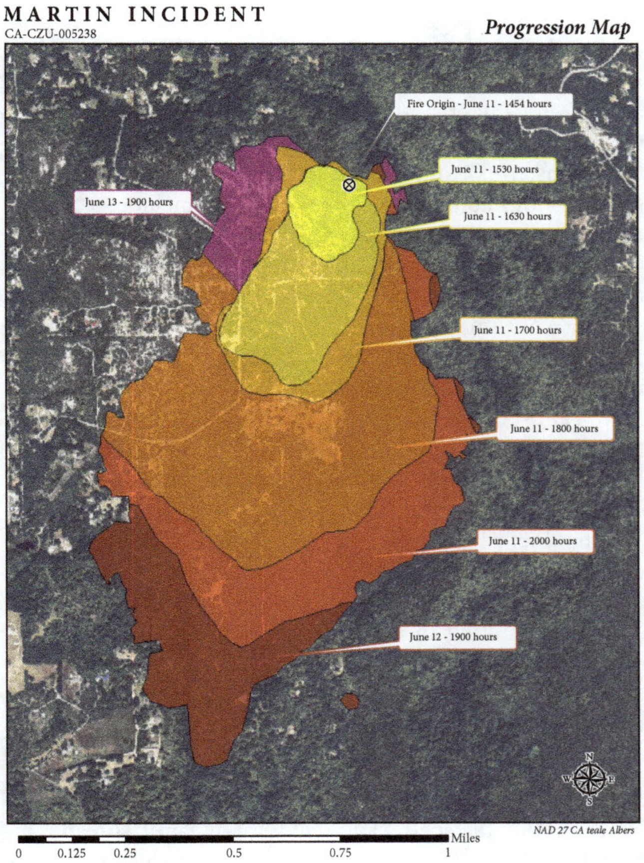

FIGURE 1: Progression map of the Martin Incident. The image is processed **after** the incident and relies on number of sources (from CAL FIRE).

are a perfect example of why a solution needs to be developed. Wildfires (and all natural and national threats phenomena) require urgent attention and an efficient response to monitor and contain their spread.

Figure 1 shows an example of a progression map of a wildfire incident. The periphery grows in time, and the boundary of the wildfire spreads. Knowing that the terrain is a dominant factor helps to understand the outcome in relation to the vegetation type, and knowing the actual weather explains the direction of propagation. The incident last for almost two days and spread out over a distance of 2km. It points to the necessity for a real-time monitoring system.

1.1. Existing Monitoring Systems. Previous studies have examined two different solutions: one on the ground and one in the air. In the first case, ground vehicles are used to explore the area. Use of ground vehicles depends on how passable is the area which needs to be explored. Ground vehicle capability is not necessarily suitable for scenarios with difficult terrain. In the second case, the deployment is in the air, the motion is smooth, and the area can be observed much more efficiently. In addition, most of those scenarios have a critical time factor. The systems phenomena are dealing with time, and because the existing system capabilities are limited, they cannot collect all spatial/temporal information at once. Whenever the observing vehicle is positioned in any one place, the system necessarily misses events in other places within the search area.

Ref. [6] describes existing projects that support disaster management in real time mainly systems that are space-based (e.g., GlobalStar), or high altitude and long duration (e.g., Global Hawk). The projects reinforce the importance of tracking events like floods and earthquakes, and how the tracking events help to monitor the incidents and handle them effectively from the ground control segment.

Ref. [7] describes remote sensing techniques and sensor packages that have been used on UAVs (Unmanned Air Vehicles). The author argues that these techniques can serve as the main solution for various disasters. Reviewing the literature on the development of UAVs, including projects with different types of sensors (IR/Visual) and platforms (fixed-wing/rotary-wing), he concludes that Multi-UAVs can be used to avoid the drawbacks of approaches that are based on either land or space.

1.2. Multi-UAV. The ideal mission for a UAV is monitoring a wide area and searching for the source of emission of a hazardous substance or expansion of a fire front. There has been a rising interest in increasing UAV efficiency and reliability. Autonomous vehicles have been used in a variety of applications for surveillance, search, and exploration [8]. In surveillance problems, the target space needs to be surveyed continuously. It differs from a coverage problem, which involves optimal deployment of the sensors for complete coverage of the target area. It also differs from an exploration problem, which deals with gaining more information about the bounded area [9]. This exploration research moves in two directions. The first focuses on how to pinpoint the source of an odor [10, 11]. In this area of research, robots are tasked with detecting, tracking and seeking the odor source efficiently.

The second direction of this exploration research focuses on how to establish the boundary or perimeter of a spreading phenomenon in order to monitor the area and prevent human exposure to risk [12]. Because of the spatial limitations of a single UAV, most research currently focuses on how to monitor large areas by operating multiple inexpensive simple UAVs simultaneously [13].

Though the studies mentioned above are significant, they focus on exploring the environment using clues (e.g., aerosol diffusion) for tracing emission sources. Moreover, the techniques used to detect the plume or periphery are strongly dependent on the spatial gradient change of the underlying tracked phenomenon. The research presented in our work proposes to explore the area by using approximate inference methods [14] and statistical reasoning [15]. The developed method takes into consideration the operational aspect of the mission in addition to the statistical characteristics of the underlying phenomenon.

1.3. Coordination. Most of the multi-UAV systems are designed to address problems related to specific research in a particular environment of interest. The UAVs cooperate and share data to obtain information on a certain aspect of the environment. Regardless of the number of UAVs and size of the AOI (Area Of Interest), cooperative systems deliver an improved overall picture of the environment through coordination.

The design of cooperative systems mainly discusses the control strategy (e.g., centralized or decentralized) and the level of autonomy. Framework design inspired by a biological system has been a popular concept of research for some time (to name a few [16–18]). Nevertheless, achieving such complexity through control techniques is considerably challenging.

There are many studies on multi-UAV cooperative control systems that address coordination issues. These focus on designing a system to control and monitor a region. One of the earliest studies proposed using aerial photographs to monitor fires in order to combat them [19]. The objective was to use aerial photographs to map the fire and then coordinate the team on the ground. In the past few years, the literature has included more and more research of systems utilizing a team of small cooperating UAVs to get better surveillance; that is, better response time in missions where time is critical.

Recent studies have focused on special missions that can be efficiently performed with multi-UAV systems. Some address the problems of formation flight and some the problems of coordination. Fewer studies have been done on reconfiguring the coordination [20] or on coordination where the assigned tasks have uncertainty. This paper demonstrates that if the guidance system accounts for real-time events and is able to adjust the flight formation to incorporate changes, then the trajectories are more effective than traditional methods.

Closely related is the work that has been done on multi-UAV coordination for tracking missions for search and rescue or surveillance [21]. It presents a concept that relies on low altitude and short endurance (UAVs). The work explores tracking a fire line by using a team of UAVs following the perimeter of the wildfire area. The UAVs return periodically to the ground station location for downloading the collected data. The research focused on how to minimize the latency associated with the fire perimeter measurement when it is transferred to the ground station.

In [22], the design includes a coordination scheme to control a rotary-wing platform (Quadrotor) for a similar mission to the one above. Essentially the motion of the UAV patrols the propagating perimeter. Whenever one UAV approaches another UAV (rendezvous point), deconflict the rendezvous and resolve each UAV next flight direction. That research assumes, however, that the perimeter of the fire is circular. These studies (and similar ones) examine a specific scenario where the focus is on directly tracking the periphery point. This is limited, however, by focusing on the connection between the uncertainty of the spreading perimeter and the maneuverability of the fleet needed to maintain knowledge of the complete perimeter.

1.4. Observation. In coordination, one of the basic operations is observation sharing. Most of the recent studies in multi-UAV address the problem of partial information. It reflects the "real-world" problem where the UAV has limited communications (range or bandwidth). One UAV can communicate with one that is close by, but not with another that is far away. Ref. [23] presents a variety of research problems in which multivehicle systems agree on the value of observed data (consensus), and explores control strategies and a set of solutions for implementing them. Ref. [24] includes a chapter that suggests various deployment algorithms. They consider a distributed algorithm to address the physical limitations of the communication system for observation sharing.

If a coordination algorithm for an environment with uncertainty is available, the overall system still relies on individual sensing capabilities. Even if the system uses the best or most advanced sensors, the sensors can be restricted by environmental conditions, i.e., the sensors carried by the UAV do not have sufficient range [25], and the data measured can only be local and quantized.

The inefficiency of current systems with high-level-control creates significant timing difficulties for achieving the mission objectives. The ongoing mission can leave one vehicle loitering, resulting in a high latency of updates. Based on different studies [21], this represents a large time loss during a mission, with fewer updates, which in some cases can cause the mission to fail in its tracking objective.

1.5. Propagated Periphery Modeling. Disaster growth models, which predict the spatial and temporal dynamic spread rate, may help in evaluating the situation and deciding on a suitable response in a real-time deployment [3]. Appropriate representation and estimation of the spatial uncertainty can improve the prediction or help in developing a simplified model [26]. A mission with an uncertainty model for the AOI stands to benefit substantially from the predicted confidence envelope approach. For example, in expected high rate of spread (RoS) segments along the AOI perimeter, the allocation can use the availability and priority of the segment to get better results than if it were to assume that all segments along the perimeter are identical. Available UAVs can be redirected to new areas instead of merely loitering.

In one of the biggest wildfire research projects done by the *Joint Fire Science Program*, the researchers developed fire behavior models for operational use. Their main objective was to develop a detailed dynamic model to predict the physical behavior of the ground phenomenon. They considered two simple fire modeling approaches. In both models, the assumption was that the local spread at a point on the perimeter is perpendicular to the fire perimeter into an unburned environment and that the fire has a local RoS normal to the fire line.

2. Problem Definition

This research proposes a system design and implementation for quick deployment of a low-cost, low-power fleet of UAVs with a high-level ground control system. The results of this research project introduce the new methods which can serve as high-level-control for operational multiagent systems: a method to estimate a propagated boundary [5] and a scheme

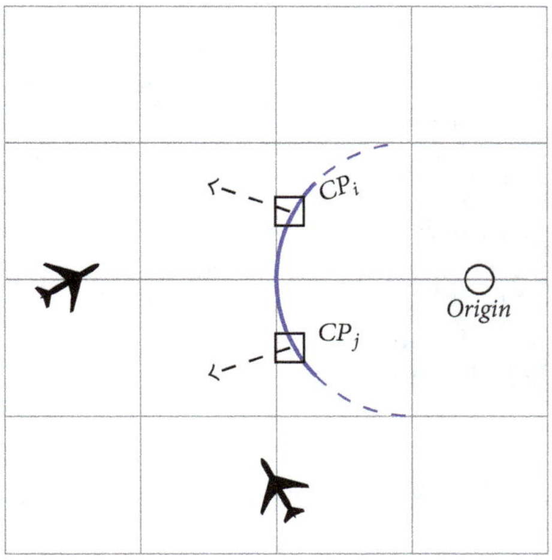

FIGURE 2: Setup of the boundary representation approach. CP_i and CP_j (square) are two of many grid points representing the close predicted periphery (in blue). The Origin (circle) is the starting point of the propagated phenomenon, and the UAVs are used to collect observations.

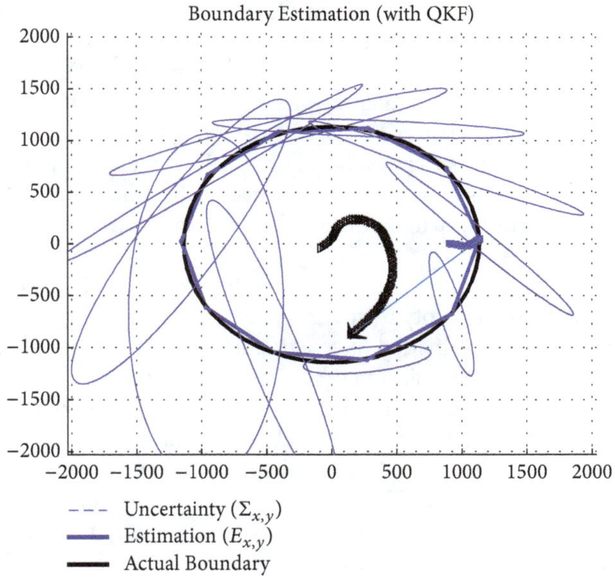

FIGURE 3: A periphery estimation with a single UAV for an autonomous mission is illustrated. The red line represents the actual periphery, and the blue line represents the estimated one. The UAV flies over the explored area autonomously. The line of sight to one of the CPs illustrates the directional effect of arbitrary CP. The QKF method is employed on all the CPs simultaneously, and the UAVs identify the current highest uncertainty to approach next (originally presented in [5]).

for optimal deployment of a fleet of agents in an exploration mission [27].

The problem is one of optimization with respect to time with sparse measurements detected by a fleet of UAVs. The UAVs have a dynamic process to monitor, as quickly as possible, a periphery represented by a set of Control Points (CPs). The complete system design considers the uncertainty of the bounded phenomenon, where each UAV fleet member carries an on-board sensor to distinguish between inside and outside areas.

Figure 2 illustrates the approach taken to represent the boundary with a set of CPs connected by straight lines. Each CP has a nominal spread rate that is considered relative to the origin point of the propagated phenomenon. That is, the spread rate is always pointed outward. The information is being gathered by a UAV to provide the observations that are noted as *IN* or *OUT* relative to the enclosed periphery. The optimum policy is derived from the decision of which CP the UAV should approach first to reduce uncertainty.

3. Periphery Estimation Methods

The estimation methods approach in this research project relies on an earlier developed technique for estimation of propagated boundary with quantized measurements [5]. The monitoring system involves large numbers of possibly randomly distributed inexpensive sensors, with limited sensing and processing. The estimator incorporates observations gathered by multiple observers and uses the Quantized Kalman Filter (QKF) estimation method [28] to update the expected location and unobserved spread rate. This technique has been extended and laid out the groundwork for the Greedy Uncertainty Suppression (GUS) strategy [27].

The estimation is meaningless in a situation where the available sensors are located inefficiently (e.g., considerably far or colocated). The GUS strategy searches for trajectories that improve the estimated boundary of a propagated phenomenon. The methodology designed for on-line look-ahead approaches to rerouting the UAVs is computationally intractable [8]. To improve the performance, even more, one can use the new approach to further suppress uncertainty. The UAV trajectory is changed to continuously reduce the uncertainty in the biggest covariance among all CPs, by flying directly to the tip of the major axis of that ellipse. Figure 3 illustrates the basic concept for reducing the uncertainty autonomously. Each associated uncertainty is represented as an ellipse (of 95% of confidence area). In previous work, it has been observed that reducing uncertainty is related to measuring distance as well as the arbitrary approach angle direction to an arbitrary analyzed CP. Moreover, uncertainty depends on measurements availability. Hence uncertainty grows with time when no significant observations have been incorporated. The UAV can approach a CP along the direction of its maximal uncertainty axis (direction of major axis of the covariance) and reduce the one-dimensional uncertainty. Interestingly the observed property affecting the uncertainty of CP is the line-of-sight direction.

4. Guidance Logic

4.1. Overview. The trajectory addresses the monitoring problem by adopting common principles. The first is a bookkeeping. The mission planner keeps track of representative quantities of interest. Reference [29] suggests discretizing the target space. Space gridding is typically performed after initial space decomposition which divides the target space into subspaces and later assigns them to different UAVs.

The second fundamental principle is the mission plan. The method considers the UAV deployment stage and pre-planned search patterns (e.g., zigzag, spiral, and alternating) in case there is a need to switch back. The third principle is the replanning which adjusts the uncertainty models with observations.

The suggested strategy considers the perpendicular direction of the uncertainty major axis as the next search direction. The periphery is divided into segments based on the number of UAVs and the resolution of the periphery (number of CPs). The planner chooses important CPs which are distributed along the perimeter. Assigning a UAV to one of the CPs will influence the others and reduce the uncertainty accordingly.

Deploying the UAVs is based on the number of available resources. For example, for two UAVs the deployment is to two segments of the predicted polygon where one UAV is assigned to the highest look-ahead uncertainty (weighted by variance and time-to-go) and the second direction is the highest uncertainty on the remaining segments.

Three benefits are gained from that allocation policy: First, the solution avoids flyby trajectories and potential collision. Second, UAVs are not allocated to the same or a close area. Third, the trajectories are being evaluated for dynamic motion feasibility to be carried out by the assigned UAV.

4.2. Greedy Uncertainty Suppression (GUS). GUS strategy tends to minimize the maximum uncertainty over all CPs by incorporating observations over a long period. The policy achieves longer look-ahead with an on-line rerouting logic for the fleet members' task.

The implementation includes two main parts; coordination and allocation. The basic operation leads to coordination involves sharing information of the assigned tasks. The UAVs share their observations with a centralized entity. The observations incorporated sequentially in the estimation process. GUS algorithm is a step by step procedure to determine the best task to each UAV. The notation uses superscript j to label UAV and i as an index of an arbitrary CP.

The first step relies on a previously developed algorithm [5] (QKF). This procedure includes system coordinates transformation, scalar probability evaluation, and Kalman Filter to estimate the state ($\widehat{x}_{x,y}^{CP_i}$) and covariance ($\widehat{P}_{x,y}^{CP_i}$) of the CPs in the original coordinates frame.

The following steps determined the new policy. Step 2, sort the CPs estimation by their major variances. After correcting the state ($\widehat{x}_{x,y}^{CP_i}$) and updating the covariances ($\widehat{P}_{x,y}^{CP_i}$) the procedure continues to evaluate the major-axes of the projected uncertainties. By sorting the variances and adjusting the waypoints along the compass line the algorithm generates candidate destinations.

To evaluate all the alternatives, Step 3, run Dubins Vehicle algorithm which also provides the length of feasible trajectories and evaluates the time-to-go for each UAV. The associated trajectory for UAV_j weighted by the cost function and step 4 assigns the UAV to its best feasible task. The weights of the cost function address the need to consider additional restrictions (for example, deploying the UAVs to one side of the periphery) or tasks.

The propagation of the error between predicted and actual boundary can increase with no control. An additional task assigned where UAV had not cross the actual boundary for a long duration. The policy includes a special allocation mode, rerouting toward the origin point and searches for a crossing point. When that step is done the algorithm is back to the default allocation mode.

5. Monitoring System Architecture

5.1. Background. For many experimental and operational applications, UAVs can enable or enhance the efforts available to researchers or operational teams. Much work has been done to make UAVs useful in a myriad of scenarios. In some scenarios, operating in the environment requires special skills or training that operational teams do not have; here an autonomous system can enable access that was previously difficult to obtain. In recent years, there has been a rapidly increasing interest in UAVs where the operational problem requires an airborne platform.

Technological progress has made it possible to use inexpensive autopilots on small UAVs. The development of high-density batteries, long-range and low-power radios, cheap airframes, high-performance microprocessors, and powerful electrical motors all make experimental research or operational team with UAVs more practical than ever [30]. The availability of UAVs as a fast deployable resource allows teams to explore many new kinds of scenarios such as wildfire. The flexibility of the system design further allows for quick changes, reducing the project workload.

A modern UAV system consists of an on-board control system (i.e., autopilot) and Ground Control Station (GCS). The autopilot utilizes various sensors, communication modules, a power supply unit, and embedded software to control the UAV. The autopilot software is the real-time implementation of the guidance, navigation and control algorithm; one of the demands on designing a rapid prototype testbed is to enable control algorithms, discussed briefly in [31].

Autopilots control and guide the UAVs in flight. They rely on data gathered by various sensors and on a central processing unit (CPU), which carries out the instructions of the program. The objective of an autopilot system is to consistently guide the UAVs to follow reference paths or navigate through several waypoints. A UAV autopilot system is a closed-loop control system consisting of two parts: the state observer and the controller. A typical observer is designed to estimate the state (e.g., attitude) from sensor measurements (gyro, accel); advanced control techniques are used in the UAV autopilot systems to guarantee smooth, desirable trajectory navigation.

This paper focuses on the design of a multi-UAV system that is used in this research project and future projects. The emphasis is on the need for multi-UAV coordination and high-level-control. Ref. [31] provides a review of the existing

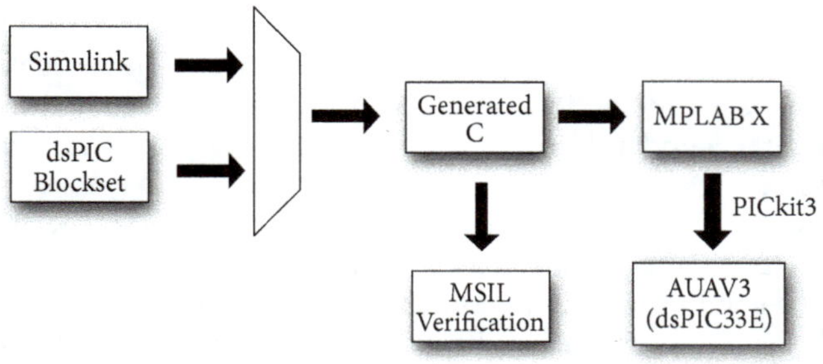

FIGURE 4: SLUGS II code generation workflow, with a new verification step, Multi-UAV Software in the loop (MSIL).

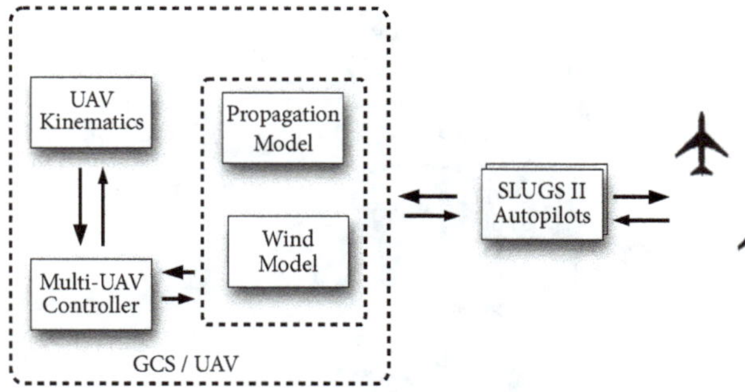

FIGURE 5: Multi-UAV Monitoring System, software block diagram.

autopilot and the migration process from the previous successful rapid prototyping concept to a new design.

SLUGS (Santa Cruz Low-cost Unmanned Aerial Vehicle Guidance, Navigation & Control System) is a platform that includes autopilot software and hardware components that enable a flexible environment for research in GNC applications [32]. The SLUGS was designed primarily for GNC research, and it has already been used in many flight tests. It is also part of the experimentation with fixed-wing UAV systems, presented in the following section.

The SLUGS II design improves on previous SLUGS design because it provides rapid prototyping control for multi-UAV systems. The control design process is made up of many iterations that can be verified and validated through both simulation in the Simulink environment and with autocode generation.

5.2. Software Design. The complete autopilot algorithm is implemented in Simulink using block diagrams and Matlab toolboxes (MPLAB X - Microchip Integrated Development Environment (IDE), and dsPIC - digital signal controllers support for Embedded Coder). Simulink blocks and Matlab routines are effective software that can be used to modify the algorithm and verify the design. Once the model is updated in the Simulink environment, it then generates the new code with the updated features. The R&D work in a model based environment makes the programming phase easier. Simulink includes tools that automatically generate and compile the code. The code is then deployed directly to the autopilot hardware [33].

SLUGS II modifies the design process to add a verification step for the generated code in a flexible and friendly environment that is committed to the sequence of events in software rather than to guarantee strong real-time performance execution of the code. Figure 4 demonstrates the code generation process and the design validation is discussed in detail in the implementation section.

The software design, as presented in Figure 5, introduces the first order constraints of a dynamical system where the vehicles are mobile and the environment domain changes.

The models are software oriented implementations in which the execution processes guarantees to reconstruct the outcome; hence, the process is deterministic. The UAV model

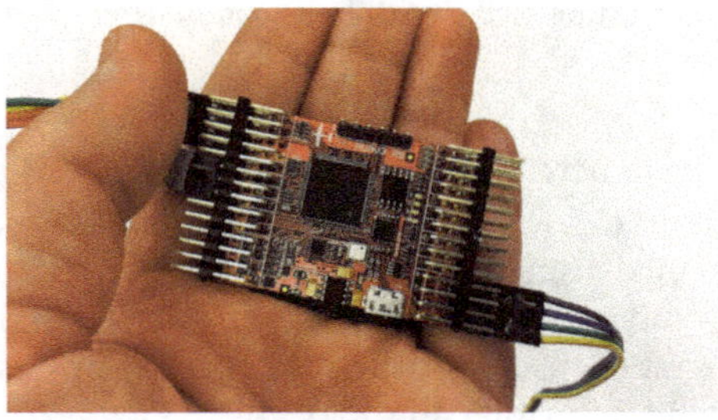

FIGURE 6: AUAV3 board.

in the software is a mathematical representation of the actual motion of nonholonomic systems.

The propagation model helps anticipate the boundary location in time. The developed model represents the firefront propagation of a wildland fire. The implementation of this model is a greedy evaluation; each point out of a set of grid points along the periphery is evaluated. Wind velocity and ground slope are incorporated into the propagation model. The model combines the wind velocity and ground slope directly to each grid point of the boundary.

5.3. Hardware Design. The literature on COTS autopilots suggests that the minimum requirements for a research autopilot are robustness and attitude accuracy, enough for low altitude flight surveillance. Hardware must include sensors on-board and software for an attitude solution [34]. This hardware design makes an important contribution to the research framework because it introduces a new design. The SLUGS embedded system features two Microchip dsPIC33F microcontrollers. That design allows SLUGS to implement more complex and effective Guidance, Navigation and Control (GNC) algorithms. It provides a high level of safety and fault tolerance features, and it is designed such that the autopilot system would have more than enough processing power. However, it means more maintenance for the research autopilot Integrated Development Environment (IDE), and increased cost.

SLUGS II simplifies the existing design by using on a reliable commercial-off-the-shelf (COTS) hardware. The AUAV3 is a commercial open-hardware development board (all PCB layouts are provided) [35]. It features a single Microchip dsPIC33EP with twice the clock rate of the dsPIC33F. The AUAV3 board (see Figure 6) comprises peripheral circuits for

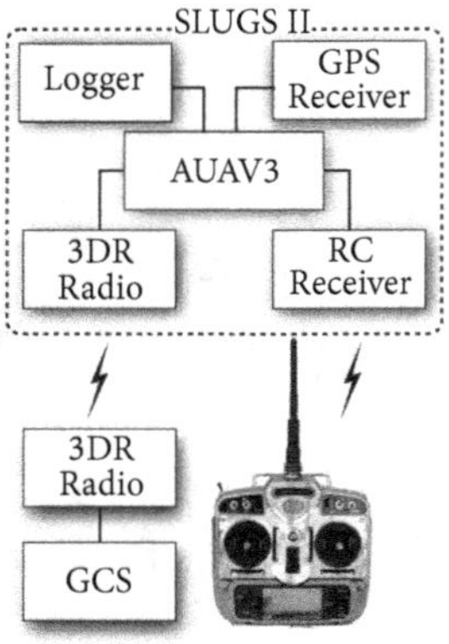

FIGURE 7: SLUGS II basic components.

IMU, Magnetometer, Barometer and the standard communication interfaces (SPI, CAN, UART and I^2C). Researchers have examined using the AUAV3 to replace the in-house SLUGS hardware. Ref. [31] discusses the SLUGS II Simulink model migration process in more detail, and the steps for verifying and validating the performance with a series of flight tests.

The design of SLUGS II is such that it can be adapted to various different scales. It provides a solution for multiple UAVs in the testing environment when they are needed for research or development. The major challenges of research in multi-UAV are handling duplicate systems with low maintenance cost, reliability, and researchers' insufficient skills. The UAV is linked with continually changing technology so that new infrastructure needs to be assessed and adopted in order to improve the existing system. The system design can help with this maintenance by enhancing the new R&D autopilot environment, eliminating the need to maintain multiple platforms can reduce overhead costs and development difficulties.

Figure 7 presents the basic hardware configuration per any single UAV employed by the monitoring system. The AUAV3 autopilot controls the GPS receiver, telemetry (recorded by a logger and transmitted to the GCS), remote-control (RC) inputs for a pilot safe mode and radio transceivers (3DR Radio) for manual and autonomous flight modes.

6. Monitoring System Implementation

6.1. SLUGS II Components. The AUAV3 addresses the issue of the skills needed to develop or maintain in-house hardware. Commercial hardware is constantly being updated, and for the R&D autopilot, this is an opportunity to put all the efforts into developing GNC algorithms and utilizing low-cost COTS hardware. The old hardware is difficult to integrate with newer sensors and sensing technology. Complex applications require a flexible and adaptive R&D autopilot to keep up with a dynamic environment.

The UAV model integrated within the Simulink development model is another challenging component. It needs specialized skills to tune and adjust to different platforms. Porting the X-Plane simulator improves the development effort and further reduces costs. Different airplane models can be found on the local simulator database instead of tuning the aerodynamic coefficients of a six-degree-of-freedom (6DOF) model by hand.

Two components are migrated as part of SLUGS II design. The benchmark configuration takes the MatrixPilot open-source autopilot and deploys the code on the AUAV3 board. Performance benchmarking ensures that the migrated components perform as well as or better than the old components. The new configuration is then evaluated in multi-UAV software in the loop (MSIL) simulation and in real flight tests.

Once the assessment of the AUAV3 board is completed, the Simulink model is then modified. The model adjusts to the new dsPic configuration. This integration phase includes eliminating the blocks that handle communication between the separate processors, improving the modeling style, optimization, removing dead code, and identifying incompatible porting issues. Configuring the Simulink model to the new AUAV3 board is based on the Microchip dsPIC toolbox (a new revision of the Lubins Blockset [36]). Although the complete process requires significant manual work, the main intellectual property (IP) of the R&D autopilot remains almost untouched.

In the final phase, the newly migrated autopilot is subjected to rigorous testing using test cases applied to the original design (SLUGS) and MatrixPilot [37]. Apart from the functional load testing, testing is carried out to ensure that the necessary performance level is achieved. The migrated autogenerated code is deployed, and parameters are fine-tuned for the new airframe (BixlerII).

6.2. Ground Control Station. The GCS is one of the most important components in a UAV system. It provides an operational interface to monitor and control the assigned task to the multiple UAVs. It presents any additional information that does not require the autopilot to complete its task; however, it supports the user who monitors the mission to coordinate with other systems for better decision making. The GCS includes indications for the mission showing the relevant spatial data (i.e., geodetic coordinates) associated with the map of the area of interest (see Figure 8).

The GCS communicate with the UAVs using a bidirectional data link (X-Bees transceivers). It runs on a mobile laptop computer that can easily be transported to the test site.

A complete process that supports a multi-UAV configuration is needed to be considered by the autopilot system for real-time identification and task allocation. To support a multi-UAV configuration, the SLUGS II design extended the tools for software verification. The multi-UAV IDE offers code verification with complete software in the loop (SIL) simulation.

6.3. Multi-UAV Software in the Loop. MSIL simulation is a higher level of fidelity for the final steps of developing the high-level-controller. MSIL simulation allows running the SLUGS II research autopilot on a computer before running

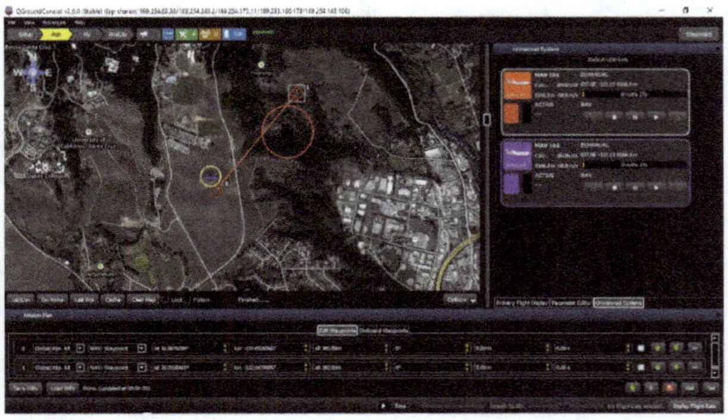

FIGURE 8: The graphical user interface (GUI) of the GCS is presented. The open-source software (Qt-Ground-Control: QGC) is adopted and extended to support the design of a multi-UAV monitoring system. The software supports the planning and visualization of the UAVs' trajectories in real time.

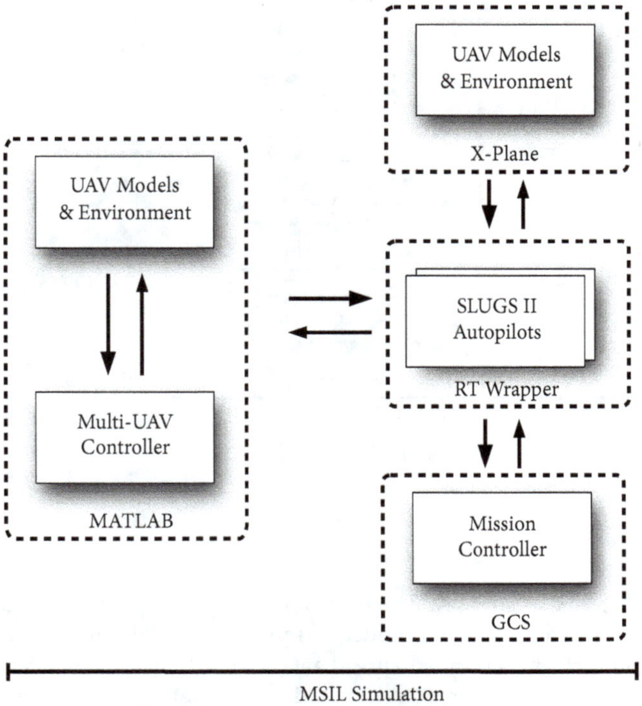

FIGURE 9: MSIL block diagram.

it on the target processor. It communicates with a simulator for simulating high-fidelity flight dynamics (X-Plane). The MSIL simulation is meant to run a single or multi-UAV configuration and support the external interfaces and built-in internal calls (for example, memory, timing and peripheral libraries) of every instance of the SLUGS II autopilot code.

The MSIL software includes the generated code, which is compiled together with a handling layer (real-time wrapper software). The RT Wrapper interfaces with the external software through a User Datagram Protocol (UDP) socket or a serial port. The MSIL simulation controls the simulated GPS, telemetry and remote-control (RC) inputs for a real RC controller (training mode). The autopilot researcher benefits from the ease of integrating the original generated code and having an easy, friendly environment for debugging.

The GCS unit controls the UAVs through a communication bridge to ensure two-way communication between the GCS and the SLUGS II autopilot. The autopilot can directly manage information from the serial port (or in case of MSIL from the buffer of the serial port). The RT Wrapper (in Figure 9) is responsible for managing the buffers and for distributing the MAVlink messages between real UAVs or simulated modules.

The coordination algorithm is executed in Matlab and works as an extension of the GCS. The RT Wrapper creates a tunnel between Matlab and the SLUGS II software through a physical communication link (UDP) using the MAVlink protocol.

6.4. Multi-UAV Hardware in the Loop. The Multi-UAV hardware In the Loop (MHIL) simulation runs the SLUGS II software stack on the AUAV3 flight controller using raw sensor data fed in from the simulated environment running on the desktop PC. HIL simulation replaces the UAV and the environment with a simulator (the simulator has a high-fidelity aircraft dynamics model and environment model for wind, turbulence, etc.) The physical autopilot hardware (AUAV3) is configured exactly as for flight and connects to a computer running the simulator rather than the aircraft. In this sense, the AUAV3 does not know it is flying a simulation.

Figure 10 shows the MHIL setup. The involved units in the MHIL configuration are depicted along with their associated interfaces. The AUAV3 and the GCS are connected physically by a telemetry link. The autopilot is connected to a computer running the simulator. The simulator is fed by the servo commands and responds with sensory values from the simulated airplane model. The generated sensor values are similar to the IMU output and injected to the navigation algorithm as the UAV autopilot flies the high-fidelity flight situation.

In the end, all of the various functionalities must work both as individual subsystems, but also integrated as part of the entire system: experiment with the UAV design, the basic multi-UAV flight formation, and the monitoring system control. Each one is a step in validating the complete system design which addresses the full multi-UAV monitoring problem.

The system architecture can be utilized in a centralized or a decentralized scheme of operation to enable coordination and information sharing. In a centralized system configuration, the UAVs relay real-time information between each other through the GCS. Alternatively, the UAVs could transmit real-time information between group members (a decentralized scheme configuration).

7. Simulation Results

7.1. Periphery Estimation Evaluation. The simulation is designed to evaluate all major components which involve

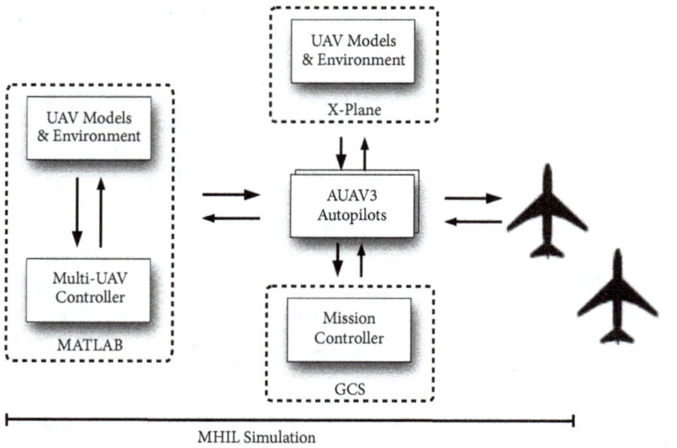

FIGURE 10: MHIL block diagram.

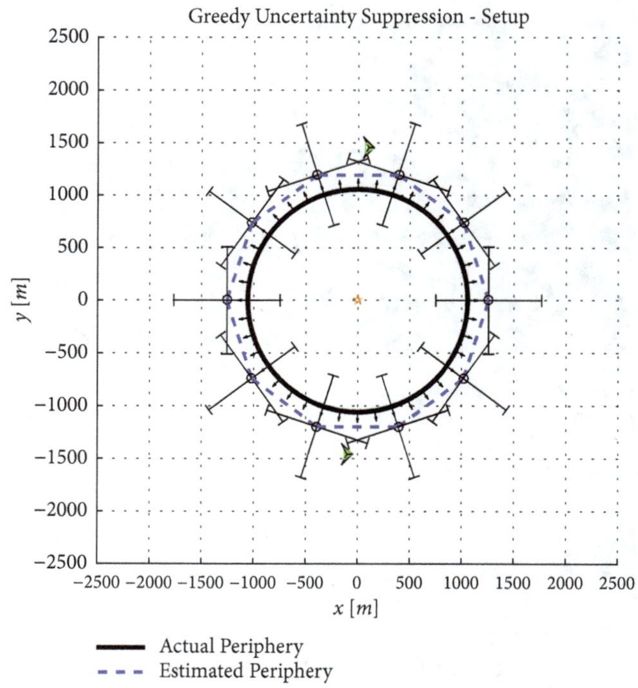

FIGURE 11: Initial setup. The UAVs are at the final stage of the deployment phase and located on the opposite sides of the boundaries. The actual periphery is a solid red line, and the predicted periphery is a dashed blue line. The error bar associated with an arbitrary CP represents its current perpendicular uncertainty (1σ). Note that the error bar is equal and results in a predetermined prediction that is based on a maximal spread rate.

in the GUS strategy. The environment conditions are being simulated based on a model of a propagated wildfire with a random and bounded spread rate (3 ± 0.1 $[m/sec]$). UAVs allocation is being implemented in a separate component which incorporates the observations gathered by the simulated UAVs.

The UAV dynamics model subject to constant-speed of $20[m/sec]$ which is the approximated speed of the platform that has been developed and examined for the experimental stage of this research. Moreover, the centralized controller comprises the QKF estimator which fuses the observations, and it is based on a previously derived technique. The following simulated scenarios explored the efficiency of the suggested concept.

The initial setup attempts to adhere to the real problem, and therefore a real-time data received from the CAL FIRE (San Mateo Santa Cruz Unit for the Martin Incident) is used. For example, the initial AOI is large ($1Km \times 1Km$) and the time scale is long (i.e., hours). Figure 11 shows the actual periphery with two UAVs deployed from both sides of it.

The propagation model used in the simulation is simplified. However, it allows investigating the major properties of fire spreading. The dynamic expansion of the boundary, the environmental effects (i.e., wind and slope) and the feasibility are all considered in the implementation and are utilized for different scenarios.

Figure 12 demonstrates the scenario with running GUS. A local error bar represents the uncertainty of each CP. The size of an error bar is correlated with the size of the perpendicular and tangential variances.

The performance measure offered in [5] accounts for two performance indicators: errors and uncertainty. The errors indicator comprised the mean-square-error, where the errors are between the predicted and the actual periphery. The uncertainty indicator is simply taking the mean of the CPs' major variances. Both indicators are weighted equally in the combined performance measure, $\sqrt{\sum_i^N (err_{CP_i}^2 + \sigma_{CP_i}^2)}$. In the figures, uncertainty is represented by an error bar in the global coordinates system, and the performances are evaluated relative to the perpendicular component of the local predicted periphery.

Figure 13 shows the combined performance measure with its two performance indicators.

If there were no errors and no uncertainty, then the traditional periphery tracking was an optimal approach. In practice, the uncertainty grows with time, and although errors are reduced to a minimum when the UAV crosses the CP, the spread rate is not observable, and the errors continue to grow shortly after updating the location with the nearby CP. The resulting trajectories and performance improve the benchmark strategy results (see [27]). Figure 14 demonstrates that the GUS and the benchmark have very different performance for a scenario with a wind. The GUS reduces the uncertainty much more over the time of the mission.

8. SLUGS II Validation

8.1. System Configuration. The platform used for the first flight tests was a Phoenix R/C aircraft; later the platform was changed to a Hobby King Bixler 2 (demonstrated in Figure 15). Both of the planes are low-cost foam kits and have a flying weight of approximately 2 lbs. The Phoenix and the Bixler 2 both feature a pusher propeller configuration that reduces vibration and increases overall robustness for a belly landing (neither aircraft has landing gear). The wings

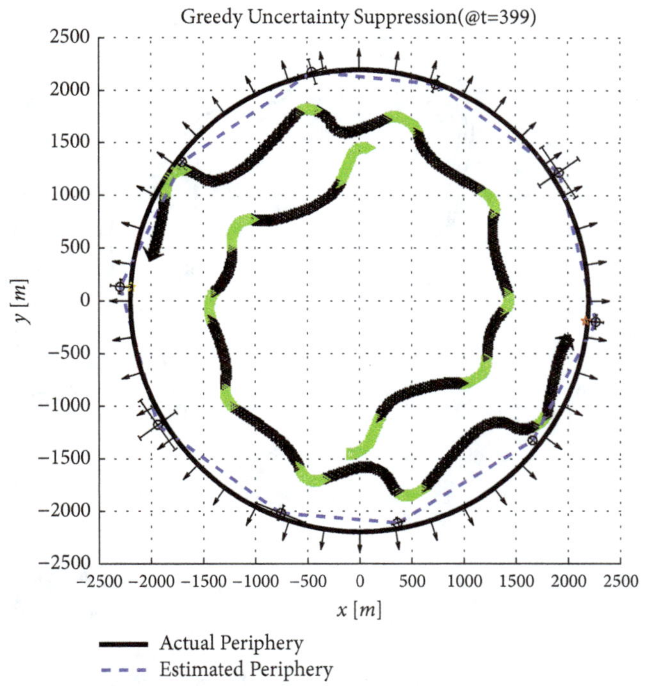

FIGURE 12: Estimation and coordination with the GUS method. The UAVs switched from the deployment phase to track the highest uncertainties. The actual periphery is a solid red line, and the predicted periphery a blue dashed line. The UAV trail is in green where the UAV is OUT and in black where the UAV is IN. The error bar associated with each CP represents its current uncertainty. Note that the error bar decreases as the UAV approaches a CP and that the observations cause the directional uncertainty of the other CPs to decrease.

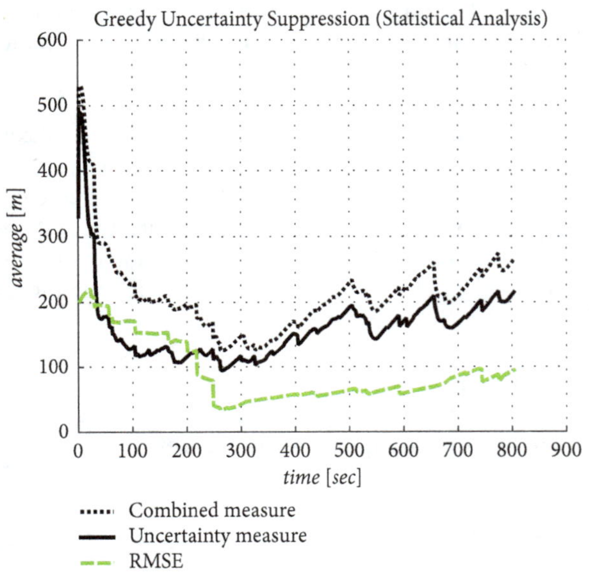

FIGURE 13: Performance analysis. The solid red line represents the average perpendicular standard deviation, the dashed green line shows the cumulative root mean squared error, and the dashed red line is the combined performance measure. Note that the mean value of the uncertainty is reduced during the mission, and the error increases as the periphery evolves since the number of crosses per AOI get smaller.

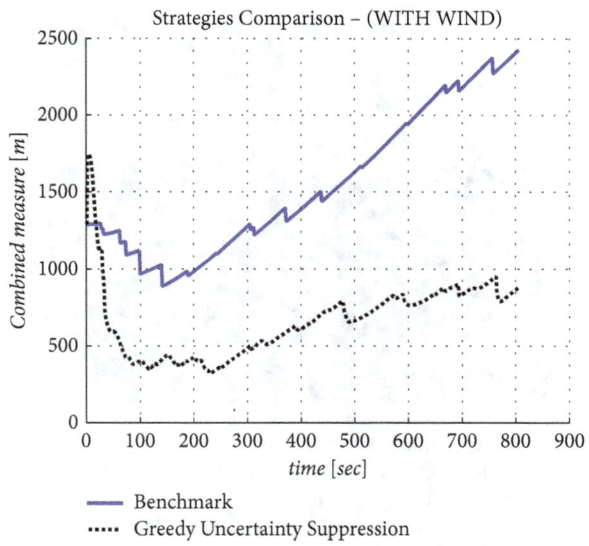

FIGURE 14: A comparison of strategies with a southwest wind. The solid blue line and the dotted red line represent the combined RMSE performance measure over time for the benchmark and the GUS strategies accordingly.

FIGURE 15: RC model plane: Hobby King Bixler 2.

and fuselage are reinforced with carbon fiber tubes that provide ample rigidity to the airframe [38]. The aircraft is hand launched for take-off. The Bixler 2 wings are almost an elliptical platform with curved winglets for increased flight efficiency. The power plant for the Bixler 2 aircraft is a 1200 kV brushless DC electric motor. The power source used is a 2200mAh Lithium Polymer battery. This battery provides sufficient current for the electric motor, servo, and the AUAV3 autopilot board, through the Electronic Speed Controller (ESC). The ESC provides a 5.0 volt supply to the servos and the AUAV3 autopilot through the Battery Eliminator Circuit (BEC) and also provides a control signal and power to the brushless motor. The BEC is designed to keep servos R/C receiver running while the baking has dropped too far in voltage to power the motor. The SLUGS II autopilot, like most other autopilots, uses a Proportional-Integral-Derivative (PID) control method for the low-level control loops [32]. The flight controller is developed as a Simulink model, and although it is relatively easy to alter its structure, it requires extensive knowledge about the inner and

FIGURE 16: Experiment hardware is shown. On the right, two airplanes model that have been used during the field test. On the left, the GCS deployed on the field.

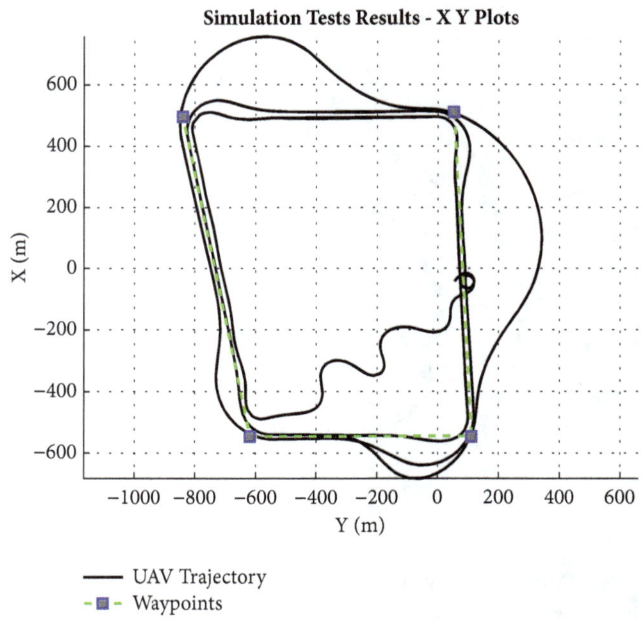

FIGURE 17: Simulated scenario with a single UAV is presented. The UAV trajectory is in X Y Cartesian coordinate frame and is relative to the Home position. The first segment of the trajectory started from take-off controlled manually by the safety-pilot (RC) and switched to autonomous mode after 23 seconds. Three laps were tested with different PID gains for tuning the roll command.

outer loop structure to redesign the controls. The simulation tests were devoted to validating the viability of the flight controller as flyable. This part of the testing covers the tuning process of the PID gains for the various autopilot control loops.

Figure 16 shows the outcome of the integration with the basic real-time components; autopilot, the UAV platforms, and the GCS a moment before performing a field test.

8.2. Flight Test. The goal of the SLUGS II validation is to support the R&D monitoring system development. The validation relies on several factors, including flight controller and path following performances. The flight controller has been extensively tested within the simulation. The environment supports parameter tuning which can accommodate hardware changes and flight mode extensions.

The most important feature of the SLUGS II autopilot for the R&D monitoring system is its autonomous waypoint navigation capabilities. The ground operator, through the GCS interface, can specify a sequence of waypoints to define the path the vehicle should follow. Figure 17 describes an example of a running scenario with four waypoints and shows how the vehicle follows the desired path while tuning PID gain parameters. Figure 17 shows that initially, the gains were too low, and the system had a slow response.

9. Conclusions

In summary, this paper has presented the design for monitoring system with a core methodology for coordinating a fleet of UAVs to suppress the uncertainty of a generic ground phenomenon. The coordination technique integrated with a R&D monitoring system which was designed carefully to improve the estimation of a propagated periphery supports decision making in an operational scenario.

The system design comprises the major components of a R&D monitoring system: high-level-controller, flight control, and ground control. The development process of the UAV flight controller (autopilot) has been improved with COTS board and a new development environment for software validation. The SLUGS II autopilot obtains the same functionality in the migrated Simulink model as found in the original model. The generated code uses on average a 60% CPU loading. The reserve computation time leaves enough computational resources for further enhancement and evolution.

MSIL simulation tests the generated code in a flexible and friendly environment that is committed to the sequence of events in the software rather than to guarantee strong real-time performance execution of the code. The system is designed to be agnostic as to the type of phenomenon that is being tracked and can be made to work well for a number of different scenarios.

Wildfire incidents are an example of a stochastic phenomenon, and knowing the fire boundary with high certainty would improve decision making by the ground team.

Conflicts of Interest

The authors declare that they have no conflicts of interest.

References

[1] National Interagency Fire Center. Federal Firefighting Costs (Suppression only). page 1987, 2016.

[2] Fire. NIFC, "Initial Attack," Tech. Rep., 2016.

[3] a Nalyses P Art, Forest Service, Martin E Alexander, David a Thomas, and Dale Bosworth. Fire Management W ILDLAND F IRE S TUDIES AND. Management, 63(3):1-96, 2003.

[4] NIFC, "National Interagency Coordination Center Wildland Fire Summary and Statistics," Technical report, 2013.

[5] S. Rabinovich, R. E. Curry, and G. H. Elkaim, "A methodology for estimation of ground phenomena propagation," in *Proceedings of the 2018 IEEE/ION Position, Location and Navigation Symposium (PLANS)*, pp. 1239–1244, Monterey, CA, April 2018.

[6] UAV Over-the-Horizon Disaster Management Demonstration Projects Project Manager: Steve Wegener February 2000 Contents. (February), 2000.

[7] C. Yuan, Y. Zhang, and Z. Liu, "A survey on technologies for automatic forest fire monitoring, detection, and fighting using unmanned aerial vehicles and remote sensing techniques," *Canadian Journal of Forest Research*, vol. 45, no. 7, pp. 783–792, 2015.

[8] N. Nigam, "The Multiple Unmanned Air Vehicle Persistent Surveillance Problem: A Review," *Machines*, vol. 2, no. 1, pp. 13–72, 2014.

[9] X. Cui, T. Hardin, R. K. Ragade, and A. S. Elmaghraby, "A swarm-based fuzzy logic control mobile sensor network for hazardous contaminants localization," in *Proceedings of the 2004 IEEE International Conference on Mobile Ad-Hoc and Sensor Systems*, pp. 194–203, USA, October 2004.

[10] G. Kowadlo and R. A. Russell, "Robot odor localization: a taxonomy and survey," *International Journal of Robotics Research*, vol. 27, no. 8, pp. 869–894, 2008.

[11] B. Ristic, D. Angley, B. Moran, and J. L. Palmer, "Autonomous multi-robot search for a hazardous source in a turbulent environment," *Sensors*, vol. 17, no. 4, 2017.

[12] B. Yamauchi, "Frontier-based exploration using multiple robots," in *Proceedings of the 2nd International Conference on Autonomous Agents*, pp. 47–53, ACM, May 1998.

[13] W. Burgard, M. Moors, and F. Schneider, "Collaborative Exploration of Unknown Environments with Teams of Mobile Robots," in *Advances in Plan-Based Control of Robotic Agents*, vol. 2466 of *Lecture Notes in Computer Science*, pp. 52–70, Springer Berlin Heidelberg, Berlin, Heidelberg, 2002.

[14] S. Russell and P. Norvig, *Artificial Intelligence A Modern Approach*, 2013.

[15] R. Elaine, K. Knight, and B. Shivashankar, *Artificial Intelligence-Rich-Knight-Nair.pdf*, 2009.

[16] M. T. Rashid, M. Frasca, A. A. Ali, R. S. Ali, L. Fortuna, and M. G. Xibilia, "Artemia swarm dynamics and path tracking," *Nonlinear Dynamics*, vol. 68, no. 4, pp. 555–563, 2012.

[17] M. T. Rashid, A. A. Ali, R. S. Ali, L. Fortuna, M. Frasca, and M. G. Xibilia, "Wireless underwater mobile robot system based on ZigBee," in *Proceedings of the 2012 International Conference on Future Communication Networks, ICFCN 2012*, pp. 117–122, Iraq, April 2012.

[18] A. A. Ali, L. Fortuna, M. Frasca, M. T. Rashid, and M. G. Xibilia, "Complexity in a population of Artemia," *Chaos, Solitons & Fractals*, vol. 44, no. 4-5, pp. 306–316, 2011.

[19] K. Arnold, "Uses of aerial photographs in control of forest fires," *Journal of Forestry*, vol. 49, pp. 26–31, 1951.

[20] A. Richards, J. Bellingham, M. Tillerson, and J. How, "Coordination and control of multiple UAVs," in *Proceedings of the AIAA Guidance, Navigation, and Control Conference and Exhibit 2002*, USA, August 2002.

[21] D. W. Casbeer, D. B. Kingston, R. W. Beard, and T. W. Mc lain, "Cooperative forest fire surveillance using a team of small unmanned air vehicles," *International Journal of Systems Science*, vol. 37, no. 6, pp. 351–360, 2006.

[22] K. Alexis, G. Nikolakopoulos, A. Tzes, and L. Dritsas, "Coordination of Helicopter UAVs for Aerial Forest-Fire Surveillance," *Applications of Intelligent Control to Engineering Systems*, vol. 39, Article ID 169193, pp. 169–193, 2009.

[23] W. Ren and R. W. Beard, *Distributed consensus in multi-vehicle cooperative control*, 2008.

[24] F. Bullo, J. Cortés, and S. Martínez, *Distributed Control of Robotic Networks: A Mathematical Approach to Motion Coordination Algorithms*, 2009.

[25] N. Dadkhah and B. Mettler, "Survey of motion planning literature in the presence of uncertainty: Considerations for UAV guidance," *Journal of Intelligent and Robotic Systems: Theory and Applications*, vol. 65(1-4), pp. 233–246, 2012.

[26] R. C. Smith and P. Cheeseman, "On the representation and estimation of spatial uncertainty," *International Journal of Robotics Research*, vol. 5, no. 4, pp. 56–68, 1986.

[27] S. Rabinovich, E. Renwick, and H. Gabriel Elkaim, "Multiple Unmanned Air Vehicle Coordination for Monitoring of Ground Phenomena Propagation," in *Proceedings of the ION GNSS+ 2018. IEEE*, 2018.

[28] C. E. Renwick, *Estimation and control with quantized measurements*, MIT press, 1970.

[29] A. Girard, A. Howell, and J. Hedrick, "Border patrol and surveillance missions using multiple unmanned air vehicles," in *Proceedings of the 2004 43rd IEEE Conference on Decision and Control (CDC) (IEEE Cat. No.04CH37601)*, pp. 620–625 Vol.1, Nassau, Bahamas, December 2004.

[30] H. Chao, Y. Cao, and Y. Chen, "Autopilots for small unmanned aerial vehicles: A survey," *International Journal of Control, Automation, and Systems*, vol. 8, no. 1, pp. 36–44, 2010.

[31] S. Rabinovich, *Multi-UAV Coordination for Uncertainty Suppression of Natural Disasters*, Phd, UC Santa Cruz, 2018.

[32] I. Mariano and Lizarraga F., *DEsign, Implementation and Flight Verification of A Versatile and Rapidly Reconfigurable Uav Gnc Research Platform*, 2009.

[33] E. Geaney, UC Santa Cruz Electronic Theses and Dissertations.

[34] R. Garcia and L. Barnes, "Multi-UAV simulator utilizing x-plane," *Journal of Intelligent & Robotic Systems*, vol. 57, no. 1–4, pp. 393–406, 2010.

[35] Nick Arsov. AUAV3.

[36] Lubin. Kerhuel, Simulink - embedded target for pic.

[37] MatrixPilot. MatrixPilot Auto-Pilot, 2016.

[38] P. Kumar, J. E. Steck, and S. G. Hagerott, "System identification, HIL and flight testing of an adaptive controllerona small scale unmanned aircraft," in *Proceedings of the AIAA Modeling and Simulation Technologies Conference 2015*, USA, January 2015.

16

Concept Design and Analysis of a Novel Steamer-Filling Robot

Enhong Xing,[1] Bin Li,[1,2] and Xinhua Zhao[1]

[1] *Tianjin Key Laboratory for Advanced Mechatronic System Design and Intelligent Control, School of Mechanical Engineering, Tianjin University of Technology, Tianjin 300384, China*
[2] *National Demonstration Center for Experimental Mechanical and Electrical Engineering Education, Tianjin University of Technology, Tianjin, China*

Correspondence should be addressed to Bin Li; cnrobot@163.com

Academic Editor: Gordon R. Pennock

Steamer-filling operation is a crucially important process in the liquor-making process, directly related to liquor yield and liquor quality. But so far, this process is still dominated by manual operation. In view of working environment and labor shortages in this industry, a novel exclusive steamer-filling robot is proposed in this paper. Firstly, the steamer-filling operation process is described, and the structure composition and function realization of the robot are particularly introduced. Secondly, the kinematics problems in terms of position analysis and workspace of the robot are analyzed in detail. Thirdly, experimental analyses are made to prove the validity and efficiency of the robot system. Finally, some conclusions and the future developing direction are prescribed.

1. Introduction

Liquor-making industry is a very important industry in China at present. According to the statistics, Chinese liquor production reached 960 million liters in 2015. Compared with automobile industry, electronics industry, food industry, and so on, the liquor-making industry is still dominated by manual operation. With growing shortage of labors and increasing labor costs, the demands on the automation equipment for liquor-making process are increasing, and these demands will generate a very large market.

Recently, a large number of wineries upgrade their automation equipment for steamer-filling operation. These technical methods can be divided into two categories, namely, nonrobot method and robot method [1–4]. For the nonrobot method, some conventional conveying devices, such as conveyor belt, are used to fill fermenting grains into fermentation containers. Since this method cannot fill fermenting grains in an automatic and well-distributed way, a worker is necessary, as shown in Figure 1; in other words, nonrobot method is just a semiautomatic method. The other method is robot method; in this way, the work system is composed of a six-degree-of-freedom manipulator and a hopper, which is fixed on the end of the manipulator. The working procedure is that, firstly, fermenting grains are filled in the hopper; then the manipulator is applied to simulate manual operation to toss fermenting grains into fermentation containers, as shown in Figure 2. Compared with the nonrobot method, this method can realize full automation, which is of vital significance to the liquor-making industry.

In this paper, a novel exclusive steamer-filling robot is proposed. Firstly, the steamer-filling operation process is described, and the structure composition and function realization of the robot are particularly introduced in Section 2. Secondly, the kinematics problems in terms of position analysis and workspace of the robot are analyzed in detail in Section 3. Thirdly, numerical experimental analyses are made to prove the validity and efficiency of the robot system in Section 4. The dynamic analysis of the robot is carried out in Section 5. Then the static stiffness of the robot is described in Section 6. Finally, some conclusions and the future developing direction are prescribed in Section 7.

2. Concept Design of Steamer-Filling Robot

In order to design the structure of the steamer-filling robot, it is necessary to introduce to the steamer-filling operation process. In general, steamer-filling operation means sprinkling

FIGURE 1: Nonrobot method for steamer-filling.

FIGURE 2: Robot method for steamer-filling.

fermenting grains into fermentation containers. In order to maximize liquor yield and liquor quality, the sprinkling process needs some skill requirements, such that operating should be quick, the fermenting grains should be porous, and the steamer should be well-distributed. For most wineries, the fermentation containers have a diameter of 2 meters and a height of 1.3 meters. A liquor worker should fill up a fermentation container within thirty-five minutes, which means hard-manual labor.

As mentioned above, a robot method has been proposed for steamer-filling operation to replace liquor workers. For this method, fermenting grains are filled in the hopper, which is fixed on the end of the robot; then the manipulator is applied to simulate manual operation to toss fermenting grains into fermentation containers. Through actual working experience, these existing steamer-filling robots have several downsides: for example, fermenting grains throwing process is a discontinuous process, which has important impact on production efficiency; fermenting grains throwing is not sufficiently homogeneous, which has an adverse effect on liquor yield and liquor quality.

Focusing on these problems of these already existing steamer-filling robots, a novel exclusive steamer-filling robot is proposed in this paper, as shown in Figure 3. The whole robot system consists of the mechanical structure part, the infrared vision part, and the control system part. A three-degree-of-freedom serial cylindrical type robot is designed to form the mechanical body, the infrared vision is chosen to detect the leakage of steam, and the PLC is used to form the control system of the robot. Due to limited space, the mechanical structure part of the whole robot system is the main research object in this paper.

When viewed as a whole, the exclusive steamer-filling robot proposed in this paper is a three-degree-of-freedom cylindrical robot, as shown in Figures 3 and 4. The robot can be seen as a R-P-R cylindrical type robot where, R and P stand for revolute and prismatic joints, respectively; firstly, the driving motor 1 and gearbox form a rotational motion around z-axis; thus the whole body of the robot can realize a rotational motion around the z-axis. Then the driving motor and ball screw form a translational motion along the z-axis. Finally, the driving motor 3 and gearbox form another rotational motion around z'-axis, which is parallel with the z-axis.

The working process of the exclusive steamer-filling robot is designed as follows: firstly, the fermenting grains are carried to the manipulator arm 1 though the hopper of the robot; then, these fermenting grains are transformed from point A to point B by using screw conveyors arranged in the manipulator arm 1. Next, these fermenting grains are pushed into the manipulator arm 2 at point B. Finally, by the same way, these fermenting grains are pushed out the manipulator arm 2 at point C by using screw conveyors arranged in manipulator arm 2, so as to complete the whole operation process. From the perspective of degree of freedom, the proposed steamer-filling robot has same 3-DOFs with SCARA type robots, namely, two rotational motions and one translational motion. But as far as function is concerned, the proposed steamer-filling robot is totally different from SCARA type robots, because the arms of the steamer-filling robot are used as multifunctional bodies, that is, not only used to transmit motion as robot arms but also used to carry the fermenting grains as actuating mechanism, that is, a noticeable difference between the proposed steamer-filling robot and SCARA type robots.

Furthermore, compared with existing steamer-filling robot, this type of robot can fill fermenting grains in a continuous mode without being standstill in the operation process, which is of great importance regarding liquor yield and liquor quality.

3. Kinematics Problem Analyses of the Robot

In this section, the kinematics problems in terms of position analysis, workspace, and dexterity of the robot are analyzed in detail [5–11]. First, the inverse and the forward positon analyses of the mechanism are calculated by explicit solution; then the workspace of the robot is determined based on the forward position analysis algorithm. Finally, the dexterity and the transmission performance of the manipulator are carried out by the conception of transmission angles.

3.1. Position Analysis. In order to increase liquor-making efficiency, one steamer-filling robot fills fermenting grains into two fermentation containers as shown in Figure 5. For analysis purpose, a plane coordinate system Axy is founded. Point A is coordinate origin; points O_1, O_2 are the center points of two fermentation containers, setting x-axis parallel to line O_1O_2; and two fermentation containers are

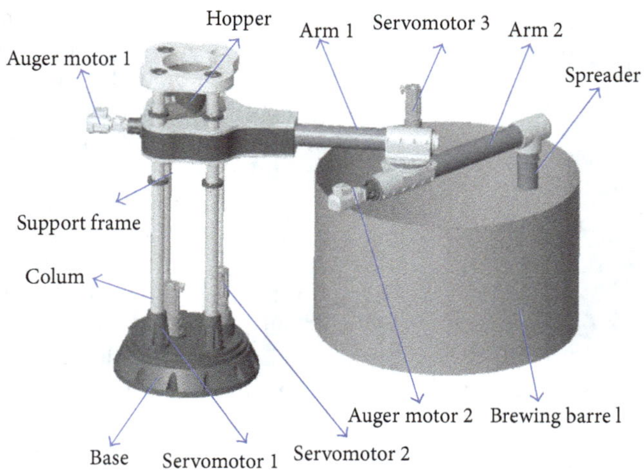

FIGURE 3: CAD model of the robot.

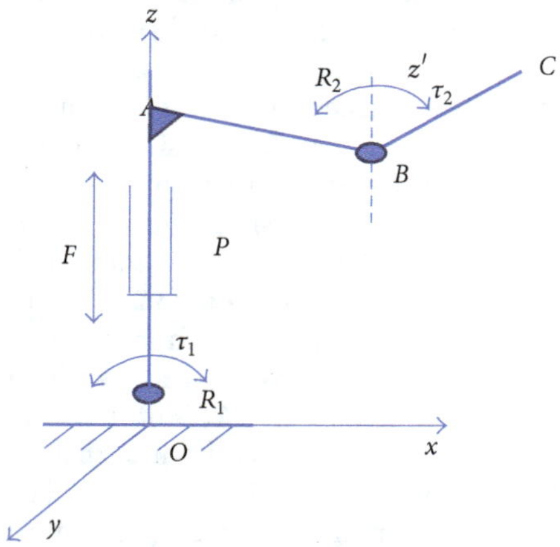

FIGURE 4: Schematic diagram of the robot.

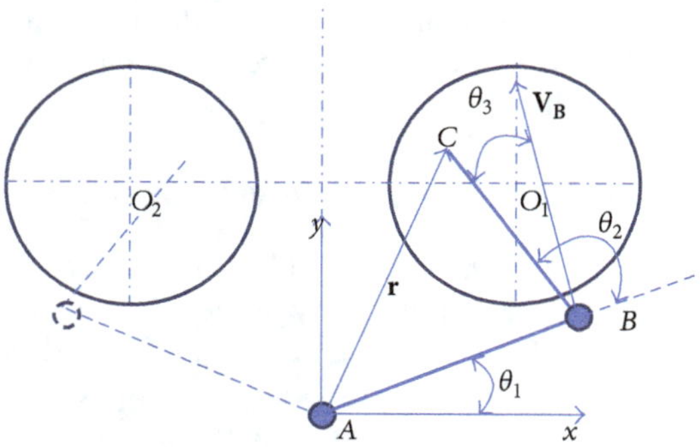

FIGURE 5: Schematic diagram of the robot (vertical view).

symmetrically distributed about y-axis. According to these geometric relationships, closed-loop position equation of the robot can be expressed as

$$r = \overrightarrow{AB} + \overrightarrow{BC}. \tag{1}$$

Expanding (1) yields

$$\begin{aligned} x &= l_1 \cos(\theta_1) + l_2 \cos(\theta_1 + \theta_2), \\ y &= l_1 \sin(\theta_1) + l_2 \sin(\theta_1 + \theta_2). \end{aligned} \tag{2}$$

According to (2), given a set of (θ_1, θ_2), the position coordinate (x, y) of the end of the robot can be calculated directly, which is the forward position solution of the mechanism. Here, $l_1 = \|AB\|$, $l_2 = \|BC\|$.

In order to get the inverse positon analysis of the robot, from (2), an equation about the relation between the parameters of θ_1, θ_2 and x, y can be deduced as follows:

$$(x - l_1 \cos\theta_1)^2 + (y - l_1 \sin\theta_1)^2 = l_2^2. \tag{3}$$

Expanding (3) yields

$$x^2 + y^2 + l_1^2 - l_2^2 = 2l_1 (x\cos\theta_1 + y\sin\theta_1). \tag{4}$$

In order to simplify the calculation, we assume that

$$\tan\theta = \frac{y}{x}, \quad \text{namely } \theta = \arctan\frac{y}{x}. \tag{5}$$

Substituting (5) into (4) yields

$$x^2 + y^2 + l_1^2 - l_2^2 = \frac{2l_1 x}{\cos\theta}\cos(\theta_1 \pm \theta). \tag{6}$$

According to (7), two set solutions of θ_1 and θ_2 can be expressed by the following:

$$\begin{aligned} \theta_1 &= \theta + \arccos\left(\frac{\cos(\theta)\left(x^2 + y^2 + l_1^2 - l_2^2\right)}{2l_1 x}\right), \\ \theta_2 &= \arccos\left(\frac{x - l_1\cos(\theta_1)}{l_2}\right) - \theta_1, \end{aligned} \tag{7}$$

$$\begin{aligned} \theta_1 &= \arccos\left(\frac{\cos(\theta)\left(x^2 + y^2 + l_1^2 - l_2^2\right)}{2l_1 x}\right) - \theta, \\ \theta_2 &= \arccos\left(\frac{x - l_1\cos(\theta_1)}{l_2}\right) - \theta_1. \end{aligned} \tag{8}$$

TABLE 1: Forward position analysis for the robot.

Case (*a*)			
Inputs		Outputs	
θ_1 (°)	θ_2 (°)	x (mm)	y (mm)
30	60	1558.8	2400

According to (7) and (8), given a set of the position coordinate (x, y) of the end of the robot, two sets of the joint locations (θ_1, θ_2) of the manipulator arm of the robot can be calculated directly, which is the inverse positon solution of the mechanism.

3.2. Jacobian Matrix and Singularity of the Robot. In order to get the Jacobian matrix of the robot, derived from both sides of (2), rewriting the equation in matrix form yields

$$\mathbf{v} = \mathbf{J}\boldsymbol{\omega}, \quad (9)$$

where $\mathbf{v}$ stands for the linear velocity of the end of the robot, $\boldsymbol{\omega}$ stands for the rotation speed of these two rotational joint, and $\mathbf{J}$ is the Jacobian matrix of the robot.

Here,

$$\mathbf{v} = \begin{bmatrix} \dot{x} \\ \dot{y} \end{bmatrix},$$
$$\boldsymbol{\omega} = \begin{bmatrix} \dot{\theta}_1 \\ \dot{\theta}_2 \end{bmatrix}, \quad (10)$$
$$\mathbf{J} = \begin{bmatrix} -l_1 \sin(\theta_1) - l_2 \sin(\theta_1 + \theta_2) & -l_2 \sin(\theta_1 + \theta_2) \\ l_1 \cos(\theta_1) + l_2 \cos(\theta_1 + \theta_2) & l_2 \cos(\theta_1 + \theta_2) \end{bmatrix}.$$

When the robot is singular,

$$\det(\mathbf{J}) = 0. \quad (11)$$

Expanding (11) gives the following result:

$$l_1 l_2 \sin(\theta_2) = 0. \quad (12)$$

From (12), when $\theta_2 = 0$ or π, namely, the two manipulator arms are collinear, the robot is singular.

Since Jacobian matrix inherently combines information about the position, orientation, and joint limits of the end-effector, the dexterity analysis of the robot can be carried out based on the Jacobian matrix, and the dexterity measure can be defined as

$$\lambda = \sqrt{\det(JJ^T)}. \quad (13)$$

3.3. Transmission Angel and Transmission Performance of the Robot. In order to evaluate the transmission performance of the mechanism, a transmission angle is designed. At a particular moment, the linear velocity of the end of the manipulator arm 1 is defined as $\mathbf{V}_B$ as shown in Figure 5; at this time, the axis of the manipulator arm 2 and the direction vector of $\mathbf{V}_B$ form an included angle θ_3. Then the angel is defined as the transmission angel θ_3 of the robot between the two manipulator arms, and the cosine value of the angel is defined as the transmission index of the two manipulator arms.

$$k = \cos(\theta_3). \quad (14)$$

Here, $k \in [0, 1]$. When $\theta_3 = 0$, $\cos(\theta_3) = 1$; this means that, in this situation, the transmission performance of the manipulator arms is best. When $\theta_3 = \pi$, $\cos(\theta_3) = 0$; this means that the two manipulator arms are collinear in this situation, the robot is in singular configuration, and this configuration is the worst situation.

4. Numerical Examples

In this section, the positon analysis, workspace, dexterity, and transmission performance of the mechanism are enumerated by corresponding numerical simulations. For consistency, the length parameters of two arms are defined as follows: $l_1 = 1800$ mm, $l_2 = 1500$ mm. The central points of fermentation containers are set as follows: $O_1 = (1390\,\text{mm}, 2055\,\text{mm})$, $O_2 = (-1390\,\text{mm}, 2055\,\text{mm})$. The swing angle ranges of the two manipulator arms are set as follows: $\theta_1 \in [-25°, 230°]$, $\theta_2 \in [30°, 150°]$.

4.1. Position Analysis Simulations. For the forward positon analysis, given one random group of inputs (θ_1, θ_2), the position coordinate (x, y) of the end of the robot can be calculated, and the corresponding parameters are listed in Table 1; meanwhile, the corresponding configurations relative to the computed solutions are described in Figure 6.

For the inverse positon analysis, given a set of the position coordinate (x, y) of the end of the robot, the joint location (θ_1, θ_2) of the manipulator arm of the robot can be calculated; the corresponding parameters are listed in Table 2; meanwhile, the corresponding configurations relative to the computed solutions are described in Figure 7.

4.2. Workspace Determination Simulations. The workspace determination of the robot can be carried out based on the forward position analysis algorithm, and a meshless method is applied to determine the reachable workspace of the robot. The reachable workspace of mechanism is shown in Figure 8.

4.3. Dexterity Analysis Simulation of the Robot. The dexterity of the robot can be thought as the ability of the manipulator to arbitrarily change its position and orientation or apply forces and torques in arbitrary directions. In this section, the dexterity of the robot can be obtained by using (13), and the simulation of the mechanism can be calculated as shown in Figure 9.

Table 2: Inverse position analysis for the robot.

Case (a)			
Inputs		Outputs	
x (mm)	y (mm)	θ_1 (°)	θ_2 (°)
1558.8	2400	30	60.0017
		83.9926	−60.0017

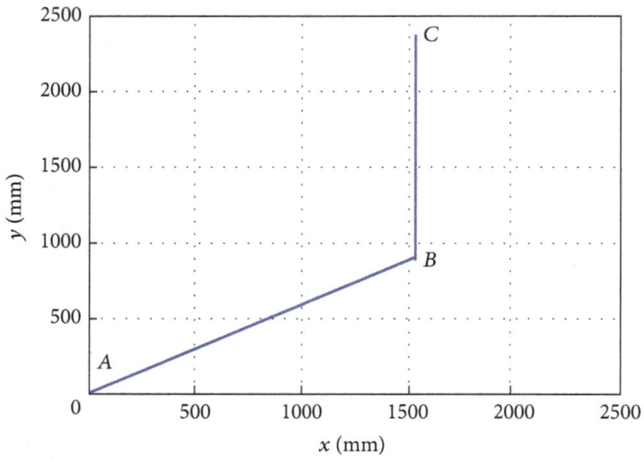

Figure 6: Different configurations for forward position.

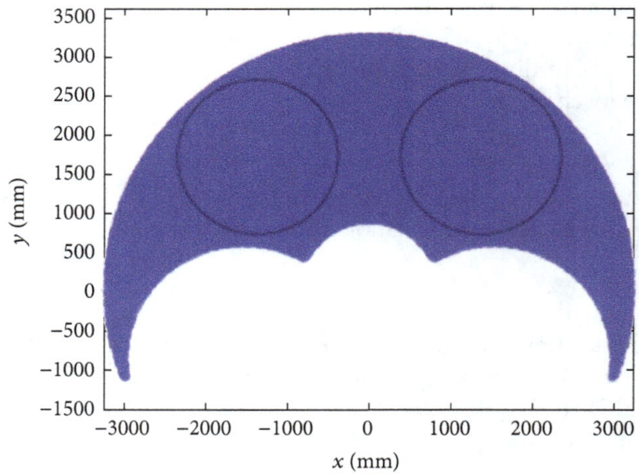

Figure 8: The reachable workspace of the mechanism.

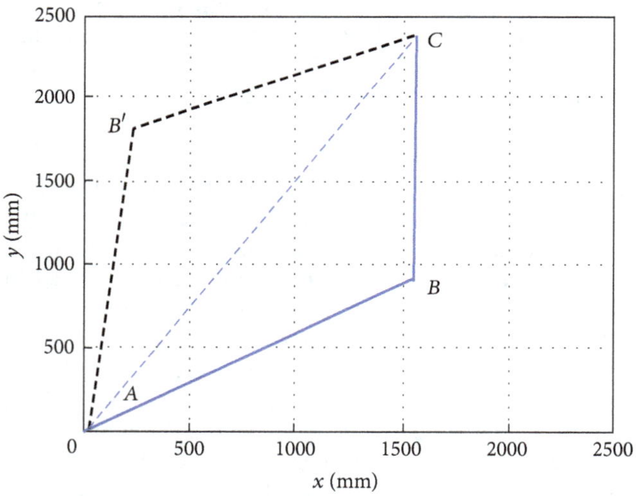

Figure 7: Different configurations for inverse position.

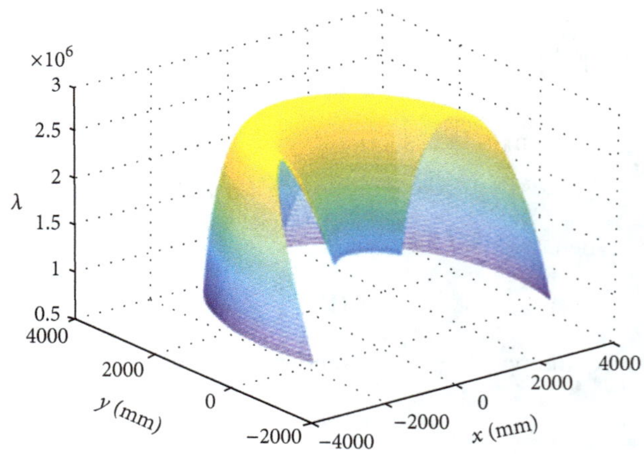

Figure 9: The dexterity analysis simulation of the robot.

4.4. Transmission Performance of the Robot. Transmission performance of the robot will be carried out through the pressure angles between two arms. The transmission performance of the robot can be obtained based on (14), and the solutions are described in Figures 10 and 11.

5. Dynamics Analysis of the Robot

In this section, the dynamics analysis of the robot is carried out on the base of the kinematics modeling and Lagrange dynamics equation.

According to the dynamic model of multilink mechanism, the dynamic equation of the robot can be founded:

$$\boldsymbol{\tau} = \mathbf{H}(\mathbf{q})\ddot{\mathbf{q}} + \mathbf{C}(\mathbf{q}\ \ \dot{\mathbf{q}})\dot{\mathbf{q}} + \mathbf{G}(\mathbf{q}), \tag{15}$$

where $\boldsymbol{\tau}$ stands for the joint drive torque of robot, $\mathbf{q}$, $\dot{\mathbf{q}}$, $\ddot{\mathbf{q}}$ stand for position, velocity, and acceleration vectors of robot joints, $\mathbf{H}(\mathbf{q})$, $\mathbf{C}(\mathbf{q}\ \ \dot{\mathbf{q}})$, $\mathbf{G}(\mathbf{q})$ represent inertial matrix, Coriolis force and centripetal force matrix, and gravity matrix of the manipulator, respectively.

As shown in Figure 4, It should be noted that the force $\mathbf{F}$ along the z-axis can be calculated directly by mechanical formula:

$$\mathbf{F} = m\mathbf{g} + m\mathbf{a}, \tag{16}$$

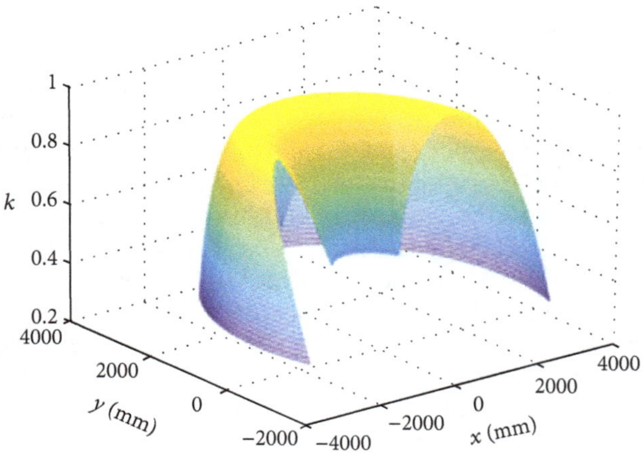

FIGURE 10: Transmission performance of the robot.

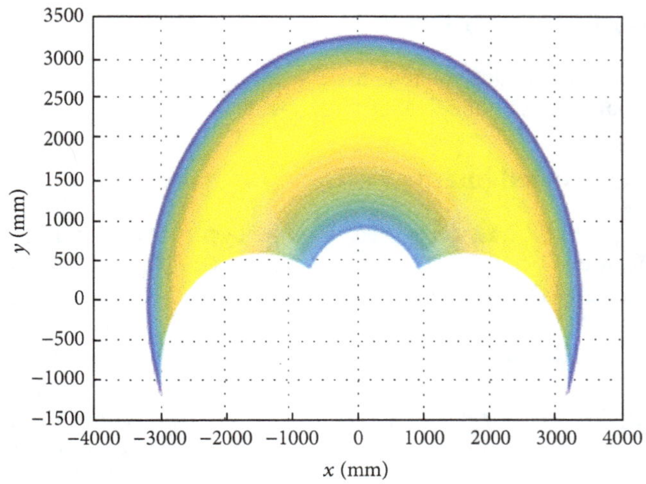

FIGURE 11: Transmission performance of the robot in x-y plane.

where m is the all weight of robot arm including fermenting grains and **g** and **a** are gravitational acceleration and translational motion acceleration of robot arm along the z-axis.

The drive torques of two revolute joints can be calculated by Lagrange dynamics equation:

$$\begin{bmatrix} \tau_1 \\ \tau_2 \end{bmatrix} = \begin{bmatrix} H_{11} & H_{12} \\ H_{21} & H_{21} \end{bmatrix} \begin{bmatrix} \ddot{q}_1 \\ \ddot{q}_2 \end{bmatrix} + \begin{bmatrix} C_{11} & C_{12} \\ C_{21} & C_{21} \end{bmatrix} \begin{bmatrix} \dot{q}_1 \\ \dot{q}_2 \end{bmatrix}, \quad (17)$$

where

$$\begin{aligned} H_{11} &= I_{z1} + I_{z2} + m_1 x_1^2 + m_2 \left(l_1^2 + x_2^2\right) \\ &\quad + 2 l_1 x_2 m_2 \cos(\theta_2), \\ H_{12} &= I_{z2} + m_2 x_2^2 + l_1 x_2 m_2 \cos(\theta_2), \\ H_{21} &= H_{12}, \\ H_{22} &= I_{z2} + m_2 x_2^2, \\ C_{11} &= -2\dot{\theta}_2 \sin(\theta_2) l_1 x_2 m_2, \end{aligned}$$

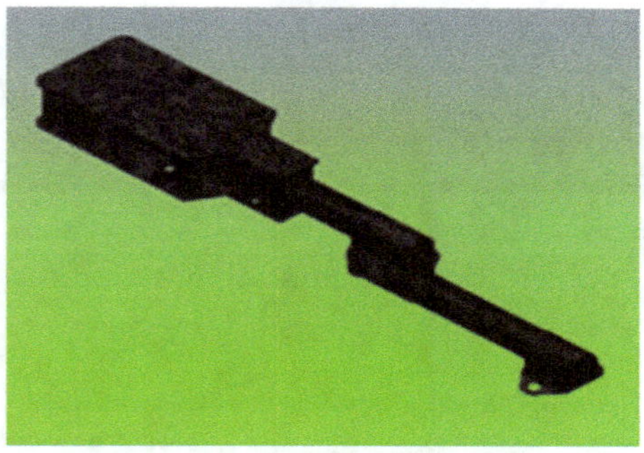

FIGURE 12: Finite element analysis model of steamer-filling robot.

$$\begin{aligned} C_{22} &= 0, \\ C_{21} &= \dot{\theta}_1 \sin(\theta_2) l_1 x_2 m_2, \end{aligned} \quad (18)$$

where m_1 and m_2 stand for the weights of arm 1 and arm 2 including fermenting grains, l_1 and l_2 stand for the lengths of arm 1 and arm 2, and x_1 and x_2 stand for the gravity center positions of arm 1 and arm 2.

For the inverse dynamic analysis, the drive torques of two revolute joints, and the drive force of the translate joint can be calculated by (16) and (17), given the position, velocity, acceleration, weights, and dimensional parameters of the robot. For the forward dynamic analysis, the weights of fermenting grains can be obtained by (16) and (17), given the position, velocity, acceleration, dimensional parameters, and drive forces of the robot.

6. Static Stiffness Analysis of the Robot

Taking into consideration the high positioning accuracy and slow rate requirements of work situation for the steamer-filling robot, the static stiffness analysis of the robot is studied by means of finite element method in this section. First, according to the original geometric model, a finite element analysis model was established using commercial software, as shown in Figure 12. Then, the actual working constraint conditions are given to the finite element analysis model. Finally, the static stiffness of the robot in different working positon is calculated by software, and the results are shown in Figures 13 and 14.

7. Experimental Results

The physical prototype of the steamer-filling robot is shown in Figures 15 and 16, which is used in the distillery. According to the production data comparison, liquor output was about 210 KG from 500 KG fermenting grains by using the steamer-filling robot, compared with about 190 KG by manual process, and the robot method has a wide application prospect.

8. Conclusions

The major contribution of this research is proposing a novel exclusive steamer-filling robot, compared with existing

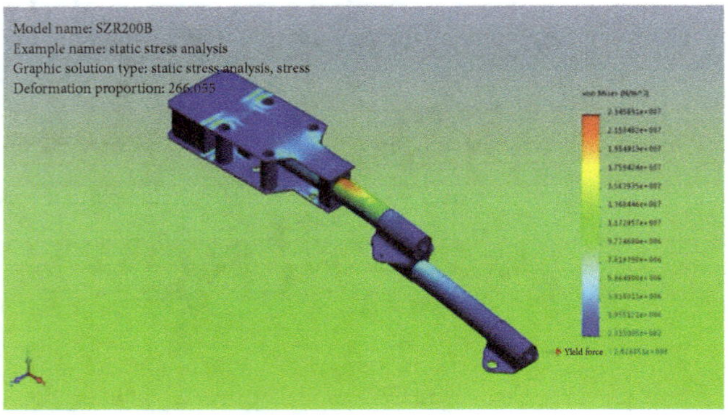

FIGURE 13: Finite element stress analysis of the robot.

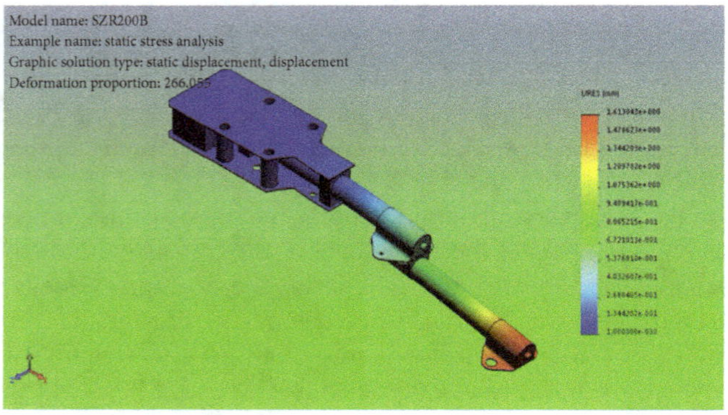

FIGURE 14: Finite element displacement analysis of the robot.

FIGURE 15: Experimental prototype of steamer-filling robot.

FIGURE 16: Steamer-filling robot used in field.

steamer-filling robot; this type robot can fill fermenting grains in a continuous mode without being standstill in the operation process, which is of great importance to liquor yield and liquor quality. Firstly, the steamer-filling operation process is described, and the structure composition and function realization of the robot are particularly introduced. Secondly, the kinematics problems in terms of position analysis, workspace analysis, dexterity analysis, and transmission performance of the robot are analyzed. Thirdly, experimental analyses are made to prove the validity and efficiency of the robot system. On the basis of the kinematics study of the robot, the inverse and forward dynamics analysis and stiffness performance and kinematic optimization of the robot will be investigated in future work, and the application and spread of this novel steamer-filling robot is of vital significance to the liquor-making industry.

Conflicts of Interest

The authors declare that there are no conflicts of interest regarding the publication of this paper.

Acknowledgments

This work was supported by Intelligent Manufacturing Technology Major Project of Tianjin (15ZXZNGX00040, 15ZXZNGX00270, and 16ZXZNGX00070).

References

[1] B. Li, Y. Li, and X. Zhao, "Kinematics analysis of a novel over-constrained three degree-of-freedom spatial parallel manipulator," *Mechanism and Machine Theory*, vol. 104, pp. 222–233, 2016.

[2] B. Li, Y. M. Li, X. H. Zhao, and W. M. Ge, "Kinematic analysis of a novel 3-CRU translational parallel mechanism," *Mechanical Sciences*, vol. 6, no. 1, pp. 57–64, 2015.

[3] Q. Yan, B. Li, Y. Li, and X. Zhao, "Comparative study of two 2-RPU+SPR parallel manipulators," in *Proceedings of the 12th World Congress on Intelligent Control and Automation, WCICA 2016*, pp. 1975–1980, chn, June 2016.

[4] Y. Li and Q. Xu, "Kinematic analysis of a 3-PRS parallel manipulator," *Robotics and Computer-Integrated Manufacturing*, vol. 23, no. 4, pp. 395–408, 2007.

[5] X.-J. Liu, X. Tang, and J. Wang, "HANA: A novel spatial parallel manipulator with one rotational and two translational degrees of freedom," *Robotica*, vol. 23, no. 2, pp. 257–270, 2005.

[6] J. Wang and X.-J. Liu, "Analysis of a novel cylindrical 3-DoF parallel robot," *Robotics and Autonomous Systems*, vol. 42, no. 1, pp. 31–46, 2003.

[7] B. Bihari, D. Kumar, C. Jha, V. S. Rathore, and A. K. Dash, "A geometric approach for the workspace analysis of two symmetric planar parallel manipulators," *Robotica*, vol. 34, no. 4, pp. 738–763, 2016.

[8] D. Sen and B. N. Singh, "A geometric approach for determining inner and exterior boundaries of workspaces of planar manipulators," *Journal of Mechanical Design*, vol. 130, no. 2, Article ID 022306, 2008.

[9] S. Yahya, M. Moghavvemi, and H. A. F. Mohamed, "Geometrical approach of planar hyper-redundant manipulators:

Inverse kinematics, path planning and workspace," *Simulation Modelling Practice and Theory*, vol. 19, no. 1, pp. 406–422, 2011.

[10] M. Gallant and R. Boudreau, "The synthesis of planar parallel manipulators with prismatic joints for an optimal, singularity-free workspace," *Journal of Robotic Systems*, vol. 19, no. 1, pp. 13–24, 2002.

[11] A. M. Hay and J. A. Snyman, "The determination of nonconvex workspaces of generally constrained planar Stewart platforms," *Computers & Mathematics with Applications*, vol. 40, no. 2000, pp. 1043–1060, 2000.

17

Galloping Trajectory Generation of a Legged Transport Robot based on Energy Consumption Optimization

Yaguang Zhu and Tong Guo

Key Laboratory of Road Construction Technology and Equipment of MOE, Chang'an University, Xi'an 710064, China

Correspondence should be addressed to Yaguang Zhu; zhuyaguang@chd.edu.cn

Academic Editor: Gordon R. Pennock

Legged walking robots have very strong operation ability in the complex surface and they are very suitable for transportation of tools, materials, and equipment in unstructured environment. Aiming at the problems of energy consumption of legged transport robot during the fast moving, a method of galloping trajectory planning based on energy consumption optimization is proposed. By establishing transition angle polynomials of flight phase, lift-off phase, and stance phase and constraint condition between each state phase, the locomotion equations of the ellipse trajectory are derived. The transition angle of each state phase is introduced into the system energy consumption equations, and the energy optimization index based on transition angles is established. Inverse kinematics solution and trajectory planning in one gait cycle are applied to genetic algorithm process to solve the nonlinear programming problem. The results show that the optimized distribution of transition angles of state phases is more reasonable, and joint torques and system energy consumption are reduced effectively. Thus, the method mentioned above has a great significance to realize fast operation outdoors of transport robot.

1. Introductions

With the development of technology, upgrading the traditional construction machinery products by using the computer control technology and robot technology and so forth has become an inevitable trend. By increasing the high-tech content, the purposes of improving operational efficiency and work quality and reducing fuel consumption and construction cost can be achieved. Due to the fact that legged robot has much difference with wheeled and tracked robot in structure, which only needs some continuous and discrete foot tip during the motion, it has stronger operation ability in complex environments such as striding over obstacles and passing the rugged, unstructured, or muddy surface easily. That is very suitable for transportation of tools, materials, and equipment in unstructured environment. Because legged robot has the characteristics of multi-DOF, construction complexity, uneasy control, and easy overturn, most of legged robot ensures stability during the motion by keeping contact with surface with one or more legs [1]. However, high-speed operation will be limited by this type of exercise, which caused the fact that primary control strategies currently can only achieve fast gait such as work or jog. The speed can be improved effectively if the foot tip of robot can be made off the ground for a period of time in the process of moving. That will enable the biped robot and multilegged robot to achieve the running in the real meaning [2].

In recent years, many researchers have engaged in research on multilegged robot running [3–5]. Kim and Park [6] proposed a leg trajectory that is generated by elliptic motion based on lip joint and keen joint. Meanwhile, the gallop was achieved by analyzing the movement rules of the robot during the contact with surface. More than that, all kinds of galloping styles can also be generated by controlling parameters such as the center of long and short axis of elliptical trajectory and rotation velocity. Step control and force redistribution strategy were proposed by Palmer III and Orin [7], which was used for controlling the locomotion of the robot. That allowed the quadruped to reach a speed of 4.75 m/s and turn at a rate of 20 degrees/s. Focchi et al. [8] were successful in trotting gait with elliptical trajectory generated by Central Pattern Generation (CPG) and controlled by

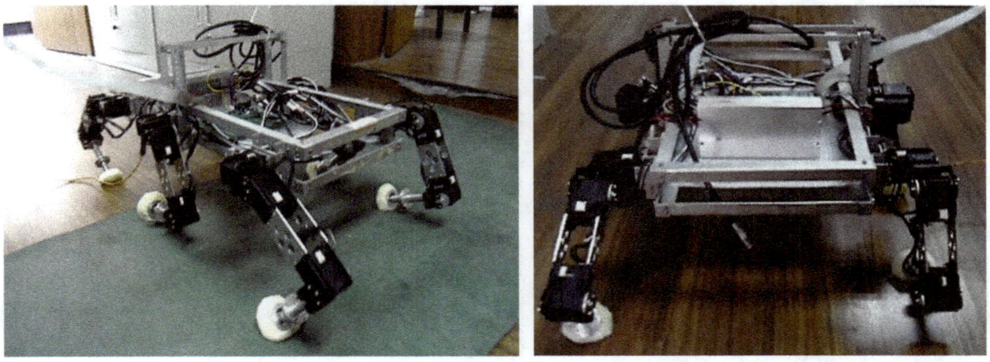

FIGURE 1: Mechanical model of hexapod transport robot.

reflection trajectory, which guarantees the stability combined with body posture controller. Marhefka et al. used fuzzy control techniques to control both bounding and galloping of multirobot [9], and they mapped inverse kinematics to a fuzzy system using intuitive heuristics. According to the question that the time of touch-down phase and swing phase are constant within the existing strategy of galloping trajectory, Capi et al. [10] reestablished the polynomial of trajectory and motion state of the robot, which was optimized by using genetic algorithm, and thus the more reasonable galloping trajectory was obtained. Roh and Park [11] proposed a simple and efficient galloping algorithm with elliptic trajectory. This method can effectively keep body height and body balance in the process of gallop based on angular momentum and fuzzy logic control, in which the desired reaction forces are obtained by angular momentum analysis and the fuzzy logic control used for keeping galloping stability.

Generally, during the running and galloping, flight phase when the robot is completely off the ground and stance phase when the robot is in contact with the ground are repeated in sequence. During the fight phase, the robot is completely governed by the gravity. This indicated that the motion of the robot in flight is decided by the robot motion during the previous stance phase [12]. Of course, another issue is the energy consumption during the moving process. The battery life is an important indicator when transport robots work outdoors independently. However, energy consumption will increase with the speed increased. So, if we make energy consumption an optimization index, it can be effectively reduced by dividing duration of each state and range of motion reasonably. To sum up, aiming at the problems about energy consumption in trajectory generation of multilegged robot, this paper proposes a galloping trajectory strategy based on energy consumption optimization. By deriving the function of elliptic trajectory, we introduce transition angles of state phase into system energy consumption equation as variables. And this issue of nonlinear planning can be solved by using genetic algorithm. This method enables making energy consumption optimization as objective to divide each state phase reasonably. The advantages of this method are that it can reduce energy consumption effectively and increase cruising ability of robot.

2. Modeling and Galloping Trajectory Generation

2.1. Transport Robot Model with Mammalian Configuration. Due to the fact that transport robot needs capacity of loading, we adopt a kind of legged structure with mammalian configuration. On the other hand, the advantages of all the legs with the same design in terms of design cost, replacement, modularity, and so on are also considered [13]. Compared with the structure of imitated insects, the workspace and flexibility are reduced, but the advantages of low energy and high load are more suitable for transportation in the construction. The mechanical model of hexapod transport robot is shown in Figure 1. Each of the legs consists of three parts which are connected by a hip joint and a knee joint, and the leg mechanism is attached to the body via a root joint. Each of the legs has three joints, and the length of each link has been optimized [14], which can be regarded as a manipulator with three rotating joints attached to a stationary base (the robotic body). Hence, the establishment of kinematic model and the derivation of the kinematic equation can follow the traditional method of robot technology, which can be obtained by defining the reference coordinate system according to the Denavit–Hartenberg model. The body of the robot is made of aluminum-alloy material, which is used to load controller, power supply module, and load. Material of sponge is adhered to the foot tip, which can effectively prevent skid and reduce the impact force from ground.

2.2. Galloping Trajectory Generation. A cycle of animal movements is divided into two states: [15] swing phase and touch-down phase. But the position and speed during the gallop are the most important factors in effecting robot motion. Here, the swing phase is further divided to two phases: flight phase and lift-off phase. In this paper, we adopt elliptic trajectory to generate galloping trajectory [16]. The entire trajectory is made of three transition angles ψ_{flight}, $\psi_{\text{touch down}}$, and $\psi_{\text{lift off}}$, which represent flight phase, touch-down phase, and and lift-off phase, respectively. ψ_1, ψ_2, and ψ_3 are initial angles of each state, which are shown in Figure 2. Those transition angles are obtained by Hermite interpolation algorithm and interconnected by continuous conditions. Thus, the gallop is achieved.

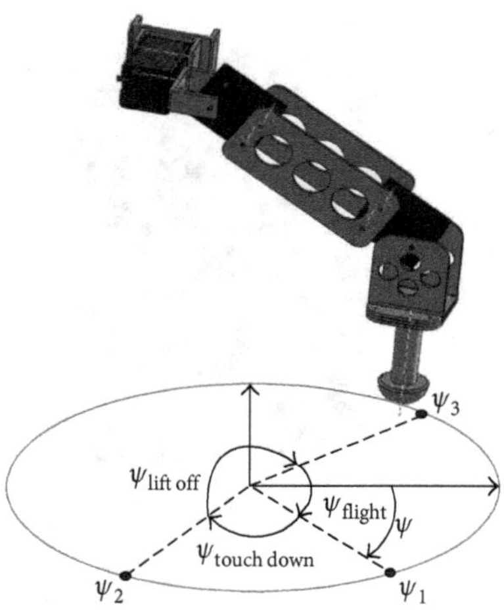

FIGURE 2: State phases and transition angles of galloping.

We assume the whole body of the robot is a mass. According to free fall conditions,

$$y = \frac{1}{2}gt^2, \qquad t_{\text{free fall}} = \sqrt{\frac{2y}{g}}, \tag{1}$$

in which $t_{\text{free fall}}$ is the time of free fall and y is falling height of body. The condition of ψ_{flight} section is from contact off to contact on, and the COG velocity is $\dot{z}_{\text{COG}} < 0$. The ellipse angle is determined by a cubic polynomial, and the 4 unknown parameters are determined by the following 4 constraints:

$$\psi_{\text{init}}^1 = \psi_1, \quad \psi_{\text{final}}^1 = \psi_2, \quad \dot{\psi}_{\text{init}}^1 = 0, \quad \dot{\psi}_{\text{final}}^1 = \frac{v_{\text{body}}}{b}. \tag{2}$$

$$t_{\text{final}_1} = t_{\text{free fall}}$$

in which ψ_{init}^i and ψ_{final}^i are origin angle and end angle belonging to i state, $\dot{\psi}_{\text{init}}^i$ and $\dot{\psi}_{\text{final}}^i$ are origin velocity and end velocity belonging to i state, and v_{body} is the motion velocity of body. ψ and $\dot{\psi}$ of flight state are all determined by cubic polynomial, which is shown in (3). According to the 4 constraints, (4) can be obtained, and then to solve it,

$$\psi_1(t) = a_3t^3 + a_2t^2 + a_1t + a_0, \qquad \dot{\psi}_1(t) = 3a_3t^2 + 2a_2t + a_1, \tag{3}$$

$$\begin{bmatrix} t_{\text{initial}}^3 & t_{\text{initial}}^2 & t_{\text{initial}}^1 & 1 \\ 3t_{\text{initial}}^2 & 2t_{\text{initial}} & 1 & 0 \\ t_{\text{final}}^3 & t_{\text{final}}^2 & t_{\text{final}}^1 & 1 \\ 3t_{\text{final}}^2 & 2t_{\text{final}} & 1 & 0 \end{bmatrix} \begin{bmatrix} a_3 \\ a_2 \\ a_1 \\ a_0 \end{bmatrix} = \begin{bmatrix} \psi_i(t_{\text{initial}}) \\ \dot{\psi}_i(t_{\text{initial}}) \\ \psi_i(t_{\text{final}}) \\ \dot{\psi}_i(t_{\text{final}}) \end{bmatrix}. \tag{4}$$

However, in the $\psi_{\text{touch down}}$ section, the feet tip of robot is keeping contact with ground all the time, and the COG velocity is $\dot{z}_{\text{COG}} < 0 \rightarrow \dot{z}_{\text{COG}} > 0$. The ellipse angle is determined by a first-order polynomial, and the 2 unknown parameters are determined by the equation of ψ and $\dot{\psi}$ of flight section when the foot first contacts ground:

$$\psi_2(t) = \dot{\psi}_{\text{final}}^1 \cdot t_2 + \psi_{\text{final}}^1, \qquad \dot{\psi}_2(t) = \dot{\psi}_{\text{final}}^1 \tag{5}$$

$$t_2 = t_{\text{final}}.$$

In $\psi_{\text{lift off}}$ section, the condition of the foot tip of robot is from contact on to contact off, and the COG velocity is $\dot{z}_{\text{COG}} > 0$. The elliptic angle of this section is similar to the section of flight, which is determined by a cubic polynomial, and the 4 unknown parameters are determined by the following 4 constraints:

$$\psi_{\text{init}_3} = \psi_{\text{final}_2}, \quad \psi_{\text{final}_3} = \psi_{\text{initial}_1}, \quad \dot{\psi}_{\text{init}_3} = \dot{\psi}_{\text{final}_2}, \quad \dot{\psi}_{\text{final}_3} = \frac{v_{\text{body}}}{b} \tag{6}$$

$$t_{\text{final}_3} = t_{\text{free fall}}.$$

On ψ_{flight} section, each ψ and $\dot{\psi}$ are similar to $\psi_{\text{lift off}}$, which are determined by a cubic polynomial equation (7), and it can be solved by following 4 constraints (6):

$$\psi_3(t) = a_3t^3 + a_2t^2 + a_1t + a_0, \qquad \dot{\psi}_3(t) = 3a_3t^2 + 2a_2t + a_1. \tag{7}$$

To sum up, each polynomial based on analysis above can be solved by corresponding constraints one by one, and the whole curve of the ellipse angle is determined, which is shown in Figure 3. According to the change of ellipse angle and following ellipse equation, the ideal location of foot tip of the robot leg can be obtained:

$$X_i(t) = \lambda_1 \cdot \cos\psi_i + x_{\text{center}}, \qquad Z_i(t) = \lambda_2 \cdot \sin\psi_i + z_{\text{center}} \tag{8}$$

in which λ_1 and λ_2 are coefficients of long and short axis of ellipse, X and Z are front direction and vertical direction, and x_{center} and z_{center} are abscissa and ordinate of elliptical center.

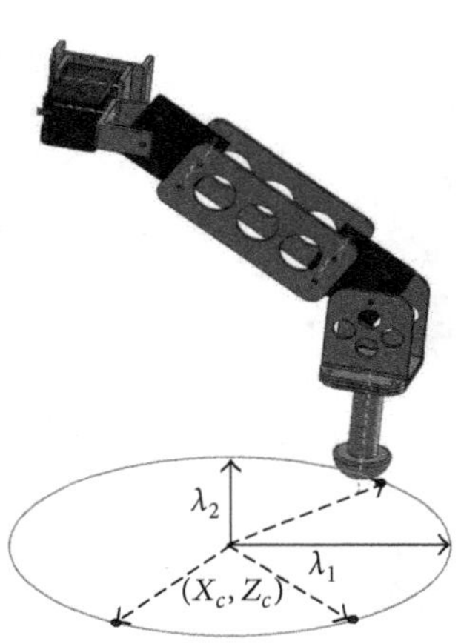

FIGURE 3: Ellipse trajectory of galloping.

3. Energy Consumption Model during Gallop

It can be known from above that as long as ψ_1, ψ_2, and ψ_3 are determined, the whole galloping trajectory is determined too. Taking the condition of transport robots working outdoors independently into consideration, energy consumption is a very important factor. So, in this paper, energy consumption is made as the optimization index, and the degrees of each state are determined by minimum system. Firstly, we establish the model of energy consumption during the gallop.

In this paper, the rotation of each joint of the robot leg is realized by brushless DC motor and reducer with corresponding motor controller. The related functions can be expressed as below [13].

Voltage balance function of DC motor is

$$U_a(t) = R_a i_a(t) + L_a \frac{di_a(t)}{dt} + U_g(t), \quad U_g(t) = k_E \dot{\theta}_m(t). \tag{9}$$

According to static characteristics of DC motor, we have

$$\tau_M(t) = k_M i_a(t). \tag{10}$$

For output torque of motor and joint driving torque, we have

$$\tau_M(t) = \frac{\tau(t)}{\eta N_G}. \tag{11}$$

According to speed of motor and joint speed, we obtain

$$\dot{\theta}_m(t) = N_G \dot{\theta}(t) \tag{12}$$

in which k_E is the EMF constant, k_M is the torque constant, R_a and L_a are the linear resistance and inductance, U_g and i_a are the armature voltage and current, N_G is the gear reduction ratio, $\dot{\theta}_m(t)$ and $\dot{\theta}(t)$ are the speed of motor and joint, η is the motor and gear efficiency, $\tau_M(t)$ is the electromagnetic torque, and $\tau(t)$ is the joint torque.

The instantaneous power of the motor is

$$P_a(t) = U_a(t) i_a(t). \tag{13}$$

Substituting (9) into (13), we obtain

$$P_a(t) = R_a i_a^2(t) + L_a i_a(t) \frac{di_a(t)}{dt} + k_E i_a(t) \dot{\theta}_m(t). \tag{14}$$

Substituting (10) and (11) into (14), we obtain

$$P_a(t) = \frac{R_a}{k_M^2 \eta^2 N_G^2} \tau^2(t) + \frac{\dot{\theta}(t) \tau(t)}{\eta} \tag{15}$$

in which inductance L_a is the energy storage element and if the energy is invariant from the beginning to the end of a period, the energy loss of the inductance can be neglected. Then, the instantaneous power of the motor becomes [16]

$$P_a(t) = \frac{R_a}{k_M^2 \eta^2 N_G^2} \tau^2(t) + \Lambda\left(\frac{\dot{\theta}(t) \tau(t)}{\eta}\right). \tag{16}$$

Taking all joints in the leg into consideration, the total energy of a single leg in a walking gait cycle becomes [16]

$$E_p = \int_0^T \sum_{j=1}^{3} \left(\Lambda\left(\frac{\dot{\theta}_j(t) \tau_j(t)}{\eta}\right) + \frac{R_a}{k_M^2 \eta^2 N_G^2} \tau_j^2(t) \right) dt \tag{17}$$

in which $\Lambda(x)$ is defined as

$$\Lambda(x) = \begin{cases} x & \text{if } x > 0 \\ 0 & \text{if } x \leq 0. \end{cases} \tag{18}$$

Here, T is cycle time, the duration of each leg's one step cycle in a periodic gait. R_a, k_M, η, and N_G are constant. The value of $\dot{\theta}_j(t)$ depends on the trajectory planning. $\tau_j(t)$ is derived by Lagrange's dynamical equations.

Since the transition angles of each state phase ψ_1, ψ_2, ψ_3 are the parameters to be optimized and they determine the final leg trajectory and ideal location of foot tip ${}^{B}P_F$, according to the inverse kinematics function, we have

$$\mathbf{q} = \left[\theta_1\left({}^{B}P(\overline{\psi}), t\right) \ \theta_2\left({}^{B}P(\overline{\psi}), t\right) \ \theta_3\left({}^{B}P(\overline{\psi}), t\right)\right]^T \in \mathbf{R}^{3\times 1} \tag{19}$$

in which

$$ {}^{B}P = \left[{}^{B}P_x \ {}^{B}P_y \ {}^{B}P_z\right]^T, \quad \overline{\psi} = \left[\psi_1 \ \psi_2 \ \psi_3\right]. \tag{20}$$

Usually, ${}^{B}P_F$ is the position vector of foot tip in the leg reference frame and is given by the gait and trajectory planning.

Joint angle, joint angular speed, and angular acceleration can be solved by inverse kinematics function, expressed as

$$\mathbf{q}(\overline{\psi},t)=\begin{bmatrix}\theta_1(\overline{\psi},t)\\ \theta_2(\overline{\psi},t)\\ \theta_3(\overline{\psi},t)\end{bmatrix},$$
$$\dot{\mathbf{q}}(\overline{\psi},t)=\begin{bmatrix}\dot\theta_1(\overline{\psi},t)\\ \dot\theta_2(\overline{\psi},t)\\ \dot\theta_3(\overline{\psi},t)\end{bmatrix}, \tag{21}$$
$$\ddot{\mathbf{q}}(\overline{\psi},t)=\begin{bmatrix}\ddot\theta_1(\overline{\psi},t)\\ \ddot\theta_2(\overline{\psi},t)\\ \ddot\theta_3(\overline{\psi},t)\end{bmatrix}.$$

Joint angle, joint angular speed, and angular acceleration can be calculated through the initial and final state of the locomotion trajectory, but it is still a function with reference to $\overline{\psi}$ and t.

For the joint torque,

$$\widehat{\tau}_p(\overline{\psi},t)=\widehat{\mathbf{D}}\cdot\ddot{\mathbf{q}}(\overline{\psi},t)+\widehat{\mathbf{H}}\cdot\dot{\mathbf{q}}(\overline{\psi},t)+\widehat{\mathbf{C}}, \tag{22}$$

$$\widehat{\mathbf{D}}=\begin{bmatrix}0 & 0 & 0\\ 0 & \frac{1}{4}m_2L_2^2+I_2+m_3L_3^2 & \frac{1}{2}m_3L_2L_3C_{3/2}\\ 0 & \frac{1}{2}m_3L_2L_3C_{3/2} & \frac{1}{4}m_3L_3^2+I_3\end{bmatrix}, \tag{23}$$

$$\widehat{\mathbf{H}}=\begin{bmatrix}0\\ -\frac{1}{4}m_3L_2L_3\dot\theta_3^2S_{3/2}\\ -\frac{1}{2}m_3L_2L_3\dot\theta_2\dot\theta_3S_{3/2}\end{bmatrix}, \tag{24}$$

$$\widehat{\mathbf{C}}=\begin{bmatrix}-\frac{1}{2}m_2gL_2C_2S_1-m_3gL_2C_2S_1-\frac{1}{2}m_3gL_3C_{23}S_1\\ -\frac{1}{2}m_2gL_2C_1S_2-m_3gL_2C_1S_2-\frac{1}{2}m_3gL_3C_1S_{23}\\ -\frac{1}{2}m_3gL_3C_1S_{23}\end{bmatrix}. \tag{25}$$

$\widehat{\mathbf{D}}$ is a symmetric matrix associated with acceleration, which involves the inertia tensor, $\widehat{\mathbf{H}}$ is the Coriolis and centrifugal matrix, and $\widehat{\mathbf{C}}$ is the gravity matrix. According to the derivation above, they are all the function with $\overline{\psi}$ and t. Hence, the power equation (16) and total energy consumption equation (17) of the three-jointed leg system can be formulated as

$$\breve{P}(\overline{\psi},t)=\sum_{j=1}^{3}P_j(\overline{\psi},t)=\widehat{\tau}_p(\overline{\psi},t)^T\widehat{\mathbf{Q}}_p\widehat{\tau}_p(\overline{\psi},t)+\max\left\{c_p^T\widehat{\tau}_p(\overline{\psi},t),0\right\}, \tag{26}$$

$$\breve{E}_p(\overline{\psi})=\int_0^T P(\overline{\psi},t)\,dt=\int_0^T\sum_{j=1}^{3}P_j(\overline{\psi},t)\,dt=\int_0^T\left(\widehat{\tau}_p(\overline{\psi},t)^T\widehat{\mathbf{Q}}_p\widehat{\tau}_p(\overline{\psi},t)+\max\left\{\widehat{c}_p^T\widehat{\tau}_p(\overline{\psi},t),0\right\}\right)dt, \tag{27}$$

where $\widehat{\mathbf{Q}}_p=(2R/k_M^2\eta^2N_G^2)\mathbf{E}_3$ and $\mathbf{E}_3$ is a unit matrix of 3 by 3,

$$\widehat{c}_p=\left[\frac{\dot q_1(\overline{\psi},t)}{\eta}\ \ \frac{\dot q_2(\overline{\psi},t)}{\eta}\ \ \frac{\dot q_3(\overline{\psi},t)}{\eta}\right]^T. \tag{28}$$

It can be known from above that the whole galloping trajectory can be determined as long as ψ_1, ψ_2, and ψ_3 are confirmed. According to the inverse kinematics function, the rotation trajectory of each joint can be determined, and then the joint torque and systematic energy consumption are obtained. The function with $\overline{\psi}$ and t can be obtained by substituting unknown transition angles of state phase into systematic energy consumption equation. ψ_1, ψ_2, and ψ_3 can be confirmed as long as the minimum energy consumption is obtained by optimization. Based on the equation above, taking the energy consumption as optimization index, the state phase transition angle that is meeting the requirement of energy consumption can be obtained by solving (29) which means that the galloping trajectory planning is completed:

$$\text{Object Function}\longrightarrow\min_{\overline{\psi}}\ \vartheta=\int_0^t\breve{E}_p(\overline{\psi}) \tag{29}$$

$$\text{Subject to}\quad C(\overline{\psi})\le 0, \tag{30}$$

where $C(\overline{\psi})\le 0$ is initial range of state phase transition angle.

4. Optimal Solutions with Genetic Algorithm

The genetic algorithm is used for solving this optimization problem that has linear or nonlinear constrained condition. It is based on natural selection, solved by biological evolution. It is widely used in various optimization problems [17]. This method is described in Figure 4.

(1) Firstly, we create a population with n size and then select the new state phase transition angle within the constraint and then put it into the galloping trajectory planner to calculate. In one gait period, the running speed and stride sizes are all determined by gait generator [16]. For each individual, the initial position and the end position are obtained by the

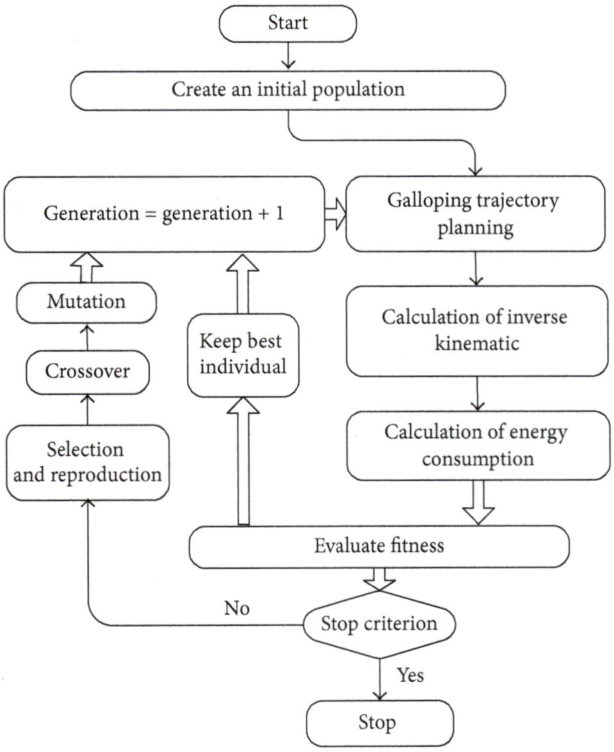

Figure 4: Process of GA.

trajectory planning. According to the initial and end condition, the initial and end joint angle, angular speed, and angular accelerations are obtained by inverse kinematics equation (21).

(2) Joint torques can be obtained by joint angle, angular speed, angular accelerations, and (22), and total energy consumption in one gait period can be obtained by (27). According to constraint condition, punishment factors are given and fitness is valuated. The inferior individuals are removed and superior individual with a number of m are to be saved directly to the next generation.

(3) If the stopping criterion is satisfied after evaluating the fitness, the optimization will stop. Otherwise, the new individual will be produced continually and create the next generation by selection, rebuilding, cross, and variation. During the optimization, cross probability is ξ, variation probability is β, and penalty factor is ρ. And the variation of mean fitness is shown in Figure 5.

In this paper, the parameters of the system are shown in Table 1. The length of each joint is L_1, L_2, L_3, the long and short axis coefficient are λ_1, λ_2, and the boundary of the variations is

$$\begin{bmatrix} 0 & \frac{\pi}{2} & \frac{3\pi}{2} \end{bmatrix} \leq \overline{\psi} \leq \begin{bmatrix} \frac{\pi}{2} & \pi & 2\pi \end{bmatrix}. \tag{31}$$

The change of the final fitness is smaller than 10^{-6} after 51 generations, and the final result is shown below:

$$\overline{\psi} = \begin{bmatrix} 0.754 & 2.23 & 4.737 \end{bmatrix}. \tag{32}$$

Table 1: Parameters of system.

Parameters	Value	Unit
k_E	13.4	mNm/A
R_a	9.17	Ω
L_a	0.219	mH
N_G	200	—
L_1	0.05	m
L_2	0.103	m
L_3	0.142	m
λ_1	0.08	—
λ_2	0.025	—
η	0.7	—
k_M	13.4	mV/rpm
I_2	0.375	10^{-3} Kg·m^2
I_3	0.375	10^{-3} Kg·m^2
n	20	—
m	4	—
ξ	0.6	—
β	0.2	—
ρ	2	—

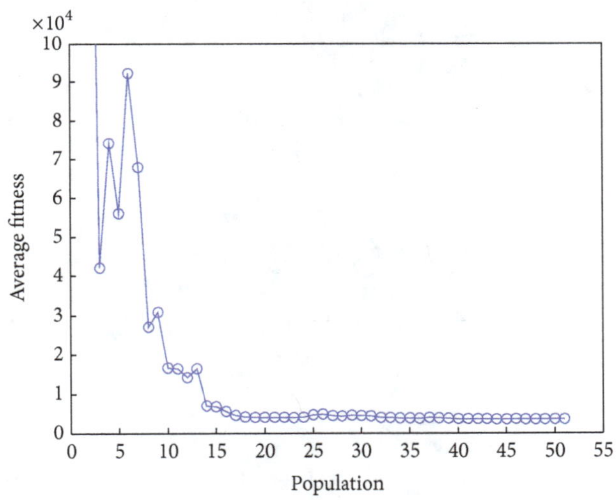

Figure 5: Variation of mean fitness.

5. Simulations

In this section, the ellipse trajectory planning during the gallop is simulated by the software Matlab/Simulink. It is assume that the robot goes forward along the x-axis on the horizontal ground, the length of a single step is 0.15 m in a cycle time, the relative height between the root joint of leg and ground is 0.17 m, and the gait cycle is 2 s. Galloping trajectory is generated based on the parameters of D-H model established in Matlab/Simulink, which is shown in Table 1.

According to the original and optimized state phase transition angle and galloping trajectory of the robot, the objective trajectory of foot tip based on root-joint coordinate is generated, which is shown in Figure 6. The parameter $\overline{\psi} = \begin{bmatrix} 0.628 & 2.512 & 5.338 \end{bmatrix}$ is used before the optimization, which

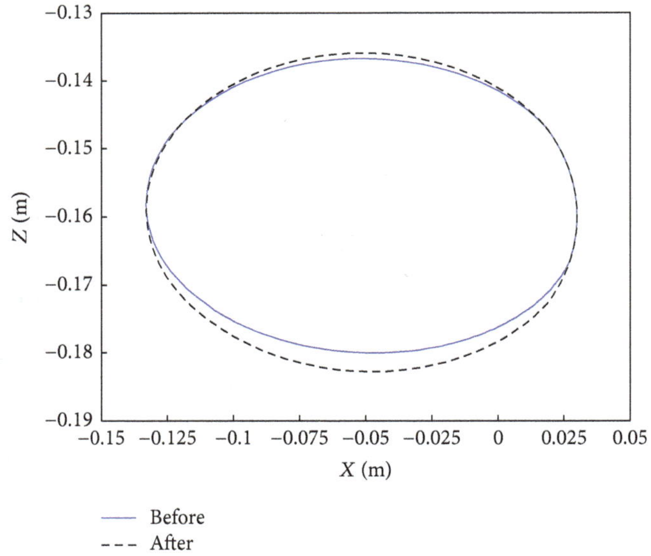

FIGURE 6: Ellipse trajectory of galloping.

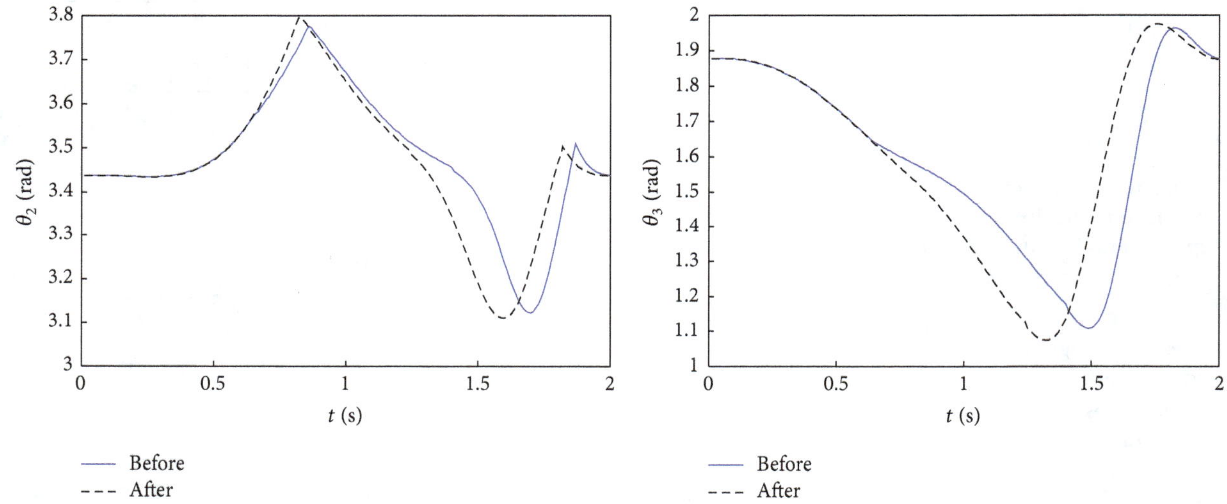

(a) Hip joint (b) Knee joint

FIGURE 7: Variations of joint angles.

is obtained by deriving the ideal step-size. Compared with the optimized parameter $\overline{\psi}$, both of the ellipse trajectories are similar basically and just differ in the division of each phase. After optimization, the angles of each state phase have a significant difference, flight phase increased, touch-down phase decreased, and the range of each state phase changed, which can be seen from the hip and knee joint movement angle (Figure 7). In the range of 0 to 0.65 seconds approximately, regardless of the optimization or not, the joint angles are all in the lift-off phase and unchanged basically. However, when near to 0.56 seconds, the optimized trajectory begins to transform with large degree and enters into the touch-down phase early. After that, the optimized trajectory becomes bigger than initial. It makes the distribution of work more reasonable during the motion.

The torques curves of each joint are shown in Figure 8. Joint torque almost does not work in 0–0.65 s, which is corresponding with the locomotion. The main reason of it is that the left-off phase is free fall and do not need to overcome its own weight to do work. However, after 0.65 s, the optimized trajectory enters the flight phase firstly, and do more work than before. But in the lift-off phase later, the work of optimized trajectory regardless of hip joint and knee joint is much smaller than nonoptimization.

This is also reflected in the system energy consumption during the movement, which is shown in Figure 9. Due to the existing of touch-down phase, the system energy consumption of optimized trajectory is larger than nonoptimization during 0.65 seconds to 1.5 seconds. But the optimized parameters are more reasonable obviously

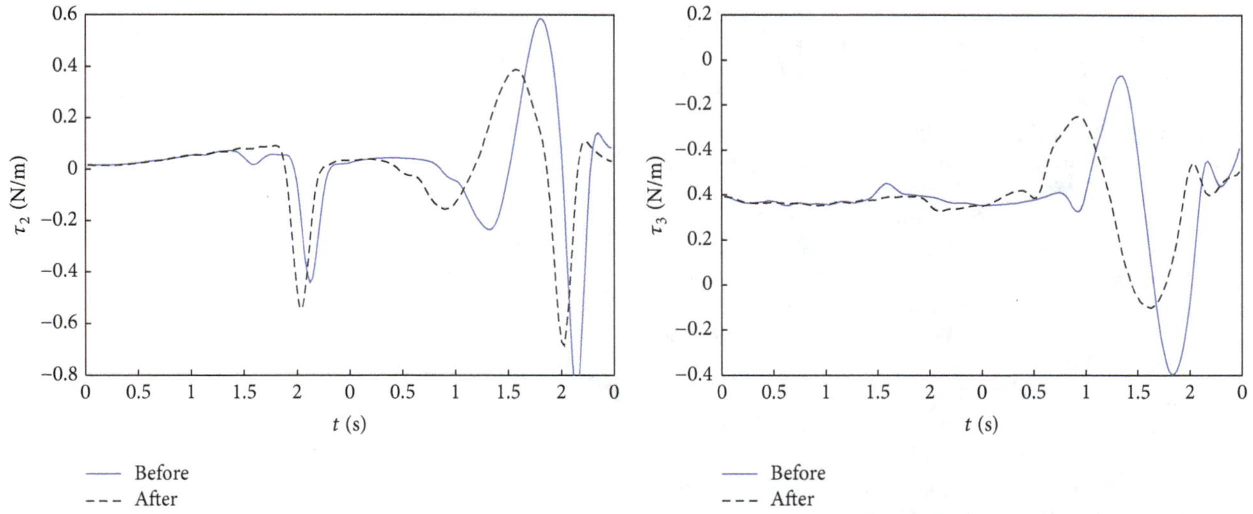

(a) Hip joint

(b) Knee joint

FIGURE 8: Variations of joint torques.

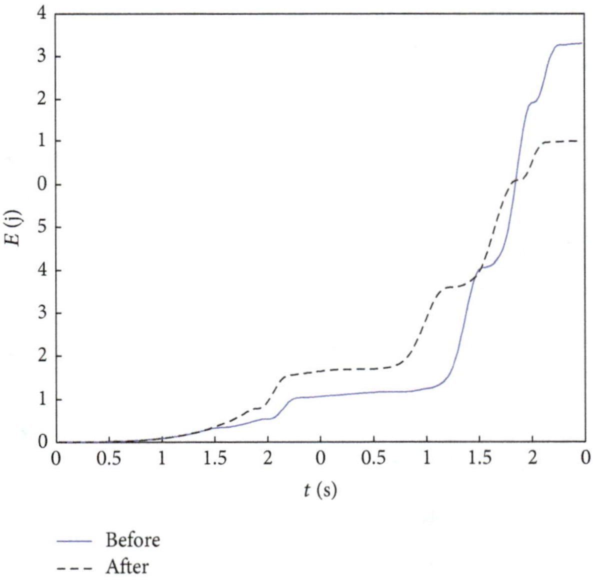

FIGURE 9: System energy consumption of galloping.

in the lift-off phase. After optimization, the system energy consumption is reduced by 23% in one gait period. It illustrates that the method of trajectory planning based on energy consumption optimization is reasonable and effective.

6. Conclusions

In this paper, using the existed transportation robot model, the ideal foot tip function based on elliptic galloping trajectory is derived by establishing the transition angles of flight phase, touch-down phase, and lift-off phase and constraint condition of each phase. On the basic analysis of the robot leg dynamics and the joint drive motor model, the formula of joint torque is derived, system energy consumption model is established, and it introduces the galloping state phase transition angle of the robot into that model, and then the index of energy consumption optimization is proposed based on the transition angle. The genetic algorithm is used for optimization. We introduce the inverse kinematics solution and the galloping trajectory planning into the process of the genetic algorithm to solve. The result shows that the distribution of optimized state phase transition angles is more reasonable, and it can reduce joint torque and energy consumption effectively. Although the algorithm proposed in this paper is based on the motor model, it can also apply to other legged robots.

Competing Interests

The authors declare that they have no competing interests.

Acknowledgments

The project supported by Natural Science Basic Research Plan in Shaanxi Province of China (Program no. 2016JQ6066); project funded by China Postdoctoral Science Foundation (no. 2016M592728); Fundamental Research Funds for the Central Universities (no. 310825151041) are acknowledged.

References

[1] A. Irawan and K. Nonami, "Optimal impedance control based on body inertia for a hydraulically driven hexapod robot walking on uneven and extremely soft terrain," *Journal of Field Robotics*, vol. 28, no. 5, pp. 690–713, 2011.

[2] A. Goswami and V. Kallem, "Rate of change of angular momentum and balance maintenance of biped robots," in *Proceedings of the IEEE International Conference on Robotics and Automation (ICRA '04)*, vol. 4, pp. 3785–3790, New Orleans, La, USA, May 2004.

[3] S. G. Roh and J. H. Park, "A 3D galloping of a quadruped robot based on force tracking control," in *Proceedings of the IASTED International Conference on Modelling, Identification, and Control*, pp. 1166–1171, 2011.

[4] A. Kawamura, K. Sugio, K. Suzuki, and C. Zhu, "Simulation study on one leg jumping for biped running," in *Proceedings of the 8th IEEE International Workshop on Advanced Motion Control (AMC '04)*, pp. 433–438, Kawasaki, Japan, March 2004.

[5] I. Poulakakis, J. A. Smith, and M. Buehler, "On the dynamics of bounding and extensions: towards the half-bound and gallop gaits," in *Adaptive Motion of Animals and Machines*, H. Kimura, K. Tsuchiya, A. Ishiguro, and H. Witte, Eds., pp. 79–88, Springer, Tokyo, Japan, 2006.

[6] K. Y. Kim and J. H. Park, "Ellipse-based leg-trajectory generation for galloping quadruped robots," *Journal of Mechanical Science and Technology*, vol. 22, no. 11, pp. 2099–2106, 2008.

[7] L. R. Palmer III and D. E. Orin, "Force redistribution in a quadruped running trot," in *Proceedings of the IEEE International Conference on Robotics and Automation (ICRA '07)*, pp. 4343–4348, IEEE, Rome, Italy, April 2007.

[8] M. Focchi, V. Barasuol, I. Havoutis, J. Buchli, C. Semini, and D. G. Caldwell, "Local reflex generation for obstacle negotiation in quadrupedal locomotion," in *Proceedings of the International Conference on Climbing Walking Robots (CLAWAR '13)*, pp. 443–450, 2013.

[9] D. W. Marhefka, D. E. Orin, J. P. Schmiedeier, and K. J. Waldron, "Intelligent control of quadruped gallops," *IEEE/ASME Transactions on Mechatronics*, vol. 8, no. 4, pp. 446–456, 2003.

[10] G. Capi, S. Kaneko, K. Mitobe, L. Barolli, and Y. Nasu, "Optimal trajectory generation for a prismatic joint biped robot using genetic algorithms," *Robotics and Autonomous Systems*, vol. 38, no. 2, pp. 119–128, 2002.

[11] S. G. Roh and J. H. Park, "Gallop with speed change for quadruped robots," in *Proceedings of the IEEE International Conference on Robotics and Biomimetics (ROBIO '11)*, pp. 2787–2792, Phuket, Thailand, December 2011.

[12] K. Y. Kim, O. Kwon, J. S. Yeon, and J. H. Park, "Elliptic trajectory generation for galloping quadruped robots," in *Proceedings of the IEEE International Conference on Robotics and Biomimetics (ROBIO '06)*, pp. 103–108, IEEE, Kunming, China, December 2006.

[13] T. Zielinska and J. Heng, "Development of a walking machine: mechanical design and control problems," *Mechatronics*, vol. 12, no. 5, pp. 737–754, 2002.

[14] Y.-G. Zhu, B. Jin, W. Li, and S.-T. Li, "Optimal design of hexapodwalking robot leg structure based on energy consumption andworkspace," *Transactions of the Canadian Society for Mechanical Engineering*, vol. 38, no. 3, pp. 305–317, 2014.

[15] J. A. Galvez, J. Estremera, and P. G. De Santos, "A new legged-robot configuration for research in force distribution," *Mechatronics*, vol. 13, no. 8-9, pp. 907–932, 2003.

[16] B. Jin, C. Chen, and W. Li, "Power consumption optimization for a hexapod walking robot," *Journal of Intelligent & Robotic Systems*, vol. 71, no. 2, pp. 195–209, 2013.

[17] G. Capi, Y. Nasu, L. Barolli, K. Mitobe, and K. Takeda, "Application of genetic algorithms for biped robot gait synthesis optimization during walking and going up-stairs," *Advanced Robotics*, vol. 15, no. 6, pp. 675–694, 2001.

18

Dynamic Decoupling Control Optimization for a Small-Scale Unmanned Helicopter

Rui Ma,[1] **Li Ding**,[2] **and Hongtao Wu**[1]

[1]*College of Mechanical and Electrical Engineering, Nanjing University of Aeronautics and Astronautics, Nanjing 210016, China*
[2]*College of Mechanical Engineering, Jiangsu University of Technology, Changzhou 213001, China*

Correspondence should be addressed to Li Ding; nuaadli@163.com

Academic Editor: Huosheng Hu

This article presents design and optimization results from an implementation of a novel disturbance decoupling control strategy for a small-scale unmanned helicopter. Such a strategy is based on the active disturbance rejection control (ADRC) method. It offers an appealing alternative to existing control approaches for helicopters by combining decoupling and disturbance rejection without a detailed plant dynamics. The tuning of the control system is formulated as a function optimization problem to capture various design considerations. In comparison with several different iterative search algorithms, an artificial bee colony (ABC) algorithm is selected to obtain the optimal control parameters. For a fair comparison of control performance, a well-designed LQG controller is also optimized by the proposed method. Comparison results from an attitude tracking simulation against wind disturbance show the significant advantages of the proposed optimization control for this control application.

1. Introduction

In recent years, rotary-wing Unmanned Aerial Vehicles (UAVs) including quadrotors, helicopters, and ducted fans are attractive to industries and academia [1, 2]. With the unique features such as hovering, good maneuverability, and low costs, they have been applied to diverse domains by installing different sensors, cameras, or other payloads on the platform [3, 4]. However, due to the complexity of flight dynamics, it is still challenging to design an appropriate flight control system that satisfies the requirement for autonomous flight.

Small-scale unmanned helicopter is a representative of the rotary-wing UAVs. It is considered as an inherently unstable, highly nonlinear, and underactuated system with significant dynamic coupling. With the small size and agile maneuverability, it is more susceptible to gust disturbance than those full-sized counterpart. Furthermore, the dynamic parameters change with the load and flight conditions. These factors cause serious challenges in dealing with concerns about the robustness, disturbance rejection, decoupling, and other control problems. To address the above problems, the classical single-input/single-output (SISO) feedback control methods are implemented in [5–7]. In these researches, the controller is optimized based on the identified model established firstly for a small helicopter. Nevertheless, many researchers have recognized that the complexity nature of helicopter and better control capability requires more advanced technologies. Cases include direct adaptive neural command controller [8], adaptive control methods [9], nonlinear control methods [10], vision-based guidance control techniques [11], and intelligent control methods like fuzzy logic approach [12] and neural network [13]. And the most pervasive choice in practical applications is the robust control approach: for example, the Kalman filter-based linear quadratic integral (LQI) approach [14], linear quadratic regulator (LQR) [15], and the H_∞ control approach [16, 17]. These methods provide a reasonable countermeasure for both disturbances and multivariable effects of the helicopter. However, the effectiveness of the prevailing model-based control approaches is exceedingly dependent on the exact model and aerodynamic coefficients of the plant. Facing the complicated control problem of the helicopter, the effective solution is to compensate the disturbance immediately and reduce the dependence of plant model.

Table 1: Physical descriptions of the state and input variables of the helicopter dynamical model.

Variable	Physical significance	Unit
$P = [X, Y, Z]^T$	Position in Earth-fixed coordinate frame	m
$V = [u, v, w]^T$	Velocity vector along body frame X-, Y-, and Z-axes	m/s
$\Omega = [p, q, r]^T$	Roll, pitch, and yaw angular rates	rad/s
$\Theta = [\varphi, \theta, \psi]^T$	Euler angles	rad
a_s, b_s	Longitudinal and lateral main rotor flapping angle	rad
r_{fb}	Intermediate state in yaw rate feedback controller dynamics	N/A
u_{lat}	lateral cyclic rotor control input	N/A
u_{lon}	longitudinal cyclic rotor control input	N/A
u_{col}	collective pitch control input	N/A
u_{ped}	tail rotor pedal control input	N/A

The main contribution of this paper is introducing a dynamic decoupling control (DDC) strategy [18] and its optimization method to a small-scale helicopter. This strategy is rooted in active disturbance rejection control (ADRC) that was recently proposed by Han [19]. The key idea of the method is to treat the total disturbance (incorporating the interactions among control loops and the unknown external disturbances) as a state variable, which can be estimated by an extended state observer (ESO) through the input-output data of the plant in real time. Consequently, unlike most existing model-dependent control methods, very little information of the model is required for ADRC [20]. ADRC offers a practical solution to decoupling control problems in the presence of large uncertainties and has been successfully applied in many engineering applications, e.g., aircraft flight control [21] and the chemical processes [22]. Moreover, in order to simplify the implementation of ADRC, Gao proposed the linear active disturbance rejection control (LADRC), which offers much better performance and needs few parameters to tune, and detailed comparison studies can be found in [23].

Since the performance of LADRC depends on the convergence speed of state observer, the bandwidth of which is the most important tuning parameter. Obviously, the trade-offs between the robustness and performance have always been difficult, especially for the helicopter. This problem can be solved by using a multiobjective optimization algorithm in the simulated environment. Artificial bee colony algorithm (ABC) was first proposed by Karaboga in 2005 [24] and successfully applied to control optimization, including optimal tuning of PID controller in [25], optimized LQR controller in [26], and robust fuzzy PSS design [27]. As we have known, usual optimization algorithms conduct only one search operation in one iteration, but ABC algorithm can conduct both local search and global search in each iteration; as a result, the probability of finding the optimal parameters is significantly increased, which efficiently avoids local optimum to a large extent.

This paper considers a design and optimization of the DDC controller used in our TREX-600 helicopter. As a controlled plant, the dynamical model is obtained through the system identification method in previous work [26]. The main idea can be characterized as follows: (i) in the design of decoupling control, all the information needed is only the predetermined input-output pairing of helicopter's dynamic model, (ii) the effect of one input to all other outputs that is not paired with, namely, the cross channel interference is viewed as a 'disturbance' to be actively estimated and canceled out in DDC framework, and (iii) the parameter tuning problem is transformed into a functional optimization problem defined by a combination of different control performance indexes, and ABC algorithm is introduced to calculate the optimal solution. Using this approach, we can optimize controllers with different control requirement.

The paper is organized as follows. In Section 2, the helicopter dynamical model under consideration is briefly introduced. Section 3 describes how to use DDC strategy to decouple the helicopter. Section 4 formulates the parameter optimization problem. Simulation tests on the model are shown in Section 5. Finally, the main conclusions are summarized in Section 6.

2. System Model for Unmanned Small-Scale Helicopter

Based on the first principle approach, the dynamics of helicopter is regarded as a six-degrees-of-freedom rigid-body dynamics augmented with a simplified main rotor flapping dynamics and a factory-installed yaw rate gyro dynamics. As illustrated in Figure 1 and summarized in Table 1, this model contains fifteen states and four inputs. Detailed information of the physical parameters and modeling structure can be found in [28]. A brief overview of the flight dynamical model is presented next.

The translational motion and rotational motion of the helicopter [15] are described as

$$\begin{aligned} \dot{P} &= R(\Theta) V, \\ \dot{\Theta} &= S(\Theta) \Omega, \\ \dot{V} &= \frac{\boldsymbol{F}_b}{m} + \frac{\boldsymbol{F}_g}{m} - \Omega \times V \\ \dot{\Omega} &= \boldsymbol{I}^{-1} \left[\boldsymbol{M}_b - \Omega \times (\boldsymbol{I}\Omega) \right] \end{aligned} \tag{1}$$

where $\boldsymbol{F}_g = [-mg\sin\theta, mg\sin\phi\cos\theta, mg\cos\theta\cos\phi]^T$ is the gravity force vector projected onto the body frame (BF); m

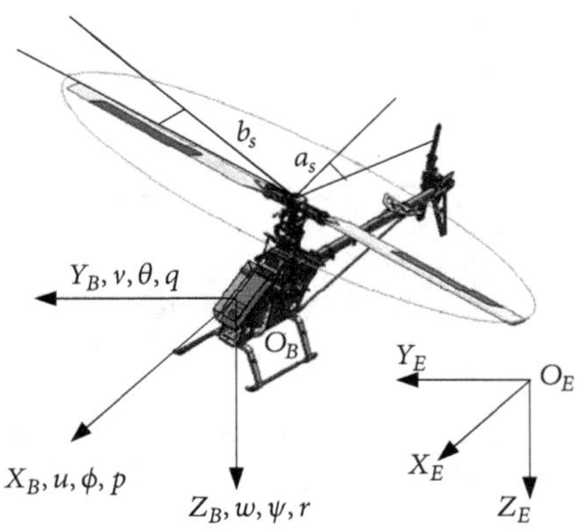

FIGURE 1: Illustration for helicopter states and coordinates.

is the helicopter mass; $\boldsymbol{I} = \text{diag}\{I_{XX}, I_{YY}, I_{ZZ}\}$ is the inertial moment matrix about the reference axes; $\boldsymbol{F}_b$ and $\boldsymbol{M}_b$ denote the combined aerodynamic force and moment vectors acting on the helicopter center of gravity (CG), respectively. The transformation matrices R and S are, respectively, given as

$$R = \begin{bmatrix} c_\theta c_\psi & s_\phi s_\theta c_\psi - c_\phi s_\psi & c_\phi s_\theta c_\psi + s_\phi s_\psi \\ c_\theta s_\psi & s_\phi s_\theta s_\psi + c_\phi c_\psi & c_\phi s_\theta s_\psi + s_\phi c_\psi \\ -s_\theta & s_\phi c_\theta & c_\phi c_\theta \end{bmatrix}, \quad (2)$$

$$S = \begin{bmatrix} 1 & t_\theta s_\phi & t_\theta c_\phi \\ 0 & c_\phi & -s_\phi \\ 0 & \dfrac{s_\phi}{c_\theta} & \dfrac{c_\phi}{c_\theta} \end{bmatrix}, \quad (3)$$

in which the compact notations $s_{(*)}$, $c_{(*)}$, and $t_{(*)}$ denote $\sin(*)$, $\cos(*)$, and $\tan(*)$, respectively.

In the helicopter system, $\boldsymbol{F}_b$ and $\boldsymbol{M}_b$ are generated by the aerodynamic forces of the fuselage and the control forces which originate from the main rotor thrust and tail rotor thrust. Generally, we calculate $\boldsymbol{F}_b$ and $\boldsymbol{M}_b$ without the consideration of the aerodynamic forces of fuselage due to the relatively small influence on the model. Therefore, the force and moment components in the BF are, respectively, given by

$$\boldsymbol{F}_b = \begin{bmatrix} F_X \\ F_Y \\ F_Z \end{bmatrix} = \begin{bmatrix} -T_{mr} \sin a_s \\ T_{mr} \sin b_s - T_{tr} \\ -T_{mr} \cos a_s \cos b_s \end{bmatrix}, \quad (4)$$

$$\boldsymbol{M}_b = \begin{bmatrix} L \\ M \\ N \end{bmatrix} = \begin{bmatrix} \left(K_\beta + T_{mr} H_{mr}\right) \sin b_s - T_{tr} H_{tr} \\ \left(K_\beta + T_{mr} H_{mr}\right) \sin a_s \\ N_{mr} + T_{tr} D_{tr} \end{bmatrix} \quad (5)$$

where T_{mr}, N_{mr}, and T_{tr} are the main rotor thrust and moment and the tail rotor thrust, respectively; D_{tr} is the distance from the CG to the tail rotor hub, along the x direction; H_{mr} and H_{tr} are the distance from the CG to the main rotor and tail rotor hub, along the z direction, respectively, and K_β is the main rotor spring constant.

The main rotor flapping dynamics, which are common to all small-scale helicopters, is described by the following two coupled first-order differential equations [29]:

$$\begin{bmatrix} \dot{a}_s \\ \dot{b}_s \end{bmatrix} = \begin{bmatrix} -q - \dfrac{a_s}{\tau_f} + A_b b + A_{lon} u_{lon} \\ -p + B_a a_s - \dfrac{b_s}{\tau_f} + B_{lat} u_{lat} \end{bmatrix} \quad (6)$$

where τ_f is the main rotor time constant; A_b and B_a are cross coupling derivatives that influence the longitudinal and lateral flapping motions; and A_{lon} and B_{lat} are effective linkage gains.

Since the high sensitivity of the bare yaw channel dynamics, a feedback yaw rate controller is widely used in small-scale helicopters. The UAV system reserved this feature for the convenience of manual control. Accordingly, the augmented yaw dynamics are modeled as a first-order bare airframe dynamics with a yaw rate feedback represented by a simple first-order low-pass filter [30]. The corresponding differential equations are given as

$$\begin{bmatrix} \dot{r} \\ \dot{r}_{fb} \end{bmatrix} = \begin{bmatrix} N_r r + N_{ped}\left(u_{ped} - r_{fb}\right) \\ -K_{rfb} r_{fb} + K_r r \end{bmatrix} \quad (7)$$

where N_r, N_{ped}, K_r, and K_{rfb} are the parameters to be identified.

3. Dynamic Decoupling Control (DDC) of the Helicopter

3.1. LESO Based Dynamic Decoupling Control Method. Linear Active Disturbance Rejection Controller (LADRC) is a novel control method which is parameterized from ADRC [20] to simplify the tuning process. In this work, LADRC-based DDC approach [18] is implemented to tackle the decoupling problem for helicopter attitude dynamics. Define f_i as the combined effect of the internal coupling dynamics and external disturbances in each channel:

$$f_i = h_i\left(x, x^{(1)}, \cdots x^{(n_i-1)}, w_i\right) + \sum_{j=1}^{m} b_{ij} u_j - b_{ii} u_i \quad (8)$$

Then, the helicopter model can be seen as a set of coupled input-output equations with a predetermined relationship:

$$\begin{aligned} y_1^{(n_1)} &= f_1 + b_{11} u_1 \\ &\vdots \\ y_i^{(n_2)} &= f_i + b_{ii} u_i \\ &\vdots \\ y_m^{(n_m)} &= f_m + b_{mm} u_m \end{aligned} \quad (9)$$

where w_i and x are external disturbances and state vector, respectively; u_i and y_i are the dominant input and output of the ith loop ($i = 1, 2, \cdots, m$), respectively; b_{ij} is the input gain; superscript (n_i) denotes the n_i th order derivative. Assuming the order n_i are given, the numbers of inputs and outputs are the same.

Most existing decoupling control approaches assume the knowledge of the elaborate plant model or disturbance model, which is a considerable challenge in practice. LADRC makes a breakthrough that realistically estimates f_i in real time from input-output data instead of identifying an accurate mathematical model. The idea is introduced next.

Define an enlarged state vector $\overline{\boldsymbol{x}}_i = (y_i, \dot{y}_i, \cdots, y_i^{(j-1)}, f_i)$, $j = 1, 2, \cdots, n_i$, in which f_i is added as an extended state. Assume f_i is differentiable and $v_i = \dot{f}_i$ is bounded. Then the augmented state-space form of ith loop in (9) is represented as

$$\begin{aligned} \dot{\overline{\boldsymbol{x}}}_i &= \boldsymbol{A}\overline{\boldsymbol{x}} + \boldsymbol{B}u_i + \boldsymbol{G}v_i \\ y &= \boldsymbol{C}\overline{\boldsymbol{x}}_i \end{aligned} \tag{10}$$

where

$$\begin{aligned} \boldsymbol{A} &= \begin{bmatrix} 0 & 1 & 0 & \cdots & 0 \\ 0 & 0 & 1 & \cdots & 0 \\ \vdots & \vdots & \vdots & \ddots & \vdots \\ 0 & 0 & 0 & \cdots & 1 \\ 0 & 0 & 0 & \cdots & 0 \end{bmatrix}, \\ \boldsymbol{B} &= \begin{bmatrix} 0 \\ 0 \\ \vdots \\ b_{0,i} \\ 0 \end{bmatrix}, \\ \boldsymbol{G} &= \begin{bmatrix} 0 \\ 0 \\ \vdots \\ 0 \\ 1 \end{bmatrix}, \\ \boldsymbol{C} &= \begin{bmatrix} 1 \\ 0 \\ \vdots \\ 0 \\ 0 \end{bmatrix}^T. \end{aligned} \tag{11}$$

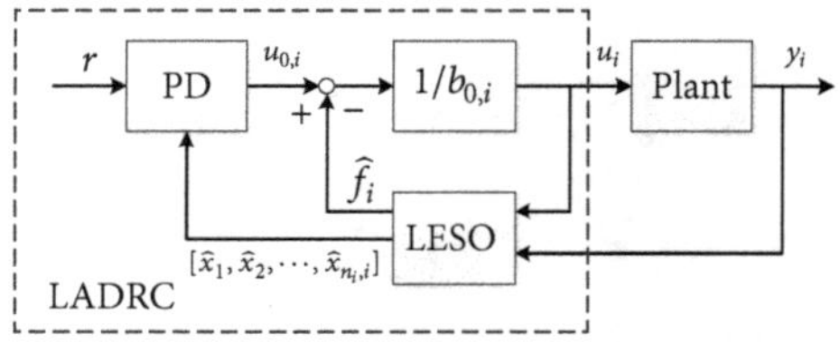

FIGURE 2: The block diagram of LADRC.

Based on the state-space model, a linear extended state observer (LESO) is designed to estimate $\overline{\boldsymbol{x}}_i$:

$$\begin{aligned} \dot{\widehat{\boldsymbol{x}}}_i &= \boldsymbol{A}\widehat{\boldsymbol{x}}_i + \boldsymbol{B}u_i + \boldsymbol{L}_i\left(y_i - \widehat{y}_i\right) \\ \widehat{y}_i &= \widehat{\boldsymbol{x}}_i \end{aligned} \tag{12}$$

where $\boldsymbol{L}_i = [l_{1,i}, l_{2,i}, \cdots, l_{n_i+1,i}]^T$ is the observer gain needed to be chosen.

With a properly selected observer gain, the system states and f_i will be accurately estimated by the LESO in real time. The following control law for i^{th} loop can be designed to reduce the closed-loop system approximately to a unit gain cascaded integrator plant $y_i^{(n_i)} = f_i + b_{ii}u_i \approx u_{0,i}$:

$$u_i = \frac{u_{0,i} - \widehat{f}_i}{b_{i,i}} \tag{13}$$

It is a relatively simple control problem, which is solved by using a PD controller with a feedforward term:

$$u_{i,0} = k_{1,i}\left(r_i - \widehat{x}_{1,i}\right) + \cdots + k_{n_i,i}\left(r_i^{(n_i-1)} - \widehat{x}_{n_i,i}\right) + r_i^{(n_i)} \tag{14}$$

where $(k_{1,i}, k_{2,i}, \cdots, k_{n,i})$ are the controller gains to be selected and r_i is the trajectory reference. The structure of LADRC is shown in Figure 2.

3.2. Parameterization of LADRC. For simplicity and practicality, both of the LESO and PD controller are parameterized in a special case as suggested in [18], where all the observer poles and controller poles are placed at $-\omega_{0,i}$ and $-\omega_{c,i}$, respectively. The characteristic polynomials of (12) and (14) are constituted, respectively, as

$$\lambda_{o,i}(s) = s^{n_i+1} + l_{1,i}s^{n_i} + \cdots + l_{n_i+1,i} = \left(s + \omega_{0,i}\right)^{n_i+1} \tag{15}$$

$$\lambda_{c,i}(s) = s^{n_i} + k_{n_i,i}s^{n_i} + \cdots + k_{1,i} = \left(s + \omega_{c,i}\right)^{n_i} \tag{16}$$

with

$$l_{j,i} = \frac{(n_i + 1)!\omega_{0,i}^j}{j!\left(n_i + 1 - j\right)!}, \quad j = 1, 2, \cdots, n_i + 1 \tag{17}$$

$$k_{j,i} = \frac{n_i!\omega_{c,i}^{n_i-j+1}}{(j-1)!\left(n_i - j + 1\right)!}, \quad j = 1, 2, \cdots, n_i \tag{18}$$

It makes $\omega_{0,i}$, $\omega_{c,i}$ the bandwidth and the only tuning parameters for LESO and the PD controller, respectively. In

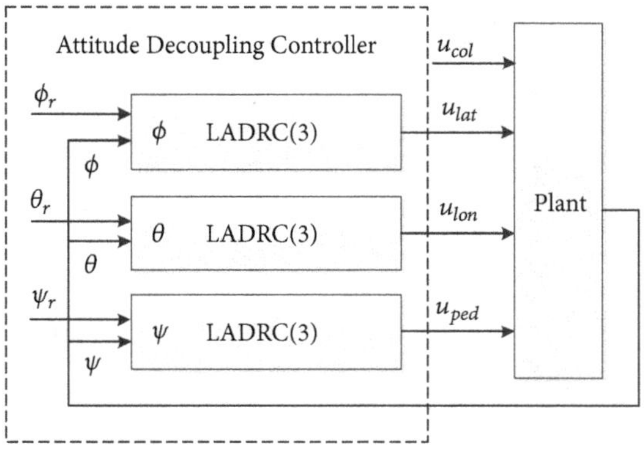

FIGURE 3: The block diagram of the attitude decoupling controller.

general, higher bandwidth corresponds to better transient response, disturbance estimation, and rejection. However, too large a value of $\omega_{0,i}$ would cause oscillation in states. The measurement noise and excessive increase of $\omega_{c,i}$ make the control signal oversize in magnitude and change rate. On the other hand, an appropriate selection of $\omega_{0,i}$ and $\omega_{c,i}$ should be subjected to physical limits and dynamic characteristics of the plant.

As seen above, the LADRC approach is a practical method for decoupling control of MIMO system. The satisfied performance will be obtained by tuning only two parameters $\omega_{0,i}$ and $\omega_{c,i}$. Furthermore, it works without the detail model of the original plant, except the orders of each input-output pairs and input gains b_{ii}. The proofs of stability are given in [31, 32].

3.3. Attitude Controller Design for the Helicopter. As shown in Figure 3, the decoupling controller is designed to have the form of three LADRC controllers. Moreover, we selected their orders according to the relative degrees of the dynamical model. It is assumed that the controlled output $\Theta = [\varphi, \theta, \psi]^T$ can be measured directly and that the trim value $\boldsymbol{r} = [\varphi_r, \theta_r, \psi_r]^T$ is within the physical limitation of helicopter flight.

To use the DDC approach, the order of input-output pairs in the model must be explicit. Figure 4 displays the interconnection of the helicopter subsystems, which offers a more physically meaningful design. Note that the helicopter attitude dynamics can be separated in two interconnected subsystems [6], i.e., the lateral and longitudinal subsystem and yaw dynamics. The cyclic commands u_{lon} and u_{lat} control the pitch and roll moment, and the pedal command u_{ped} manipulates the heading of the helicopter. In this case, we set the lateral and longitudinal subsystem as two third-order systems and the yaw dynamics as a second-order system.

According to aforementioned discussion and analysis, we define f_1, f_2, and f_3 as the total disturbance in each channel and rewrite (9) as

$$\begin{aligned} \phi^{(n_1)} &= f_1 + b_{01} u_{lat} \\ \theta^{(n_2)} &= f_2 + b_{02} u_{lon} \\ \psi^{(n_3)} &= f_3 + b_{03} u_{ped} \end{aligned} \quad (19)$$

where b_{01}, b_{02}, and b_{03} are the input gains of lateral cyclic, longitudinal cyclic, and tail rotor collective pitch, respectively. In the LADRC design, these input gains are treating as another tuning parameter besides $\omega_{0,i}$ and $\omega_{c,i}$ to improve the performance of the reduced order closed-loop system. Note that the orders of each loop are $n_1 = n_2 = 3$ and $n_3 = 2$, and the LADRC-based DDC controller can be realized by designing the LESO and PD controller for each loop, accordingly.

4. Optimization Problem Formulation

4.1. The Objective Function. The proposed LADRC tuning method using ABC is schematically shown in Figure 5, where the plant is the identified model of the TREX-600 helicopter. As stated above, the primary concern in the implementation of LADRC is maximizing the bandwidth $\omega_{0,i}$ and $\omega_{c,i}$, and identifying a suitable value of b_{ii} while satisfying the system constraints and design objective. It can be accomplished by forming a functional optimization problem. Also, the design specifications are comprehensively represented by a new objective function. In the optimization procedure, by changing the closed-loop step responses according to its automatically selected controller parameters and calculating the objective function value at every generation, the iterative algorithm searches the optimal parameters for the controller subjected to the design specifications.

In the tuning of the controller, the objective function can be formed by different performance index that considers the step responses of the entire system. Typical performance index in the time domain includes integral square error (ISE), integral of absolute error (IAE), integral time absolute error (ITAE) [33], rise time (T_r), settling time (T_s), overshoot (OS), and steady-state error (E_{ss}). The selection of these factors and form of the function can be determined depending on the design requirements. In this work, the desired control performance should have a small or no overshoot in the step response with a minimal settling time, and the control signal should be smooth within the physical limit. Hence, we defined the objective function F in this work as a linear combination of the ISE, integral of the square of the control signal, the overshoot OS, and the one percent settling time T_s [34]:

$$F = \sum_{i=1}^{3} \alpha \int (y_i - r)^2 + \beta \int u_i^2 + \sigma (OS_i) + \varepsilon (T_{s,i}) \quad (20)$$

where the variables of α, β, σ, and ε are the adjustment parameters. The values of these parameters are generally selected by using trial-and-error method. During the minimization of the objective function, all of the performance indexes are minimized and all of the disadvantaged controller parameters caused to system unstable or a poor performance will be eliminated by the algorithm.

Using the proposed objective function (20), the parameter tuning for the controller becomes a function optimization problem. This method combines a variety of performance indices which can be selected and weighted as required. Then the desired control performance and its parameter setting

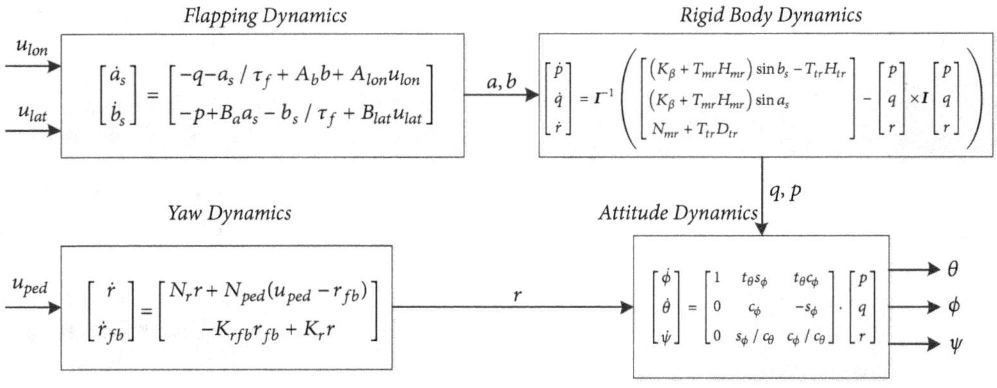

FIGURE 4: Interconnection of the helicopter dynamics model. The terms associated with the gravity force are disregarded.

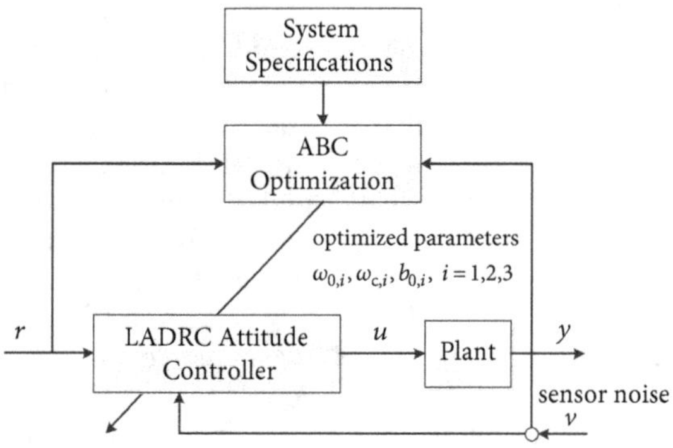

FIGURE 5: LADRC tuning scheme with ABC.

can be found by minimizing the value of (20). It makes this method different to the traditional optimal control method. Since (20) is nonlinear and discontinuous, simple search methods are usually lost in local optimum, as shown in [35]. Advanced search methods like GA, PSO, and ABC provide us with efficient solutions for solving this problem.

4.2. ABC Algorithm. In order to introduce the search mechanism of ABC algorithm, we should define three essential components: employed bees, unemployed bees, and food source [36]. And the unemployed bees are divided into the following bees and scout bees. The population of the colony bees is N_s, the number of employed bees is N_e, and the number of unemployed bees is N_u, which satisfies the relation $N_s = 2N_e = 2N_u$. We also define D as the dimension of solution vector, i.e., the number of the unknown parameters. ABC algorithm treats each solution vector as a food source and combines the global search of unemployed with the local search of employed bees. The detailed procedure of executing the proposed algorithm is described as follows.

Step 1. Randomly initialize a set of possible solutions $(x_1, \cdots, x_{N_s})$, and the particular solution x_i can be governed by

$$x_i^j = x_{\min}^j + \text{rand}(0,1)\left(x_{\max}^j - x_{\min}^j\right) \tag{21}$$

where $j \in \{1, \cdots, D\}$ denotes the j th dimension of the solution vector. $x_{\min}^j$ and $x_{\max}^j$ mean the lower and upper bounds, respectively.

Step 2. Apply a specific function to calculate the fitness of the solution x_i according to the following equations and select the top N_e best solutions as the number of the employed bees:

$$fit_i = \frac{1}{(1+F_i)} \tag{22}$$

where fit_i is the fitness function and F_i is objective function depicted in (20).

Step 3. Each employed bee searches new solution in the neighborhood of the current position vector in the n th iteration as follows:

$$v_i^j = x_i^j + \lambda_i^j\left(x_i^j - x_k^j\right) \tag{23}$$

where $k \in \{1, \cdots, D\}$, $k \neq i$, both k and j are randomly generated, and λ_i^j is a random parameter in the range from -1 to 1. In order to ensure that the algorithm evolves to the global optimal, we apply the greedy selection equation (22) to choose the better solution between v_i^j and x_i^j into the next generation:

$$x_i^j = \begin{cases} v_i^j, & fit\left(v_i^j\right) > fit\left(x_i^j\right) \\ x_i^j, & fit\left(v_i^j\right) \le fit\left(x_i^j\right) \end{cases} \tag{24}$$

Step 4. Each following bee selects an employed bee to trace according to the parameter of probability value. The formula of the probability method is described as

$$p_i = \frac{fit_i}{\sum_{i=1}^{Ne} fit_i} \tag{25}$$

Step 5. The following bee searches in the neighborhood of the selected employed bee's position to find new solutions. Update the current solution according to their fitness.

Step 6. If the search time *trial* is larger than the predetermined threshold *limit* and the optimal value cannot

TABLE 2: Tuning performance of trial-and-error method, GA, PSO, and ABC algorithm.

Applied process	Time domain performance				
	T_r/s	$OS/\%$	T_s/s	E_{ss}/rad	F
trial-and-error	0.70	2.33	1.17	0	1.2351
	0.55	1.30	0.77	0	
	0.37	1.86	0.65	0	
GA	0.596	2.13	0.97	0	1.013
	0.514	1.74	0.80	0	
	0.303	1.27	0.48	0	
PSO	0.62	2.01	0.94	0	0.9713
	0.53	1.82	0.81	0	
	0.30	1.11	0.40	0	
ABC	0.60	2.08	0.96	0	0.9315
	0.57	1.60	0.86	0	
	0.35	0.72	0.31	0	

be improved, the location vector can then be reinitialized randomly by scout bees according to the following equation:

$$x_i(n+1) = \begin{cases} x_{\min} + \mathrm{rand}(0,1)\left(x_{\max} - x_{\min}\right), & \textit{trial} > \textit{limit} \\ x_i(n), & \textit{trial} \le \textit{limit} \end{cases} \quad (26)$$

Step 7. Output the best solution parameters achieved at the present time, and go back to Step 3 until termination criteria $T_{\max}$ are met.

5. Simulation Tests

For best performance of optimization, we compare the results of ABC with the existing search iterative algorithm methods, including the trial-and-error method, GA, and PSO. We set the population size as 20 and iteration numbers as 50 for each algorithm. The adjustment parameters (α, β, σ, ε) are selected as 1, 1, 0.1, and 0.2, respectively. The step command with the value of 0.2618 rad is applied to each of the input channels. 0.1% measurement white noise is added to the plant. For GA, the crossover probability and mutation probability are chosen as 0.8 and 0.2, respectively. For PSO, the optimal parameters, i.e., social, individual and inertia weight, are set to 2, 2, and 0.8, respectively. Finally, for ABC the threshold is set to *limit* =5. The results are presented in Table 2 and Figure 8.

Figure 6 shows the evolution curves of ABC, GA, and PSO. The figure demonstrates that the objective function reduces as the generation iterates with time, gradually converging to an optimal result. Compared with GA and PSO, ABC achieves a better result with smaller objective function after 28 iterations. Table 2 indicates that all of the controllers have no steady-state error and that the trial-and-error method gets the largest time domain index. The ABC-optimized LADRC responds to the input and stabilizes the system faster than other three methods. In summary, the results suggest that our proposed method outperforms

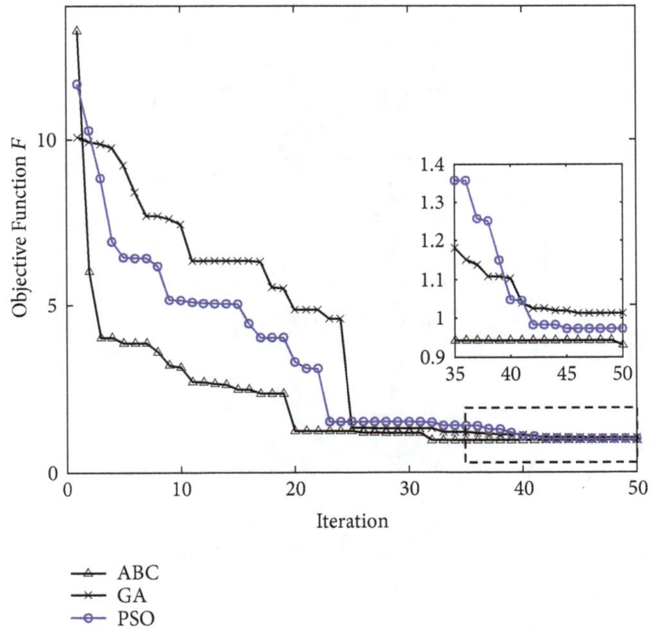

FIGURE 6: Evolutionary curves of the three algorithms.

other techniques in terms of rising time, settling time, and quadratic performance index.

To assess the improvements of the proposed controller, the closed-loop performance of helicopter attitude control with ABC-based LADRC and LQG is compared and analyzed by attitude tracking test under wind disturbance. The LQG controller is designed based on the linear model of Trex-600 helicopter; it is also implemented experimentally in [13]. As shown in Figure 7, the LQG controller consists of a state estimator based on Kalman filter and a MIMO state-feedback controller, which ensures that the output Θ tracks the reference command $\boldsymbol{r}$ and rejects process disturbances and measured output noise. The Kalman filter produces estimates $\hat{\boldsymbol{x}}$ of the plant. The observer gain $\boldsymbol{L}_f$ and optimal state feedback gain $\boldsymbol{K}_f$ are achieved by solving two independent Riccati equations [14]

TABLE 3: Parameters of LADRC and LQG.

Results	Controllers	
	LADRC	LQG
Optimal parameters	$w_{0,1} = 126.50$ $w_{c,1} = 11.81$ $b_{01} = 658.66$ $w_{0,2} = 116.67$ $w_{c,2} = 12.40$ $b_{02} = 638.15$ $w_{0,3} = 160.45$ $w_{c,3} = 19.32$ $b_{03} = -163.33$	$\boldsymbol{Q} = \mathrm{diag}(0, \cdots, 0, 79.11, 242.98, 72.78)$ $\boldsymbol{R} = \mathrm{diag}(1, 1, 1)$ $\boldsymbol{Q}_f = \mathrm{diag}([0.01, 0.01, 0.01]);$ $\boldsymbol{R}_f = \mathrm{diag}([0.01, 0.01, 0.01]);$
F	1.1355	1.7086

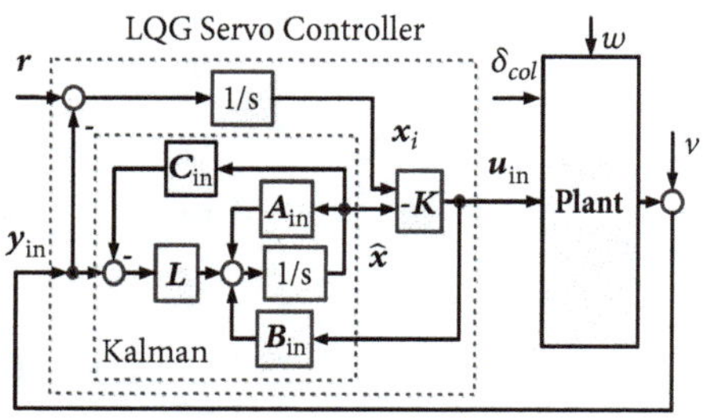

FIGURE 7: LQG controller.

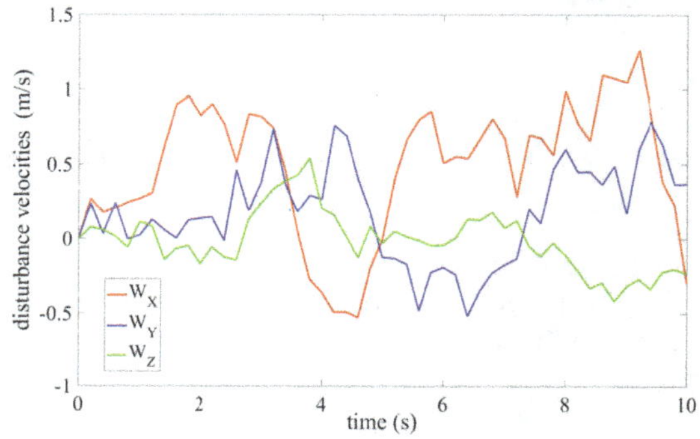

FIGURE 8: Wind disturbance vector.

$$\boldsymbol{K}_f = \boldsymbol{R}^{-1}\overline{\boldsymbol{B}}^T\boldsymbol{P}$$
$$\overline{\boldsymbol{A}}^T\boldsymbol{P} + \boldsymbol{P}\overline{\boldsymbol{A}} - \boldsymbol{P}\overline{\boldsymbol{B}}\boldsymbol{R}^{-1}\overline{\boldsymbol{B}}^T\boldsymbol{P} + \boldsymbol{Q} = 0 \tag{27}$$

$$\boldsymbol{L}_f = \boldsymbol{P}_f\boldsymbol{C}_{in}^T\boldsymbol{R}_f^{-1}$$
$$\boldsymbol{P}_f\boldsymbol{A}_{in}^T + \boldsymbol{A}_{in}\boldsymbol{P}_f - \boldsymbol{P}_f\boldsymbol{C}_{in}^T\boldsymbol{R}_f^{-1}\boldsymbol{C}_{in}\boldsymbol{P}_f + \boldsymbol{G}\boldsymbol{Q}_f\boldsymbol{G}^T = 0 \tag{28}$$

where $\boldsymbol{Q}$ and $\boldsymbol{R}$ are symmetric weight matrices; $\mathbf{Q}_f$ and $\mathbf{R}_f$ are covariance matrices of process disturbances w and measured output noise v, respectively. It is obvious that the choice of weighting matrices $(\boldsymbol{Q}, \boldsymbol{R})$ dominates the closed-loop performance.

Using the proposed method to optimize LQG, Table 3 summarizes the optimal parameters and objective values of these two controllers. It is observed that objective value of LQG is 1.5 times larger than that of LADRC. Hence, LADRC achieves better performance than LQG.

To simulate the measurement noise of the helicopter, the white noise is included in the output of the plant. The wind turbulence disturbances (W_X, W_Y, W_Z), as shown in Figure 8, are also injected to the velocity vector $\boldsymbol{V}$ along body frame X_B-, Y_B-, and Z_B-axes. Here, a shaping filter [37] modeled by independently exciting of the correlated Gauss-Markov processes is chosen for the wind components:

$$\begin{bmatrix} \dot{W}_X \\ \dot{W}_Y \\ \dot{W}_Z \end{bmatrix} = \begin{bmatrix} -\frac{1}{\tau_s} & 0 & 0 \\ 0 & -\frac{1}{\tau_s} & 0 \\ 0 & 0 & -\frac{1}{\tau_s} \end{bmatrix} \begin{bmatrix} W_X \\ W_Y \\ W_Z \end{bmatrix} + \rho\boldsymbol{B}_W \begin{bmatrix} d_X \\ d_Y \\ d_Z \end{bmatrix} \tag{29}$$

where τ_s is the correlation time of the wind; ρ is the scalar weighting factor; $\boldsymbol{B}_W$ is the turbulence input identity matrix; d_X, d_Y, and d_z are independent with zero mean.

Figures 9 and 10 show the roll and pitch responses and tracking error of the two control systems. We can observe that LADRC controller has more advanced performance in attitude tracking and disturbance resisting as compared to its counterpart. Figure 11 shows that LADRC controller responds faster and has smaller interfere between channels. For the LQG controller, however, the oscillation of u_{lon} increases apparently when there is a change in u_{lat}. Figures 12 and 13 show the angular velocities versus their estimates of both controllers. LESO has more effective estimation performance than Kalman filter, which means faster compensating for the

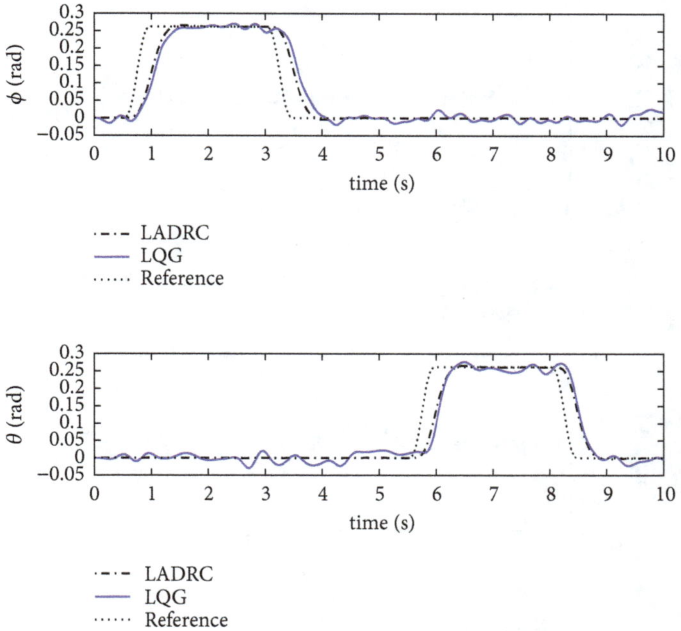

Figure 9: Attitude responses of both controllers with wind disturbance.

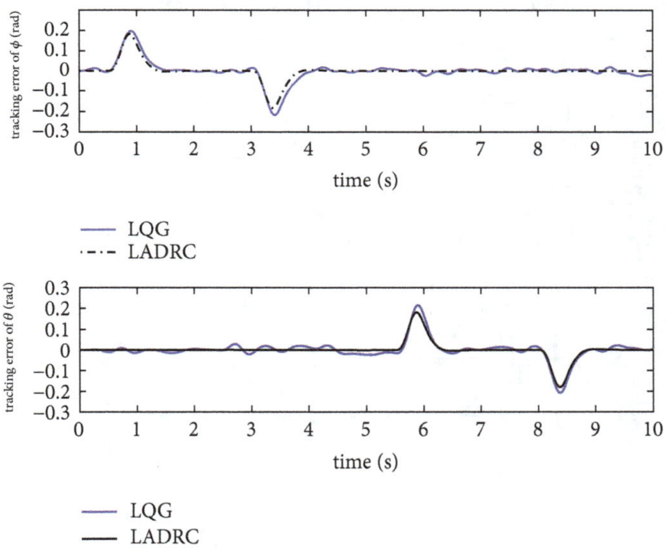

Figure 10: Tracking error of both controllers.

coupling effects and disturbances. All the simulation results indicate that the proposed ABC-optimized LADRC is the perfect control optimization strategy in terms of both the control performance and efficiency of design and parameter tuning. It obtains the lowest objective value and fastest convergence speed. But above all, LQG relies on the precise linear model of the plant, while LADRC only needs the input gains b_{ii} that can be even considered as the tuning parameter.

6. Conclusion

In this paper, the ABC algorithm is first applied to tune the controller parameters of LADRC-based DDC controller for a small-scale unmanned helicopter. With the proposed method, the decoupling control of small helicopters is reformulated as a disturbance rejection one, with only the orders of each input-output pairs of the system. The controller optimization is formulated as a function optimization problem and an objective function is proposed for multiple conflicting performance specifications. Four different optimization algorithms are investigated and evaluated in the search of global optimum. The proposed controller is also compared with the traditional LQG technique on the performance of state estimation and disturbance rejection. The simulation results verify the robustness and effectiveness of the ABC-optimized DDC strategic. As future works, the presented strategic will be utilized to design a path following controller for our helicopter and test its reliability in real flight experiments.

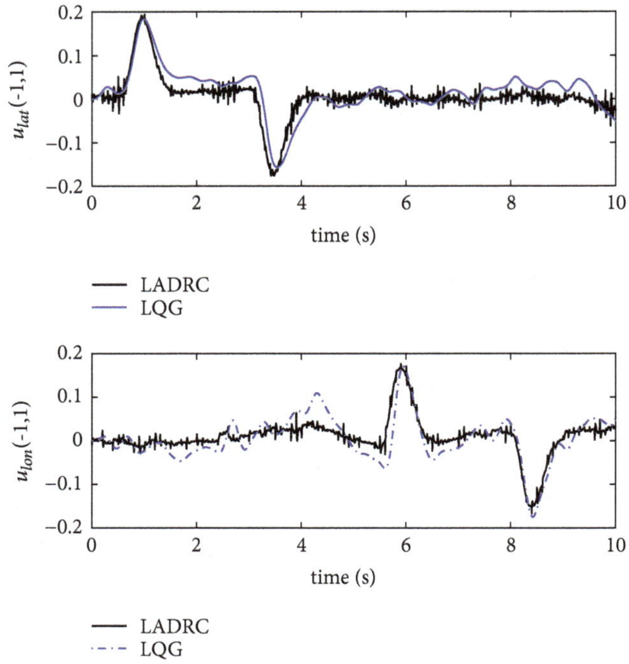

FIGURE 11: Control signals of both controllers.

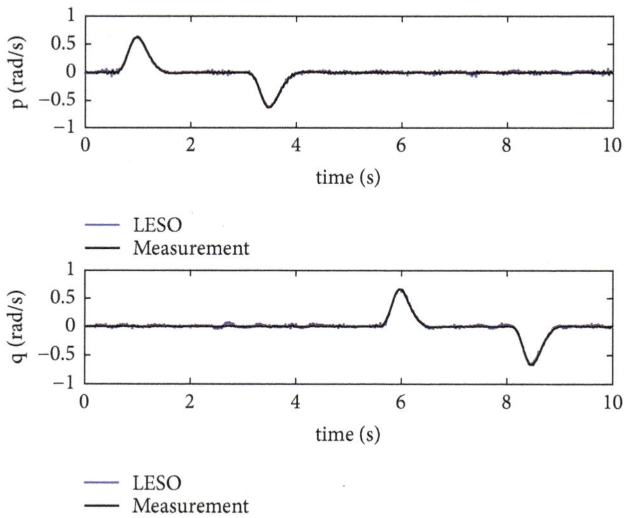

FIGURE 12: Performance of the LESO.

Acronyms

ABC: Artificial bee colony
ADRC: Active disturbance rejection control
BF: Body frame
CG: Center of gravity
DDC: Dynamic decoupling control
ESO: Extended state observer
GA: Genetic algorithm
IAE: Integral of absolute error
ISE: Integral square error
ITAE: Integral time absolute error
LADRC: Linear active disturbance rejection control
LESO: Linear extended state observer
LQG: Linear quadratic Gaussian
LQI: Linear quadratic integral
LQR: Linear quadratic regulator
MIMO: Multi-input multi-output
PD: Proportional derivative
PID: Proportional integral derivative
PSO: Particle swarm optimization
SISO: Single-input/single-output
UAVs: Unmanned aerial vehicles.

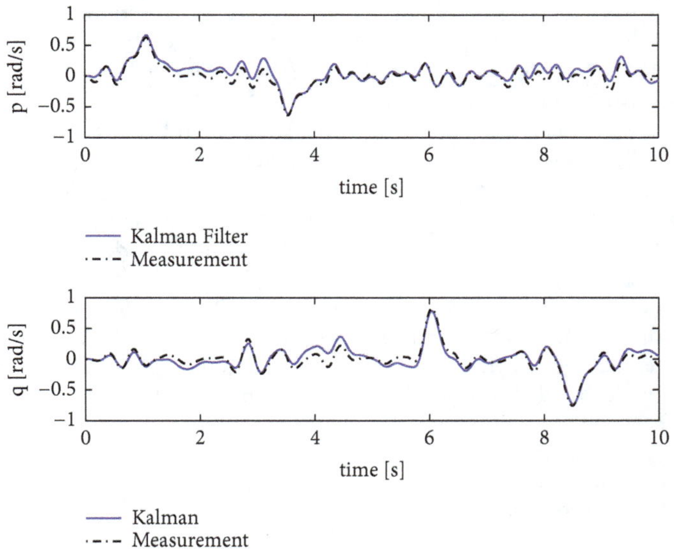

FIGURE 13: Performance of the Kalman filter.

Conflicts of Interest

The authors declare that they have no conflicts of interest.

Acknowledgments

This work was partially supported by the Foundation Research Project of Jiangsu Province (the Natural Science Fund no. K20170315) and Changzhou Sci&Tech Program of China (Grant no. CJ20179017).

References

[1] H. Xie and A. F. Lynch, "State transformation-based dynamic visual servoing for an unmanned aerial vehicle," *International Journal of Control*, vol. 89, no. 5, pp. 892–908, 2016.

[2] L. Huang, H. Qu, P. Ji, X. Liu, and Z. Fan, "A novel coordinated path planning method using k -degree smoothing for multi-UAVs," *Applied Soft Computing*, vol. 48, pp. 182–192, 2016.

[3] J. Zhou, L. R. Khot, T. Peters, M. D. Whiting, Q. Zhang, and D. Granatstein, "Efficacy of unmanned helicopter in rainwater removal from cherry canopies," *Computers and Electronics in Agriculture*, vol. 124, pp. 161–167, 2016.

[4] C. Goerzen, Z. Kong, and B. Mettler, "A Survey of Motion Planning Algorithms from the Perspective of Autonomous UAV Guidance," *Journal of Intelligent & Robotic Systems*, vol. 57, no. 1-4, pp. 65–100, 2010.

[5] H. Shim, T. Koo, F. Hoffmann, and S. Sastry, "A comprehensive study of control design for an autonomous helicopter," in *Proceedings of the 37th IEEE Conference on Decision and Control*, pp. 3653–3658, Tampa, FL, USA.

[6] B. Mettler, *Identification Modeling and Characteristics of Miniature Rotorcraft*, Springer Science & Business Media, New York, NY, USA, 2013.

[7] H. J. Kim and D. H. Shim, "A flight control system for aerial robots: algorithms and experiments," *Control Engineering Practice*, vol. 11, no. 12, pp. 1389–1400, 2003.

[8] M. Vijaya Kumar, S. Suresh, S. N. Omkar, R. Ganguli, and P. Sampath, "A direct adaptive neural command controller design for an unstable helicopter," *Engineering Applications of Artificial Intelligence*, vol. 22, no. 2, pp. 181–191, 2009.

[9] C.-T. Lee and C.-C. Tsai, "Adaptive backstepping integral control of a small-scale helicopter for airdrop missions," *Asian Journal of Control*, vol. 12, no. 4, pp. 531–541, 2010.

[10] H. Voos, "Nonlinear control of a quadrotor micro-uav using feedback-linearization," in *Proceedings of the IEEE International Conference on Mechatronics (ICM '09)*, pp. 1–6, IEEE, April 2009.

[11] S. Ahrens, D. Levine, G. Andrews, and J. How, "Vision-based guidance and control of a hovering vehicle in unknown, GPS-denied environments," in *Proceedings of the 2009 IEEE International Conference on Robotics and Automation (ICRA)*, pp. 2643–2648, Kobe, May 2009.

[12] B. Kadmiry, P. Bergsten, and D. Driankov, "Autonomous helicopter control using fuzzy gain scheduling," in *Proceedings of the 2001 ICRA. IEEE International Conference on Robotics and Automation*, pp. 2980–2985, Seoul, South Korea.

[13] R. Enns and J. Si, "Helicopter trimming and tracking control using direct neural dynamic programming," *IEEE Transactions on Neural Networks and Learning Systems*, vol. 14, no. 4, pp. 929–939, 2003.

[14] J. Shin, K. Nonami, and D. Fujiwara, "Model-based optimal attitude and positioning control of small-scale unmanned helicopter," *Robotica*, vol. 23, no. 1, pp. 51–63, 2005.

[15] A. Budiyono and S. S. Wibowo, "Optimal Tracking Controller Design for a Small Scale Helicopter," *Journal of Bionic Engineering*, vol. 4, no. 4, pp. 271–280, 2007.

[16] J. Gadewadikar, F. L. Lewis, K. Subbarao, K. Peng, and B. M. Chen, "H-infinity static output-feedback control for rotorcraft,"

Journal of Intelligent & Robotic Systems, vol. 54, no. 4, pp. 629–646, 2009.

[17] G. Cai, B. Wang, B. M. Chen, and T. H. Lee, "Design and implementation of a flight control system for an unmanned rotorcraft using RPT control approach," *Asian Journal of Control*, vol. 15, no. 1, pp. 95–119, 2013.

[18] Q. Zheng, Z. Chen, and Z. Gao, "A practical approach to disturbance decoupling control," *Control Engineering Practice*, vol. 17, no. 9, pp. 1016–1025, 2009.

[19] J. Han, "Nonlinear design methods for control systems," *IFAC Proceedings Volumes*, vol. 32, no. 2, pp. 1531–1536, 1999.

[20] Z. Gao, Y. Huang, and J. Han, "An alternative paradigm for control system design," in *Proceedings of the 40th IEEE Conference on Decision and Control (CDC '01)*, vol. 5, pp. 4578–4585, Orlando, Fla, USA, December 2001.

[21] . Yi Huang, . Kekang Xu, . Jingqing Han, and J. Lam, "Flight control design using extended state observer and non-smooth feedback," in *Proceedings of the 40th Conference on Decision and Control*, pp. 223–228, Orlando, FL, USA.

[22] Q. Zheng, Z. Chen, and Z. Gao, "A Dynamic Decoupling Control Approach and Its Applications to Chemical Processes," in *Proceedings of the 2007 American Control Conference*, pp. 5176–5181, New York, NY, USA, July 2007.

[23] Q. Zheng and Z. Gao, "Motion control design optimization: Problem and solutions," *Int. J. Intell. Control Syst*, vol. 10, pp. 269–276, 2005.

[24] D. Karaboga, "An idea based on honey bee swarm for numerical optimization," 2005.

[25] M. Abachizadeh, M. R. Yazdi, and A. Yousefi-Koma, "Optimal tuning of PID controllers using Artificial Bee Colony algorithm," in *Proceedings of the 2010 IEEE/ASME International Conference on Advanced Intelligent Mechatronics (AIM)*, pp. 379–384, Montreal, QC, Canada, July 2010.

[26] C. Sun and D. Haibin, "Artificial bee colony optimized controller for unmanned rotorcraft pendulum," *Aircraft Engineering and Aerospace Technology*, vol. 85, no. 2, pp. 104–114, 2013.

[27] O. Abedinia, B. Wyns, and A. Ghasemi, "Robust fuzzy PSS design using ABC," in *Proceedings of the 2011 10th International Conference on Environment and Electrical Engineering (EEEIC)*, pp. 1–4, Rome, Italy, May 2011.

[28] L. Ding, H. T. Wu, and Y. Yao, "Chaotic artificial bee colony algorithm for system identification of a small-scale unmanned helicopter," *International Journal of Aerospace Engineering*, vol. 2015, Article ID 801874, 11 pages, 2015.

[29] V. Gavrilets, B. Mettler, and E. Feron, "Nonlinear model for a small-size acrobatic helicopter," in *Proceedings of the AIAA Guidance, Navigation, and Control Conference and Exhibit*, pp. 6–9, August 2001.

[30] B. Mettler and T. Kanade, "System Identification Modeling of a Model-Scale Helicopter," *Robotics*, vol. 1, p. pp, 2000, http://citeseerx.ist.psu.edu/viewdoc/download.

[31] Z. Gao, "Scaling and bandwidth-parameterization based controller tuning," in *Proceedings of the American Control Conference*, pp. 4989–4996, Denver, Colo, USA, June 2003.

[32] Z. Gao, "Active disturbance rejection control: a paradigm shift in feedback control system design," in *Proceedings of the American Control Conference (ACC '06)*, p. 7, IEEE, Minneapolis, Minn, USA, June 2006.

[33] L. dos Santos Coelho, "Tuning of PID controller for an automatic regulator voltage system using chaotic optimization approach," *Chaos, Solitons & Fractals*, vol. 39, no. 4, pp. 1504–1514, 2009.

[34] A. Bagis, "Determination of the PID controller parameters by modified genetic algorithm for improved performance," *Journal of Information Science and Engineering*, vol. 23, no. 5, pp. 1469–1480, 2007.

[35] F. J. Goforth, "On motion control design and tuning techniques," in *Proceedings of the 2004 American Control Conference (AAC)*, pp. 716–721, Boston, MA, USA, July 2004.

[36] D. Karaboga and B. Basturk, "A powerful and efficient algorithm for numerical function optimization: artificial bee colony (ABC) algorithm," *Journal of Global Optimization*, vol. 39, no. 3, pp. 459–471, 2007.

[37] W. E. Hall and A. E. Bryson, "Inclusion of rotor dynamics in controller design for helicopters," *Journal of Aircraft*, vol. 10, no. 4, pp. 200–206, 1973.

19

The Design and Evaluation of an Ergonomic Contactless Gesture Control System for Industrial Robots

Gilbert Tang and Phil Webb

School of Aerospace, Transport and Manufacturing, Cranfield University, Building 83, Cranfield, Bedfordshire MK43 0AL, UK

Correspondence should be addressed to Gilbert Tang; g.tang@cranfield.ac.uk

Academic Editor: Keigo Watanabe

In industrial human-robot collaboration, variability commonly exists in the operation environment and the components, which induces uncertainty and error that require frequent manual intervention for rectification. Conventional teach pendants can be physically demanding to use and require user training prior to operation. Thus, a more effective control interface is required. In this paper, the design and evaluation of a contactless gesture control system using Leap Motion is described. The design process involves the use of RULA human factor analysis tool. Separately, an exploratory usability test was conducted to compare three usability aspects between the developed gesture control system and an off-the-shelf conventional touchscreen teach pendant. This paper focuses on the user-centred design methodology of the gesture control system. The novelties of this research are the use of human factor analysis tools in the human-centred development process, as well as the gesture control design that enable users to control industrial robot's motion by its joints and tool centre point position. The system has potential to use as an input device for industrial robot control in a human-robot collaboration scene. The developed gesture control system was targeting applications in system recovery and error correction in flexible manufacturing environment shared between humans and robots. The system allows operators to control an industrial robot without the requirement of significant training.

1. Introduction

Many high-value low-volume manufacturing tasks are labour intensive, and workers are often required to work in poor conditions and uncomfortable positions. Due to the physical nature of medium and large-scale manufacturing, workers spend considerable amount of time traveling around the factory floor and sizable work pieces to retrieve tools and to carry out work on various locations, which prolong manufacturing cycle time. On the other hand, the use of full automation on these applications is impeded by the complex nature of the processes and the environments. However, by combining the strengths of humans and robots, human-robot collaborative systems could be viable manufacturing solutions for these applications [1]. Collaborative robots could operate as assistants in shared workspaces to reduce manual handling and improve ergonomic of manual tasks by minimising the requirements for human to work in awkward positions and to lift heavy parts, which also improve productivity [2, 3].

Interaction is an integral part of human-robot collaboration that requires effective communication for seamless performance [4]. Industrial robots are often programmed and controlled using robot teach pendants, which require trained users who have knowledge of the user interface and its proprietary programming languages. These training requirements have negative implications on the overall implementation cost and the complexity of the controls could reduce the seamlessness of the interaction. On the other hand, the investment cost of effective human-machine interfaces could be recovered by savings generated by a more efficient working system which will turn into profit once the cost is broke even [5].

In human work teams, verbal and gesture communications are practically inseparable. Hand gestures are often used to reinforce the speaker's ideas, and people sometimes communicate only through hand gestures [6]. Thus, it is logical to consider using similar methods for communication in human-robot cooperative systems. Gesture communication is sometimes preferred in industrial settings due to the restrictions of working environments; a prime example is hand signals for crane operation where communication between operators could subject to high background noise and over a long distance. It shows that gesture communication could be valuable in human-robot cooperative manufacturing systems, especially those in unstructured open workspaces.

In large-scale manufacturing, assistant industrial robots could manipulate large components but the position of components may require frequent adjustment due to misalignments. In this case, gesture control systems enable workers to guide robots into correct positions using simple hand gestures to continue the tasks. Furthermore, gesture control could be used anywhere within a work cell. In a large scale manufacturing environment the robot controller can be separated from operator by a significant distance; the ability to command the robot without the need to relocate will have positive impact on process cycle time.

The feasibility of performing robot teleoperation using gestures interfaces has been studied before. These studies have used both contact and noncontact digital devices. Handheld accelerometer-based input device [7, 8] and photogrammetry based data gloves [9–11] are examples of handheld and wearable interfaces. These devices can track position and orientation of hands but they can hinder human-limb movements and cause discomfort for prolonged use [12]. Furthermore, the use of wearable devices is less preferable in industrial contexts due to the time required for wearing, removing and calibrating the devices, and wear and tear will increase maintenance cost. Contactless gesture user interfaces enable human-machine interaction using natural hand movements in a noninvasive manner to users' motions and the surroundings which makes it a preferable option for this research [13].

Existing research on robot teleoperation using contactless gesture control could be classified into two main types and these are imitation of human movements [12, 14–18] and gesture recognition [19–26]. These researches have focused on technological development with a lack of application and ergonomic considerations; these could affect users' health and experience for practical long-term use. Some research has shown that gesture interfaces designed without human factors considerations are prone to cause negative health implication such as muscle fatigue after prolonged usage [27–29]; apart from intuitiveness of the system, biomechanical constraints of humans are also important design elements. In terms of intuitiveness and control precision, Norman (2010) has pointed out that a pure gestural system makes it difficult to discover the accurate dynamics of execution, but the problem could be overcome by adding conventional interface elements such as menus, operations, tutorials, and other forms of feedback and guides. It was also suggested that because gesturing is a natural, automatic behaviour, the system must be adjusted to avoid false responses to movements that were not intended to be system inputs [30]. It is particularly important when industrial robots could potentially cause damage to people and surrounding objects by false triggers.

The novelty of the research presented in this paper is the human-centred approach used in the development of a hand gesture robot control system which enables the user to perform motion control and simple programming of an industrial robot. The user can control the robot via joint movement as well as linear movement in Cartesian space. The system was designed using a user-centred design approach with the use of human factor analysis tool in the design of gesture commands. The system has undergone functionality test and usability test with participants to compare physical workload associated with using the system and to gather suggestion for future improvements.

The rest of the paper is organised as follows. First, the characteristics of the system and its functionality are described in System architecture. In this chapter the system is described in detail as well as the design process of the gesture and design constraints. In the method section the system testing method and details are described. The findings of the experiment are explained in results and discussion. Finally the paper is concluded with future work suggestions.

2. System Architecture

The system consists of a small industrial robot with six degrees of freedom (DoF), control system with software written in C#, and a hand-tracking sensor. A Leap Motion sensor is used to capture the user's hands positions and orientations and the control software processes the input and sends signals to robot controller. A decoder programme is embedded in the robot controller which receives signals from the control software through TCP server and actuate robot to perform actions. The system architecture is illustrated in Figure 1.

2.1. Hands and Fingers Tracking. The Leap Motion Controller is a low cost consumer hand tracking device designed as a human-computer interface. The controller is a small device with surface area of 24 cm^2 and it is based on stereo vision which has three Infrared Light emitters and two Infrared cameras. The field of view of the controller is approximately 150° and the effective range is approximately 25 to 600 mm above the device. The controller is capable of tracking position of hands, fingers, and finger-like tools with submillimetre accuracy. The manufacturer claimed that the accuracy of fingertip detection is approximately 0.01 mm and the frame rate is up to 300 fps. However, Weichert et al., 2013, have performed a series of tests to measure Leap Motion's accuracy with an industrial robot and they have discovered different results as the manufacturer has stated. It is summarised that it is not possible to achieve the claimed accuracy of 0.01 mm under real condition, but the overall average accuracy of 0.7 mm from the experiments is still relatively good for gesture-based user interfaces [31]. On the other hand, Guna et al. (2014) have performed a series of test on the Leap Motion with the aid of a professional fast and high-accuracy motion tracking system, and by measuring a plastic arm with pointing finger they have measured accuracy with standard deviation of 0.5 mm and best case at less than 0.01 mm. However, they found a significant increase in the standard deviation when moving towards the edge of the controller's

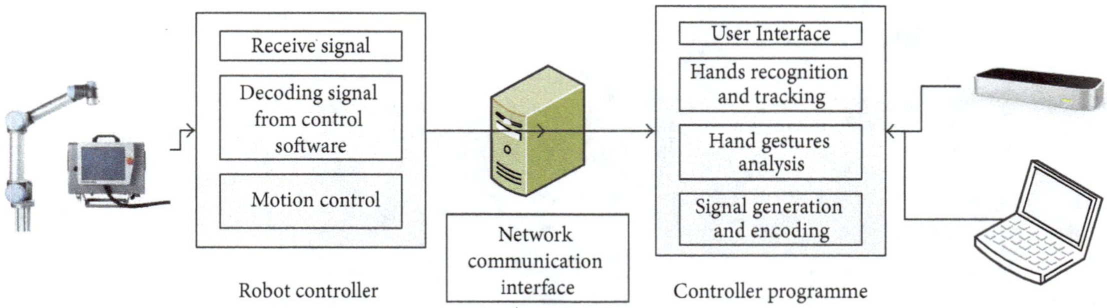

FIGURE 1: System architecture.

working area [32]. Furthermore, the set of measurements in the dynamic scenario have revealed some inconsistent performance of the controller which is limited by its inconsistent sampling frequency. The accuracy of tracking data can be improved by applying filtering algorithms in the programme. Du and Zhang (2015) have incorporated both particle filters and Kalman filters to improve the tracking accuracy of Leap Motion. However, the recorded tracking error is still significant compared to typical manufacturing tolerance in high-value production [33]. The developed system utilise a gesture control method based on hand positional thresholds which is not affected by the reported tracking error. Furthermore, the Leap Motion can measure hands positional data which cannot be performed with other markerless and noncontact off-the-shelf device so it is a suitable sensor to be used in this system.

2.2. Gesture Control for Robotic Applications. Hand and body gestures are broadly recognised as a natural way of communication, but when designing gestures as machine inputs or control it is important to consider the constraints of both the technology and the human. The main design objectives are to create a natural human-machine interface for users to perform robot motion control without significant training and to allow operators to work in comfortable postures. One of the many causes of work-related musculoskeletal disorders is the adoption of static or constrained postures. Using a control system may involve placing a load on the musculoskeletal system and discomfort, pain, and fatigue will be influenced by the amount, duration, and distribution of this load [34]. Thus, a gesture control system used in prolonged periods should be designed with ergonomic considerations that reduce risks to these injuries.

Gesture control should be responsive where the user can alter robot motion with immediate intended response. This provides instant feedbacks which reassures users that the robot is operating according to the their input. In this case, dynamic gestures are less preferred due to their completion time of typically over half a second that cause delayed system responses; therefore, it cannot provide a continuous input to control the robot motion. The actuation method used in the developed system is based on relative hand positions so the robot responses when user's hands reach the actuation positions.

The BS ISO/IEC 30113-1 and BS ISO 9241-960 provide guidelines for the design of gesture-based interfaces. The standard highlighted a number of requirements and recommendations and these include activating/finishing a gesture, performing a gesture, feedback for confirming a gesture, feedforward, cancelling a gesture, criteria of gesture size, controlling the criteria, changing correspondence of a gesture command, and descriptions of individual gestures within part. A number of these recommendations have been considered during the design of the developed system [35, 36].

2.3. Technological Constraints and Solutions. Machines perceive human gestures through sensors by using algorithms to recognise and to track a human body and its components, so it is important to know the limitation of the technology to design a gesture control system that is reliable and robust to use. The developed system uses the leap motion as an input device. The device API provides tracking information of the user's hands which include the pitch, roll, yaw, and the x, y, z positions relative to the sensor's centre point. The API can also recognise pinch or grab motions and return values between zero to one where one is a full pinch or grab, and zero is none. As explained previously, the Leap Motion device has tracking errors which can be compensated by incorporating appropriate filters as demonstrated in [12]. However, the errors have remained relatively high compared to typical high-value production manufacturing tolerance [33]. For this reason, real-time robot teleoperation involving mimicking human hand movement is not feasible for critical tasks that require precise positioning because precise hand positional data is one of the key enablers for this method. However, for gesture recognition the Leap Motion has sufficient accuracy when applied with appropriate filtering and a well-designed gesture set. The developed system uses a mean filter to remove noises from the sensor data stream prior to gesture recognition to stabilise the hand tracking. For recognition of the hand gestures, the user's hands must be arranged within a set of positional thresholds which are relative to a set of calibrated hands position values recorded prior to motion control. The thresholds could be adjusted to minimise the ambiguity between different gestures and to reduce sensitivity to tracking uncertainties which improve the system robustness.

2.4. Robot Control. The robot used is a Universal Robots UR5 as shown in Figure 2. It has a common layout with six joints: base, shoulder, elbow, wrist 1, wrist 2, and wrist 3. This is a common layout for industrial robots for its flexibility,

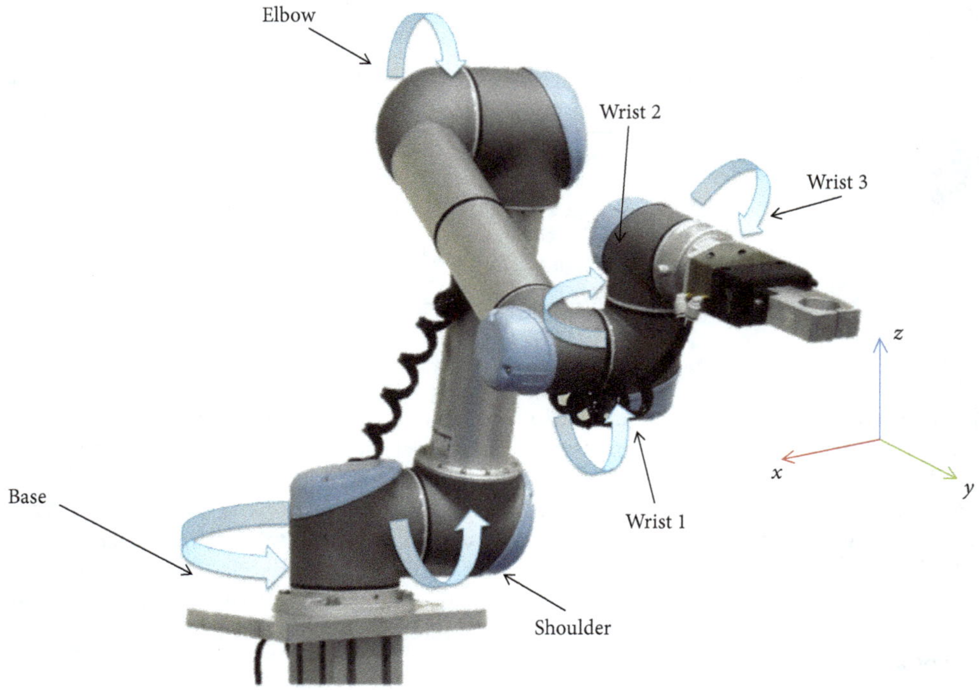

FIGURE 2: The Universal Robot UR5 has 6 dof which can be utilised with the gesture control system.

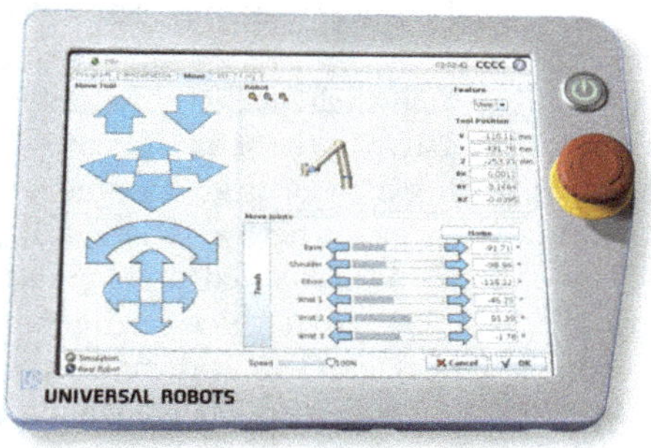

FIGURE 3: The Universal Robots teach pendant.

spherical working envelop, and adaptability to numerous applications. This robot is controlled similarly to most industrial robots using a touch screen teach pendant as shown in Figure 3.

The purpose of the developed system described in this paper is to enable users to control robot movements precisely without the use of a teach pendant. This novel gesture control system enables users to move the robot by joint and linear movements. These different control strategies are integrated into one control programme by segregating different control methods into modes where the users can switch between modes with a simple pinch motion of their left hand.

The developed software consists of two joint modes and a Cartesian mode. In joint mode 1, the user can take control of the robot's base, shoulder, and elbow. Joint mode 2 enables control of wrist 1, wrist 2, and wrist 3. The linear mode enables user to move the robot in x, y, and z directions of the base frame (Figure 2 and Table 1). The list of modes, functions, and triggers are shown in Table 1. The level of control is expandable to include rotational movements and actuation of auxiliary systems.

The hands recognition and tracking modules process data from the sensor, passing the tracking data on to the data enhancing module before forwarding to the gesture recognition module. If the identifier recognises a hand gesture, the system sends a signal to the robot controller. Each gesture has a threshold which could be adjusted to determine the sensitivity for recognition. For example, the pitch of the right hand has to reach a certain degree to trigger a signal as illustrated in Figure 4. The base value of these thresholds is defined using the results of the ergonomic analysis.

The control PC communicates with the robot controller via TCP/IP network. The control software sends encoded signals to the robot controller and a decoder within the robot programme interprets the signals and drives the robot according to the input. Each encoded signal is an array which contains variables representing the movement mode, joints speeds, velocity vectors, and rotation angles. The robot controller decoder interprets this array and sets the speed of the robot depending on the mode of movement and directional values. For movements in Cartesian space, the controller sets the speed in the format of $(x, y, z, \alpha, \beta, \gamma)$ where (x, y, z) represent the linear movement and (α, β, γ) represent the rotation movement about the x-, y-, z-axes. Joints movements are assigned by rotational speed in the format of (J1, J2, J3, J4, J5, J6). The robot movement is configured according to the acceleration, speed, and response time of the robot. The design of this control system is to ensure the user interface is comfortable and intuitive to use

TABLE 1: User input and robot control mode.

Mode	Function	Trigger
Off	No movement	None
Joint 1	Base movement	Right hand roll
Joint 1	Shoulder	Right hand pitch
Joint 1	Elbow	Right hand yaw
Joint 2	Wrist 1 movement	Right hand pitch
Joint 2	Wrist 2 movement	Right hand yaw
Joint 2	Wrist 3 movement	Right hand roll
Cartesian	Movement in x-axis	Right hand horizontal movement to left and right
Cartesian	Movement in y-axis	Right hand movement to forward and backward positions
Cartesian	Movement in z-axis	Right hand movement to upward and downward position

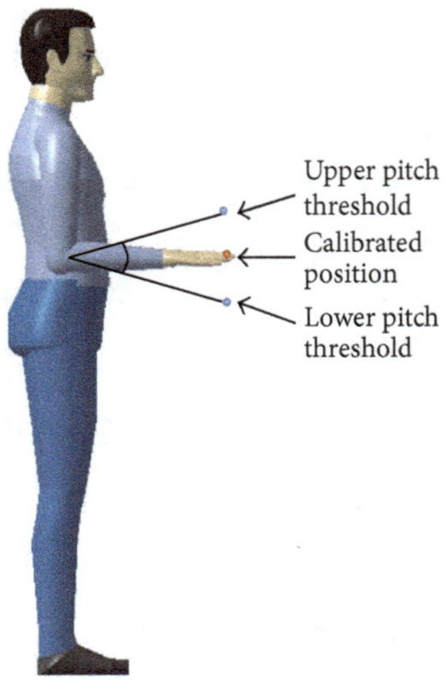

FIGURE 4: An illustration of calibrated hand position and trigger thresholds for pitch movements.

without the requirement of significant training and the user can produce input with simple hand movements.

Due to individual differences, people have varied neutral positions of the wrists, which affects radial deviation and flexion of the wrists. For adapting to different users, the neutral hands positions are stored in temporary memory during initiation and this calibration occurs at the beginning of every mode change cycle. Furthermore, some system characteristics were included purposely to enhance robustness and safety. For example, the robot movement is progressive and the speed is limited to less than 250 mm/s so the user should notice any abnormal movement of the robot and react to it. The system only operates when both hands are present above the sensor, if the user lifts one or both hands away from the sensor; then the robot will stop immediately. The system also features an "Off mode" for when the system is not being used and this feature avoid false trigger by movements near the sensor.

A simple graphical user interface has been designed to provide information about robot mode and the hands tracking state of the programme as shown in Figure 5. An image of a pair of hands appears when both hands are detected above the sensor. Directional arrows become visible when user's hands go above or below any positional thresholds; this reassures the user that the robot is engaged in the appropriate mode and provides the status of the hand tracking module. The interface is shown on a screen in front of the sensor and the user.

2.5. Human Constraints. The developed system receives input from the users in the form of hand gestures which detects the arrangements of arms, hands, and fingers. Thus, the posture analysis focuses on musculoskeletal movements involved in changing the hands' positions and orientations. The control hand movements are broken down into potential risks of musculoskeletal injury. For instance, changing the hands' pitch, roll, and yaw requires wrist flexion, radial deviation, and ulnar deviation, respectively. Changing the x, y, z position of the hand requires upper arm movement or lower arm movement and sometimes both. When designing a control system, these required body movements should be identified and the implication on health should be assessed.

The Rapid Upper Limb Assessment (RULA) [37] postural analysis is used to aid the design of gestures and assess the risk of musculoskeletal injury associate with system usage. It is an effective method for assessing the risk level of task requires movement of the upper limbs. The tool is chosen based on its simplicity and reliability compared to other assessment methods [38].

The RULA assessment method is used in the assessment of human constraint during the system design phase. The assessment has a score system with two parts which include part A, arm and wrist analysis, and part B, neck, trunk, and leg analysis. The scores accumulate at each step and the final score is calculated using a combination of three score tables.

The arm and wrist analysis indicates that the upper arm position should stay within ±20 degrees from the vertical axis or the score increases. A raised shoulder or abducted arm also increases the score which should be avoided. The lower arm position should stay within 60–100 degrees from the vertical axis; otherwise, the risk increases. Any deviation of the lower arms from the midline of the body will increase the risk, but it is inevitable if one of the hand positions is required to change in the x-axis of the sensor. Wrist flexion within ±15 degrees

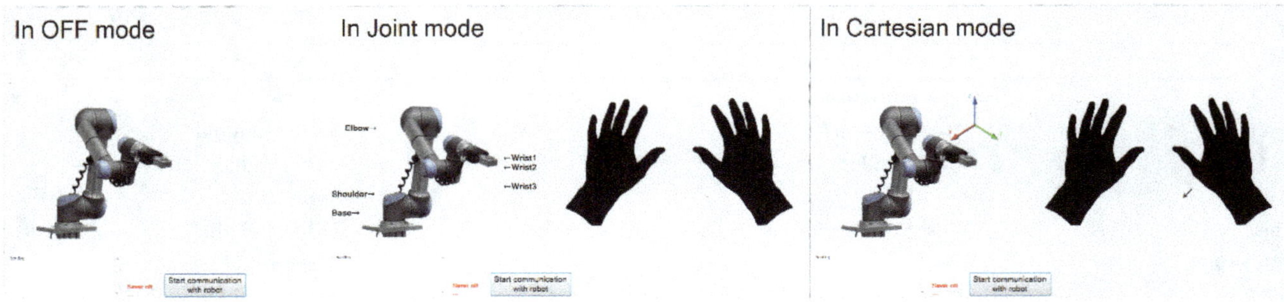

FIGURE 5: User interface in different modes.

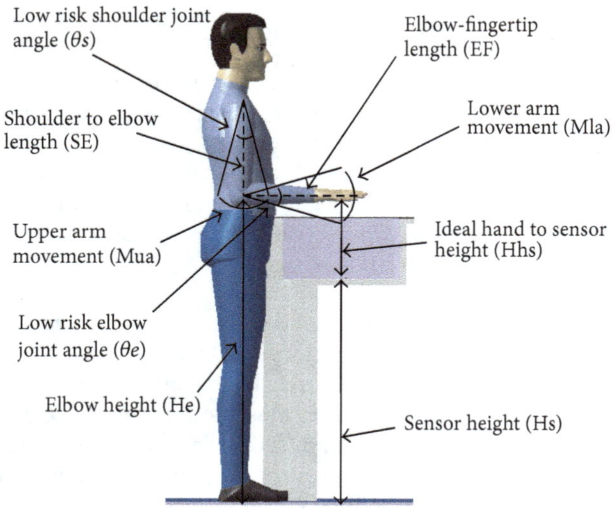

FIGURE 6: Model showing the upper arm constraint.

will add two points, but any movement over the range will add three points. Any ulnar deviation will add another one point to the score, but it is avoidable by using lower arm movement instead of changing hand yaw in sensor values. Wrist twist within midrange will only add one point to the wrist and arm score. Finally, highly repeatable posture or prolonged static posture will increase the risk as well as added force or load larger than 4.4 lbs on the hand. The developed system has received minimum risk score from the neck, trunk, and leg analysis as it is design to be used in comfortable standing posture as described in BS EN ISO 9241-5:1999 [39].

The RULA assessment is the core of the human-centred design process of this development. The human constraints highlighted from the assessment are considered at the initial gesture design stage as well as each design iterations. It was established at the beginning that in order to achieve intuitive gesture control the user's hands will be moving within a Cartesian space and some wrist flexion movement will be required. In this case, the threshold for low risk hands movements can be calculated using the joint angles provided in RULA.

The Leap Motion sensor was positioned flat on a solid surface and in front of the user as originally intended. The x, y, z positions of one of the user's hands will be used to actually control input which require shoulder and elbow joint movements. It is known from the assessment that neither joints should be required to move over ±20 degrees from a default position. Using this information combined with anthropomorphic data the ideal sensor height, working area, and threshold for actuation are calculated. Figure 6 is a model in the operating posture; the parameters used in the calculation for the system setup are labelled. Maximum joints movement angles for low risk classification are obtained from the RULA assessment. Limb lengths can be obtained from national anthropomorphic data according for use in different region.

The sensor height Hs is calculated using the elbow height He, the neutral lower position angle θen, the elbow to fingertip length EF, and the ideal hand to sensor height Hhs which can be expressed as

$$\tan \theta en \bullet \mathrm{EF} + \mathrm{Hhs} = \mathrm{Hs}. \tag{1}$$

The lower arm movement threshold Mla is calculated using the elbow to fingertip length EF and the maximum low risk elbow joint angle θe which can be expressed as

$$\frac{\theta e \pi}{180} \bullet \mathrm{EF} = \mathrm{Mla}. \tag{2}$$

The upper arm movement threshold Mua is calculated using the shoulder to elbow length SE and maximum low risk shoulder joint angle θs which can be expressed as

$$\frac{\theta s \pi}{180} \bullet \mathrm{SE} = \mathrm{Mua}. \tag{3}$$

3. Method

The evaluation of the contactless robot control system is described in this section. Traditionally the motion of a robot manipulator is controlled using a teach pendant. However, in a shared human-robot collaborative workspace a more flexible and intuitive way of communicating with robots is required. Using the omnipresent gesture control, the user can control a robot without having to hold any devices, which reduces physical workload and enables seamless cooperation. Evaluations were carried out during the design process and after the development to address potential ergonomic issues at design phase and to assess the user experience of the developed system by comparing it with a standard robot teach pendant.

3.1. RULA Study. A comparison study was carried out at the development stage to compare the risk of using the developed system against a touch screen teach pendant as the benchmark using RULA. RULA has a final score scale from one to seven where a final score of one to two means the posture is acceptable, three to four means change may be needed, five and six require further investigation and change required soon, and a score of seven leads to investigation and implements change.

3.2. Exploratory Usability Study. In addition to the RULA assessment, an exploratory ergonomic study was carried out.

3.2.1. Participants. Eight people from the general population of Cranfield University have participated in the usability experiment. Four of them were males and four of them were females, and all participants were right-handed. Their age ranged from 23 to 51 years with mean of 30 (SD = 9.27). Five of the participants have some experience in robot control and three of the participants are inexperienced robot operator.

3.2.2. Design. This experiment followed a within-subject design. The two independent variables are interface type and usability attribute, and the dependent variable is the usability score. Participants had to complete one time in each of the two conditions, which are the gesture control and teach pendant. The testing orders for the two control interfaces were counterbalanced to mitigate the risk of carryover effects. Each experiment took around 35 minutes.

3.2.3. Experiment Setup. The experiment was conducted in a 4 m × 4 m laboratory area surrounded by four sides of wall. The gesture control system was tested with a Universal Robot UR5 robot positioned on top of a stand at a height of 1 m. The robot has 850 mm reach with a circular working envelop and features a touch-screen teach pendant, which was used in the test. The Leap Motion device and a laptop were positioned next to the edge of the robot's working envelop on a table with adjustable height. The robot was equipped with a two-finger gripper that was deactivated. Three black plastic pipes were suspended within the working envelop of the robot on a worktop which were used as obstacles during the test.

3.2.4. Procedure. Informed consent was sought from participants prior to participating in the experiment. Participants were informed regarding their right to withdraw at any time, confidentially and anonymously. They were asked to provide their demographic information (age, sex, and job title) in a questionnaire. A scripted experiment briefing was given to each participant to explain about the purpose of the study and the expectations of him or her.

Participants were given a five-minute brief about the basic of industrial articulated robot with 6DoF which covers motion control techniques. Before the test of each control interface, participants were given a five-minute training and practicing period. The actual testing time of each system lasted 10 minutes, participants were given the option to take a five minutes' break after the first test before continuing to the second test. Each participant was asked to fill out a system usability questionnaire (SUS) [40] immediately after each test. Participants were asked to score the system against three ergonomic criterions which are physical workload, intuitiveness, and enjoyableness. These criteria are similar to those described by Bhuiyan and Picking in [41], but the questionnaire used in this test was simplified. Each criterion is assessed based on a seven-point Likert scale where seven is positive and zero is negative. The purpose of the questionnaire is to measure their experience of using the system.

The experiment focuses on evaluating ergonomic aspects of the systems and participants were not given specific task to perform. They were instructed to first try all possible movements with each system and then manipulating the robot gripper around the suspended plastic pipes without hitting them. They were told that each test will end in 10 minutes, but they can stop at any time if they are too tired to continue.

Table 2: RULA analysis scores for teach pendant and gesture control.

	Teach pendant	Gesture control
	Arm and wrist analysis	
Wrist posture score	3	3
Muscle use score	1	1
Force load score	1	0
Arm and wrist score	5	4
	Teach pendant	Gesture control
	Neck, trunk, and leg analysis	
Trunk posture score	3	1
Muscle use score	1	1
Force load score	1	0
Neck, trunk, and leg score	5	2
Final RULA score	6	3

4. Discussion

In the RULA test a higher score indicates an increased risk of musculoskeletal injury after prolonged usage. In this particular assessment gesture control scored three points and teach pendants scored four points. The score of the teach pendant can further increases if the device weighs over two kilograms. It was identified that the wrist movement in Leap Motion Robot Control poses the biggest problem in terms of posture, but the risk is relatively insignificant. The analysis highlighted potential issue with the neck when using a teach pendant, because its operating angle suggests a high load which could lead to musculoskeletal discomfort when used for prolonged periods. Furthermore, the lower arms have to be raised to a significant angle when using a teach pendant which may cause tiredness after a long period of usage. It is also found that prolonged usage of a teach pendant causes tiredness in the wrist of the hand holding the teach pendant. The RULA results are summarised in Table 2.

The ergonomic test results are summarised in Table 3 and presented in Figure 7. Two participants verbally reported it was tiring to hold the robot teach pendant during the

TABLE 3: Usability test results.

Participant ID	Less tiring to use		Intuitiveness		Enjoyable to use	
	Teach pendant	Gesture control	Teach pendant	Gesture control	Teach pendant	Gesture control
1	3	4	4	5	6	6
2	2	3	6	5	6	6
3	5	5	3	5	5	6
4	3	4	4	5	5	5
5	3	4	4	3	5	6
6	4	3	5	4	5	5
SD	1.03	0.75	1.03	0.84	0.52	0.52
Mean	3.33	3.83	4.33	4.50	5.33	5.67

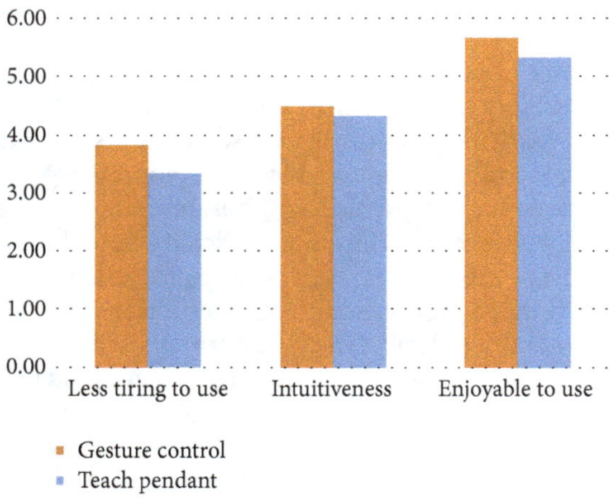

FIGURE 7: Mean ratings given to the two systems in three criteria.

test which lead them to leave the device on the worktop in majority of the test. However, in the questionnaire they have provided a positive score for the level of tiredness which were contradictory. Thus, their results were deemed as inaccurate and discarded.

The mean scores for physical workload were 3.83 (SD = 0.75) for gesture control and 3.33 (SD = 1.03) for teach pendant. For intuitiveness, gesture control received a mean score of 4.5 (SD = 0.84) and teach pendant received a mean score of 4.33 (SD = 1.03). The enjoyability mean scores were 5.67 (SD = 0.52) for gesture control and 5.33 (SD = 0.52) for teach pendant. A Wilcoxon Signed-Rank test showed that there are no significant difference between the usability scores in all categories: physical workload ($Z = -1.342$, $p = 0.18$), intuitiveness ($Z = -0.333$, $p = 0.739$), and enjoyability ($Z = -1.414$, $p = 0.157$). Nonetheless, the developed gesture control system is comparable with the off-the-shelf teach pendant in terms of usability with some addition advantages in human-robot collaboration scenarios.

Some participants have reported the visual feedback of the GUI and the tactile feedback of the touchscreen have made it easier to use, which indicates the potential benefit of incorporating visual feedback and tactile feedback into gesture control system. Half of the participants reported that the teach pendant is heavy to hold.

5. Conclusion and Future Work

The purpose of the developed system is not to completely replace a traditional robot teach pendant, but to operate as a complementary input device in an interactive environment. The system is designed to be a secondary input device to reduce travel distance of operator to location of robot teach pendant as well as to reduce risk of musculoskeletal injury from frequent usage. The developed gesture control system was targeting applications in system recovery and error correction in large-scale manufacturing environment. The system allows operators to control an industrial robot without the requirement of significant training. The experiment shows that the developed gesture control system has potential to be used as an input device for industrial robot control in a human-robot collaboration scene. However, addition features could be incorporated in the future to improve the ease of use and intuitiveness. The usability of gesture control for robot can be enhanced when complemented by a graphical interface with conventional elements which provide user with visual feedback. It is also suggested that tactile feedback can improve user experience. Visual feedback can be achieved by using augmented reality or virtual reality technology to provide real-time information and instructions. Contactless tactile feedback can be provided by integrating ultrasonic force fields devices around the control area. Such implementations are in the scope of future work.

Conflicts of Interest

The authors declare that there are no conflicts of interest regarding the publication of this paper.

Acknowledgments

The authors would like to thank Teegan Johnson who provided practical advice on the use of RULA assessment that greatly helped the research. This work was funded by the

EPSRC Centre for Innovative Manufacturing in Intelligent Automation under the Grant Reference no. EP/1033467/1.

References

[1] M. Walton, P. Webb, and M. Poad, "Applying a concept for robot-human cooperation to aerospace equipping processes," SAE Technical Papers 2011-01-2655, 2011.

[2] British Standards Institution, "Robots and robotic devices—Safety requirements for industrial robots Part 2: Robot systems and integration," BS ISO 10218-2, 2011.

[3] British Standards Institution, "Robots and robotic devices—Collaborative robots," ISO/TS 15066, 2016.

[4] S. A. Green, M. Billinghurst, X. Chen, and J. G. Chase, "Human-robot collaboration: a literature review and augmented reality approach in design," *International Journal of Advanced Robotic Systems*, vol. 5, no. 1, pp. 1–18, 2008.

[5] J. Löwgren, *Human-Computer Interaction: What Every System Developer Should Know*, Studentlitteratur, 1993.

[6] T. I. Cerlinca, S. G. Pentiuc, and V. Vlad, "Real-Time 3D Hand Gestures Recognition for Manipulation of Industrial Robots," *Elektronika ir Elektrotechnika*, vol. 19, no. 2, pp. 3–8, 2012.

[7] P. Neto, J. N. Pires, and A. P. Moreira, "High-level programming and control for industrial robotics: using a hand-held accelerometer-based input device for gesture and posture recognition," *Industrial Robot*, vol. 37, no. 2, pp. 137–147, 2010.

[8] T. Schlömer, B. Poppinga, N. Henze, and S. Boll, "Gesture recognition with a Wii controller," in *Proceedings of the 2nd International Conference on Tangible and Embedded Interaction (TEI '08)*, pp. 11–14, Bonn, Germany, February 2008.

[9] N. Karlsson, B. Karlsson, and P. Wide, "Glove equipped with finger flexion sensors as a command generator used in a fuzzy control system," in *Proceedings of the 1998 IEEE Instrumentation and Measurement Technology Conference (IMTC '98)*, vol. 1, pp. 441–445, IEEE, May 1998.

[10] K. Kuklinski, K. Fischer, I. Marhenke et al., "Teleoperation for learning by demonstration: data glove versus object manipulation for intuitive robot control," in *Proceedings of the 6th International Congress in Modern Telecommunications and Control Systems and Workshops (ICUMT '14)*, pp. 346–351, IEEE, October 2014.

[11] L. Paredes-Madrid and P. Gonzalez De Santos, "Dataglove-based interface for impedance control of manipulators in cooperative human-robot environments," *Measurement Science and Technology*, vol. 24, no. 2, Article ID 025005, 2013.

[12] G. Du and P. Zhang, "A markerless human-robot interface using particle filter and kalman filter for dual robots," *IEEE Transactions on Industrial Electronics*, vol. 62, no. 4, pp. 2257–2264, 2015.

[13] J. Kofman, X. Wu, T. J. Luu, and S. Verma, "Teleoperation of a robot manipulator using a vision-based human-robot interface," *IEEE Transactions on Industrial Electronics*, vol. 52, no. 5, pp. 1206–1219, 2005.

[14] G. L. Du, P. Zhang, L. Y. Yang, and Y. B. Su, "Robot teleoperation using a vision-based manipulation method," in *Proceedings of the International Conference on Language and Image Processing (ICALIP '10)*, pp. 945–949, IEEE, November 2010.

[15] G. Du and P. Zhang, "Markerless human-robot interface for dual robot manipulators using Kinect sensor," *Robotics and Computer-Integrated Manufacturing*, vol. 30, no. 2, pp. 150–159, 2014.

[16] D. Kruse, J. T. Wen, and R. J. Radke, "A sensor-based dual-arm tele-robotic system," *IEEE Transactions on Automation Science and Engineering*, vol. 12, no. 1, pp. 4–18, 2015.

[17] K.-C. Nguyen and V. Perdereau, "Arm-hand movement: Imitation of human natural gestures with tenodesis effect," in *Proceedings of the 2011 IEEE/RSJ International Conference on Intelligent Robots and Systems: Celebrating 50 Years of Robotics (IROS '11)*, pp. 1459–1464, IEEE, September 2011.

[18] C. Stanton, A. Bogdanovych, and E. Ratanasena, "Teleoperation of a humanoid robot using full-body motion capture, example movements, and machine learning," in *Proceedings of the 2012 Australasian Conference on Robotics and Automation (ACRA '12)*, pp. 3–5, Victoria University of Wellington, December 2012.

[19] T. Kornuta and C. Zieliński, "Behavior-based control system of a robot actively recognizing hand postures," in *Proceedings of the IEEE 15th International Conference on Advanced Robotics: New Boundaries for Robotics (ICAR '11)*, pp. 265–270, IEEE, June 2011.

[20] Y. V. Parkale, "Gesture based operating system control," in *Proceedings of the Advanced Computing & Communication Technologies (ACCT '2012)*, pp. 318–323, IEEE, January 2012.

[21] P. M. Yanik, J. Manganelli, J. Merino et al., "Use of kinect depth data and growing neural gas for gesture based robot control," in *Proceedings of the 6th International Conference on Pervasive Computing Technologies for Healthcare and Workshops (PervasiveHealth '12)*, pp. 283–290, IEEE, May 2012.

[22] T. Wan, Y. Wang, and J. Li, "Hand gesture recognition system using depth data," in *Proceedings of the 2nd International Conference on Consumer Electronics, Communications and Networks (CECNet '12)*, pp. 1063–1066, IEEE, Yichang, China, April 2012.

[23] A. A. Bhurane and S. N. Talbar, "Vision-based authenticated robotic control using face and hand gesture recognition," in *Proceedings of the 3rd International Conference on Electronics Computer Technology (ICECT '11)*, vol. 1, pp. 64–68, IEEE, April 2011.

[24] M. Hasanuzzaman, T. Zhang, V. Ampornaramveth, and H. Ueno, "Gesture-based human-robot interaction using a knowledge-based software platform," *Industrial Robot*, vol. 33, no. 1, pp. 37–49, 2006.

[25] M. Hasanuzzaman, T. Zhang, V. Ampornaramveth, H. Gotoda, Y. Shirai, and H. Ueno, "Adaptive visual gesture recognition for human-robot interaction using a knowledge-based software platform," *Robotics and Autonomous Systems*, vol. 55, no. 8, pp. 643–657, 2007.

[26] G. Tang, S. Asif, and P. Webb, "The integration of contactless static pose recognition and dynamic hand motion tracking control system for industrial human and robot collaboration," *Industrial Robot*, vol. 42, no. 5, pp. 416–428, 2015.

[27] B. Feng and X. Zhang, "Study on t-test, analysis of variance and multiple comparisons," *Journal of Taiyuan Normal University (Natural Science Edition)*, vol. 11, no. 4, pp. 46–49, 2012.

[28] M. H. van Beurden, W. Ijsselsteijn, and K. Hopf, "User centered design of gesture-based interaction technology," in *Proceedings of the 3DTV Conference: The True Vision-Capture, Transmission and Display of 3D Video (3DTV-CON '11)*, pp. 1–4, IEEE, May 2011.

[29] M. Nielsen, M. St, T. B. Moeslund, E. Granum, and M. Störring, "A procedure for developing intuitive and ergonomic gesture interfaces for HCI," in *Gesture-Based Communication in Human-Computer Interaction*, pp. 409–420, Springer Berlin Heidelberg, 2004.

[30] D. A. Norman, "Natural user interfaces are not natural," *Interactions*, vol. 17, no. 3, pp. 6–10, 2010.

[31] F. Weichert, D. Bachmann, B. Rudak, and D. Fisseler, "Analysis of the accuracy and robustness of the leap motion controller," *Sensors*, vol. 13, no. 5, pp. 6380–6393, 2013.

[32] J. Guna, G. Jakus, M. Pogačnik, S. Tomažič, and J. Sodnik, "An analysis of the precision and reliability of the leap motion sensor and its suitability for static and dynamic tracking," *Sensors*, vol. 14, no. 2, pp. 3702–3720, 2014.

[33] N. Jayaweera and P. Webb, "Automated assembly of fuselage skin panels," *Assembly Automation*, vol. 27, no. 4, pp. 343–355, 2007.

[34] A. Smith, "Upper limb disorders—time to relax?" *Physiotherapy*, vol. 82, no. 1, pp. 31–38, 1996.

[35] British Standards Institution, "Information technology—Gesture-based interfaces across devices and methods," BS ISO/IEC CD 30113-1, 2015.

[36] British Standards Institution, "Ergonomics of human-system interaction, Part 960: Framework and guidance for gesture interactions," Draft BS ISO 9241-960, 2015.

[37] L. McAtamney and E. N. Corlett, "RULA: a survey method for the investigation of work-related upper limb disorders," *Applied Ergonomics*, vol. 24, no. 2, pp. 91–99, 1993.

[38] D. Kee and W. Karwowski, "A comparison of three observational techniques for assessing postural loads in industry," *International Journal of Occupational Safety and Ergonomics*, vol. 13, no. 1, pp. 3–14, 2007.

[39] British Standards Institution, "Ergonomic requirements for office work with visual display terminals (VDTs). Workstation layout and postural requirements," BS EN ISO 9241, 1999.

[40] J. Brooke, "SUS-A quick and dirty usability scale," *Usability Evaluation in Industry*, vol. 189, no. 194, pp. 4–7, 1996.

[41] M. Bhuiyan and R. Picking, "A gesture controlled user interface for inclusive design and evaluative study of its usability," *Journal of Software Engineering and Applications*, vol. 4, no. 09, p. 513, 2011.

20

Advanced Methods for Robot-Environment Interaction towards an Industrial Robot Aware of Its Volume

Fabrizio Romanelli

Comau Robotics S.p.A., Via Rivalta 30, 10095 Turin, Italy

Correspondence should be addressed to Fabrizio Romanelli, fabrizio.romanelli@comau.com

Academic Editor: Gordon R. Pennock

A fundamental aspect of robot-environment interaction in industrial environments is given by the capability of the control system to model the structured and unstructured environment features. Industrial robots have to perform complex tasks at high speeds and have to satisfy hard cycle times while maintaining the operations extremely precise. The capability of the robot to perceive the presence of environmental objects is something still missing in the real industrial context. Although anthropomorphic robot producers have faced problems related to the interaction between robot and its environment, there is not an exhaustive study on the capabilities of the robot being aware of its volume and on the tools eventually mounted on its flange. In this paper, a solution to model the environment of the robot in order to make it capable of perceiving and avoiding collisions with the objects in its surroundings is shown. Furthermore, the model will be extended to take also into account the volume of the robot tool in order to extend the perception capabilities of the entire system. Testing results will be showed in order to validate the method, proving that the system is able to cope with complex real surroundings.

1. Introduction

Robots are meant to become part of everyday life, as assistants at home, as our appliances, and in the industrial environments; they can take part to the work chain as coworkers, completing hard and demanding jobs. In this context, creating autonomous robots that can learn to act in unpredictable environments has been a long-standing goal of robotics, artificial intelligence, and cognitive sciences. An important step towards the autonomy of robots is the need to provide them with a certain level of independence and dynamic behaviour in order to face quick changes in the environment surrounding them; to get robots operating outside rigidly structured environments, such as research centres or universities facilities and beyond the supervision of engineers or experts, it is necessary to face different technological challenges, amongst them, the development of strategies that allow robots to learn from their own experiences and interaction with the environment. The interaction is the generic relation between two (or more) subjects where each subject modifies repeatedly its behaviours in relation to the behaviours of other subjects. In order to extend this definition to robotics, it is important to understand the basis of human interaction with the environment in order to make the robot acting in a similar manner to what people do while moving and interacting across their space. The ways the robot interacts with the environment are several: it acquires data from the surroundings through its sensors to provide the necessary input signals to the controller and it performs actions in order to achieve desired tasks. The entire robot-environment interaction can be described completely using two models, under the assumption that the robot controller is reactive where the output of the controller does not depend on its internal states, but only on the current input signals provided to it; moreover, another as-sumption must be that the robot works and operates in a controlled environment with no other external factors which influence its surroundings (i.e., the change in robot perception is only dependent on the actions of the robot). Given these assumptions, the robot-environment interaction model is described both by the robot controller model, which computes the desired motor responses of the manipulator according to its perception, and the perception model, which emulates how the perception of the robot is affected by its own actions.

We will focus on the robot controller model in order to describe the robot-environment interaction, starting from the modelling of the static environment through the use of elementary geometrical regions as in [1]. In regard to the nature of the interaction between a robot and its environment, robotic applications can be categorizes in two classes. The first class is referred to noncontact tasks (unconstrained motion in a free space, without any environmental influence on the robot). In these tasks, the robot dynamics is the most important aspect as regards its performance: several industrial applications such as pick-and-place, spray painting, gluing, and arc or spot welding belong to this category. In contrast to these tasks, many complex advanced robotic applications (packaging, assembling, or machining) require the manipulator to be coupled with other objects which can move. The contact tasks can be furthermore divided into two subclasses: essential force tasks and compliant motion tasks, as described in [2]. The first subclass requires the end-effector to establish a physical contact with the objects in the environment and exert a process-specific force. In these tasks, a synergy between the control of the end-effector position and interaction forces is required; some examples of this kind of tasks are deburring, roughing, bending, polishing, and so forth. In these tasks, the force has to be controlled in relation to the particular process in order to prevent overloading or damaging the tool or the objects to be manufactured. In the second subclass, the tasks focus on the end-effector motion, which has to be realized close to the constrained surfaces, and it must be compliant (i.e., capable to reacting to the interaction forces). In this subclass, the problem of controlling the robot is joined to the problem of accurate positioning (as in part-mating process). These processes are often characterized by the presence and occurrence of contact with constrained surfaces, and then the control must cope with reaction forces. The measurement of interaction force here provides information for error detection and for the identification of the parameters to modify the prescribed robot motion. In the future of robotics, the interaction with the environment is fundamental and more and more tasks will include and require interaction. In this paper, we will focus on the noncontact task class; in this context, a lot of considerations can be done, taking into account the feasibility of these applications in a real industrial environment.

As a first step, we have developed a system which can manage a set of different geometrical shapes (spheres, cylinders, and parallelepipeds) which define regions where movement and access are forbidden or allowed; besides this feature, a warning zone is defined as a thickness from the region of interference, useful in order to control the general speed override of the robot Tool Centre Point (TCP for now on) in fact, in these areas the speed override is controlled according to a control law which changes in function of the robot TCP position, adjusting the speed proportionally to the distance between the TCP and the avoided zone. At this stage, the algorithm checks only the TCP position and the spatial checks are performed on it (and not on other parts of the robot). The following step will take into account the modelling of volumes interactions in order to extend the method and the control system to include collision detection paradigms between volumes. This will allow the definition of an effective model of the static features inside a real industrial cell, which must be checked against the volume of part of the robot. At first, an overview of the theory of collision between solids has to be analyzed in order to introduce in the interference regions that model the possibility to check minimum distances between volumes and to prevent collision between the same volumes. Subsequently, the critical points which have to be passed to the interference regions control have to be detected by the engine. The entire volumetric collision detection engine will be engineered in order to take into account the bulk of the tool mounted on the robot, without limiting the capabilities of the interference regions control system of testing the only tool centre point against the elementary geometrical regions. The theory of collisions between volumes is here studied, in order to detect if a volume is in contact with others; furthermore, in order to correctly implement the methods of volumetric interference regions, a further module of com-putation of minimum distance between volumes has to be taken into account. The study of such architecture is a step towards the realization of a more intelligent robot that is aware of its volume, starting from the tool mounted on its flange and giving the opportunity to extend this model to the whole robot arm. The final step is to create a general architecture where volume collision detection and closest point detection are integrated together with the interference regions control system. Test results will be shown to prove the effectiveness of the presented architecture. These tests refer to the robot while it is moving at its maximum speed and while performing technological tasks in a real industrial cell. The algorithm that has been tested here is the interference regions control system without the management of volumetric con-trol paradigm. We are confident that these results can be easily obtained with the presented volumetric method, since the following tests refer to the collision avoidance core which takes as input a set of tool centre point positions which are the same output of the closest point detection block previously described; therefore, we think that these results can be generalized for the presented solution.

A side aspect of the robot-environment interaction is the interaction of the robot with human operators; this is an important social aspect that has to be taken into account when developing new robotics theories; in the last century, the growth in automation inside the factories and industries was exponential and this has led to a rapid change in the conditions of human operators. In particular for muscular fatigue, technology has substituted tension and mental effort; for the more advanced automated plants, the transformation of physical energy into technical and mental skills is even emphasized [3]. Another product of the increase in automation is the sense of alienation which has to be faced by the human operators who have to work in tight connection with robots or other mechanical instruments, being actors of repetitive tasks without complex interactions with their "mechanical coworkers." The interaction between robot and environment has been studied in this paper in order to also take into account the role of the human operator in the programming and maneuvring of robots, making his role closer to the productive process and less alienating; this is possible, considering the overall increase in the interaction between robot and environment which is the basic idea of this paper.

2. Model of Static Environment

In this section, a model of the static environment will be described in order to define a correct and effective model of static features inside a real industrial robotized cell. The first step consists of the definition of a general model that can be analyzed in order to cover all the possible scenarios inside the industrial environment; keeping this in mind, it is necessary to synthesize a control system capable of taking into account the elementary geometrical regions in order to provide the robot with the fundamental tools to face a basic level of interaction. The study of this subject is then a good start to realize a more intelligent integrated robotized cell, where the strong interaction between robots and humans becomes closer and closer.

2.1. State of the Art. In this paragraph, the modelling of static environment will be shown; this topic is still at a basis level, and there are several starting points of study in particular in the industrial automation background. PILZ developed a new control system for areas crowded with robots, moving machines, and human operators [4]; with this option, it is possible to broadcast a safe output in order to immediately halt the machines operating in those areas, avoiding potentially harmful situations. With ABB world zone software option [5], it is possible to define zones where the robot tool centre point cannot enter. If the robot end effector ends in an off-limits area, previously defined by the programmer, the control cuts the power and quickly stops the robot. This system lets the user program world zone software which is especially useful when two robots work close in order to prevent collision and establish working protocols and policy. However, this solution is quite limited as the model of the static environment is made taking into account the robot tool centre point rather than the complex structure of the robot arm; anyway, this solution represents a valid starting point to go through the development of a wider and more complex control paradigm to take into account volume versus volume collision, as will be shown later on. In the literature, there are several different approaches to the study of space occupancy and cooperation and to cope with collision avoidance problem. In particular, path planning is strongly associated with the problem of forbidden zones [6–8]. The management of operative space is a matter of study and development in the field of telemanipulation and robot assisted tasks [9], where security of avoiding forbidden zones is the main objective of the work. Amongst the previous studies in collision detection, there are some works to be mentioned as the one [10] where given two general polyhedra of complexity n, where one is moving while translating or rotating about a fixed axis, determine the first collision, if any, between the two objects. Another important aspect of collision detection and control of forbidden areas is presented in several works [11, 12], where the dynamics of complex bodies is simulated over a system equipped with a collision detection algorithm. In the industrial field, other approaches aimed to reach a greater level of automation in robot-environment interaction. Fuzzy logic allows controlling a system in order to avoid access to dangerous areas [13]. There is a further approach that will be presented later on this paper, where a system uses Bayesian occupancy grid and a fuzzy logic controller in order to avoid the collision between robot and other objects or humans moving around the cell [14]. In the following paragraphs, a new method to synthesize a control system capable of managing a set of predefined geometrical areas will be shown; with this paradigm, the advantages of taking into account a space model will be shown as well. This method will be the basis to extend the control from point versus volumes, to volume versus volumes collision detection as described later on in the paper.

2.2. Modelling Static Environment with Elementary Geometrical Volumes. The modelling of robot surroundings is a crucial problem in order to cope with management of robot working space, especially when this space is shared with other robots or machinery. The basis of the robot-environment interaction is the capability of the control system to define primitive geometrical areas in order to cover all the possible configuration of the objects inside a real industrial cell. The elementary objects are defined as parallelepipeds, cylinders, spheres, and planes. With this simple modelling, the system can be provided with the capability of defining multiple geometrical areas of these types in order to cover almost every object inside the cell (such as working tables, machinery, or moving objects such as rails and conveyors). Starting from this definition of elementary geometrical area, in the following paragraph, a control system will be shown. This control system allows the user to move the robot around the operating area with the certainty that it will not collide with the forbidden regions; these regions must be previously defined, declared, and activated in order to work correctly. On the other hand, the system could manage another type of regions (called monitored) linked to a digital output which can be raised when the robot TCP enters that region.

2.2.1. Control Paradigm. The developed system allows the programmer to define multiple elementary zones which can be integrated inside the standard robot control and which represent the database of spatial forbidden areas that are used in order to control both the position and speed of the robot end effector. The areas can be easily defined by the programmer, considering several simple geometrical reasoning in order to define a parallelepiped, it is sufficient to take two points (the lower left corner at the base of the shape and the upper right corner); with these two points declared, the first shape can be integrated into the forbidden zone database. The cylinder on the other hand is univocally defined considering the centre of the base circumference, the radius, and its height with this convention, the base of the cylinder is parallel to the XY plane of the world frame reference. The sphere is instead defined considering its centre and radius. Planes can also be defined into the system, declaring three points as the origin, X- and Y-axes and considering the Z-axis as perpendicular to the plane. Thanks to the possibility of defining several zones in the same operating area, obviously a lot of industrial applications can be covered by using this control paradigm. An important feature of this system is the possibility to consider the areas with two different features, concerning their *life*; they can be either constant or temporary. The first typology is programmable and modifiable from a particular class of users, and they cannot be ignored or modified by user programs these zones are active during all the cycle of the robot. These can be used, for instance, in order to define zones that cannot be covered by the robot end effector as they are occupied by fixed structures, such as pillars or other irremovable facilities. The second typology is temporary as it can be activated or deactivated from each user program, and it is a very useful function in order to manage interlocks for exchange zones between robots when a robot is inside an elementary zone, it is compulsory that the other robot is avoided to access the zone. This feature can be extended to those systems including a network of robot controllers where the information about the elementary areas in the robots cell is shared. In this case, the control system of each robot can be supervised by another controller in order to update the information of the position of the robots as regards the position of multiple elementary areas defined on the cell. Another important feature of the presented system is that it allows the user to define an area bigger than the elementary region, called warning zone in this zone, the robot can keep working but with a safe control system that checks the distance between the surface of the elementary area and the robot end effector, forcing the robot speed override to a value proportional to that distance. With this method, the robot end effector speed is reduced when the control system realizes that the robot violates the warning zone; activating this, it is also possible to avoid mechanical solicitations due to hard brakes and to allow human operator to better perceive the enabled elementary zone around the robot. This control law is applied to each geometrical area and the resulting speed overrides (one for each declared elementary area) are computed in order to find the minimal value of them and to apply it to the robot. Another important innovation of the presented method is that concerning the implementation of dynamic management of elementary areas; in particular, it is possible to program areas which can change their positions in time. This allows the elementary zones to be linked to moving objects (such as moving machines or to end effector of other robots inside the cell); in order to use this position, it is necessary that the current position of the objects has to be known since the robot controller must know this information, or it must similarly share it with other cooperative robots inside the cell. In order to fully describe this feature, a space where several robots operate can be considered in this configuration, it would be useful to define a dynamic zone on each robot end effector (linked to it). In this condition, each robot knows exactly and instantly the position of other robots end effector: if a robot draws too much to another one, the presented control is able to prevent damage between them. This allows the robot programmer to be released from the need of defining interlocks with some useless waits, while with this management it is possible to define digital outputs when a robot accesses a specified zone (not when specified in a user program); this control is parallel and acts in real-time, despite the classic management of interlocks. Thus, it is very important during the productive cycle when robot programmer has to develop applications where more robots and machines share the same working space, and the presented method helps the user to exactly bound the working areas. The definition of a shared information on the state of the elementary zones (if they are occupied by a robot or not) is very useful and innovative for what concerns monitoring a dynamical geometrical area (e.g., with conveyors). With this system, it is possible to link a dynamic zone to a moving object and this allows the definition of dynamical interlocks, which can be shared through a network between robots, giving a global visibility in the whole cell. The kinematics information of the first three robot joints and the tool allows the definition of useful information avoiding collision between robots cooperating in the same cell; this system, on the other hand, is not safe for the interaction between robots and human operators, but it is thorough in order to protect and prevent damage between robots and facilities without the need of further devices.

2.2.2. Control System Integrated in the Robot. In this section, an accurate description of the architecture of the control system will be shown. With the proposed solution, each controlled elementary zone can be programmed and defined using both the programming language (PDL2 for Comau robots), describing the complete geometry, shape, thickness of the warning zone, and its static or dynamic typology. Using this method, it is possible to program elementary areas to be controlled, in a precise way, and this solution can be very suited for off-line programming. A second method to perform the definition of an elementary zone is to use the robot in order to teach the distinctive points of a geometrical area (as shown in the previous section). With this method, the user will be asked to move the robot around the working area while locating the characteristic points of the geometrical shapes to be defined teaching these points will bring the advantage of having a direct comparison with the taught elementary-controlled zone and the real obstacle inside the working area. These methods provide the user with simple tools in order to create the database of the controlled elementary zones which allow the control system to perform complex operations of the spatial checks on the robot working area. With this solution, it is also possible to program and define for each declared zone, *channels* of shared information which can be activated automatically whenever the robot end effector enters a controlled zone or a warning zone; this also allows to have a quantization of the working area. The operator who uses the proposed solution has the possibility of tuning a set of parameters which makes the system extremely flexible and modular. It is also possible for example to define a dynamic controlled zone, linked to the end effector of another robot, in order to check the possible collisions between the robots. The control scheme is depicted in Figure 1.

As shown in the scheme the geometrical control algorithm checks if the robot end effector is inside a controlled zone or, analogously, a warning zone. This check is done on the basis of the database of geometrical areas, previously defined by the user; in this context, the dynamic objects position provides the control system with the possibility to link the geometrical areas to arbitrary moving points (as conveyors or rails), which can be read from external sensors like encoders. The speed control is performed by the geometrical area control block which detects the typology of the shape and selects the correct control law to be applied in order to modify the robot override, preventing collisions with the user-defined zones. The speed override is changed smoothly when the robot end effector comes up against a spherical elementary zone, according to the following control law:

$$
\begin{aligned}
v &= v_o \cdot \frac{d-r}{\delta}, \quad r \le d \le r+\delta, \\
v &= v_o, \quad d > r+\delta, \\
v &= 0, \quad d < r,
\end{aligned}
\tag{1}
$$

where v is the actual speed override of the robot end effector, v_o is the old override, d is the distance between the robot end-effector and the centre of the elementary spherical area, δ is the thickness of the warning zone and r is the radius of the sphere (the area is depicted in Figure 2(a)). When the robot encounters a cylindrical elementary zone its speed override is subject to the following control law:

$$
\begin{aligned}
v &= v_o \cdot \frac{d_1-r}{R-r}, \quad 0 \le z \le h, \\
v &= v_o \cdot \frac{d_2}{\delta}, \quad h < z \le h+\delta, \quad -\delta \le z < 0, \quad p \in \text{cyl}, \\
v &= v_o \cdot \frac{d_3}{\delta}, \quad h < z \le h+\delta, \quad -\delta \le z < 0, \quad p \notin \text{cyl},
\end{aligned}
\tag{2}
$$

where h is the height of the cylinder p the position of the robot end effector. The distances d_i represent the distance between the centre of the cylinder and the robot position (d_1), the distance between the cylinder top or bottom base and the robot position, when it belongs to the top/bottom cylinder with thickness δ (d_2), and the minimal distance between the robot position and the points on the circumference of the top/bottom cylinder base (d_3). The robot speed override coincides to the old speed override when the robot end effector is outside the warning zone; the cylindrical elementary area is depicted in Figure 2(b) with its warning zone.

Another geometrical shape is the plane depicted in Figure 2(c). The control law for the plane is the following:

$$
v = v_o \cdot \frac{d}{\delta}, \tag{3}
$$

where d is the distance between the plane surface and the robot position when it is inside the warning region with thickness δ. The last modelled elementary geometrical volume is represented by the parallelepiped in Figure 2(d); its control law is quite complex as its warning zone is composed by 8 half-lunes, 12 quarters of cylinder, and 6 planes. Given that the mathematical treatment of the control law for the parallelepiped warning area is not reported here, it is enough to consider that this control law smoothly covers the whole warning zone area, with an appropriate speed override for each sector of it. Each elementary zone declared inside the working space has its own control law, also depending on the thickness of the warning zone; it is fundamental when the control system has to fix the controlled speed override that the correct value will be chosen in an efficient way. It is chosen according to the following:

$$
\min(v_1, v_2, \ldots, v_M), \tag{4}
$$

where M is the number of elementary geometrical volumes defined inside the robot working space. This solution, notwithstanding its easiness, assures that the selected controlled speed override v_i follows a smooth trend when several different elementary zones are defined in the robot environment, even if they overlap. The last important feature of the presented paradigm is the possibility to manage dynamic geometrical areas, linking the position of moving objects to distinctive points belonging to the previously defined elementary shapes for the spherical area, this point is characterized by the centre of the sphere. The cylinder will have its characteristic point on the centre of its bottom base; the plane will have its characteristic point on the origin; eventually, the bottom base centre of the parallelepiped will represent its characteristic point.

3. Volume Environment Modelling

In this section, a model of the collision between volumes will be described; this will allow the definition of an effective model of the static features inside a real industrial cell, which must be checked against a part of the robot volume. First of all, an overview of the theory of collision between solids has to be analyzed in order to introduce in the presented interference regions model the possibility to check minimum distances between volumes and to prevent collision between the same volumes. Subsequently, the engine has to detect the critical points that have to be passed to the interference regions control in order to manage in the usual way the override speed of the robot. The entire volumetric collision detection engine will be engineered in order to take into account the bulk of the tool mounted on the robot, with no limit on the capabilities of the interference regions control system to test the tool centre point against the elementary geometrical regions. The theory of collisions

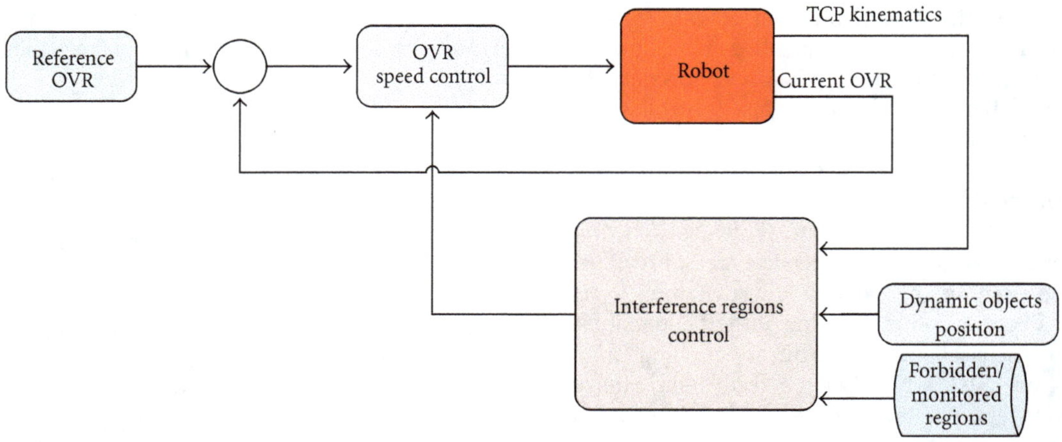

FIGURE 1: Interference Regions Control Scheme.

between volumes is here studied, in order to detect if a volume is in contact with others; furthermore, in order to correctly implement the methods of volumetric interference regions, a further module to compute the minimum distance between volumes has to be taken into account this study is strictly connected to the one related to collision detection and it will not be described here. For a deeper treatment of this topic, the reader can refer to [15]. The study of such architecture is a step towards the realization of a more intelligent robot and aware of its volume, starting from the tool mounted on its flange and giving the opportunity to extend this model to the whole robot arm.

3.1. Objects Intersection. In order to begin the analysis and study of collision detection for volumes, it is necessary to introduce some theory from computer graphics; at first we approach the method of detecting intersection between two convex shapes in two-dimensional space [16], after we will extend these results to 3D [17]. For generic geometrical objects lying in 2D space the separating axis theorem, a special case of the separating hyperplane theorem [18] states that given two convex shapes, there exists a line onto which their projections will separate if and only if they are not intersecting (i.e., colliding); a line for which the objects have disjoint projections is called a separating axis.

Another way of stating the theorem is to say that two convex shapes in the plane are not intersecting if and only if a line can be placed with one shape to one side of the line and the other shape to the other side such a separating line will be perpendicular to a separating axis (see Figure 3). In the specific case of 3D objects, the separating axis theorem can be easily extended in order to detect intersections between three-dimensional convex shapes. The separating plane theorem thus states that for any arbitrary convex disjoint polyhedra there exists at least one separating axis, where the projections of the polyhedra, which form intervals on the axis, are also disjoint.

The separating axis in 3D can be obtained considering that where a plane can be inserted between two three-dimensional objects, that plane's normal defines a separating axis (see Figure 4). This theory is the basis to describe more complex interactions between volumes.

3.2. Volume versus Volume Collision Detection. Starting from separating axis/plane theorem, we want to model the robot geometry with convex disjoint polyhedra, in order to adapt the presented theory to robotics. Initially, we want to prevent collisions between the robot tool and the elementary geometrical regions, as defined in the previous chapter. The basic idea is to include the tool mounted on the robot flange inside a bounding box, which can be both oriented and moving rigid with the robot flange. The bounding box should be analyzed to fully describe its minimum set of variables which can be used in the computation of interactions and collisions.

The OBB (oriented bounding box) $\mathbf{B}$ is depicted in Figure 5 and it can be described by the following set of variables:

$$\mathbf{B} = (\mathbf{b}^c, \mathbf{b}^u, \mathbf{b}^v, \mathbf{b}^w, h^u, h^v, h^w). \tag{5}$$

It is true that for each robot tool can be found an OBB which can contain it: this is an initial step towards the definition of a higher level of collision detection between solids. The tool centre point check of the interference regions control system is then substituted by an OBB check against interference regions. In order to proceed in this direction, it is necessary to analyze singularly the possible intersections between solids, in particular we have to describe intersections between

(i) *OBB* versus *plane*;

(ii) *OBB* versus *sphere*;

(iii) *OBB* versus *cylinder*;

(iv) *OBB* versus *OBB*.

3.2.1. OBB versus Plane Intersection. With reference to Figure 6, the analytical method to determine whether an OBB intersects a plane $\pi : \mathbf{n} \cdot \mathbf{x} + d = 0$ is to put the vertices of the OBB into the plane equation. If both a positive and

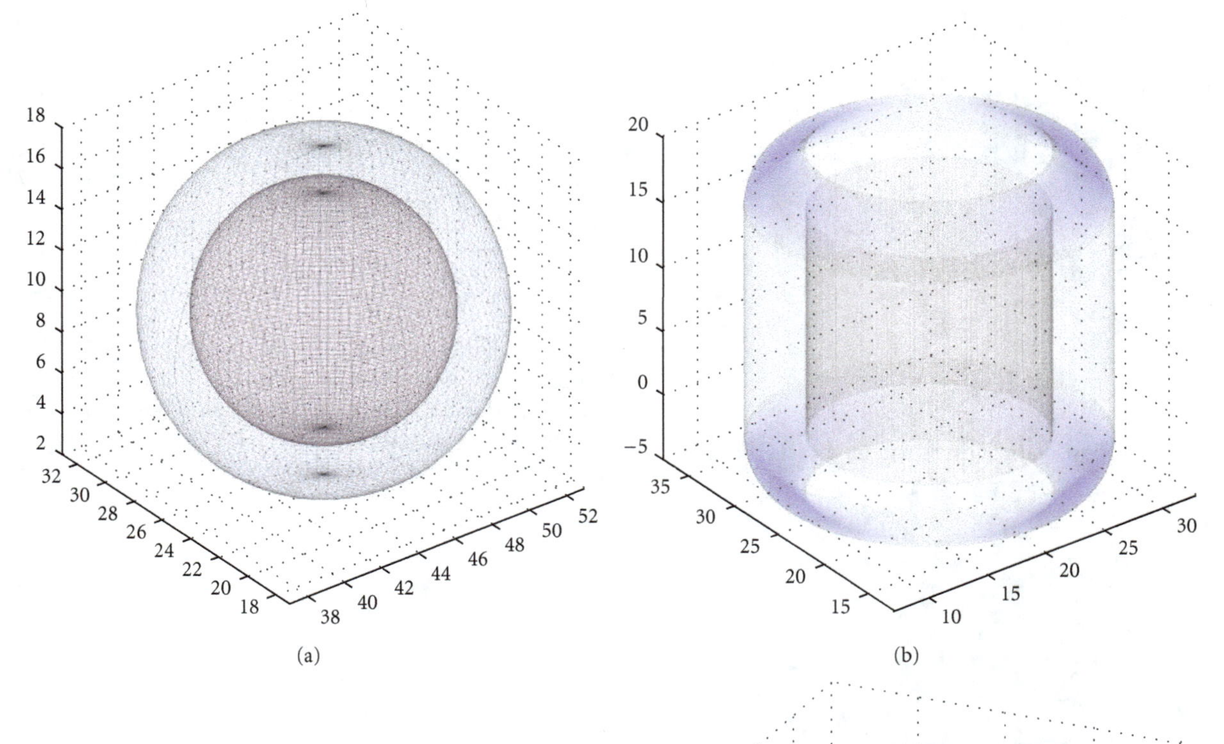

Figure 2: Elementary shapes (a) sphere, (b) cylinder, (c) plane, (d) parallelepiped (red), and their warning zones (blue).

a negative results (or a zero) are obtained, the vertices are located on both sides of (or on) the plane, and therefore, an intersection is detected. There are other smarter methods where only two points have to be inserted into the plane equation; for an OBB, there are two diagonally opposite corners on the box that are the maximum distance apart, when measured along the plane's normal. Every box has four diagonals formed by its corners. Taking the dot product of each diagonal's direction with the plane's normal, the largest value identifies the diagonal with these two furthest points. By testing these two points, the box as a whole is tested against a plane. So we can assume that we have an OBB defined by a centre point, $\mathbf{c}$ and a positive half diagonal vector, $\mathbf{h}$. The first step is to compute $\mathbf{c}$ and $\mathbf{h}$ as follows:

$$\mathbf{c} = \frac{\mathbf{b}^{\max} + \mathbf{b}^{\min}}{2}, \qquad \mathbf{h} = \frac{\mathbf{b}^{\max} - \mathbf{b}^{\min}}{2}, \tag{6}$$

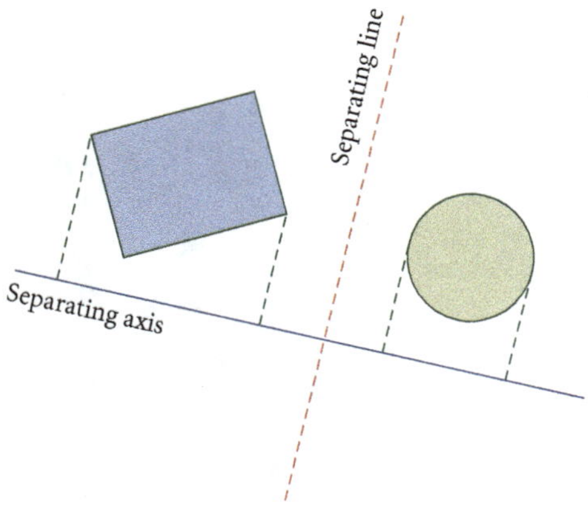

FIGURE 3: Illustration of the separating axis theorem.

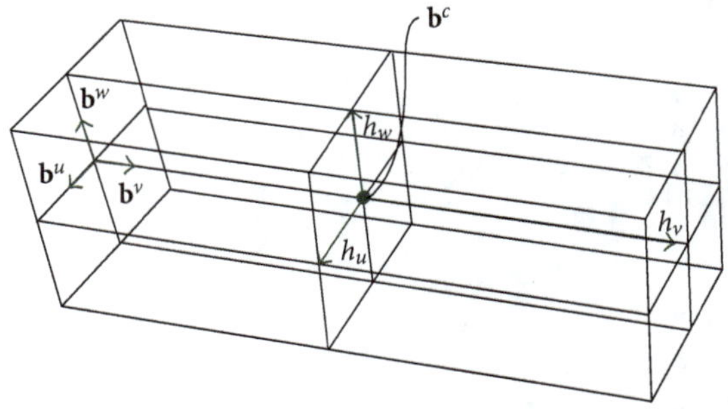

FIGURE 5: Oriented 3D bounding box: $\mathbf{b}^c$ is the centre point, together with normalized and positive-oriented side vectors $\mathbf{b}^u$, $\mathbf{b}^v$, and $\mathbf{b}^w$. The vectors h_u, h_v, and h_w are the distances from the centre of the box to the centre of its faces.

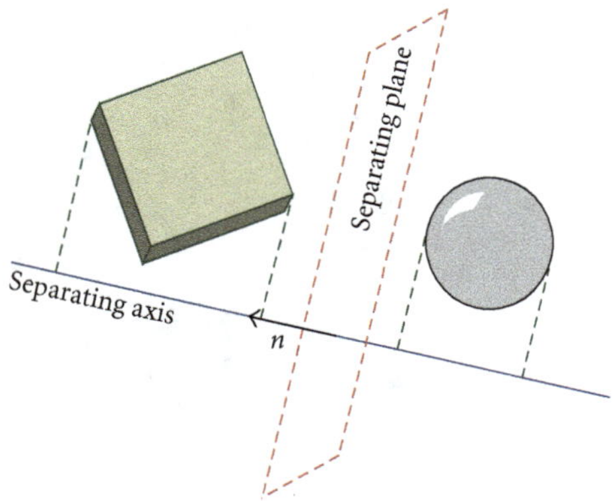

FIGURE 4: Illustration of the separating plane theorem.

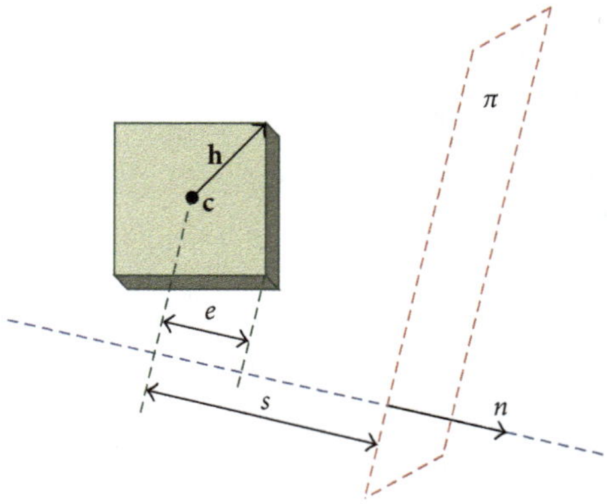

FIGURE 6: Principles of the intersection test between an OBB and a plane.

where $\mathbf{b}^{\min}$ and $\mathbf{b}^{\max}$ are the minimum and maximum corners of OBB. Now we have to test our OBB against the plane π. In order to do that in a fast way, we can consider the computation of the extent, denoted as e, of the box when projected onto the plane normal, $\mathbf{n}$. The equation of the extent is the following:

$$e = h_u |\mathbf{n} \cdot \mathbf{b}^u| + h_v |\mathbf{n} \cdot \mathbf{b}^v| + h_w |\mathbf{n} \cdot \mathbf{b}^w|. \quad (7)$$

Next, we compute the signed distance, s, from the centre point, $\mathbf{c}$, to the plane. This can be achieved by computing $s = \mathbf{c} \cdot \mathbf{n} + d$. With the signed distance and the extent of the box, the following will state if box is inside/outside the plane region:

$$\begin{aligned} &\text{if } (s - e) > 0 \quad \text{then outside,} \\ &\text{if} (s + e) < 0 \quad \text{then inside,} \end{aligned} \quad (8)$$

This simplification is explained in [18].

3.2.2. OBB versus Sphere Intersection. The principle to detect an intersection between an OBB and a sphere is depicted in Figure 7 and it consists on finding the point of the OBB that is closest to the sphere's centre $\mathbf{c}$. The tests to be used are one-dimensional one for each axis of the OBB. The sphere's centre coordinate for an axis is tested against the bounds of OBB: if it is outside the bounds, the distance between the sphere's centre and the box along this axis is computed and squared. After having executed the same calculus for each of the three axes, the sum of these squared distances is compared to the squared radius, r^2, of the sphere. If the sum is less than the squared radius, the closest point is inside the sphere, and the box overlaps. However, this algorithm is only applicable if the OBB reference frame is parallel to the sphere's frame reference, and in order to use the presented method, it is compulsory for an oriented bounding box to first transform the sphere's centre into the OBB's space; this means that the OBB's normalized axes as the basis for transforming the sphere's centre have to be used.

The algorithm can be resumed as shown in Algorithm 1.

The tests $(e < -r)$ and $(e > r)$ are used here to quickly reject the possibility of any intersection between the volumes.

3.2.3. OBB versus Cylinder Intersection. With reference to Figure 8, the principle to detect an intersection between an OBB and a cylinder is quite complex and here it will be analyzed a possible implementation. First of all, we have to transform cylinder into the OBB space in order to simplify the computation. After that, the closest points, P_{OBB} and P_{CYL}, between OBB and cylinder have to be detected; a first simplification can be done, stating that if the distance

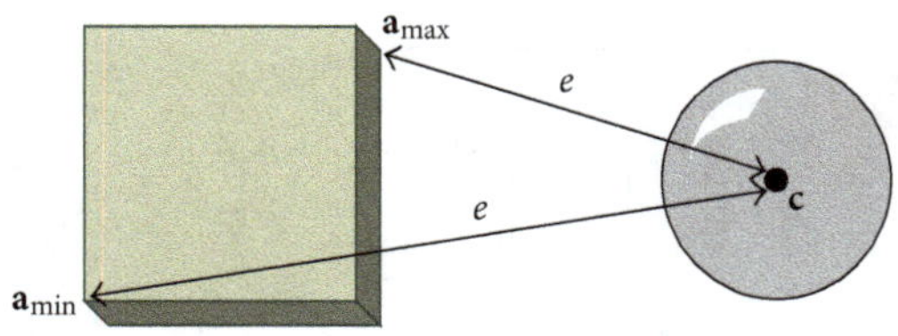

FIGURE 7: Principles of the intersection test between an OBB and a sphere.

```
d = 0
for each i ∈ {x, y, z}
  if ((e = c_i − a_i,min) < 0)
    if (e < −r) return (DISJOINT);
    d = d + e²;
  else if ((e = c_i − a_i,max) > 0)
    if (e > r) return (DISJOINT);
    d = d + e²;
if (d > r²) return(DISJOINT);
return(OVERLAP);
```

ALGORITHM 1:

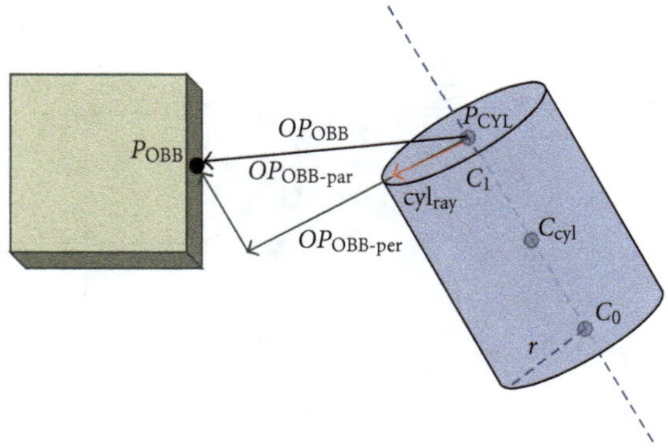

FIGURE 8: Principles of the intersection test between an OBB and a cylinder.

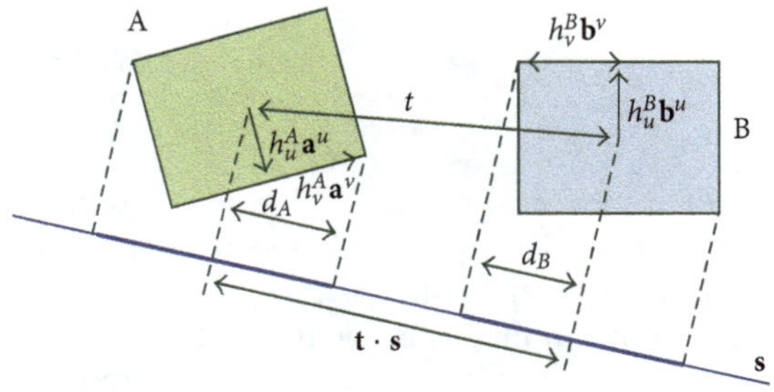

FIGURE 9: Principles of the intersection test between two OBBs.

between P_{OBB} and P_{CYL} is greater than r (the radius of the cylinder), then the two volumes do not intersect.

Otherwise, the algorithm should go forward computing:

$$C_{\text{cyl}} = \frac{C_0 - C_1}{2}, \tag{9}$$

Then

$$\begin{gathered} \text{if } \left(\text{Length}\left(OP_{\text{OBB-per}}\right) > r\right) \quad \text{then outside,} \\ \text{if } \left(\text{Length}\left(OP_{\text{OBB-per}}\right) < \text{dist}\left(C_{\text{cyl}}, C_0\right)\right) \quad \text{then inside.} \end{gathered} \tag{10}$$

The algorithm continues stating that

$$\text{if } \left(\text{Length}\left(OP_{\text{OBB-para}}\right) > \text{dist}\left(C_{\text{cyl}}, C_0\right)\right). \tag{11}$$

Then if the red vector cyl_{ray} intersects the plane and the point of intersection lies in the OBB's boundaries, then there is an intersection.

3.2.4. OBB versus OBB Intersection. The algorithm to detect intersections between two OBBs uses the separating axis theorem; the test is done in the coordinate system formed by A's centre and axes, as depicted in Figure 9. The origin then is $\mathbf{a}^c = (0, 0, 0)$ and the main axes are $\mathbf{a}^u = (1, 0, 0)$, $\mathbf{a}^v = (0, 1, 0)$, and $\mathbf{a}^w = (0, 0, 1)$. The OBB B is assumed to be located relative to A, with a translation t and a rotation matrix $\mathbf{R}$.

Based on the separating axis theorem, it is sufficient to find one axis that separates A and B to be sure that they are disjoint and so do not overlap. In order to complete the test, fifteen axes have to be tested: 3 from the faces of A, 3 from the faces of B, and $3 \cdot 3 = 9$ from the combinations of edges from A and B. Assuming that a potential separating axis is $\mathbf{s}$, the radii, d_A and d_B, of the OBBs on the axis, $\mathbf{s}$, are obtained by simple projections as stated in the following

$$\begin{aligned} d_A &= \sum_{i \in \{u,v,w\}} h_i^A \left| \mathbf{a}^i \cdot \mathbf{s} \right|, \\ d_B &= \sum_{i \in \{u,v,w\}} h_i^B \left| \mathbf{b}^i \cdot \mathbf{s} \right|. \end{aligned} \tag{12}$$

If and only if $\mathbf{s}$ is a separating axis, then the intervals on the axis should be disjoint and the following disjoint is valid:

$$|\mathbf{t} \cdot \mathbf{s}| > d_A + d_B. \tag{13}$$

Summarizing, if any of these fifteen tests is positive, the OBBs are disjoint; otherwise, if all the tests are negative, the OBBs intersect. For an extensive and more detailed theoretical treatment of this subject refer to [19].

4. Volumetric Interference Regions

The control systems' features, for the management of robot behaviour in static environments, have been described as the capability to control the robot speed override as different changes in the configuration of the robot arise. The architectural scheme of the proposed volumetric interference regions control system is depicted in Figure 10. From this scheme, it is possible to locate the blocks where the volume collision detection and closest point detection are taken; in particular, the volume collision detection (VCD) block takes as input the tool volume kinematics and the database of user-defined monitored/forbidden regions.

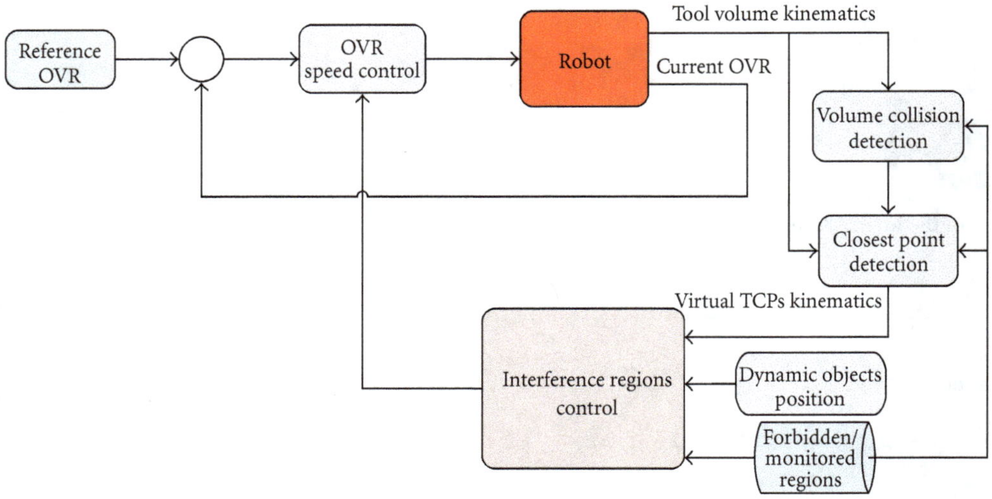

Figure 10: Volumetric interference regions control scheme.

The robot tool must be bounded inside an OBB; in this way, all the forbidden and monitored regions, previously defined by the user, can be successfully checked for collision. The information coming from the robot is then a complete kinematics set of the tool's OBB (i.e., its position, speed, and acceleration of its centre point); this information, together with some previous computation from the VCD and the information about the forbidden and monitored regions, are then used inside the closest point detection (CPD) block in order to detect for each region in the robot's workspace the closest point. This set of points is the output of the CPD which gives is an array of virtual TCPs kinematics (i.e., positions, speeds, and accelerations of the detected closest points). After the VCD and CPD block, the integrated solution showed in Figure 1 depicting the interference regions control scheme can be fully adopted in order to manage collisions between volumes and output a controlled speed override applied to the robot. This solution assures that the results obtained with the previous control architecture, concerning performances and effectiveness, are still valid, as showed in the next section.

Figure 11: The SMART NS16 industrial robot manipulator.

5. Test Results

In the present section an overview of the tests executed on a real industrial robot will be shown. These tests refer to the robot while it is moving at its maximum speed and while performing technological tasks in a real industrial cell. The interference regions control system has been analyzed during this test session, without taking into account the management of the volumetric control paradigm. We are confident that these results can be easily obtained with the presented volumetric method since the following tests refer to the collision avoidance core which takes as input a set of tool centre point positions which are the same output of the closest point detection block previously described. Therefore, we do think that these results can be generalized for the presented solution. The robot arm used in the test is a SMART NS16 manipulator (shown in Figure 11) which is a 6-axis industrial manipulator with a maximum load at wrist of 16 kilograms and a high repeatability of 0.05 mm. During this test phase, the SMART NS16 has been programmed, as usual in industrial robotics, as if it was moving over a working path (technological move of welding process) with one active cylindrical-controlled area.

In Figure 12(a), the robot end-effector position (X,Y) is plotted for two different movements: the first one where the control is not active (the blue pattern) and the second one where the control algorithm is enabled (the red pattern). The manipulator moves from the starting position (*Start*), up to a first point A. The next movement is a Cartesian linear move towards point B, parallel to Y-axis. The two circles depicted in the figure, highlight in yellow the warning zone and the forbidden zone in red. On the other hand, when the forbidden zone is not active, the end effector moves towards the start position after reaching point B (blue pattern in

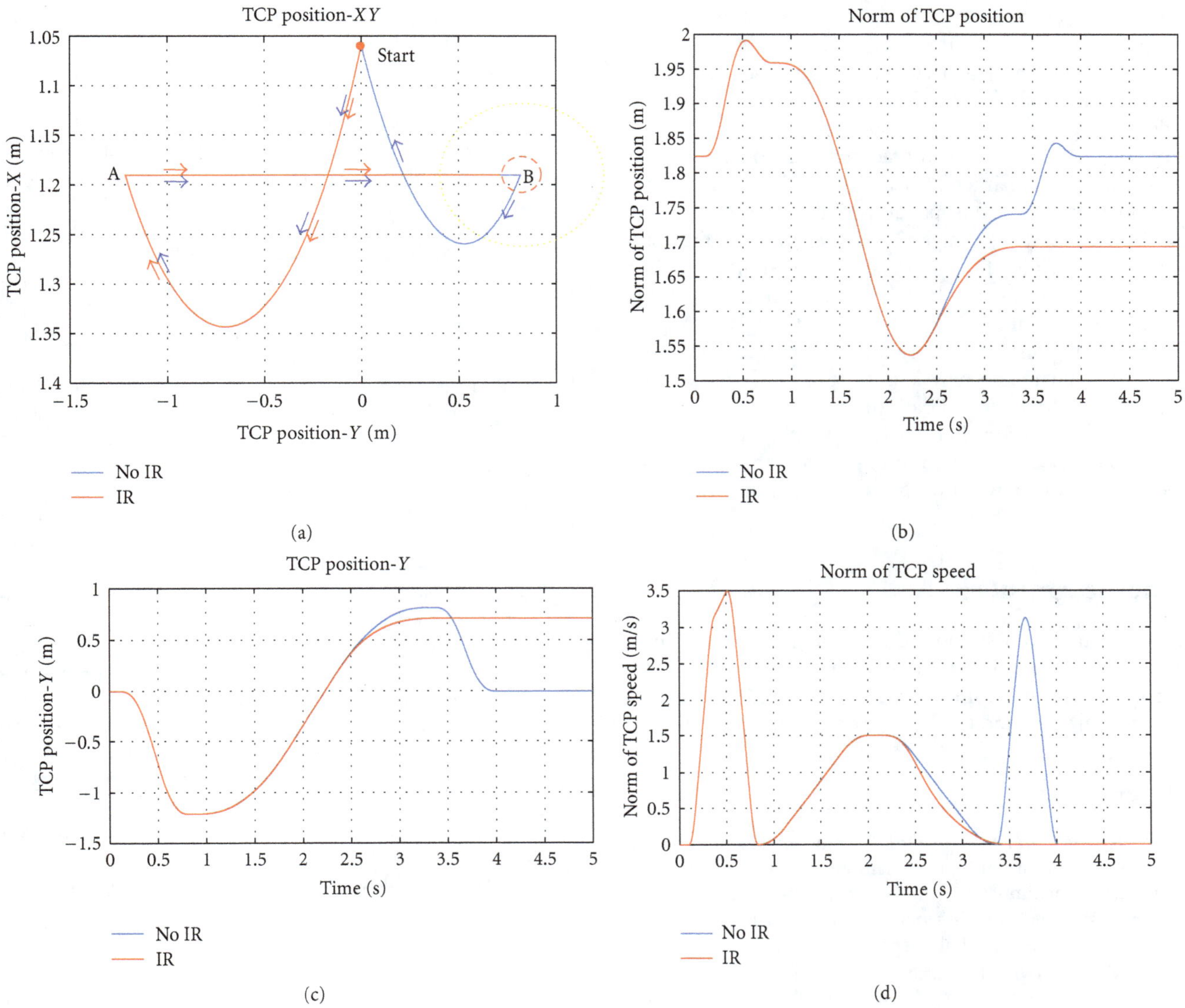

FIGURE 12: (a) Robot position (X,Y), with enabled/disabled control, (b) norm of the end-effector position over time, (c) trend of the Y end-effector position over time, and (d) norm of the robot speed over time.

the figure). The second movement is depicted in red, and represents the behaviour of the robot end effector when the forbidden zone is enabled; the robot end effector stops inside the warning zone at the boundaries of the forbidden zone when the control is enabled.

The position trend over time is depicted in Figures 12(b) and 12(c). From Figure 12(b) it is possible to see the different trends of the norm position of the end effector when the control is enabled (in red) and when it is deactivated (in blue). The robot begins moving along the trajectory parallel to Y at 1.5 seconds, and after almost one second it enters the warning zone. In Figure 12(c), it is depicted how, without enabled warning zone control algorithm, the trend along the Y-axis continues as far as the next movement (at about 3.5 seconds); when the control algorithm is enabled, it is shown that the Y position of the robot end effector becomes constant when the end effector encounters the forbidden zone. In Figure 12(d), is then depicted how the end-effector speed changes when it comes in contact with the warning zone at about 2.5 seconds. From these results, it comes out clear that the speed reduction is in inverse relation to the distance between the end effector and the geometrical elementary region surface. These results can be easily extended to a more complex environment where several different interference regions are present and where the control algorithm has to manage the override control for each area, even if they overlap. As stated before, the algorithm is still valid if the input of the interference regions control system is a vector of virtual tool centre points, coming from the closest point detection and from collision avoidance blocks.

6. Conclusion

In this paper, we presented a new method to extend the perceptive capabilities of an industrial robot. Although anthropomorphic robot producers have faced problems related to the interaction between robot and its surroundings, there is not an exhaustive study on the capabilities of a robot to be aware of its volume (in particular on the tools eventually mounted on its flange). This paper presents methods to model the space around the robot in a manner that the robot could be capable of interacting with some particular regions, here called interference regions, in its workspace. The added value of this paper is that the model of the robot surroundings is designed in order to make the robot capable of perceiving the volume of the surrounding objects through the volume of the bounding box containing its tool. The test results have been shown considering the original interference regions control architecture, and the possibility to extend them to the new architecture are remarked and expounded in order to validate the new control paradigm. The proposed solution then prove that the system is able to cope with complex real surroundings where the interactions to be checked are between volumes in order to get closer to a real industrial environment.

Acknowledgments

This paper has been supported by Comau Robotics S.p.A.; the author would like to thank the company and the control engineering team for the support and help on the experimental work and on the state of the art research.

References

[1] F. Romanelli and F. Tampalini, "A control algorithm for the management of multiple dynamical geometrical areas for industrial manipulators," in *Proceedings of the 17th International Workshop on Robotics in Alpe-Adria-Danube Region (RAAD '08)*, Ancona, Italy, September 2008.

[2] M. Vukobratović, D. Surdilovic, and Y. Ekalo, *Dynamics and Robust Control of Robot-Environment Interaction*, World Scientific, River Edge, NJ, USA, 2008.

[3] H. Marcuse, *One-Dimensional Man: Studies in the Ideology of Advanced Industrial Society*, Routledge, 1964.

[4] B. Schulz, "The gift of sight," *Austrlian Manufacturing Technology*, pp. 86–89, 2007.

[5] B. Rooks, "Precision robot welding for world's largest particle physics apparatus," *Industrial Robot*, vol. 32, no. 4, 2005.

[6] S. Chunxue, B. Yingyong, and L. Jianghui, "Mobile robot path planning in three-dimensional environment based on ACO-PSO hybrid algorithm," in *IEEE/ASME International Conference on Advanced Intelligent Mechatronics (AIM '08)*, pp. 252–256, August 2008.

[7] J. Wolfe, B. Marthi, and S. Russell, "Combined task and motion planning for mobile manipulation," in *the 20th International Conference on Automated Planning and Scheduling (ICAPS '10)*, pp. 254–257, May 2010.

[8] M. H. Kazemi, S. Ahangarian, and A. Hashempour, "Fuzzy path tracking control of a 5-DOF experimental robot," in *IEEE International Conference on Information and Automation (ICIA '08)*, pp. 436–440, June 2008.

[9] M. Matinfar, C. Baird, A. Batouli, R. Clatterbuck, and P. Kazanzides, "Robot-assisted skull base surgery," in *IEEE/RSJ International Conference on Intelligent Robots and Systems (IROS '07)*, pp. 865–870, November 2007.

[10] G. van den Bergen, *Collision Detection in Interactive 3D Environments*, Morgan Kaufmann, 2003.

[11] E. Drumwright, "A fast and stable penalty method for rigid body simulation," *IEEE Transactions on Visualization and Computer Graphics*, vol. 14, no. 1, pp. 231–240, 2008.

[12] J. U. Kuehnle, M. Danzer, A. Verl, and R. Bischoff, "Real-time 3D environment model for obstacle detection and collision avoidance with a mobile service robot," in *Intelligent Robots and Computer Vision XXVII: Algorithms and Techniques*, vol. 7539 of *Proceedings of SPIE*, January 2010.

[13] M. Shahrokhi and A. Bernard, "A fuzzy approach for definition of dangerous zone in industrial systems," in *IEEE International Conference on Systems, Man and Cybernetics (SMC '04)*, pp. 6318–6324, October 2004.

[14] F. Romanelli and F. Tampalini, "BOG and fuzzy filters based multimodal collision avoidance for industrial manipulators," in *Proceedings of the IEEE Conference on Robotics, Automation and Mechatronics*, pp. 779–784, Chengdu, China, September 2008.

[15] M. A. Abam, M. de Berg, S.-H. Poon, and B. Speckmann, "Kinetic collision detection for convex fat objects," in *the European Symposium on Algorithms*, 2006.

[16] P. K. Agarwal and N. H. Mustafa, "Independent set of intersection graphs of convex objects in 2D," in *Proceedings of the 9th Scandinavian Workshop on Algorithm Theory (SWAT '04)*, T. Hagerup and J. Katajainen, Eds., Lecture Notes in Computer Science, pp. 127–137, 2004.

[17] H. A. Sulaiman and A. Bade, "Balanced hierarchical method of collision detection in virtual environment," in *the 2nd International Conference on Software Engineering and Computer Systems (ICSECS '11)*, vol. 181, pp. 493–501, 2011.

[18] S. Boyd and L. Vandenberghe, *Convex Optimization*, Cambridge University Press, 2004.

[19] T. Akenine-Möller, E. Haines, and N. Hoffman, *Real-Time Rendering*, A K Peters, Wellesley, Mass, USA, 2008.

21

Design of a Sapling Branch Grafting Robot

Qun Sun,[1] Dongjie Zhao,[1,2] Chong Wang,[1] and Ying Zhao[1]

[1] *School of Mechanical & Automotive Engineering, Liaocheng University, Liaocheng 252059, China*
[2] *College of Engineering, China Agricultural University, Beijing 100083, China*

Correspondence should be addressed to Qun Sun; sunxiaoqun97@163.com

Academic Editor: Giovanni Muscato

The automatic sapling grafting methods and grafting robot technologies are not comprehensively studied despite the fact that they are urgently required in practice. For this reason, a sapling grafting robot is developed to implement automatic grafting for saplings. The developed grafting robot includes clipping mechanism, moving mechanism, cutting mechanism, binding mechanism, and Arduino MCU based control system, which is capable of clipping, moving, positioning, cutting, grafting, and binding saplings. Experiments show that the stock cutting efficiency is 98.4%, the scion cutting efficiency is 98.9%, the grafting efficiency is 87.3%, and the binding efficiency is 68.9%.

1. Introduction

Sapling grafting is an ancient technology that can assist breeding of good plant species, enhance their ability of anti-illness, improve the fruit quality, and strengthen their adaptabilities to harsh environments. The earliest literatures about manual citrus sapling grafting techniques appeared around 3000 years ago in China.

Japan is the first country in the world that developed vegetable grafting robots. The first commercial model of a grafting robot (GR800 series, Iseki & Co. Ltd., Matsuyama, Japan) became available for cucurbits back in 1993 [1–3]. Since then, semiautomated or fully automated vegetable grafting robots have been invented by several agricultural machine companies, and several commercial models became available in East Asia, Europe, and more recently the U.S. [4–9].

However, grafting robots for saplings have not been comprehensively studied. A semiautomatic grape grafting robot was developed by CFCL in France to improve the grape production. Till present, the sapling grafting works are still following traditional manual methods in China, which are labour intensive but less productive. Besides, the grafted saplings often exhibit low survival rates and are less adaptive to the seasonal planting characteristics, which made it difficult to achieve industrialized sapling breeding production.

Some similar literatures about sapling grafting robot could be found in China in the following areas.

(1) Sapling Force Characteristics Experiments. Li et al. studied peach saplings and carried out experiments on peach force characteristics such as the cutting resistance force in vertical, traverse, and inclined cutting manners and also studied the variations of compressive injury forces [10]. Zhao et al. investigated poplar saplings based on a stick-to-stick grafting manner and studied the varying relationships between cutting resistance force and cutting depth, as well as the impact of scion diameter on the cutting resistance force [11].

(2) Sapling Grafting Mechanism Design. Li et al. designed a cutting mechanism to carry out experiments for peach sapling grafting, and the average cutting successful rate of this mechanism reached up to 99.14% [12]. Zhao et al. created a scion cutting mechanism for the poplar sapling studies, which provided a successful rate of 96.5% [13]. Jiang designed a pure mechanical grape sapling mechanism for a stick-to-stick grafting approach. Its grafting efficiency is 15–20 times that of manual work and the grafting survival rate exceeded 85% [14].

The Harbin Forestry Machinery Institute of China's Forestry Administration Bureau developed a camellia grafting robot based on a cutting grafting method [15]. The author of this paper and his research team also initially studied

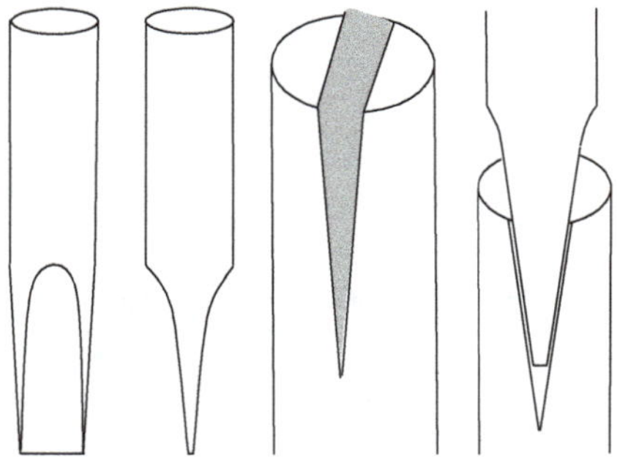

FIGURE 1: Operation flow of sapling cleft grafting by hand.

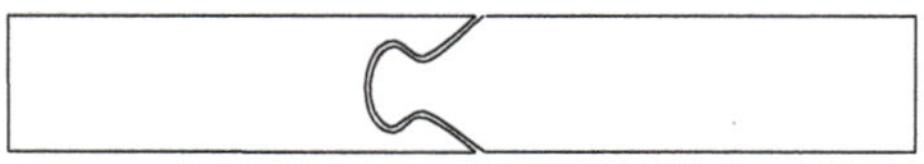

FIGURE 2: "Ω" shaped blades.

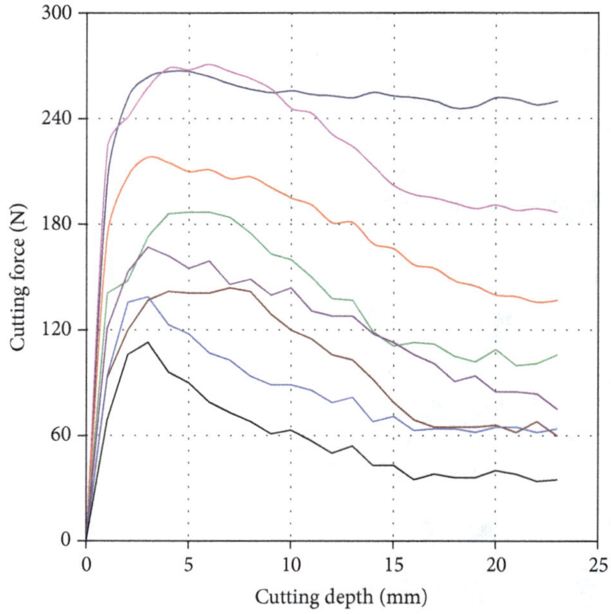

FIGURE 3: Relationship between cutting force and cutting depth.

cleft grafting mechanism for poplar saplings and associated automatic grafting devices [16].

This study developed a sapling grafting robot and carried out experiments for hard branch grafting applications, in order to investigate universal methods for grafting hard sapling branches and the associated grafting robot devices.

2. Sapling Hard Branch Grafting Method

2.1. Grafting Method. Cleft grafting is the most important method for sapling hard branch grafting. The manual operation flow diagram is shown in Figure 1. Firstly, a scion branch is cut into a double wedge shaped incision and a rootstock is cut into an "I" shaped cleft; then, force the cross section of the rootstock cleft to be tightly attached to the wound surface of the scion wedge incision, leaving no gap on the top and the bottom, and if loose further bind tightly.

In order to accurately align the cutting surfaces of the scion and the rootstock cross sections and ensure a tight binding of the saplings, people now use "Ω" shaped blades to directly cut scions and rootstocks, as shown in Figure 2, which simplifies the program of cleft grafting process and enhances grafting efficiency. Therefore, the "Ω" shaped blades grafting method is adopted in the paper.

2.2. Sapling Grafting Mechanical Characteristics Analysis. By a mechanical cold cutting manner, the sapling mechanical grafting characteristics include scion and rootstock growing geometry parameters, cutting force characteristics, cutting surface physiology, and biochemistry characteristics. In particular, the scion and rootstock cutting force characteristics are directly associated with the design of the grafting robot mechanism variables [11]. Using force measurement equipment, the relationship between cutting force and cutting depth could be obtained by 100 branch samples, as shown in Figure 3. Some key parameters are also confirmed using sensors as in the following chart [11].

3. Design of Mechanical System for Sapling Grafting Robot

3.1. Design of Mechanical System for Integrated Sapling Grafting Robot

3.1.1. Sapling Grafting Robot Working Process. The working process of a sapling grafting robot is as follows: feeding scion and rootstock saplings → moving saplings → cutting scions and rootstocks → moving saplings → attaching saplings → binding saplings → detaching saplings. Among these, feeding and finally detaching scion and rootstock saplings need manual operation. The working process of the sapling grafting robot is shown in flow diagram in Figure 4.

3.1.2. Integrated Structure of the Sapling Grafting Robot. The overall structure of the sapling grafting robot consists of a frame, a robot hand to handle rootstock, another robot hand to clamp scion, rootstock moving mechanism, scion moving mechanism, rootstock cutting mechanism, scion cutting mechanism, and binding mechanism. It adopts modular design concepts so that the functionalities are relatively independent of each other and utilizes many standard parts and mature products such as standard aluminum frame, angle parts, robot hands, and grafting blades. The overall mechanical structure of the example robot machine is shown in Figure 5.

3.2. Design of the Key Structures of the Sapling Grafting Robot

3.2.1. Scion and Rootstock Clipping Robot Hands. The scion and rootstock clipping robot hands are of the same structures; both include screw driver and motor, guiding shaft, linear bearing, robot hand, and other parts, as shown in Figure 6.

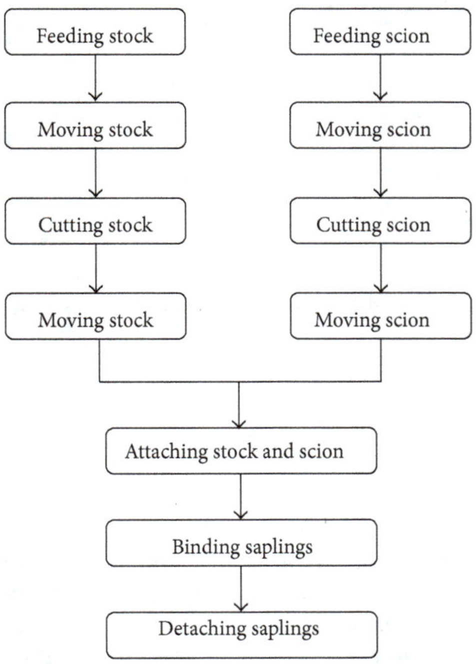

FIGURE 4: Work flow of grafting robot.

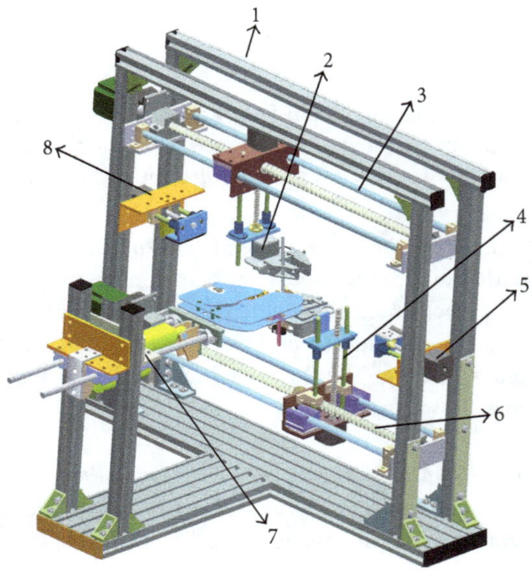

FIGURE 5: The whole mechanical structure of grafting robot. 1: frame, 2: scion clipping robot hand, 3: scion moving mechanism, 4: stock clipping robot hand, 5: stock cutting mechanism, 6: scion moving mechanism, 7: binding mechanism, and 8: scion cutting mechanism.

The robot hand is based on BigHand single steering clipping actuator, which can control clipping and releasing rootstock and scion as well as adjusting the clipping force. The screw motor adopts a BG42 step motor that can control the robot hand to bring scion and rootstock to move vertically along the bare shaft.

3.2.2. Rootstock and Scion Moving Mechanism. The moving mechanisms for rootstocks and scions have basically the same structures. They both include supporting base, transmission motor, screw, and nut as shown in Figure 7.

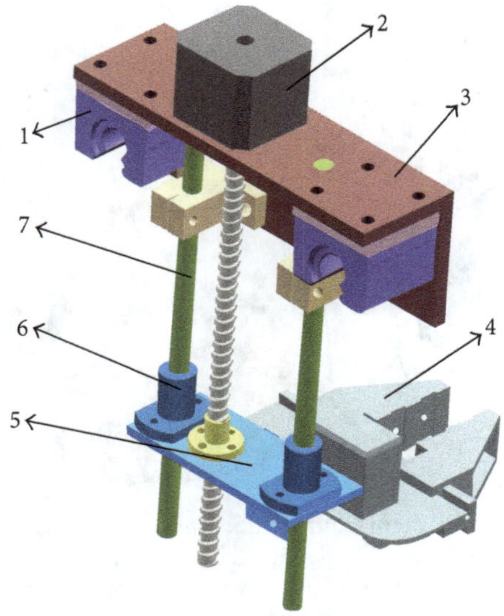

FIGURE 6: Structure of the clipping mechanism. 1: sliding unit, 2: screw motor, 3: connection base, 4: robot hand, 5: mounting base for robot hand, 6: linear bearing, and 7: bare guiding shaft.

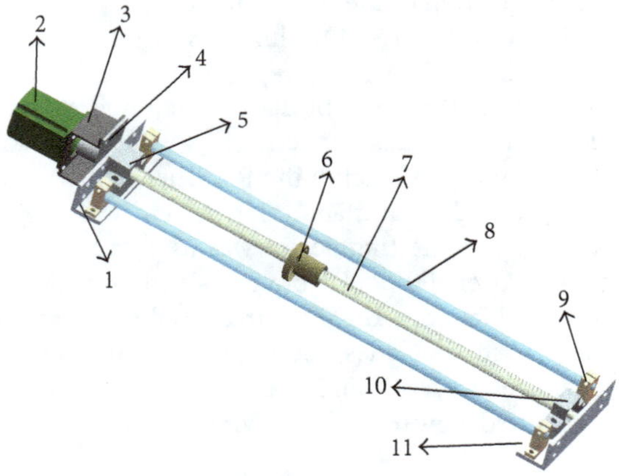

FIGURE 7: Structure of the moving mechanism. 1: left supporting base, 2: driving motor, 3: motor mounting base, 4: motor connection board, 5: front screw mounting base, 6: driving screw nut, 7: driving screw, 8: bare guiding shaft, 9: bare supporting base, 10: rear screw mounting base, and 11: right supporting base.

The driving motor uses a 57BYG250-76 step motor, which powers the screw to rotate through a coupler and drives the clipping robot hand connected on the screw nut to move horizontally along the bare shaft.

3.2.3. Rootstock and Scion Cutting Mechanism. The scion and rootstock cutting mechanisms have the same structure as they both include screw motor, bare guiding shaft, screw nut, and cutting parts, as shown in Figure 8.

The screw motor drives the cutting blade to move horizontally through connection on the blade mounting base and the mounting board, so as to cut the saplings allocated between the cutting blade and the rubber pad. The cutting

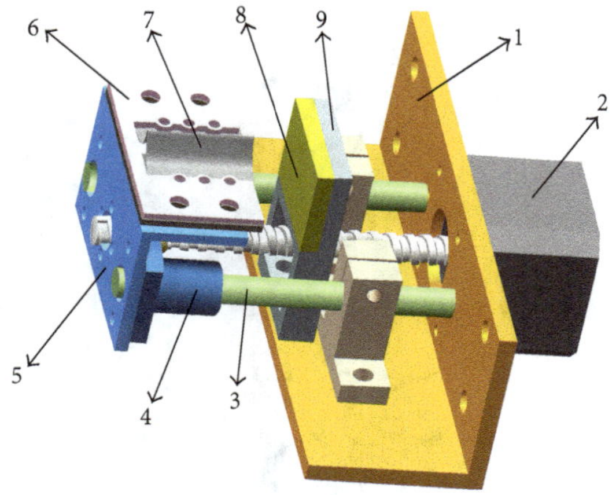

FIGURE 8: Structure of cutting mechanism. 1: mounting base, 2: screw motor, 3: bare guiding shaft, 4: linear bearing, 5: cutting blade mounting base, 6: cutting blade mounting board, 7: cutting blade, 8: rubber pad, and 9: rubber pad back base.

blade is a "Ω" shaped knife as extensively adopted in manual grafting tools.

The flow diagram of the cutting force test experiments is shown in Figure 9. Two FSR402 film pressure sensors are fixed on a YANSHEN hand-held sapling grafting device, forming half-bridge circuit through a bridge circuit junction box. The signal goes through a resistance strain gauge, a data acquisition device and then reaches the PC. Then, Vib'SYS signal gathering, processing, and analyzing system was employed for processing. That testing device was used to measure the cutting forces over 100 poplar branches and grape branches each of 6 mm thickness. The measured results indicated that, using the "Ω" shape blades, the required cutting forces for poplar branches are less than 300 N.

The required power of the driving motor is calculated from

$$P_m = \lambda P = \lambda \frac{Tn}{9550} = \frac{\lambda F p \eta \cdot (\pi/2) \cdot n}{9550}, \quad (1)$$

where F is the required cutting force, λ is the surplus coefficient, T is the motor torque, n is the screw rotational speed, p is the screw lead, and η is the screw efficiency.

According to the calculation, the adopted screw motor is Shidai Chaoqun type BG42 step motor.

3.2.4. Sapling Binding Mechanism. The binding mechanism includes base components, moving parts, and binding components. The structure is shown in Figure 10.

By adjusting the height of the moving components of this binding device, the long slot of the binding components can be aligned with the sapling intake nozzle. The binding belt should be set around the gear and locked between the gear and the claw. The motor drives the pusher to move which sends the sapling into the tooth space of the gear wheel and then moves along the long slot. The binding belt wraps the sapling end and bonds tightly under the pressure from the claw. Then, the electromagnet drives the cutting blade to cut off the binding belt. The working process is as shown in Figure 10(c).

4. Design of Sapling Robot Control System

4.1. Design of Overall Control System. The control objects include step motor, servo motor, LCD display, DC motor, and electromagnet, whilst the measured objects include micro switch signal and approach switch signal. The control system diagram is shown in Figure 11.

4.2. Design of Control System Hardware. The controller adopts Arduino MCU and the hardware design includes the following.

(1) Front Channel Configurations. The sensors of the control system are 2 micro switches and 2 approach switches; both include TTL levels. The front channel configuration includes 4-way binary input circuit.

(2) Rear Channel Configurations. The control system requires 2 servo motors, 6 step motors, 1 DC motor, 1 electromagnet, and 1 LCD module. According to modular design concepts, the design adopts 6-way step motor control module, 1-way DC motor driving module, 1-way relay control module, and 1-way LCD driving module.

(3) Exchange Channel Configurations. The control system was configured with 2-way stock processor communications and 2-way scion processor communications.

4.3. Design of Control System Software. Control system software includes two parts which are rootstock processing control software and scion processing control software.

4.3.1. Design of Main Control Module for Rootstock Processor. The functionality of the rootstock main control module is to implement stock processing, which includes stock robot hand clipping, stock moving, stock inversed "Ω" shaped cutting, and sapling binding. The flow diagram of rootstock control software is shown in Figure 12.

4.3.2. Design of Main Control Module for Scion Processor. The functionality of the scion processor main control module is to implement scion processing, including LCD display, scion robot hand clipping, scion moving, and "Ω" shaped cutting. The program diagram is similar to that of the rootstock processing main control module program. The flow diagram of scion control software is shown in Figure 13.

5. Experiment

The developed sapling grafting robot system is shown in Figure 14. Grafting experiments have been carried out in a lab environment.

5.1. Experimental Objects and Evaluation Criteria

5.1.1. Experimental Objects. The grafting experiment adopted straight poplar branches that have diameter of about 4~6 mm and length around 100 mm. Totally, 100 rootstock samples

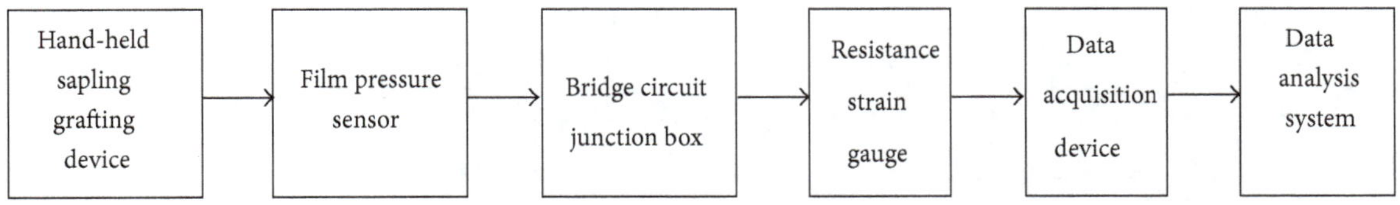

FIGURE 9: Flow diagram of the test circuit.

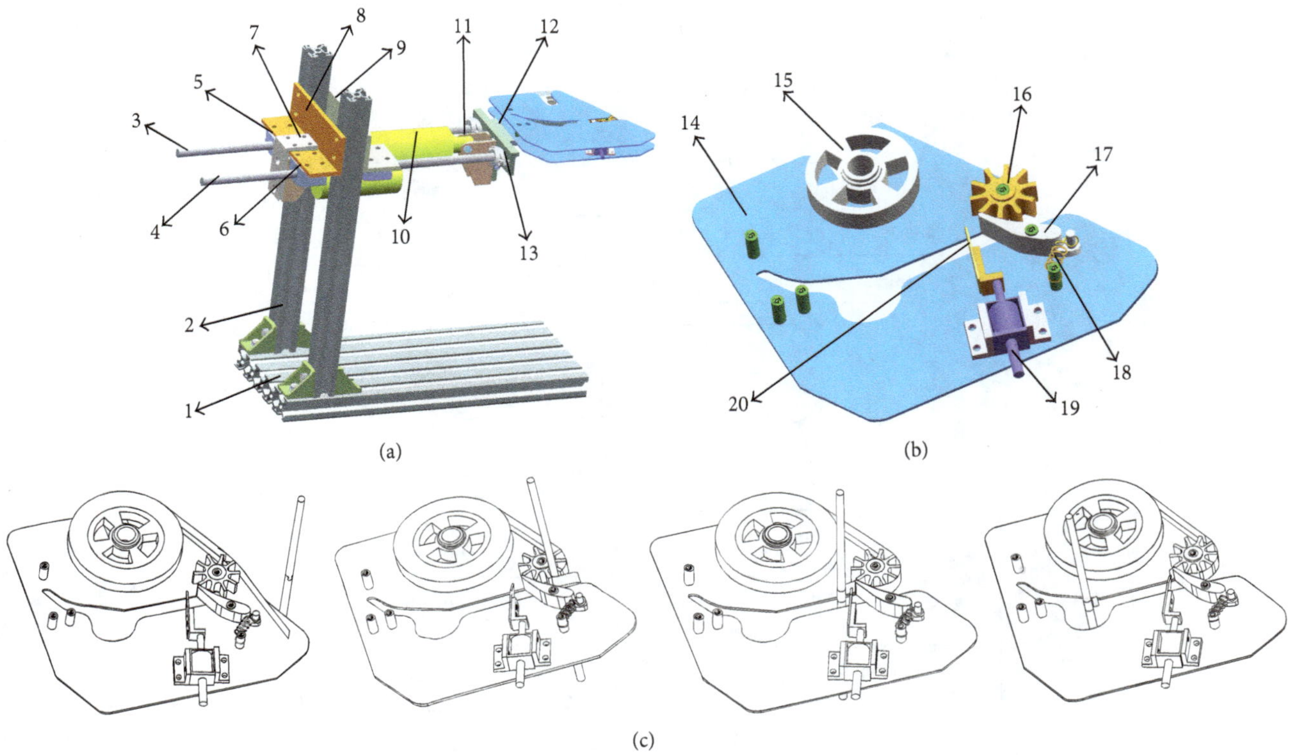

FIGURE 10: Structure of the binding mechanism. (a) Binding mechanism integrated diagram. (b) Internal diagram of the binding components. (c) Working process. 1: base board, 2: vertical pillar, 3: guiding shaft I, 4: guiding shaft II, 5: sliding unit I, 6: sliding unit II, 7: push rod mounting base, 8: left mounting base, 9: right mounting base, 10: electrical push rod, 11: guiding shaft base I, 12: moving base, 13: guiding shaft base II, 14: cover board, 15: belt wheel, 16: gear, 17: claw, 18: spring, 19: electromagnet, and 20: cutting blade.

and 100 scion samples are chosen, paired according to diameter and then divided into 10 groups.

5.1.2. Grafting Evaluation Criteria

(1) The criteria for successful rootstock cutting include inverted Ω shaped cutting cross section, giving smooth surfaces without burr.

(2) The criteria for successful scion cutting include normal Ω shaped cutting cross section, giving smooth surfaces without burr.

(3) The criteria for successful grafting include (a) good binding between rootstock and scion; that is, the two central lines are aligned, in good contact without deflections, (b) the rootstock and scion are basically correctly bound; that is, the two central lines are aligned but the contact may not be in perfect condition or there are deflections, and (c) the rootstock and scion are badly bound including bad alignment, failed contact, and deflections. Among these, both (a) and (b) are regarded as successful grafting.

(4) The criteria for successful binding: the binding belt tightly wraps the grafting saplings, without obvious misplacement of the saplings.

5.2. Experimental Results and Analysis. Based on the above-mentioned assessment criteria, experiments have been conducted using the developed grafting robot. Among the sapling success statistical data, 43.3% are good, 44% are acceptable, and 12.7% are poor. Therefore, the grafting efficiency is 87.3%.

The results of stock cutting and scion cutting and binding success rate are shown in Table 1, and corresponding curves are shown in Figure 15. Experiments show that the stock cutting efficiency is 98.4%, the scion cutting efficiency is 98.9%, and the binding efficiency is 68.9%.

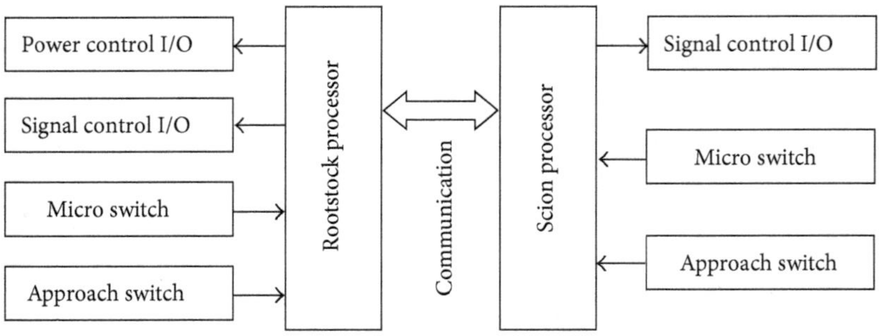

FIGURE 11: Structure of the control system.

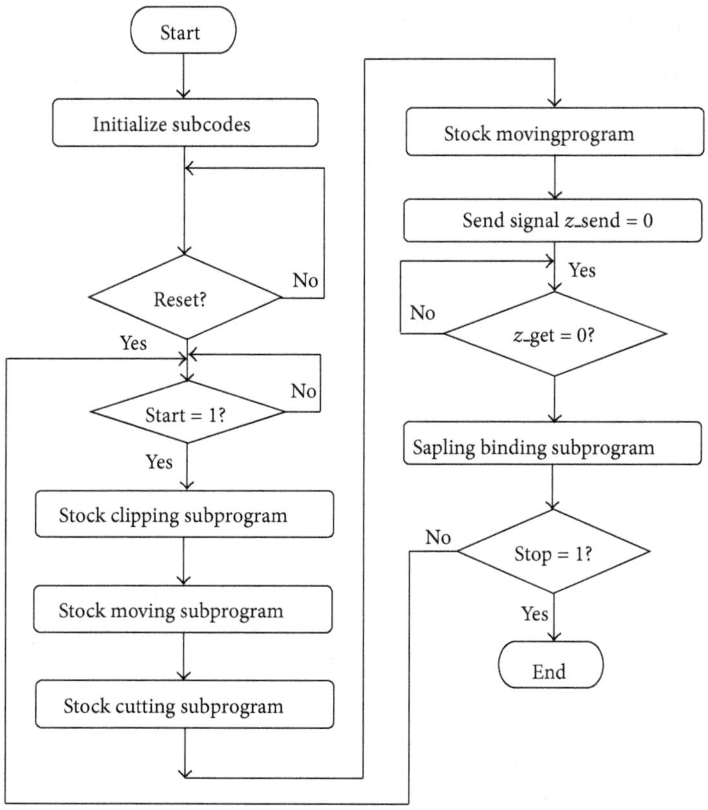

FIGURE 12: Flow diagram of stock processing main control module.

TABLE 1: Results of stock cutting, scion cutting, and binding experiments.

Group	Stock cutting success rate (%)	Scion cutting success rate (%)	Binding success rate (%)
1	100	99	63
2	98	98	71
3	99	100	76
4	98	99	71
5	99	98	73
6	98	100	73
7	98	98	70
8	97	99	62
9	98	99	61
10	99	99	69
Total	98.4	98.9	68.9

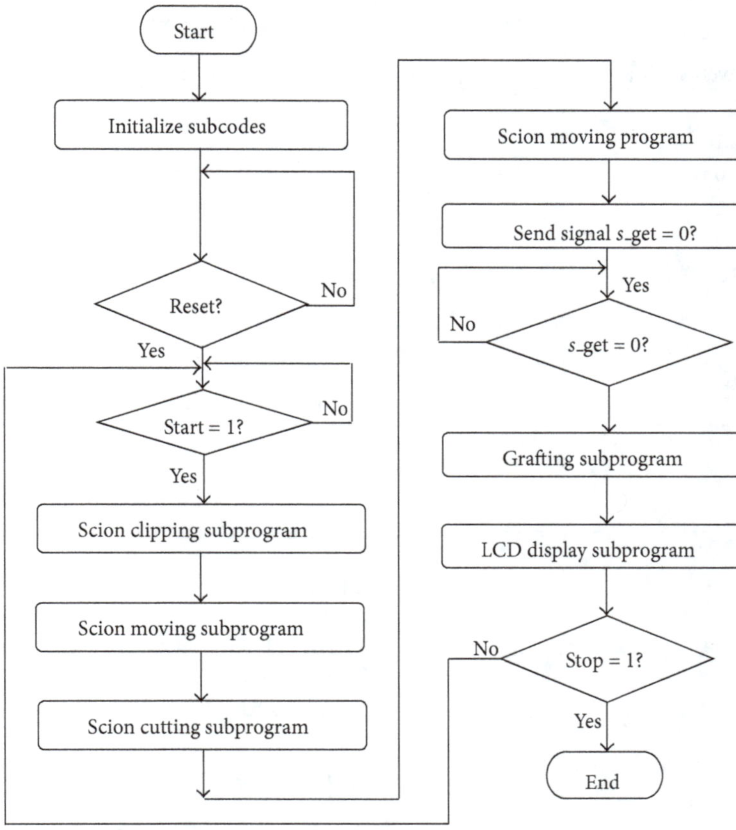

FIGURE 13: Flow diagram of scion processing main control module.

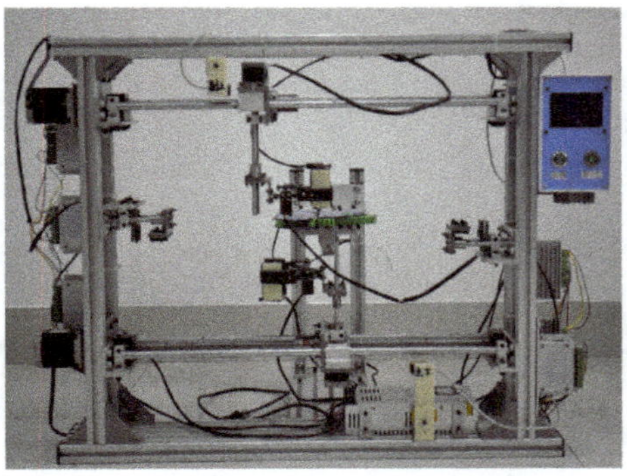

FIGURE 14: The developed sapling grafting robot.

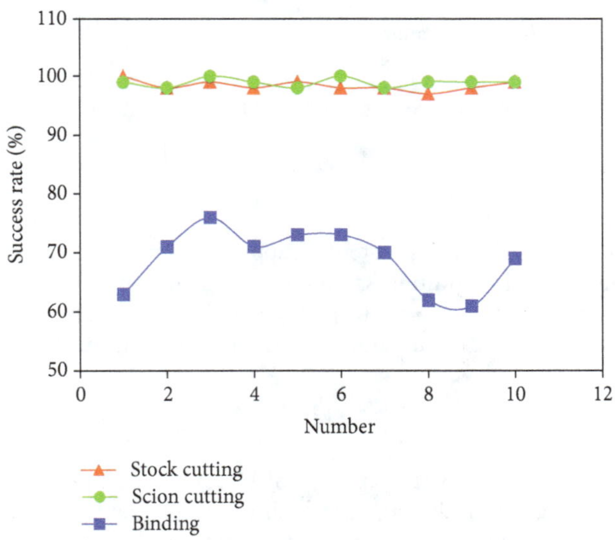

FIGURE 15: Curves of stock cutting, scion cutting, and binding success rate.

The results of consumption time for sapling grafting experiments are shown in Table 2, and corresponding curves are shown in Figure 16. Experiments show that the average time for stock cutting is 2.20 seconds, the average time for scion cutting is 2.18 seconds, and the average time for one grafting is 34.40 seconds.

The experimental results indicate that the cutting success rates for rootstocks and scions are high, but due to individual differences between stock and scion samples, errors exist in the alignment of the cutting surfaces when binding them together, which affects the grafting success rates. When there are bends near the sapling joining location, the sapling binding success rates decrease. Stock cutting, scion cutting, and sapling binding processes take relatively short time and the efficiencies are high. However, due to the large empty strokes of the main moving parts, the overall grafting process has a relatively low efficiency.

Further studies include automatically pairing rootstock and scion samples, automatically aligning rootstock and scion

TABLE 2: Results of consumption time for sapling grafting experiments.

Group	Stock cutting average time (S)	Scion cutting average time (S)	Average grafting overall time (S)
1	2.2	2.1	33
2	2.1	2.2	35
3	2.0	2.0	32
4	2.2	2.2	39
5	2.2	2.3	35
6	2.3	2.2	34
7	2.1	2.1	36
8	2.3	2.2	31
9	2.2	2.3	33
10	2.4	2.2	36
Total	2.20	2.18	34.40

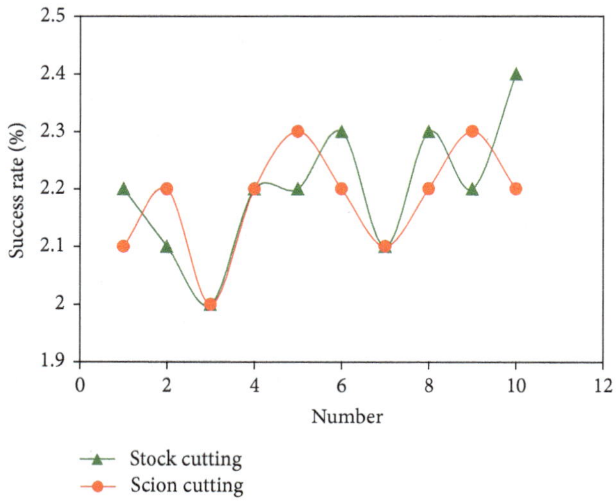

FIGURE 16: Curves of consumption time for stock cutting and scion cutting.

cutting surfaces, accurately wrapping saplings, automatically detecting grafting sapling qualities, and automatically detecting the accuracy of moving parts' motions. The main structural dimensions should also be further optimized.

6. Conclusion

(1) An automatic sapling hard branch grafting robot has been developed, which can implement clipping, moving, positioning, cutting, binding, and wrapping rootstock and scion saplings.

(2) Experimental results indicate that the grafting success rate of this robot is 87.3% and the binding success rate is 68.9%.

Conflict of Interests

The authors declare that there is no conflict of interests regarding the publication of this paper.

Acknowledgments

This work was supported by the Natural Science Foundation of Shandong Province, China (Grant nos. ZR2011EL038 and ZR2012CQ026), and A Project of Shandong Province Higher Educational Science and Technology Program (Grant nos. J11LD16 and J12LB63).

References

[1] Y. Nishiura, N. Honami, and T. Taira, "Development of a new grafting method (part 4)—robotization of grafting operation," *Journal of the Japanese Society of Agricultural Machinery*, vol. 61, no. 6, pp. 103–112, 1999.

[2] Y. Osamu and F. Akiko, "Growth adjustment technique of seedling optimal for grafting by seedling storage for cucumber full-automatic machine grafting," *Tokyo Agricultural Research*, vol. 55, no. 1, pp. 201–202, 2002.

[3] K. Ken, S. Kenta, and S. Sadao, "Study on automation of seedlings feeding for grafting robot for cucurbitaceous vegetables (part1)-necessary accuracy for cotyledon orienting mechanism," *Journal of the Japanese Society of Agricultural Machinery*, vol. 68, no. 6, pp. 117–123, 2006.

[4] C.-H. Kang, G.-S. Han, T.-H. Noh et al., "Splice grafting robot for fruit and vegetable plants: World Intellectual Property Organization," WO/2005/089532, 2005.

[5] S. Chen, Y. C. Chiu, and Y. C. Chang, "Development of a Tubing-Grafting robotic system for fruit-bearing vegetable seedlings," *Applied Engineering in Agriculture*, vol. 26, no. 4, pp. 707–714, 2010.

[6] J. M. Lee, C. Kubota, S. J. Tsao et al., "Current status of vegetable grafting: diffusion, grafting techniques, automation," *Scientia Horticulturae*, vol. 127, no. 2, pp. 93–105, 2010.

[7] Y.-C. Chang, S. Chen, Y.-C. Chiu, L.-H. Lin, and Y.-S. Chang, "Growth and union acclimation process of sweet pepper grafted by a tubing-grafting robotic system," *Horticulture Environment and Biotechnology*, vol. 53, no. 2, pp. 93–101, 2012.

[8] Z. Libin, Y. Qinghua, B. Guanjun et al., "Overview of research on agricultural robots in China," *International Journal of Agricultural and Biological Engineering*, vol. 1, no. 1, pp. 12–21, 2008.

[9] C. Kubota, M. A. McClure, N. Kokalis-Burelle, M. G. Bausher, and E. N. Rosskopf, "Vegetable grafting: history, use, and current technology status in North America," *HortScience*, vol. 43, no. 6, pp. 1664–1669, 2008.

[10] M. Li, C. Tang, F. Xie, M. Wu, and S. Sun, "Experimental study on mechanical properties of peach seedling," *Transactions e of the Chinese Society of Agricultural Engineering*, vol. 21, no. 3, pp. 29–33, 2005.

[11] Y. Zhao, T. Zhang, and L. Yang, "An experimental study on cutting resistance of root stock of branch to branch graft," *Agricultural Mechanization Research*, vol. 30, no. 7, pp. 146–148, 2008.

[12] M. Li, S. Dai, C. Tang, and Y. Xiang, "Simulation test on plantlet-cutting mechanism of grafting robot," *Transactions of the Chinese Society of Agricultural Engineering*, vol. 24, no. 6, pp. 129–132, 2008.

[13] Y. Zhao, T. Zhang, and H. Wang, "Cutting mechanism of root parental stock in automatic sapling grafting machine," *Transactions of the Chinese Society of Agricultural Engineering*, vol. 24, no. 9, pp. 79–83, 2008.

[14] X. Jiang, "Design of model PJJ-50 grape grafting machine," *Agricultural Equipment & Vehicle Engineering*, vol. 11, pp. 7–9, 2011.

[15] F. Wang, M. Liu, X. Wu et al., "Study and design of grafting robot for *Camellia oleifera* seedlings," *Forest Machine and Equipment*, vol. 39, no. 4, pp. 36–39, 2011.

[16] Q. Sun, D. Zhao, and Y. Zhao, "Mechanical design and test for a sapling grafting robot based on the Cleft method," *Journal of Henan Technology University*, vol. 34, no. 1, pp. 69–71, 2013.

22

Tracking Mobile Robot in Indoor Wireless Sensor Networks

Liping Zhang,[1] Cheng-Chew Lim,[2] Yiping Chen,[1] and Hamid Reza Karimi[3]

[1] *College of Electronic and Electrical Engineering, Shanghai University of Engineering Science, Song jiang District, Shanghai 201620, China*
[2] *School of Electrical and Electronic Engineering, The University of Adelaide, Adelaide, SA 5005, Australia*
[3] *Department of Engineering, Faculty of Engineering and Science, University of Agder, N-4898 Grimstad, Norway*

Correspondence should be addressed to Hamid Reza Karimi; hamid.r.karimi@uia.no

Academic Editor: Xiaojie Su

This work addresses the problem of tracking mobile robots in indoor wireless sensor networks (WSNs). Our approach is based on a localization scheme with RSSI (received signal strength indication) which is used widely in WSN. The developed tracking system is designed for continuous estimation of the robot's trajectory. A WSN, which is composed of many very simple and cheap wireless sensor nodes, is deployed at a specific region of interest. The wireless sensor nodes collect RSSI information sent by mobile robots. A range-based data fusion scheme is used to estimate the robot's trajectory. Moreover, a Kalman filter is designed to improve tracking accuracy. Experiments are provided to assess the performance of the proposed scheme.

1. Introduction

Mobile robots have the capability of moving around in their environment and are able to perform tasks without being fixed to one physical location. Some examples of mobile robots are domestic robots and automatic guided vehicles (AGV), which are used today in the home, industry, the military, and security environments. In fact, there is a high demand for mobile robots to use in hard-to-reach and hazardous areas [1]. In such scenarios, mobile robots are often operated semiautonomously or remote-controlled from a safe distance. In order to ensure mobile robots operate correctly, system must ensure a constant monitoring in mobile robots' motion parameters.

The problem of mobile robot tracking refers to the extraction, detection, identification, and tracking of the mobile robot from its environment to obtain motion parameters such as position, trajectory, velocity, and acceleration of the mobile robot. In many mobile robot applications, the robot should know its motion parameters in order to complete some typical tasks. For example, it is crucial for a traction control robot to know its motion parameters before changing its speed and moving direction in order to walk in a specified trajectory.

There are two solutions to tracking indoor mobile robots. One is to start at a known location and track the robot's position locally using methods such as odometry or inertial navigation [1]. Although these methods have been proven to be efficient and provide good short term position estimates, they suffer from unbounded error growth due to the integration of minute measurements to obtain the final estimate [2]. The other is to estimate the robot position globally using external sensors. GPS provides a sufficient and reliable way to track mobile robots, but it cannot be used indoors. Wireless sensor networks (WSNs) are often used to track indoor mobile robots.

WSNs deploy a large number of wireless sensor nodes which can sense, process, and communicate in a specified field. WSNs are used in monitoring and detecting specific events in specific environments. In general, WSN nodes have some characteristics as follows: (1) limited power: most wireless nodes are equipped with a battery and it is difficult to change or charge the battery; (2) fault tolerance: WSN should complete its task even if some wireless nodes malfunction or leave and the network is under enemy intrusion and interference; (3) dynamic topology: wireless nodes are often deployed at random; WSN should autonomously form a network to transfer interested information effectively; (4) ability to withstand harsh environmental conditions: wireless sensor nodes are often deployed in harsh environments such as a battle field, a nuclear plant, and outdoor environments; (5) heterogeneity of nodes: in a WSN, there are often different nodes which have different sensor equipment, battery levels, communication ability, and so on; and (6) large scale of deployment: WSN often consists of millions of inexpensive sensor nodes.

Target tracking is a typical WSN application. For example, it is especially useful in such applications as tracking emergency rescue workers, tracking military targets in modern battlefield scenarios, and monitoring traffic in transportation systems. In addition to the general advantage of tracking indoor mobile robots, WSN is able to forward the gathered tracking data from the deployed sensors to a central observation point, which is needed in most scenarios. Target tracking in wireless sensor networks has received more and more attention in recent years [3]. The accuracy of localization and saving energy are always two critical issues because of the limited power and wireless bandwidth of sensor nodes.

The main contribution of this paper is to develop a realistic tracking scheme. We consider the problem of tracking a mobile robot moving through a particular indoor target region. In our system, wireless sensor nodes equipped with a wireless transceiver are deployed in the target region. The wireless sensor nodes keep listening. If they detect a tracking request from a mobile robot, they sample the signal strength of the tracking request and relay the measured samples to a central observation point. The moving robot periodically broadcasts a tracking request as it travels, which is captured by the wireless sensor nodes that are located near the robot. A location algorithm as well as a Kalman filter is used to track the mobile robot in the central observation point.

In this work, we have conducted experiments on this tracking system to study tracking performance. The metric that we have used to evaluate the performance of the system is track estimation error, that is, the distance between the estimated and the actual location of the mobile robot.

The rest of this paper is organized as follows. Section 2 summarizes related work. In Section 3, we propose a framework for tracking mobile robots and give a brief overview of our problem and the discrete-time linear state-space model of mobile robot. RSSI (received signal strength indication) based moving robot position estimation is presented in Section 4. Section 5 details the Kalman filter which is used to improve tracking accuracy. Section 6 introduces the experiment and results. Finally, Section 7 concludes the paper.

2. Related Works

We address the data fusion algorithm in the indoor mobile robot tracking application of WSN. Mobile robot tracking is necessary when it does not follow a predictable path. Mobile robot tracking solutions require a real-time location estimation algorithm. Mobile robot tracking can be viewed as a sequential target location estimation problem. Therefore, most target localization approaches such as received signal strength (RSS), time of arrival (TOA), time difference of arrival (TDOA), signal angle of arrival (AOA), and their combinations [4, 5] are able to track mobile robots. Meanwhile, different filters such as the Kalman filter, the extended Kalman filter, and the particle filter were used to improve tracking accuracy.

Klančar et al. [6] proposed a vision system for mobile robot tracking. Ohmi et al. [7] implemented, tested, and compared different calibration techniques (linear and nonlinear methods, with and without lens distortion) in 3D-particle tracking velocimetry. Mazo Jr. et al. [8] presented a sensor fusion scheme that tracks the motion of a moving target by a team of mobile robots. Kuang and Morris [9] put forward an ultrasonic tracking system which can track high-speed robot motion. The ultrasonic tracking system estimates the frequency of the interference signal and a wavelet filter constructed with wavelet packet filter banks to improve tracking accuracy. All robots in the scheme have a directional sensor. Langerwisch and Wanger [10] proposed a system which tracks a mobile robot in a feature-based map using a 2D laser rangefinder and wheel odometry. Ramya et al. [3, 11] proposed a cooperative tracking method which each sensor node stores neighbor node identifier, intersection points of sensing circle of the node. Zhao [12] and Bhatti and Xu [13] used the detection duration of each sensor as a weight and performed line fitting in a time window to estimate target position. Astudillo et al. [14] presented an optimization method inspired on the chemical reactions to address the tracking problem for the dynamic model of a unicycle mobile robot by integrating a kinematic and a torque controller based on fuzzy logic theory. Cui et al. [15] proposed a stochastic Lagrangian model to describe the motion of a two-link planar rigid robot manipulator in random vibration environment. Moreover, they designed a state feedback controller such that the error system is 4th moment exponentially practically stable. Kim et al. [16] proposed Kalman filter (KF) based recursive algorithms for distributed state estimate based on the sign of innovations (SOI). In the scheme, each sensor is used to detect the target sound and compute the state and observation predicted estimate. Meanwhile, some authors [3, 12, 13, 16–22] used the particle filters (PFs) for target tracking in WSN environments. However, though PF can have lower computational complexity [23, 24], it has the major drawback that the error propagates through the sensor network [16–23]. In order to compensate, as an improvement over PF, Ristic et al. [24] present a variational filter (VF) for target tracking which is claimed to be able to track targets in near random motion with high accuracy. Su et al. [25] formulate WSN topology design, subject to random link failures and to a communication cost constraint, as a constrained convex optimization problem which improves significantly the convergence speed of the consensus algorithm and can achieve the asymptotic performance of a nonrandom network at a fraction of the communication cost. Wu et al. [26] proposed a sliding mode control law which is synthesized to drive the system trajectories onto the predefined sliding surface in a finite time. Zhang et al. [27–30] proposed remote proportional-integral-derivative controller, H_∞ controller, and robust observer-based controller for networked control systems which are subject to network-induced delays and missing data.

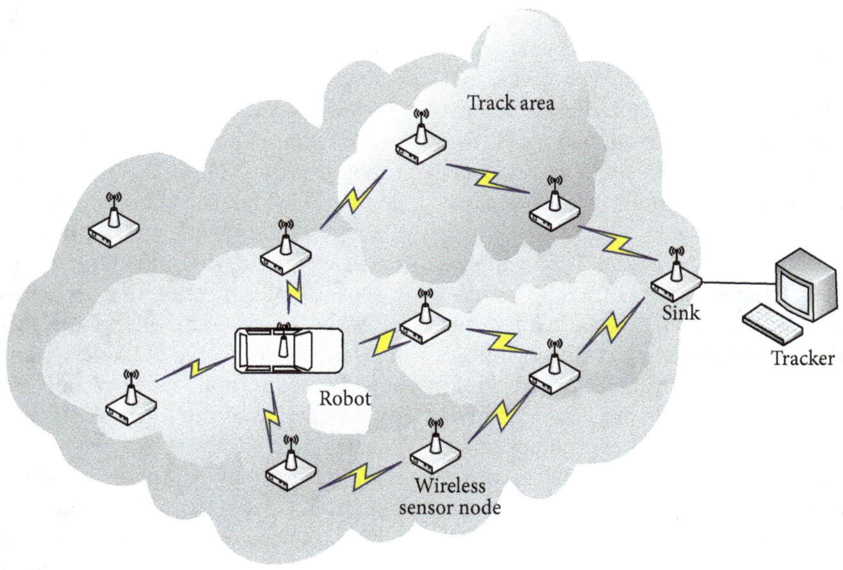

FIGURE 1: Proposed framework.

All the previously mentioned studies use the power law which excludes the Doppler effect so that their detection model and track results are applicable only for very low speed targets. Moreover, the computational complexity, especially for PF and VF, is so high that their trackers may be costly for real world WSN system.

3. System and Model

3.1. Robot Tracking Framework in Indoor Wireless Sensor Network. The framework proposed in this study is used to track mobile robots in an indoor wireless sensor network environment. As shown in Figure 1, there are some key network entities in the framework.

(1) *Wireless sensor nodes.* This is designed to realize three functions as follows. First, it gathers RSSI information of moving objects. Second, it realizes information fusion about position information of moving objects from different wireless sensor nodes in order to get more accurate position information and decrease the wireless sensor network load. Last, all of them cooperate to transfer position information of moving objects to sink. All of them are equipped with a Zigbee module.

(2) *Sink.* It is a specialized wireless sensor node. It not only has the functions as other wireless sensor nodes, but also it can transfer information to tracker.

(3) *Tracker.* It is designed to gather position and path information of moving objects and show them to interested people.

(4) *Mobile Robot.* It is a wireless sensor node located at a robot such as vehicle. It sends information frequently through a Zigbee module in order that wireless sensor nodes can receive its RSSI information.

It should be noticed that if moving objects leave track area as shown in Figure 1, the wireless sensor network may not receive RSSI information of moving objects, and the tracker will not show its position.

3.2. Problem Statement. We consider a sensor network of n anchored nodes at the positions denoted by a set of 2-dimensional vectors. The anchor nodes are deployed in known locations to provide RSSI information. A moving robot travels nearby, whose maneuver is not known in advance. The moving robot is equipped with a signal emitter whose signal transmission is measured by the m anchor sensor nodes out of all n anchored nodes. The RSSI information acquired by the m anchor nodes is sent to a central observation point. The central observation point executes the tracking algorithm and produces the estimates.

In order to describe the moving robot's kinematics and its dynamical behavior, the mobile robot tracking problem is formed in a discrete-time linear state-space approach.

3.3. Discrete-Time Linear State-Space Model. The state equation describes the time evolution of the state-vector. The system is considered to run for a total amount of time denoted as T_a, which is divided into several time steps in terms of the sampling period. The state vector x represents all the relevant information required to describe the system. Hence, at each time step k, the state vector $x_k = \begin{bmatrix} x & y & v_x & v_y \end{bmatrix}^T$ to be estimated comprises of the planar coordinates and the two axis velocities of the mobile robot. At each time step k an integer regime variable indicates which model is in use during the sampling period from $(k-1, k)$.

The model can be written as a standard discrete-time linear Gaussian state-space equation for continuous robot movement:

$$x_k = Ax_{k-1} + W_k, \tag{1}$$

where

$$A = \begin{bmatrix} 1 & 0 & T & 0 \\ 0 & 1 & 0 & T \\ 0 & 0 & 1 & 0 \\ 0 & 0 & 0 & 1 \end{bmatrix} \quad (2)$$

is state matrix and T is the time interval between samples.

W_k is the disturbance transition vector defined as a two-dimensional random speed vector with covariance matrix being $Q_v(k)$ while $W_k = [0 \ \ 0 \ \ w_x \ \ w_y]^T$.

The measurements vector contains the measurements that are made available to the system. The observations in this case correspond to the position of the mobile robot. Thus, at each time step k, the measurements equation provides the mathematical association between the observations vector y_k and the state-vector x_k:

$$y_k = Hx_k + v_k, \quad (3)$$

where $y_k = [x \ \ y]^T$ is the observation vector of the moving robot position. H is the time-invariant measurement matrix mapping the state of the system to observation vector where

$$H = \begin{bmatrix} 1 & 0 & 0 & 0 \\ 0 & 1 & 0 & 0 \end{bmatrix} \quad (4)$$

is the output matrix and v_k is a noise vector sampled from a known distribution that represents the additive measurements noise.

In order to get an estimation of y_k, a RSSI-based moving robot position estimation algorithm is used in this study.

4. RSSI-Based Moving Robot Position Estimation

There are two categories of algorithm to estimate target position in wireless sensor networks. The former is range-based algorithms that use absolute point-to-point distance estimates (range) or angle estimates for calculating location. The latter is range-free algorithms that make no assumption about the availability or validity of such information. In general, range-based algorithms are more accurate but also more complicated. Because accuracy is very important in tracking a mobile robot, a range-based algorithm is adopted in this study.

As a range-based positioning system, RSSI can be read from almost every transceiver of a wireless sensor network. This makes RSSI-based systems be of low cost and power economic since no extra hardware is needed. While "low-cost" and "power economical" are two key factors for wireless sensor network applications, however, it is difficult for RSSI-based systems to obtain accurate and high resolution position estimation. In wireless sensor networks, RSSI values are normally expressed in dBm. If the distance is very large, small variation of the RSSI value causes a large bias of distance estimation. Moreover, increasing RSSI from one quantized level to another level requires a long distance movement. Therefore, RSSI-based position systems have a good resolution at a near distance while having a bad resolution at a far distance.

As mentioned in Section 3, the m anchor nodes receive a tracking request from the moving robot at time step k. They measure RSSI and send it to a central observation point. The central observation point executes the RSSI-based localization algorithm to estimate the moving robot position.

Transmitted signals of wireless sensor networks are attenuated in the communication channel between sender and receiver. Thus, the RSSI at the receiver is lower than the emitted signal by the sender. This dependency can be described with the log-normal-shadowing model [31]:

$$P_{r,\mathrm{dB}}(d) = P_{r,\mathrm{dB}}(d_0) - \eta 10 \log\left(\frac{d}{d_0}\right) + X_{\sigma,\mathrm{dB}}, \quad (5)$$

where $P_{r,\mathrm{dB}}(d)$ is the signal strength measured at receiver (dBm), d is the reference distance between transmitter and receiver, $P_{r,\mathrm{dB}}(d_0)$ is the received power strength at a reference point d_0 (dBm), and $X_{\sigma,\mathrm{dB}}$ is the normal random variable with a standard deviation of σ. The value of σ depends on building construction and the number of partitions the signal has to travel. The range of σ is 3 dBm up to 20 dBm. η is the path loss exponent, between 2.5 and 3.0 in the indoor environment.

In order to get the reference distance between the transmitter and the receiver, (5) can be rewritten as

$$d = 10^{[P_{r,\mathrm{dB}}(d_0) - P_{r,\mathrm{dB}}(d) + X_{\sigma,\mathrm{dB}}]/(10\eta)} * d_0. \quad (6)$$

Therefore, all the distances between transmitter (moving robot) and receiver (m anchor sensor nodes) can be calculated by (6) if signal strength measured at receiver is measured.

However, RSSI measurements are highly error-prone due to the following reasons. First, several reflections of radio waves at obstacles cause multipath propagation. Second, other electrical fields in the environment interfere with the transmitted radio waves. Third, the resolution of the RSSI can be limited by the transceiver hardware. Finally, the required line of sight between transmitter and receiver cannot be guaranteed.

In order to improve location accuracy, an ML (maximum likelihood) algorithm is used to minimize the differences between the measured and estimated range between the moving robot and the m anchor node.

The difference between the measured and estimated distance is formed by

$$e_i = d_i - \sqrt{(x_i - x)^2 + (y_i - y)^2}, \quad (7)$$

where e_i is the difference between measured and estimated distance between moving robot and anchor node i, $i = 1, 2, \ldots, m$, d_i is the estimated distance between moving node and anchor node i and can be calculated by (6). (x, y) is the position of the moving robot, and (x_i, y_i) is the position of anchor nodes i.

When $e_i = 0$, (7) can be rewritten as

$$d_i^2 - x_i^2 - y_i^2 = x^2 + y^2 - 2x_i x - 2y_i y. \quad (8)$$

We can use the first $m-1$ equations to subtract the mth equation as follows:

$$\begin{aligned} & d_i^2 - x_i^2 - y_i^2 - \left(d_N^2 - x_N^2 - y_N^2\right) \\ & \quad = 2x_N x + 2y_N y - 2x_i x - 2y_i y. \end{aligned} \tag{9}$$

Equation (9) can be rewritten in matrix form as follows:

$$Y = XD, \tag{10}$$

where

$$\begin{aligned} Y &= \begin{bmatrix} d_1^2 - x_1^2 - y_1^2 - \left(d_N^2 - x_N^2 - y_N^2\right) \\ \vdots \\ d_{N-1}^2 - x_{N-1}^2 - y_{N-1}^2 - \left(d_N^2 - x_N^2 - y_N^2\right) \end{bmatrix} \\ X &= \begin{bmatrix} 2\left(x_N - x_1\right) & 2\left(y_N - y_1\right) \\ \vdots & \vdots \\ 2\left(x_N - x_{N-1}\right) & 2\left(y_N - y_{N-1}\right) \end{bmatrix} \\ D &= \begin{bmatrix} x \\ y \end{bmatrix}. \end{aligned} \tag{11}$$

Therefore, the position of the moving robot D can be calculated by

$$D = \left(X^T X\right)^{-1} X^T Y. \tag{12}$$

5. Kalman Filter

Because the algorithm mentioned before is easily subject to errors due to the RSSI measurements which are highly error-prone, a Kalman Filter is introduced to improve tracking accuracy. The Kalman Filter has made a dramatic impact on linear estimation because of its adaptivity for implementation on a digital computer for on-line estimation and usefulness of the state-space approach. It is widely applied in the fields of navigation too.

The Kalman Filter uses a form of feedback control to estimate the state of the system. The process through which we obtain the kth state is divided into two main parts, the predictor and the corrector. The predictor is responsible for letting the current state and error covariance estimates advance in time, to calculate the a priori estimates for the next time step. The corrector is the feedback and thus introduces a new measurement into the a priori estimate to obtain an improved a posteriori estimate.

The predictor is used to calculate the a priori state by (1) and the a priori estimation error covariance matrix $P_{k+1|k}$ with

$$P_{k+1|k} = AP_{k|k}A^T + Q_{k|k}. \tag{13}$$

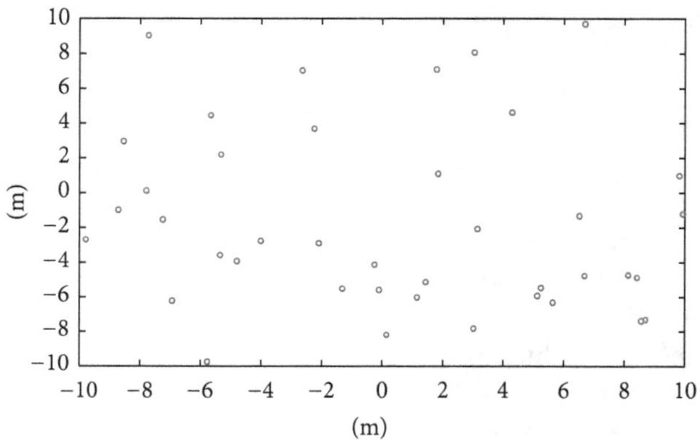

FIGURE 2: Topology of anchor nodes.

The *corrector* is used to calculate the a posteriori state estimate and the a posteriori error covariance estimate as follows:

$$\begin{aligned} \delta_k &= y_k - \hat{y}_{k|k-1} \\ Pd_{k+1|k} &= HP_{k+1|k}H^T + R_k \\ \alpha_{k+1} &= HP_{k+1|k}H^{-1} \\ P_{k+1|k+1} &= P_{k+1|k} - \alpha_{k+1}Pd_{k+1|k}\alpha_{k+1}^T \\ \hat{x}_{k+1|k} &= Ax_k \\ y_{k+1|k} &= H\hat{x}_{k+1|k} \\ \hat{x}_{k+1|k+1} &= \hat{x}_{k+1|k} + \alpha_{k+1}\delta_{k+1} \\ \hat{y}_{k+1|k+1} &= H\hat{x}_{k+1|k+1}. \end{aligned} \tag{14}$$

A detailed derivation of algorithm can be found in [32].

6. Experiment

In this section, we present and analyze the experiment results of our scheme.

The moving robot tracking system is tested in a chamber. The chamber is 20 meters × 20 meters (length × width). There are 40 anchor nodes spread over the tracking area at random. The topology of anchor nodes is shown in Figure 2.

The anchor node is a CC2530 module as shown in Figure 3. CC2530 is a system-on-chip (SoC) solution for IEEE 802.15.4, Zigbee applications. The CC2530 combines the excellent performance of a leading RF transceiver with an industry standard enhanced 8051 MCU, in-system programmable flash memory, 8 KB RAM, and many other powerful features. CC2530 is also featured in its low power and supports short range wireless communication such as Zigbee [11]. The PHY layer and MAC layer of CC2530 comply with IEEE 802.15.4. Moreover, its work spectrum located in 2.4 GHz. It supports data rates as high as 250 kbps and multipoint to multipoint communication. It is also featured by hardware encryption, reliability, flexible networking, and high destroy-resistance. The anchor node is composed of processor, wireless transceiver, and power module as shown

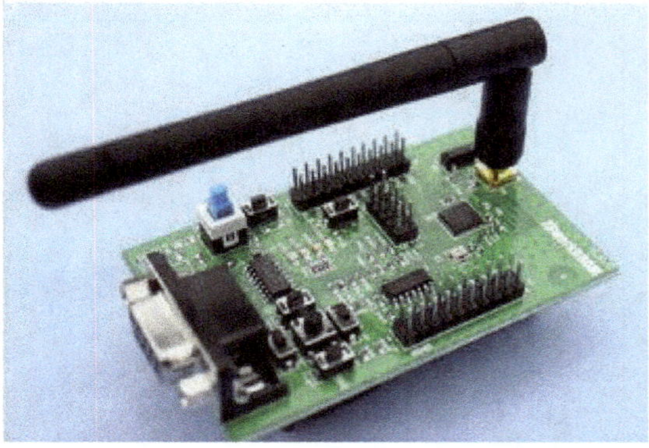

FIGURE 3: Prototype of anchor node.

FIGURE 5: Moving robot.

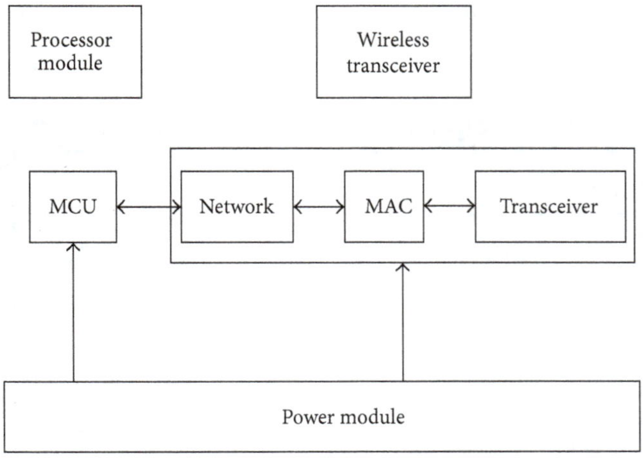

FIGURE 4: Block diagram of the prototype.

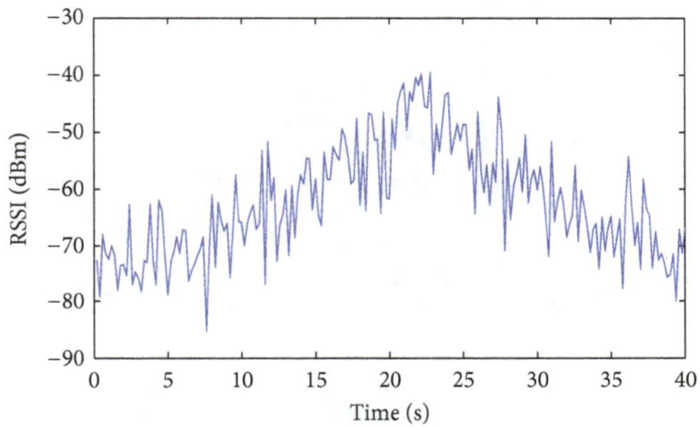

FIGURE 6: RSSI sample of an anchor node.

in Figure 4. Each module of this platform is described in Figures 3 and 4.

The moving robot is a remote programmable tracked mobile robot as shown in Figure 5. It is equipped with three wheels which are driven by three direct motors. Moreover, it is controlled by a STC11F32XE MCU which is featured in ultralow power. The MCU is connected to a CC2530 module; therefore, it can be manipulated via Zigbee to move forward, backward, turn left, and turn right. The robot is powered by 2 1.5 v batteries. It can be controlled by a touch screen control. The moving robot is moving from the start point (−10 m, 0 m) to end point (10 m, 0 m) across the tracking area.

Figure 6 is RSSI measured in an anchor node whose coordinate is (1.1, 1.2). As can be seen in Figure 6, RSSI is added by noises. Meanwhile, the peak point of RSS is −40 dBm in 21 seconds where the distance between anchor node and moving robot approaches the least. The low point of RSS is −82 dBm in 4 seconds where the distance between anchor node and moving robot approaches the largest.

In order to save computation resources, five nearest anchor nodes' RSSIs are used to estimate mobile robot locations according to (12). The result of tracking moving nodes is shown in Figure 6. The true trajectory and the positional estimates before and after applying the Kalman filter are described.

As can be seen in Figure 7, as the robot moves, the estimate with Kalman Filter converges towards the correct trajectory and ends very close to the true position compared to the estimate without Kalman filter. Moreover, the tracking accuracy with Kalman filter improves. For example, the biggest tracking estimation error without Kalman filter is 10 meters while the biggest tracking estimation error after applying Kalman filter is 7.3 meters. Therefore, the biggest tracking estimation error improves about 27%.

The tracking estimation error is shown in Figure 7 which shows the performance of estimate with and without Kalman filter. The mean tracking estimation error after applying Kalman filter is 1.10 while the mean tracking estimation error without Kalman filter is 2.23.

The true velocity of the robot and the velocity estimates before and after applying the Kalman filter are described in Figure 8. It can be seen from Figure 8 that the performance of velocity estimation improves greatly with the use of Kalman filter comparing to those without Kalman filter. Figure 9 shows the velocity estimation comparison.

The GUI of tracker is shown in Figure 10. Position and velocity are shown in the status bar.

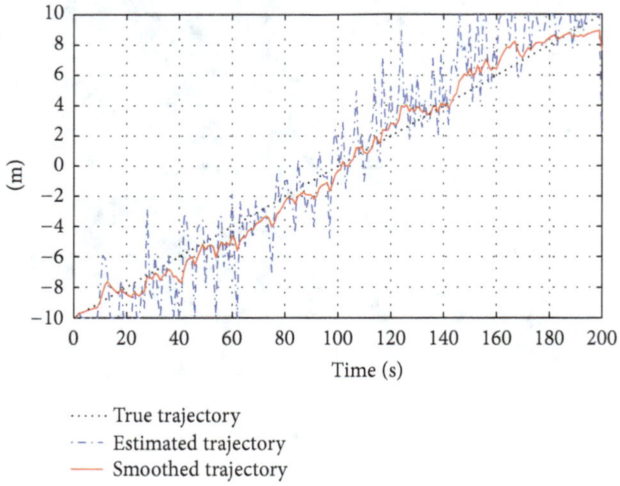

Figure 7: Trajectories comparison.

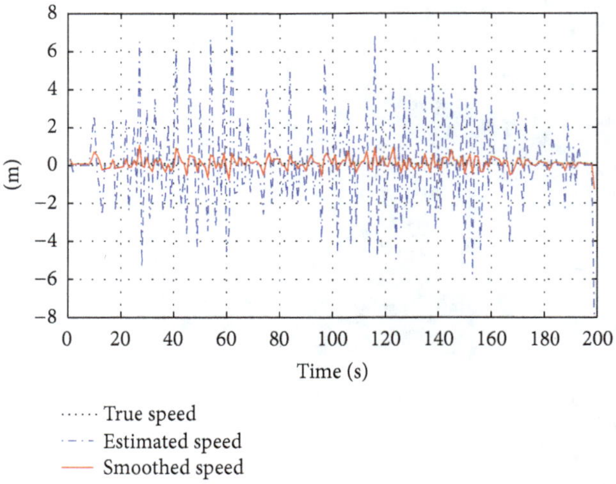

Figure 9: Velocity estimation comparison.

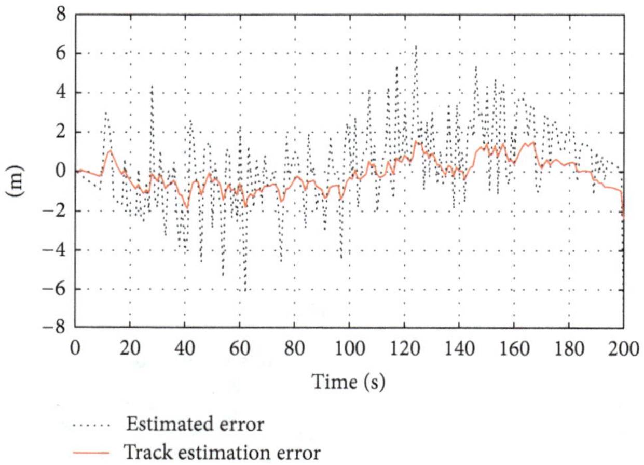

Figure 8: Tracking estimation error.

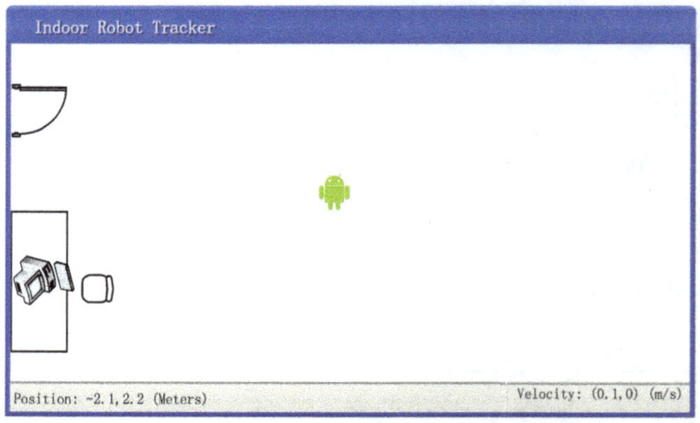

Figure 10: Tracker GUI.

7. Conclusion

In this study, an indoor moving robot tracking system implemented on a Zigbee WSN has been proposed. The experimental results prove that it is able to obtain good results at low speeds, which are typical in indoor robot tracking applications.

In fact, tracking accuracy in WSN presents several tradeoffs. If higher accuracy is desired, one has to deploy more anchor nodes or use more samples. However, a large number of anchor nodes and samples may increase energy consumption and communication payloads which may lead to shortening the WSN's lifetime. Therefore, we should deploy anchor nodes or use samples according to application requirements.

Conflict of Interests

The authors declare that there is no conflict of interests regarding the publication of this paper.

Acknowledgments

The projects have been supported by the Innovation Program of Shanghai Municipal Education Commission "Vehicle Collision Avoidance System Based on Vehicle Wireless Communication" (no. 12YZ151), "Discipline Comprehension Construction on Control Science and Engineering" (no. 13XKCZ07), and "First Class Discipline Cultivation-Mechanical Engineering" (no. YLJX12-7) and Shanghai University of Engineering Science's project "Vehicle Tracking System Based on VANET" (no. 2011XY18).

References

[1] K. S. Chong and L. Kleeman, "Accurate odometry and error modelling for a mobile robot," in *Proceedings of the IEEE International Conference on Robotics and Automation (ICRA '97)*, pp. 2783–2788, April 1997.

[2] J. Borenstein, H. R. Everett, and L. Feng, *Navigating Mobile Robots:Systems and Techniques*, A. K. Peters, 1996.

[3] K. Mechitov, S. Sundresh, Y. Kwon, and G. Agha, "Cooperative tracking with binary detection sensor networks," Tech.

Rep. UIUCDCS-R-2003-2379, University of Illinois at Urbana-Champaign, 2003.

[4] A. H. Sayed, A. Tarighat, and N. Khajehnouri, "Network-based wireless location: challenges faced in developing techniques for accurate wireless location information," *IEEE Signal Processing Magazine*, vol. 22, no. 4, pp. 24–40, 2005.

[5] N. Patwari, J. N. Ash, S. Kyperountas, A. O. Hero III, R. L. Moses, and N. S. Correal, "Locating the nodes: cooperative localization in wireless sensor networks," *IEEE Signal Processing Magazine*, vol. 22, no. 4, pp. 54–69, 2005.

[6] G. Klančar, M. Kristan, and R. Karba, "Wide-angle camera distortions and non-uniform illumination in mobile robot tracking," *Robotics and Autonomous Systems*, vol. 46, no. 2, pp. 125–133, 2004.

[7] K. Ohmi, B. Joshi, and S. P. Panday, "Comparative study of camera calibration models for 3D particle tracking velocimetry," *International Journal of Innovative Computing Information and Control*, vol. 9, no. 5, pp. 1971–1986, 2013.

[8] M. Mazo Jr., A. Speranzon, K. H. Johansson, and X. Hu, "Multi-robot tracking of a moving object using directional sensors," in *Proceedings of the IEEE International Conference on Robotics and Automation*, vol. 2, pp. 1103–1108, May 2004.

[9] W. T. Kuang and A. S. Morris, "Using short-time Fourier transform and wavelet packet filter banks for improved frequency measurement in a Doppler robot tracking system," *IEEE Transactions on Instrumentation and Measurement*, vol. 51, no. 3, pp. 440–444, 2002.

[10] M. Langerwisch and B. Wagner, "Guaranteed mobile robot tracking using robust interval constraint propagation," in *Intelligent Robotics and Applications*, vol. 7507 of *Lecture Notes in Computer Science*, pp. 354–365, 2012.

[11] K. Ramya, K. P. Kumar, and V. S. Rao, "A survey on target tracking Techniques in wireless sensor networks," *International Journal of Computer Science & Engineering Survey*, vol. 3, no. 4, pp. 93–108, 2012.

[12] C. Zhao, "Zigbee communication design based on CC2420," *Microprocessor Information*, vol. 23, no. 1, 2007.

[13] S. Bhatti and J. Xu, "Survey of target tracking protocols using wireless sensor network," in *Proceedings of the 5th International Conference on Wireless and Mobile Communications (ICWMC '09)*, pp. 110–115, August 2009.

[14] L. Astudillo, P. Melin, and O. Castillo, "Chemical optimization paradigm applied to a fuzzy tracking controller for an autonomous mobile robot," *International Journal of Innovative. Computing, Information and Control*, vol. 9, no. 5, pp. 2007–2018, 2013.

[15] M.-Y. Cui, Z.-J. Wu, and X.-J. Xie, "Stochastic modeling and tracking control for a two-link planar rigid robot manipulator," *International Journal of Innovative Computing Information and Control*, vol. 9, no. 4, pp. 1769–1780, 2013.

[16] W. Kim, K. Mechitov, J.-Y. Choi, and S. Ham, "On target tracking with binary proximity sensors," in *Proceedings of the 4th International Symposium on Information Processing in Sensor Networks (IPSN '05)*, pp. 301–308, April 2005.

[17] Z. Wang, E. Bulut, and B. Szymanski, "A distributed cooperative target tracking with binary sensor networks," in *Proceedings of the IEEE International Conference on Communications (ICC '08)*, pp. 306–310, Beijing, China, May 2008.

[18] Z. Wang, E. Bulut, and B. K. Szymanski, "Distributed target tracking with imperfect binary sensor networks," in *Proceedings of the IEEE Global Telecommunications Conference (GLOBECOM '08)*, pp. 1–5, New Orleans, Lo, USA, December 2008.

[19] A. Ribeiro, G. B. Giannakis, and S. I. Roumeliotis, "SOI-KF: distributed Kalman filtering with low-cost communications using the sign of innovations," *IEEE Transactions on Signal Processing*, vol. 54, no. 12, pp. 4782–4795, 2006.

[20] P. M. Djurić, M. Vemula, and M. F. Bugallo, "Signal processing by particle filtering for binary sensor networks," in *Proceedings of the IEEE 11th Digital Signal Processing Workshop and 2nd IEEE Signal Processing Education Workshop*, pp. 263–267, Taos Ski Valley, NM, USA, August 2004.

[21] P. M. Djurić, M. Vemula, and M. F. Bugallo, "Target tracking by particle filtering in binary sensor networks," *IEEE Transactions on Signal Processing*, vol. 56, no. 6, pp. 2229–2238, 2008.

[22] N. Ahmed, M. Rutten, T. Bessell, S. S. Kanhere, N. Gordon, and S. Jha, "Detection and tracking using particle-filter-based wireless sensor networks," *IEEE Transactions on Mobile Computing*, vol. 9, no. 9, pp. 1332–1345, 2010.

[23] J. Teng, H. Snoussi, and C. Richard, "Decentralized variational filtering for target tracking in binary sensor networks," *IEEE Transactions on Mobile Computing*, vol. 9, no. 10, pp. 1465–1477, 2010.

[24] B. Ristic, S. Arulampalam, and N. Gordon, *Beyond the Kalman Filters for Tracking Applications*, Artech House, 2004.

[25] X. J. Su, L. G. Wu, and P. Shi, "Sensor networks with random link failures: distributed filtering for T-S fuzzy systems," *IEEE Transactions on Industrial Informatics*, vol. 9, no. 3, pp. 1739–1750, 2013.

[26] L. Wu, W. X. Zheng, and H. Gao, "Dissipativity-based sliding mode control of switched stochastic systems," *IEEE Transactions on Automatic Control*, vol. 58, no. 3, pp. 785–793, 2013.

[27] H. Zhang, Y. Shi, and A. S. Mehr, "Robust $\mathscr{H}_\infty$ PID control for multivariable networked control systems with disturbance/noise attenuation," *International Journal of Robust and Nonlinear Control*, vol. 22, no. 2, pp. 183–204, 2012.

[28] H. Zhang, Y. Shi, and A. Saadat Mehr, "Robust static output feedback control and remote PID design for networked motor systems," *IEEE Transactions on Industrial Electronics*, vol. 58, no. 12, pp. 5396–5405, 2011.

[29] H. Zhang, Y. Shi, and M. X. Liu, "H_∞ step tracking control for networked discrete-time nonlinear systems with integral and predictive actions," *IEEE Transactions on Industrial Informatics*, vol. 9, no. 1, pp. 337–345, 2012.

[30] H. Zhang, Y. Shi, and J. Wang, "Observer-based tracking controller design for networked predictive control systems with uncertain Markov delays," *International Journal of Control*, vol. 86, no. 10, pp. 1824–1836, 2013.

[31] J. Goldhirsh and W. J. Vogel, "Handbook of Propagation Effects for Vehicular and Personal Mobile Satellite Systems, Over of Experimental and Modelling Results," http://www.utexas.edu/Research/mopro/index.html.

[32] P. Zarchan and H. Musoff, *Fundamentals of Kalman Filtering: A Practical Approach*, American Institute of Aeronautics & Ast, 3rd edition, 2009.

23

Cam Drive Step Mechanism of a Quadruped Robot

Qun Sun, Chong Wang, Dongjie Zhao, and Cuihua Zhang

School of Mechanical and Automotive Engineering, Liaocheng University, Liaocheng, Shandong 252000, China

Correspondence should be addressed to Qun Sun; sunxiaoqun97@163.com

Academic Editor: Shahram Payandeh

Bionic quadruped robots received considerable worldwide research attention. For a quadruped robot walking with steady paces on a flat terrain, using a cam drive control mechanism instead of servomotors provides theoretical and practical benefits as it reduces the system weight, cost, and control complexities; thus it may be more cost beneficial for some recreational or household applications. This study explores the robot step mechanism including the leg and cam drive control systems based on studying the bone structure and the kinematic step sequences of dog. The design requirements for the cam drive robot legs have been raised, and the mechanical principles of the leg operating mechanism as well as the control parameters have been analyzed. A cam drive control system was constructed using three cams to control each leg. Finally, a four-leg demo robot was manufactured for experiments and it showed stable walking patterns on a flat floor.

1. Introduction

Quadruped robots have been extensively studied as the most important branch of bionic robot applications. In 1968, General Electrics and the US Army Mobility Systems Laboratory constructed a quadruped walking machine, which used hydraulic server motors to drive [1]. It was later suggested that a large number of controllable degrees of freedom require highly efficient drives properly arranged, special design of feet to dissipate the energy of the strike, and so forth, and the problem of the control seems to be the main problem of the walking robot [2].

The first comprehensive quadruped robot, KUMO-I, was developed in 1976 by Japan's Tokyo Institute of Technology, which subsequently also produced TITAN series quadruped robots based on this study [3–6]. Currently quadruped robots are widely studied in many universities and laboratories around the world [7–13]. The most advanced quadruped robot so far is BigDog, jointly developed by Boston Dynamics and M.I.T. in 2005 for the US Army. BigDog is able to be operated in unstructured environment, with multiple capabilities such as independent walking, running, jumping, and climbing obstacles. An upgraded version, LS3, is currently in intense development [13].

Considered in a constructed environment with flat terrain, the robot performs steady walking, and it is possible to use a mechanical transmission system to replace the servomotors at the leg joints. This reduces the system weight, cost, and control complexities; thus it may be more cost beneficial for some recreational or household applications. For this reason, this study utilizes some previous research outcomes [14–16] and the bionic theories to improve the theoretical analysis and the component design and presents a novel cam drive quadruped robot step mechanism to achieve steady walking on an even floor surface.

2. Analysis of the Step Sequences of Typical Quadrupeds

2.1. Kinematic Bone Structure of Typical Quadrupeds. The animals that are most widely used in the bionic studies for quadruped robots are dogs. According to observations and analysis of dog walking, scientists investigated the dogs bone structures involved in body movement and walking, including the shoulder, elbow, hip, and knee joints [3].

The dog's shoulder can be regarded as a ball joint, which is able to tilt back and forth or yaw to the two sides, as shown in Figure 1. The dog's elbow is similar to an articulated joint structure, which can only move in a single plane, that is, tilting back and forth, to the maximum 90-degree angle as presented in Figure 2. Hip is another ball joint structure

FIGURE 1: Movements of dog's shoulder.

FIGURE 2: Movements of dog's elbow.

which can turn about three axes, as shown in Figure 3. Similar to elbow, knee is another articulated joint structure which can only tilt back and forth, as shown in Figure 4.

2.2. Step Sequence Analysis of Typical Quadrupeds. When quadrupeds steadily walk, the four legs move coordinately and each leg is in turn in two different states. The step sequence usually obeys the diagonal principle; that is, the left front leg moves first and then the right lateral, the left lateral, the right front, the left front leg, and so on in repeated cycles. When a quadruped is in the state of a supporting phase, there are at least three legs touching the ground, whilst in the hanging phase there is only one leg lifted above the ground. Therefore, regarding the motion of each leg in steady walking, the ratio between the time that the foot is on the ground and off ground, that is, the duty cycle, is 0.75.

The limiting condition of this study is that a quadruped robot walks following a steady and regular pace; that is, a robot moves in a steady manner and changing position is achieved by continuously shifting gaits following the fundamental modes. The step mechanisms described in this paper include leg mechanism and the cam drive control system that governs the motions of the leg joints. These two aspects will be presented in the following sections.

3. Theoretical Analysis of Robot Leg Mechanism

3.1. Design Requirements of Leg Mechanism. According to quadruped walking characteristics, the design requirements of the quadruped robot walking mechanism are given below [15].

(1) The movement of the foot end relative to the robot body is of a dome shape ⌓, where the straight line section is the motion trajectory of the foot supporting mechanism in the supporting phase. The curve section corresponds to the foot trajectory when it is lifted above the ground, that is, in the hanging phase.

(2) In order to avoid unnecessary energy consumption involved in body bumping when the robot moves, it is necessary to ensure certain straightness of the straight line section mentioned in (1).

(3) In order to avoid friction and unnecessary energy consumptions occurring due to uneven speed of three supporting legs when the robot moves, it is proposed to ensure that each leg moves in a steady speed within the straight line section mentioned in (1).

(4) Regarding the curve section in (1), there is a requirement about the ceiling point rather than the shape, since the maximum height determines the ability that the robot travels on an uneven surface.

(5) In addition to requirement (1), the time that a foot travels along the straight line section should be three times of that used in the curved section. This means that the phase angle of the supporting phase is $3\pi/2$ whilst for the hanging phase it is $\pi/2$. The coordinated motion of the four legs should obey the diagonal principle; that is, the difference of the phase angle between the two legs that move in turn should be $\pi/2$.

3.2. Mechanical Principle of the Leg Mechanism. The scope of this study is the step mechanism of a quadruped robot that can achieve steady walking, excluding turning. By simplifying the real leg joint structure based on the hip, elbow, and knee joints, the mechanical schematic of the leg structure can be shown in Figure 5.

In Figure 5, sections BC, O_2B, and O_1O_2, points B and O_2 correspond to the calf, thigh, hip bone, knee joint, and hip joint, respectively. The associated parameters are as follows [14–16]: (1) E is hip joint offset gap, (2) h_1 is calf length, (3) h_2 is thigh length, (4) S is step length, (5) θ_1 is hip joint turning angle, (6) θ_2 is step angle, and (7) θ_3 is knee joint turning angle. The designed leg parameters are (1) hip joint offset gap $E = 50$ mm, (2) calf length $h_1 = 220$ mm, (3) thigh length $h_2 = 180$ mm, and (4) step length $S = 150$ mm.

The leg mechanism required to achieve the movement described in Figure 5 needs the following structural criteria [15]: firstly assume the starting point is $\theta_1 = 0$ and $\theta_2 = 0$; thus if joint A moves to the right, then the foot C tends to move along a circle whose center is O_2 and the radius is $(h_1 + h_2)$. If point A maintains uniform motion in the horizontal

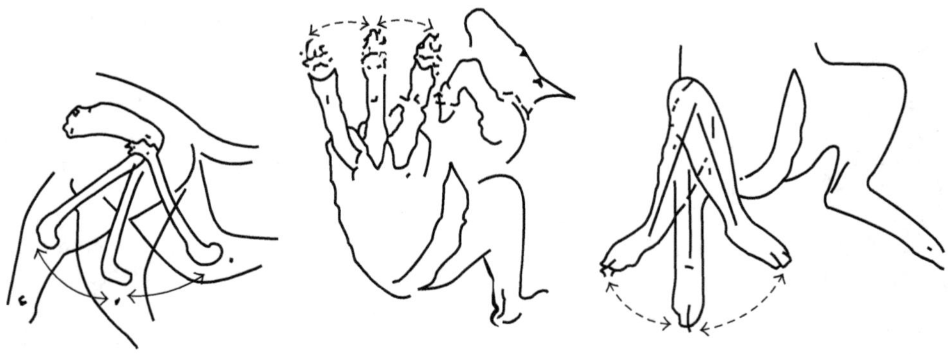

FIGURE 3: Movements of dog's hip.

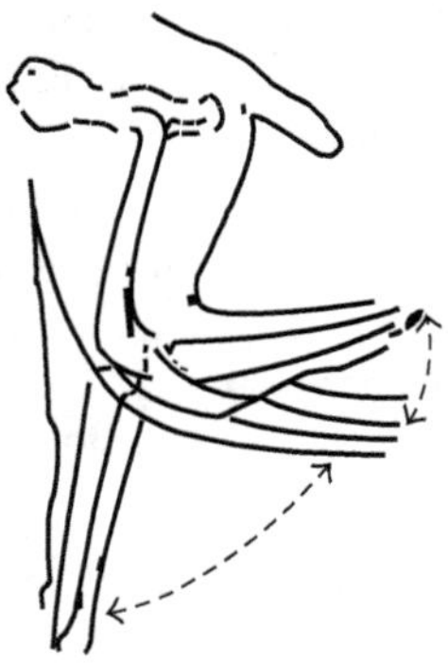

FIGURE 4: Movements of dog's knee.

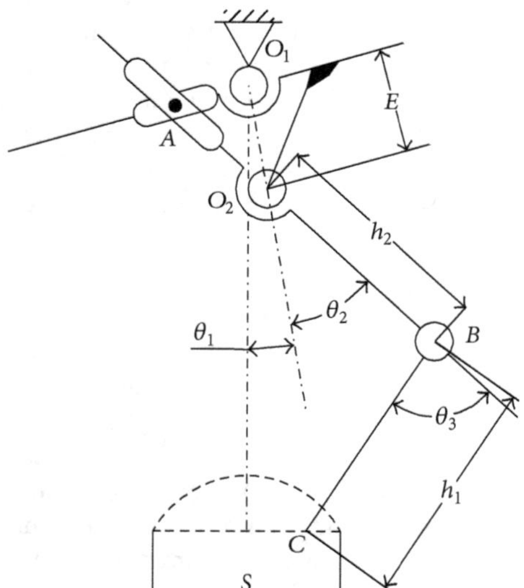

FIGURE 5: Schematic diagram of the walking mechanism.

direction, then the intersection between calf BC and the flat dash line in Figure 5 also maintains steady horizontal velocity. Therefore, if the hip joint angle θ_1 and the knee join θ_3 turn to enable the foot end C to move to this intersection point, then it satisfies the requirements (2) and (3) mentioned previously in Section 3.1. Next, move joint A quickly to the left, making sure the displacement equals that in the last procedure but the travelling time is only 1/3 of it and then enable foot end C to move along the curved dash line in Figure 5.

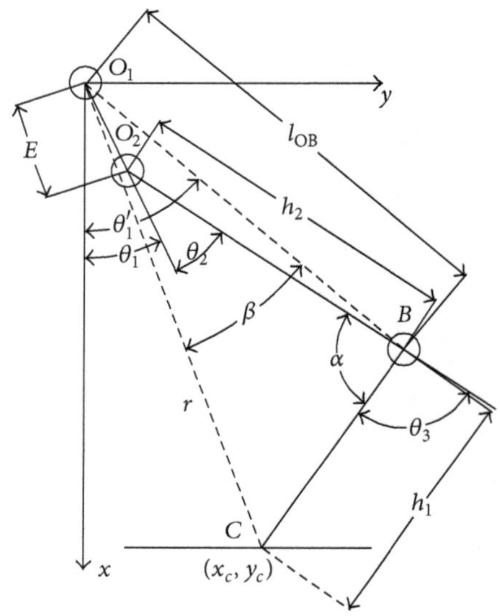

FIGURE 6: Schematic diagram of robot leg in the Cartesian coordinate system.

3.3. Joint Parameters and Drive Principle. As shown in Figure 6, in order to satisfy quadruped robot walking requirements, when a cam drive robot leg is in the hanging phase, it is only necessary to control the position of the foot (x_c, y_c). In the supporting phase, however, the relationships between the three angles θ_1, θ_2, and θ_3 should be clearly defined and controlled.

The parameter θ_2 should satisfy the following constraint [15]:

$$\theta_2 = \arctan \frac{y_c}{\sqrt{(x_c^2 + y_c^2) - (S/2)^2}}, \tag{1}$$

where S is the step length which is the same as that defined in Figure 5. The relationships between θ_1 and θ_3 relate to the inverse kinematic solution of the in-plane two-bar linkages

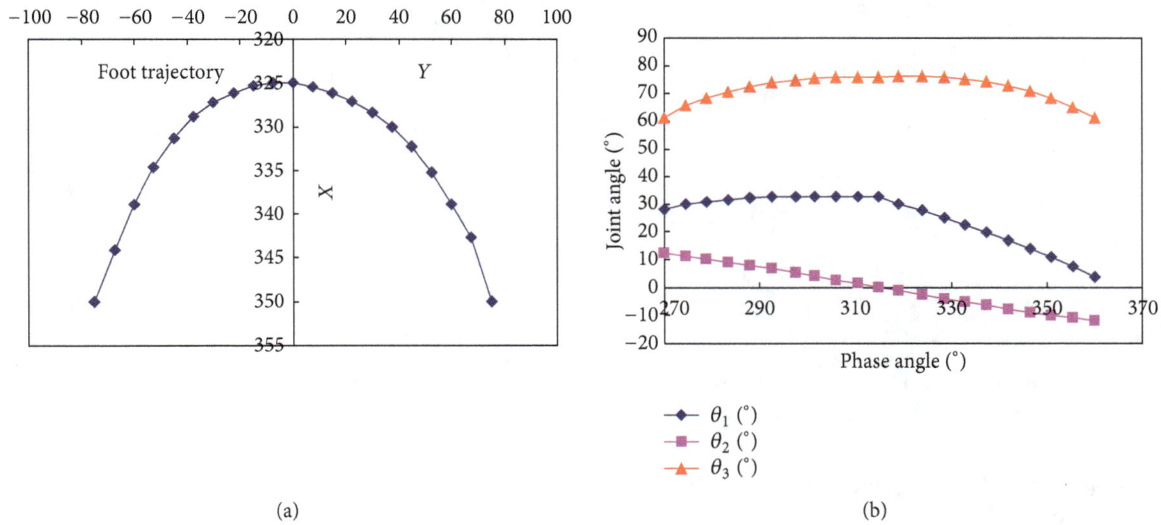

FIGURE 7: Key sample points chosen for cam control.

O_1B and BC. By analyzing in-plane geometry, the kinematics functions of the linkage are given by

$$x_c = l_{OB}\cos(\theta_1 + \angle O_2O_1B) + h_1\cos(\theta_1 + \angle O_2O_1B - \theta_3),$$
$$y_c = l_{OB}\sin(\theta_1 + \angle O_2O_1B) + h_1\sin(\theta_1 + \angle O_2O_1B - \theta_3). \tag{2}$$

The below relationship can be obtained from cosine law:

$$l_{OB}^2 = h_1^2 + E^2 - 2h_1E\cos\angle O_1O_2B, \tag{3}$$

and from law of sines the following is given:

$$l_{OB}\sin\angle O_2O_1B = h_1\sin\theta_2. \tag{4}$$

The inverse solution of the in-plane linkage kinematics function is about deriving θ_1 and θ_2 when (x_c, y_c) is given. It is possible to derive

$$\theta_1' = \arctan\left(\frac{y}{x}\right) \pm \beta, \tag{5}$$

where

$$\beta = \arccos\left[\frac{\left(r^2 + l_{OB}^2 - h_2^2\right)}{(2l_{OB}r)}\right],$$
$$r^2 = x_c^2 + y_c^2. \tag{6}$$

According to the actual situation of quadruped moving, either the calf and thigh remain in a straight line or the calf is behind the knee. However, since $\theta_1' = \arctan(y/x) - \beta$ is derived from an ideal theoretical situation without considering this aspect, (5) can be simplified as

$$\theta_1' = \arctan\left(\frac{y}{x}\right) + \beta. \tag{7}$$

Hence,

$$\theta_1 = \theta_1' - \angle O_2O_1B. \tag{8}$$

From

$$l_{O_2C}^2 = \sqrt{\left(X_c - X_{O_2}\right)^2 + \left(Y_c - Y_{O_2}\right)^2},$$
$$X_{O_2} = E\sin\theta_1,$$
$$Y_{O_2} = E\cos\theta_1, \tag{9}$$

the following can be derived:

$$\alpha = \arccos\left[\frac{\left(h_1^2 + h_2^2 - l_{O_2C}^2\right)}{(2h_1h_2)}\right]. \tag{10}$$

Therefore,

$$\theta_3 = \pi - \alpha. \tag{11}$$

Equations (1), (8), and (11) form the relationship functions between θ_1, θ_2, and θ_3. In the hanging phase, to control θ_1, θ_2, and θ_3, it is only necessary to know a few point coordinates. This paper chooses some key points for control purpose as shown in Figure 7, where each sample point was obtained by equally dividing the total phase angle $\pi/4$ of a hanging phase.

Given by the current technologies, using either electric, hydraulic, or pure mechanical drive methods can implement the control of θ_1, θ_2, and θ_3. In order to reduce the cost and control complexities, this paper uses 4 groups of cams to control the 3 joint parameters on each of the 4 legs, totally 12 parameters, which can enable quadruped robot steady walking on an even terrain.

4. Cam Drive and Control Mechanism of Quadruped Robot

4.1. The Principle of the Hip Joint Cam Kinematics. The movement principle of the hip joint cam drive control parameter θ_1 is shown in Figure 8.

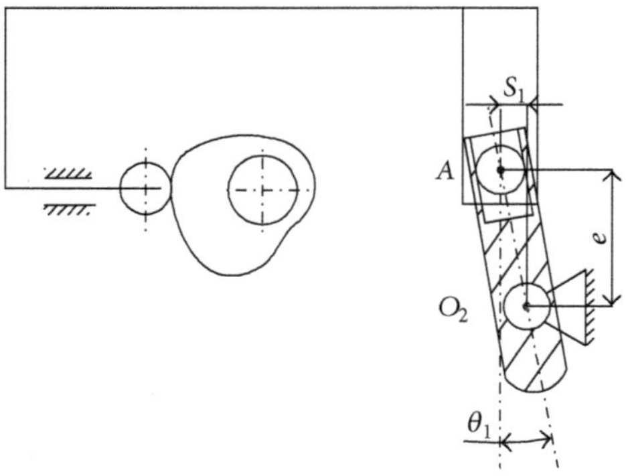

FIGURE 8: The principle of the hip joint cam kinematics.

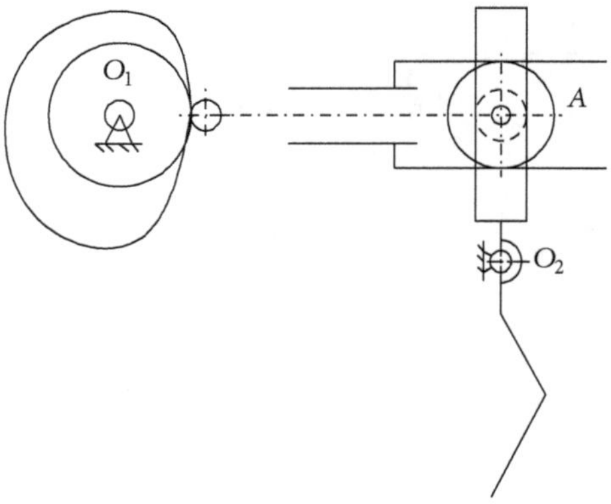

FIGURE 9: The principle of the main cam kinematics.

The moving part of the hip joint cam is subject to the below law of motion:

$$s_1 = e \cdot \left[\tan\theta_1 - \tan\theta_1^*\right]. \tag{12}$$

In this paper $e = 30$, where e is the vertical distance between points A and O_2, A is a movable pivot point, and O_2 is a fixed pivot point. The initial value of θ_1 is $\theta_1^* = 0.11768$, and in the hanging phase, the θ_1^* values are given in Figure 7. Therefore, the (x, y) coordinates of the points on the cam profile curve can be written as

$$\begin{aligned} x &= \left(r_{o1} + s_1\right)\cos\varphi, \\ y &= \left(r_{o1} + s_1\right)\sin\varphi, \end{aligned} \tag{13}$$

where s_1 is the pushing distance of the hip joint cam follower and φ is the rotational angle of the camshaft.

Equation (13) can only provide the theoretical profile curve of the hip joint cam. The actual profile curve is the inner envelop of a series of circles, whose centers are the points on the theoretical profile curve, and the radius is the radius of the rolling wheel. C programming was used to sample 50 points equally from the whole circumference of the cam, and then these points were smoothly connected to produce an approximated cam model.

4.2. The Principle of the Main Cam Kinematics. The principle of the main cam control parameter θ_2 is to use the motion of the movable joint A along AO_1 presented in Figure 5 to govern the follower. The cam structure to control θ_1 is shown in Figure 9.

The stroke length of the main cam is shown in Figure 10.

In Figure 10, a is the sum of the follower stroke and the radius of the cam base plate, when the follower is at the turning point when approaching the angle of repose. And b is the follower stroke and the radius of the cam base plate when the follower is at the angle of repose. In this paper, $a = 10.8932$ and $b = 13.6771$. The schematic diagram of the main cam follower motion cycle is presented in Figure 11.

In Figure 11, n is the radian angle of the camshaft rotation φ and $n = \varphi\pi/180$. Assume s_2 is the stroke of the main cam follower, divided into three sections corresponding to OA, AB, and BC. If the cam follower strokes corresponding to

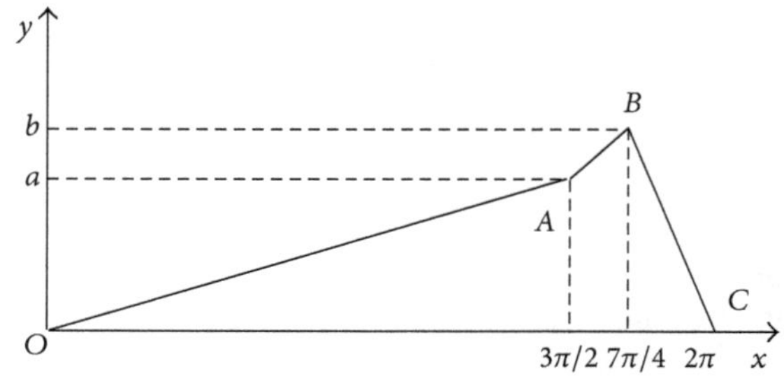

FIGURE 10: The stroke of the main cam.

sections OA, AB, and BC are s_{21}, s_{22}, and s_{23}, respectively, then the below equations can be given:

$$\begin{aligned} s_{21} &= \frac{(2 \cdot a \cdot n)}{3\pi}, \\ s_{22} &= \frac{4 \cdot (b - a) \cdot n}{(\pi + 7a - 6b)}, \\ s_{23} &= \frac{-4 \cdot b \cdot n}{(\pi + 8b)}. \end{aligned} \tag{14}$$

Therefore, the (x, y) coordinates of the points on the cam profile curve can be expressed as

$$\begin{aligned} x &= \left(r_{o2} + s_2\right)\cos\varphi, \\ y &= \left(r_{o2} + s_2\right)\sin\varphi. \end{aligned} \tag{15}$$

Similarly (15) can only give the theoretical profile curve of the main cam, and C programming was used to produce a fitted cam model.

4.3. The Principle of the Knee Joint Cam Kinematics. The principle of the knee joint cam control parameter θ_3 is given in Figure 12, and the knee joint cam follower follows the below relationship:

$$\frac{r\theta_3}{r_2} = \frac{s_3}{r_1} \tag{16}$$

which leads to

$$s_3 = \frac{r_1\theta_3 r}{r_2}. \tag{17}$$

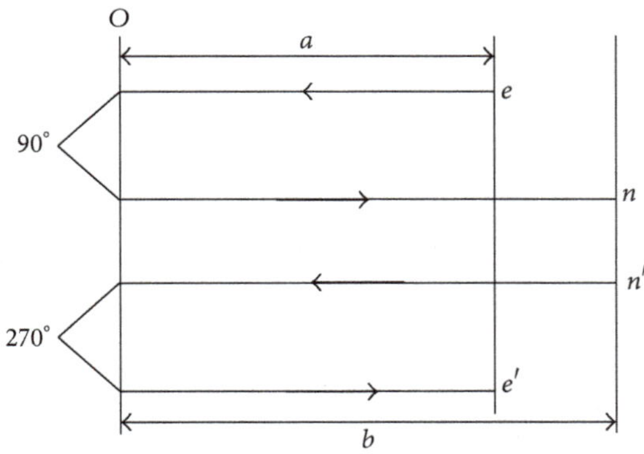

FIGURE 11: Schematic diagram of the main cam motion cycle.

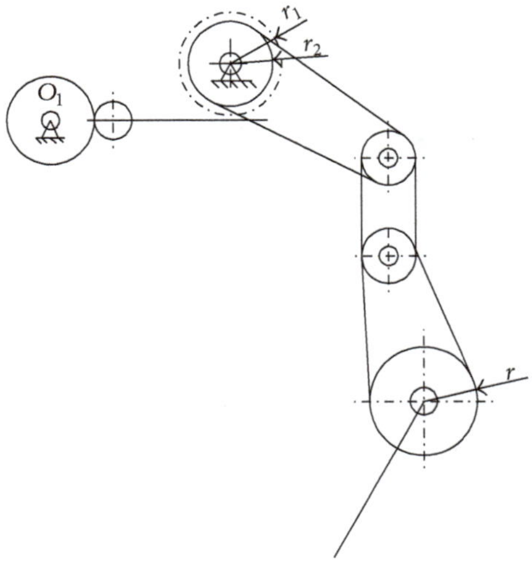

FIGURE 12: The knee joint cam control method.

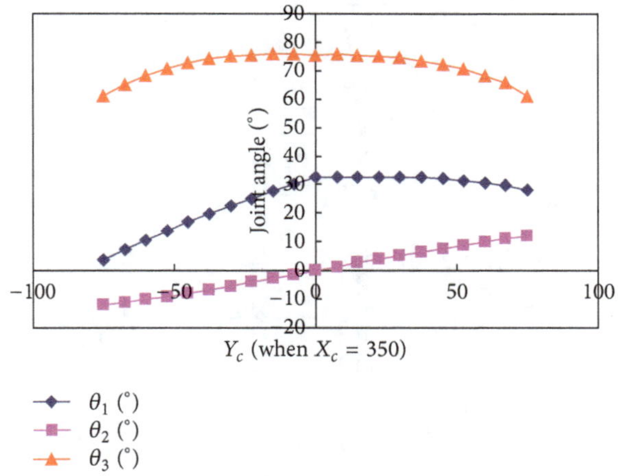

FIGURE 13: Key sample points chosen for verifying the fitted cam model.

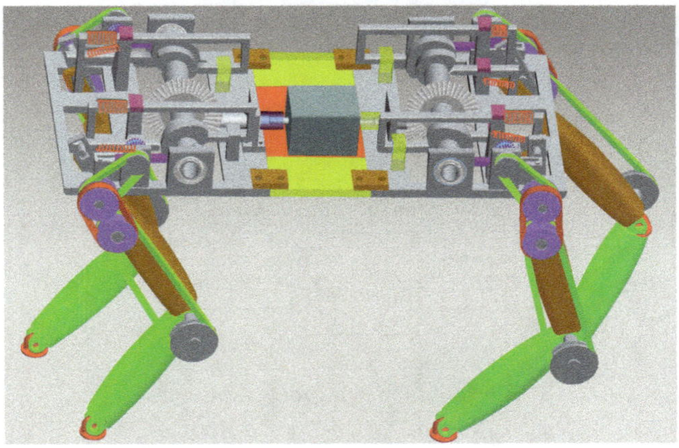

FIGURE 14: Three-dimensional model of the quadruped robot.

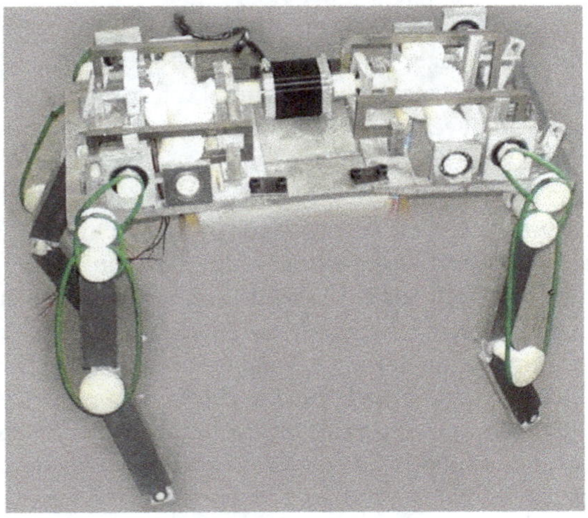

FIGURE 15: An example quadruped robot LCQR-1.

The (x, y) coordinates of the points on the cam profile curve can be written as

$$x = (r_{o3} + s_3)\cos\varphi,$$
$$y = (r_{o3} + s_3)\sin\varphi, \tag{18}$$

where the θ_3 value in the supporting phase is given by (11). In order to verify the strokes of the three joints presented correctly, some key points can be used as shown in Figure 13.

5. Design of the Quadruped Robot Step Mechanism

Virtual system design software, Pro/E, was utilized to model the overall three-dimensional structure of the step mechanism of the cam drive quadruped robot, as shown in Figure 14.

The system was assembled and integrated after different components were manufactured. An actual example of the assembled quadruped robot is shown in Figure 15.

6. Experiments

The quadruped robot prototype has been built with a controller that can perform walking tests. As shown in Figure 16, the control system employs an 80C51 MCU, a DAC0832 unit that transfers the digital control command to analog

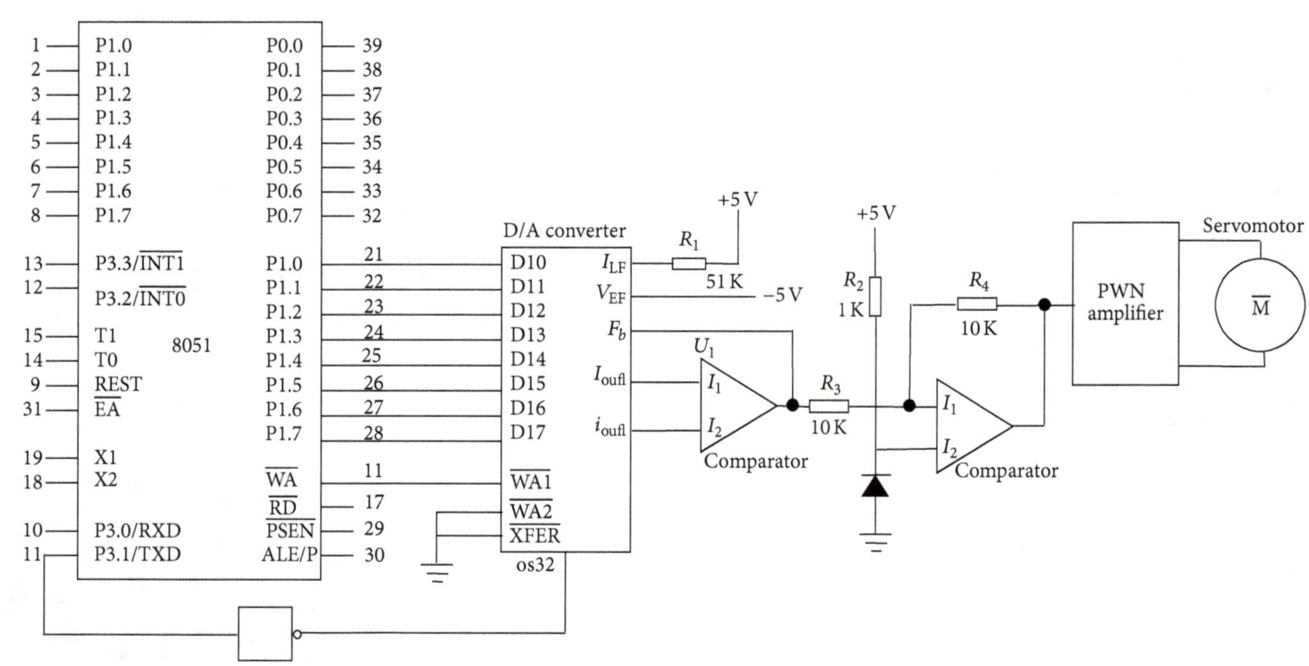

FIGURE 16: The control unit of the prototype quadruped robot.

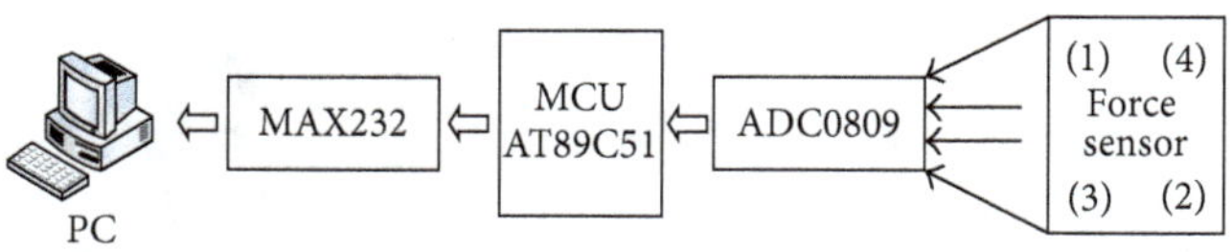

FIGURE 17: The foot supporting force measurement unit.

signals, and a PWN amplifier to drive the servomotor. Since the legs are all driven by the servomotor through the cam mechanisms, only one servomotor is needed, which simply governs the leg paces by motor rotational speed.

In order to assess the walking tests, a force measurement unit was also developed, which consists of a force sensor at the bottom of each foot, a ADC0809 unit that transfers analog force signals into digital data required by the 80C51 MCU, and the recorded data were sent to a remote PC through MAX232 interface. The four legs and the sensors are numbered (1)–(4) as shown in Figure 17, which are consistent with the quadruped robot step sequence as mentioned previously.

The supporting force of each foot was recorded with 1 KHz sampling rate and then averaged for every twenty samples to smooth out transient disturbances. Some testing results for the robot walking with a speed of two steps per second are shown in Figure 17, where the duty cycle of the step sequence is 0.75; that is, when one leg is lifted above the ground, all the other three legs are in the supporting phase. As the legs are all driven by one servomotor and a set of mechanical linkages, they do appear to hit the ground and leave the ground in a somehow rigid manner; there are clear force distinctions between the supporting phase and the hanging phase.

As shown in Figure 18, the force measured at the bottom of each foot is zero in the hanging phase and is about one-third of the system weight in the supporting phase. The time sequence of each leg force indicates that the walking paces are in a well-coordinated manner. This is ensured by the solid mechanical linkages between each leg and therefore

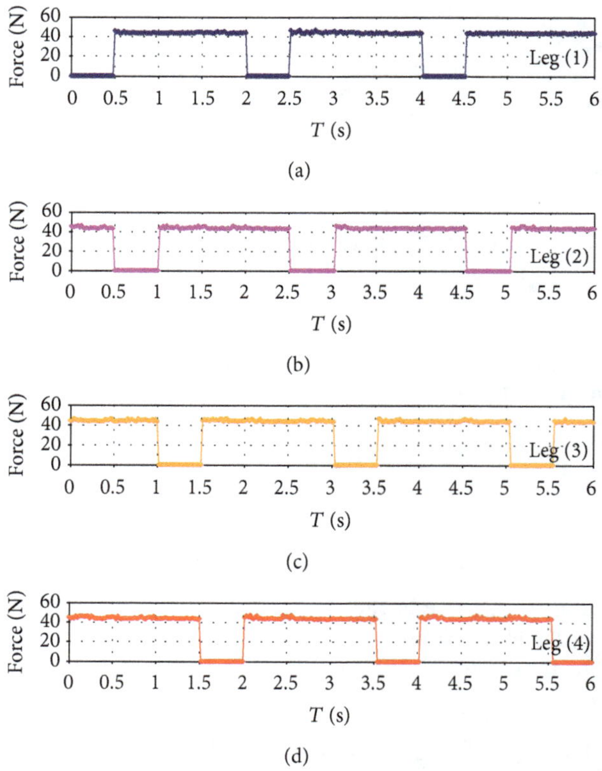

FIGURE 18: Foot supporting forces when duty cycle is 0.75 and the speed is 2 steps per second.

it is a goal much easier to achieve than using individual servomotors on each leg joint.

A limitation of the system is that the duty cycle is fixed at 0.75; therefore, it is not very suitable for fast running, in which case the hanging phase might involve two legs lifted about the ground at the same time. However, this design does demonstrate a simple and reliable solution to perform four-leg walking with steady paces on flat terrain using much lower costs, and it may be applicable to some applications such as household robots or educational devices.

7. Conclusion

This study is focused on the walking mechanism design of a quadruped robot and it covers the studies in the following areas:

(i) introducing the basic concept of quadruped robot walking and analyzing typical quadruped walking characteristics, the kinematic bone structures, and step sequences;

(ii) investigating the mechanical theories of the quadruped robot walking mechanism and selecting the required control parameters;

(iii) investigating the kinematic principles of the quadruped robot walking mechanism and using hip joint cam, knee joint cam, and main cam to control the leg governing parameters;

(iv) building the three-dimensional model of the quadruped robot walking mechanism, developing a prototype machine, initially testing walking performance, and improving the system design;

(v) performing walking tests using the prototype machine and analyzing the force characteristics.

The cam drive bionic quadruped robot study is a long term research project, needing some theoretical breakthroughs and practical issues to be better handled in reality. In the later studies, more attention should be given to the optimal layout design of the mechanical components and to better stabilize the gravity center of the system for faster walking, as well as dynamics simulation of the system using virtual system simulation software MSC.ADAMS.

Conflict of Interests

The authors declare that there is no conflict of interests regarding the publication of this paper.

Acknowledgments

This work was supported by the Natural Science Foundation of Shandong Province, China (Grants nos. ZR2011EL038 and ZR2012CQ026), and a Project of Shandong Province Higher Educational Science and Technology Program (Grants nos. J11LD16 and J12LB63).

References

[1] R. B. McGhee and A. A. Frank, "On the stability properties of quadruped creeping gaits," *Mathematical Biosciences*, vol. 3, no. 1-2, pp. 331–351, 1968.

[2] V. S. Gurfinkel, E. V. Gurfinkel, A. Y. Shneider, E. A. Devjanin, A. V. Lensky, and L. G. Shtilman, "Walking robot with supervisory control," *Mechanism and Machine Theory*, vol. 16, no. 1, pp. 31–36, 1981.

[3] S. Hirose, T. Masui, and H. Kikuchi, "TITANIII: a quadruped walking vehicle-its structure and basic characteristics," *International Journal of Robotics Research*, pp. 325–331, 1985.

[4] S. Hirose, K. Yoneda, K. Arai, and T. Ibe, "Design of prismatic quadruped walking vehicle TITAN VI," in *Proceedings of the 5th International Conference on Advanced Robotics*, pp. 723–728, Pisa, Italy, June 1991.

[5] S. Hirose, K. Yoneda, and H. Tsukagoshi, "TITAN VII: quadruped walking and manipulating robot on a steep slope," in *Proceedings of the IEEE International Conference on Robotics and Automation*, vol. 1, pp. 494–500, Albuquerque, NM, USA, April 1997.

[6] K. Kato and S. Hirose, "Development of the quadruped walking robot, TITAN-IX -mechanical design concept and application for the humanitarian de-mining robot," *Advanced Robotics*, vol. 15, no. 2, pp. 191–204, 2001.

[7] Y. Fukuoka, H. Kimura, Y. Hada, and K. Takase, "Adaptive dynamic walking of a quadruped robot "Tekken" on irregular terrain using a neural system model," in *Proceedings of the IEEE International Conference on Robotics and Automation*, pp. 2037–2042, Taipei, Taiwan, September 2003.

[8] M. Raibert, K. Blankespoor, G. Nelson et al., "The rough terrain quadruped robot," in *Proceedings of the 17th International Federation of Automation Control*, p. 10822, Seoul, Republic of Korea, 2008.

[9] L. Fei, *Research about gait planning and balance control of quadruped robot [Ph.D. thesis]*, University of Science and Technology of China, Heifei, China, 2010.

[10] Y. Fukuoka, H. Kimura, and A. H. Cohen, "Adaptive dynamic walking of a quadruped robot on irregular terrain based on biological concepts," *The International Journal of Robotics Research*, vol. 22, no. 3-4, pp. 187–202, 2003.

[11] Z. G. Zhang, H. Kimura, and K. Takase, "Adaptive running of a quadruped robot using forced vibration and synchronization," *Journal of Vibration and Control*, vol. 12, no. 12, pp. 1361–1383, 2006.

[12] I. Poulakakis, E. Papadopoulos, and M. Buehler, "On the stability of the passive dynamics of quadrupedal running with a bounding gait," *The International Journal of Robotics Research*, vol. 25, no. 7, pp. 669–687, 2006.

[13] L. Ding, R. Wang, H. Feng, and J. Li, "Brief analysis of a BigDog quadruped robot," *China Mechanical Engineering*, vol. 23, no. 5, pp. 505–514, 2012.

[14] S. Qun, S. Chunlei, and L. Baolong, "Virtual design of the mechanical system of the quadruped robot and theory analysis of the wheel mechanism," *Mechanical Design and Manufacture*, vol. 46, no. 8, pp. 183–185, 2009.

An Autonomous Omnidirectional Robot

Yanfei Liu,[1] Jiaxin Zhao,[2] Josh Apple,[1] Torrey Frank,[1] Matthew Saylor,[1] and Ted Siegel[1]

[1] *Department of Engineering, Indiana Univeristy-Purdue University Fort Wayne, 2101 E. Coliseum boulevard, Fort Wayne, IN 46805-1499, USA*
[2] *West Hills College Lemoore, 555 College Avenue, Lemoore, CA 93245-9098, USA*

Correspondence should be addressed to Yanfei Liu, liu@engr.ipfw.edu

Academic Editor: Yangmin Li

RoboCup is an international research and education initiative, which aims to foster artificial intelligence and robotics research by using competitive soccer as a standard problem. This paper presents a detailed engineering design process and the outcome for an omni-directional mobile robot platform for the Robocup Middle Size League competition. A prototype that can move omnidirectionally with kicking capability was designed, built, and tested by a group of senior students. The design included a mechanical base, pneumatic kicking mechanism, a DSP microcontroller-based control system, various sensor interfacing units, and the analysis of omnidirectional motions. The testing results showed that the system was able to move omnidirectionally with a speed of ~2 m/s and able to kick a size 5 FIFA soccer ball for a distance of at least 5 meters.

1. Introduction

In August 2007, the College of Engineering, Technology and Computer Science at Indiana University-Purdue University Fort Wayne and a local company initiated a 5-year project to promote robotics, artificial intelligence, and software engineering in the college curricula. The main goal of this project is to build a robot team to compete in the Robocup Middle Size League competition [1] by 2012. This project also aims at introducing robotics into a variety of computer science and engineering courses. As part of the first year plan, a Pioneer 3-DX robot was purchased. Using this robot as a development platform, the first task was to design and build a kicking mechanism that is connected and interfaced with this robot [2]. As the outcome of this project, the robot with the mounted kicker (Figure 1) can locate an orange soccer ball and a green goal post, compute the angle that the robot should turn in order to line up with the ball and the post, approach the ball with the computed angle, and kick the ball towards the post. The Pioneer 3-DX robot is a sturdy and reliable commercial robot. However, it does not suit the specific needs for soccer playing. One problem is its slow speed due to its relatively large weight. Also its predetermined shape and dimensions impose additional constrains to the mounting of other components, for example, the kicker, in order to meet the dimension limits for the Robocup Middle Size League. Additionally, the traditional two-wheel with a caster structure leads to a long reaction time when the robot has to turn. Therefore, on the second year the emphasis of this project switched to the development of an omni-directional mobile robot platform specifically for the Robocup competition.

A large body of robotics research in several areas related to the Robocup middle size league competition has been published during the past decade. These research areas include real-time motion planning [3], data fusion [4], robot platform design [5], communications among robots [6], and multiagent control structure [7]. Among these areas, the platform design is where a new robot team would start. However, to the best of our knowledge, there are few works published describing the initial design and building process of their basic robot platform. Usually a team's description paper [8] only discusses the components of their robot platform without explaining the process of

Figure 1: The kicking mechanism mounted on a Pioneer 3-DX robot.

the design generation. To fill this blank, in this paper, a detailed process of designing, building, and testing an omni-directional mobile robot platform for the Robocup middle size league competition is presented. This design task was carried out by a multidisciplinary team of senior engineering students.

The task of building an omni-directional mobile robot with kicking capability was carefully reviewed by the professors whose expertise is in robotics and in machine design, and determined to be an appropriate two-semester multi-disciplinary capstone senior design project. The team was formed with two electrical/computer engineering students and two mechanical engineering students. In the fall of 2008, the students started with the formulation of the problem, generation of conceptual designs, evaluation of the conceptual designs and finished with a detailed design of the final chosen design. In the spring of 2009, the students built the robot platform and conducted the experimental testing.

The remainder of this paper is organized as follows. In Section 2, we describe an overview of the robot system. In Sections 3 and 4, we present the mechanical and electrical design of the mobile robot platform. Testing procedures and results are presented in Section 5. Finally, we conclude the paper in Section 6.

2. System Overview

At the beginning of the project the design requirements and specifications were set so that the robot was to be fully autonomous and capable of moving omnidirectionally with a speed of at least 2 m/s and able to kick a size 5 FIFA soccer ball for a distance of at least 5 meters. Due to the specific intended use of this robot it has to comply with a maximum weight constraint of 40 kg and also with other rules and regulations defined by the RoboCup Middle Size League as of 2008. Students started generating conceptual designs by breaking the whole system into the following individual modules: mobility, mechanical base, power, kicking mechanism, microcontroller, and motor control. Several conceptual designs were generated for each module. Table 1 shows the design details and their ranking from best to worst. The ranking of different designs in each category is based on functional capability, easiness of implementation, modularity for interface with other components, reliability, and cost. Each criterion was assigned a specific weight factor based on its importance to the final product. A detailed description of the generated conceptual designs for each module can be found in [9]. Considering compatibility, the team combined the designs of each individual module to compose five overall conceptual designs. Figures 2, 3, 4, 5, and 6 show the 3D CAD drawings for these five conceptual designs.

Table 2 shows the individual components of each design. To choose the best design decision, matrices were generated to evaluate these five conceptual designs according to the same criteria of capability, ease of implementation, modularity, reliability, and so forth, as for the ranking in each category. The detailed evaluation process and matrices can be found in [9]. Based on this evaluation, the designs that had the top two scores, concepts 2 and 4, were picked as the primary and secondary designs, respectively.

The primary conceptual design, shown in Figure 2, utilizes the pyramid shell with a four-drive wheel. With this design, the mobility is greatly increased over the other concepts. With the four-wheel design, the robot should be able to maintain orientation with the goal while moving around the entire field. Four-wheel design also provides a larger wheel base to allow a larger resistance to tilt over by the robot as it might get hit in a game. Assuming uniform mass distribution and everything else equal (Mass M, adjacent wheel distance d), four-wheel design requires almost twice as much moment to tilt it over, compared with the three-wheel design, as shown in (1). In addition, the kicker on this robot will be very reliable compared to the other concepts due to its relatively simple design. Lack of reliability in other kicker options stems from the fact that more kicker parts in other designs led to more failure modes. The Parallel 12 V DC power supply is also advantageous due to its relative simplicity and ability to produce twice as much current to actuators/motors. The motor controller design for this concept will allow each drive motor to have complete independence by receiving independent signals from the microcontroller

$$M_{\text{tilt}} = \begin{cases} Mg\left(\dfrac{\sqrt{3}}{3}\dfrac{d}{2}\right) = 0.288\,Mgd & \text{Three Wheel Base,} \\ Mg\left(\dfrac{d}{2}\right) = 0.5\,Mgd & \text{Four Wheel Base.} \end{cases} \tag{1}$$

3. Mechanical Design

In the mechanical design, the design selection and optimization was focused on providing rigidity of the platform, reducing weight, increasing tilt-over resistance, while maintaining the mobility and functionality of the robot. The design first started by taking into consideration the size of the wheels and motors selected for the robot. With this specific information along with the diameter of the ball, the desired wheel orientation and the size constraints for

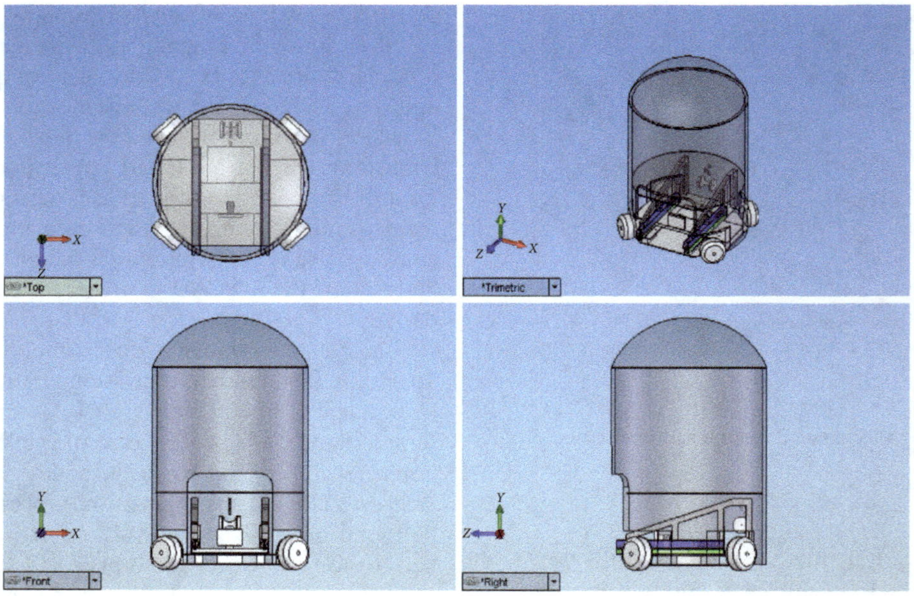

FIGURE 2: Conceptual design 1: Bullet design with a spring kicker.

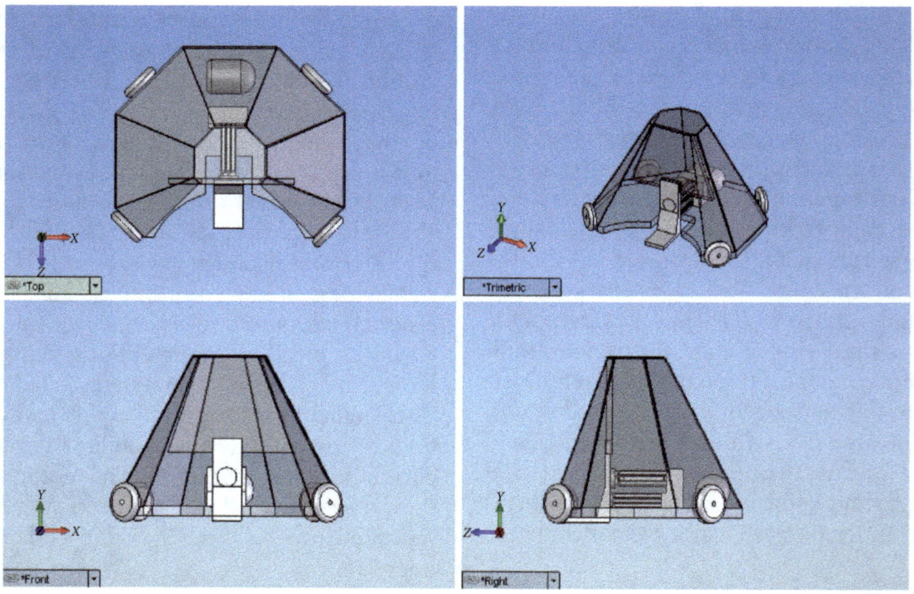

FIGURE 3: Conceptual design 2: Pyramid structure with a swing lift kicker.

TABLE 1: Ranking of the individual design modules.

	Best . . .			. . . Worst
Mobility	Four wheel	Two wheel	Three wheel	
Mechanical base	Pyramid	Bullet	Box	
Kicking mechanism	Swing lift with plunger	Spring loaded sliding	Pneumatic kicker with plate	Swing Arm
Microcontroller	DSP microcontroller	Low power microcontroller with a laptop	Networked microcontroller	Low power microcontroller with a PC
Motor control	Multiple motor controllers	Single motor controller		
Power	24 V DC	Parallel 12 V DC	Single 12 V DC	

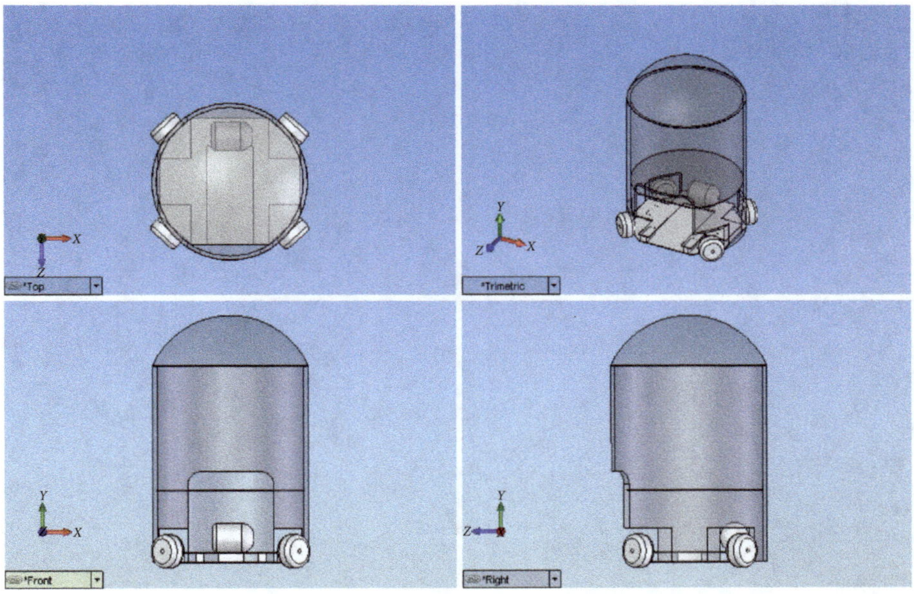

FIGURE 4: Conceptual design 3: Bullet design with a pneumatic kicker.

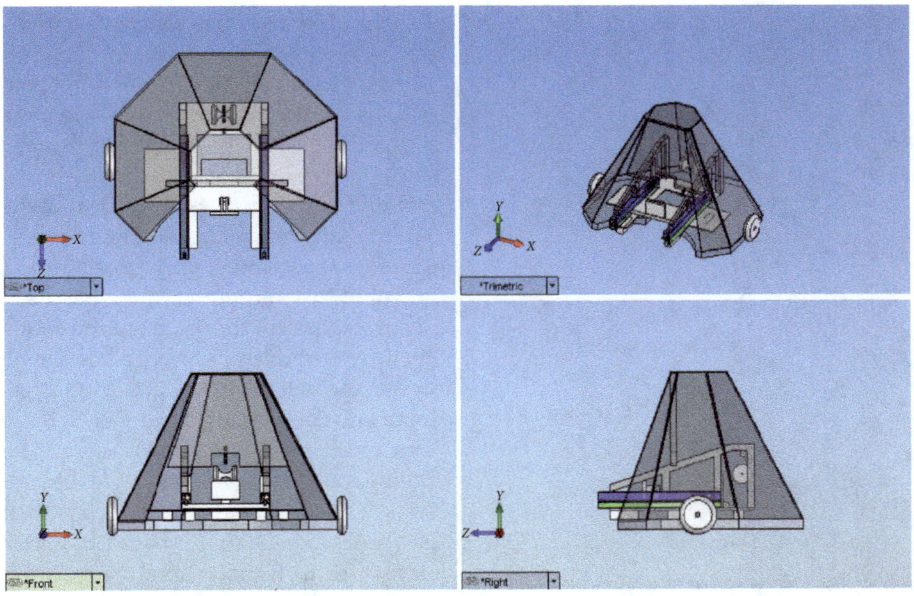

FIGURE 5: Conceptual design 4: Pyramid of two wheels with a spring-loaded kicker.

TABLE 2: Five conceptual designs.

	Concept 1	Concept 2	Concept 3	Concept 4	Concept 5
Mobility	Four wheel	Four wheel	Four wheel	Two wheel	Three wheel
Mechanical Base	Bullet	Pyramid	Bullet	Pyramid	Pyramid
Kicking mechanism	Spring loaded sliding	Swing lift with plunger	Pneumatic kicker	Spring Loaded Sliding	Swing lift with plunger
Microcontroller	Low power microcontroller with a laptop	DSP microcontroller	DSP microcontroller	Networked microcontroller	Low power microcontroller with a PC
Motor Control	Single motor controller	Multiple motor controllers	Multiple motor controllers	Single motor controller	Multiple motor controllers
Power	Parallel 12 V DC	Parallel 12 V DC	24 V DC	24 V DC	Parallel 12 V DC

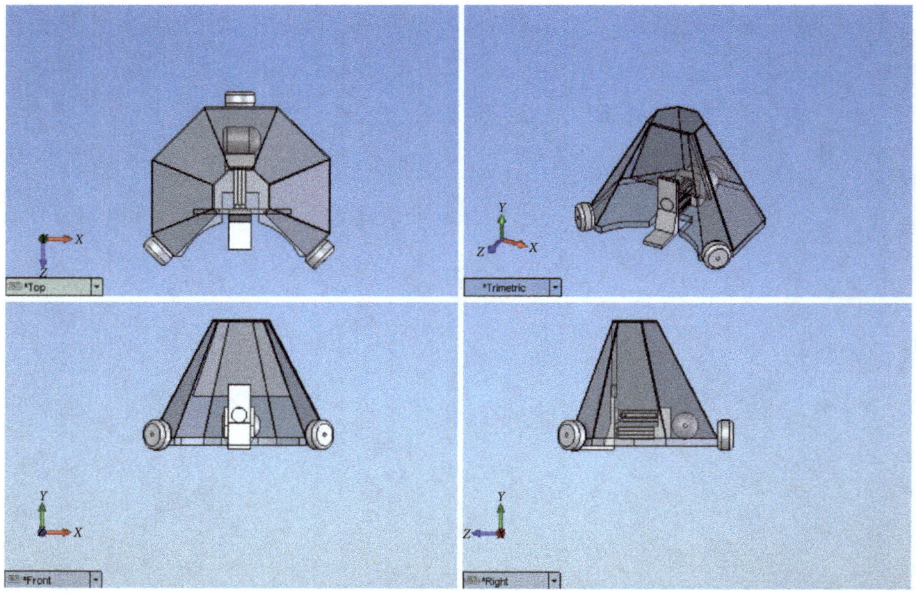

FIGURE 6: Conceptual design 5: Pyramid of three wheels with a lift kicker.

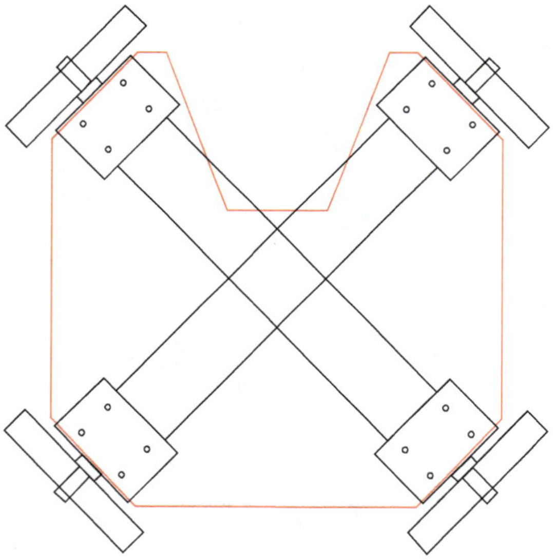

FIGURE 7: Omni-directional wheel arrangement.

the base of the platform were generated. Figure 7 shows the shape of the bottom plate and the wheel arrangement. The resistance to tilt over is further optimized by arranging to have heavier objects lower in the robot to lower the center of gravity. The capability of determining center of gravity of the CAD software used was utilized for verification of components arrangement. It was decided that the motors would be mounted on the underside of the plate. A cut-out in the base plate had to be made in order for the ball to enter into the robot one third of its diameter. The angle of the cut-out and size was determined based on the size of the ball, the motor arrangement, and the size of the plate. A detailed description of the cut-out calculations can be found in [10].

The next task was to create a mounting structure for the kicker itself as well as an overall body for the robot. During this process, it was found out that the pyramid design selected would not provide an adequate amount of space due to the dimensions of the nitrogen tank and the pneumatic components. To solve this problem, a combination of the pyramid and bullet designs was utilized. Figure 8 shows the 3D CAD drawing of the modified design. This design allowed for more space and open areas and it also kept the weight of the robot to a minimum. There was also enough room left in the overall height for a later addition of a vision system. The open areas of the robot were all designed to be small enough so that the ball could not accidentally enter the inside of the robot and damage any components. In the CAD drawing, the large mechanical components, such as the nitrogen tank, the pneumatic kicker, were modeled and added to give a better visualization of the remaining space. It shows that there was plenty of space left for the batteries, drivers, microcontrollers, and so forth. Figure 9 shows a picture of the actual robot built. Aluminum bars form the frame of the robot, providing rigidity of the structure as well as means to attach other components. Aluminum is selected as the material because of its better strength/density ratio compared with steel. Rubber bends could be wrapped around the frame to reduce the effects of impact from the ball or other robots in a game.

A final mechanical optimization was done to align the kicking plate which was pinned on top as shown in Figure 9. The pneumatic cylinder pushes the plate from below and the plate rotates up, kicking the ball. For simple trajectory motion from ground level with initial velocity magnitude of v and angle of θ from the horizon, the distance of travel until hitting ground again is given by (2), which results in a maximum distance of travel when $\theta = 45°$. In lieu of this, a rubber stop was positioned on top of the plate to stop the plate motion at approximately 45° from horizon, resulting an optimal release angle of the ball to ensure the longest travel

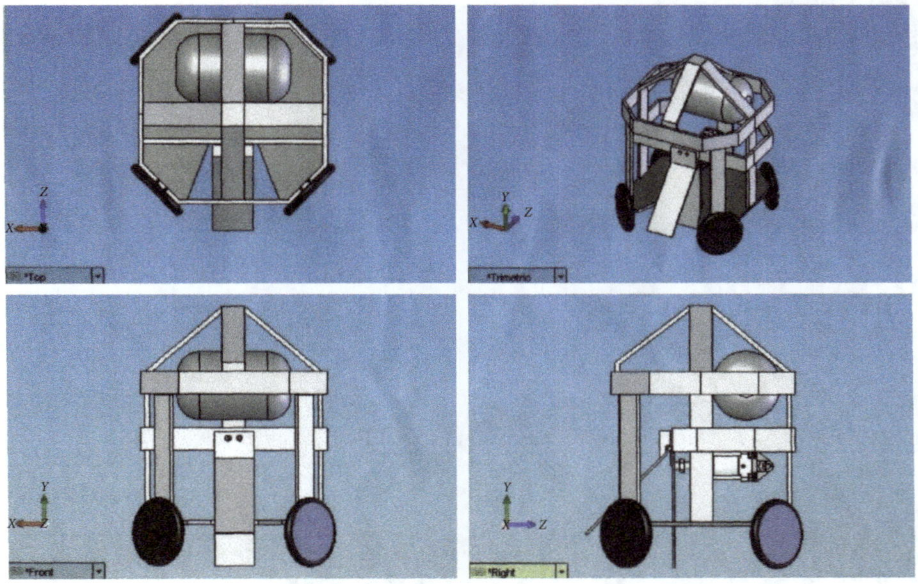

FIGURE 8: 3D CAD drawing for the final modified design.

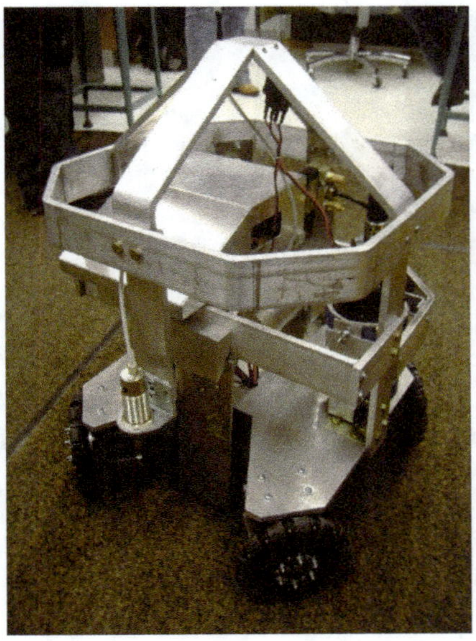

FIGURE 9: The final built robot.

possible through air to achieve the distance requirement of the kicker

$$S = \frac{v^2}{g} \sin(2\theta). \quad (2)$$

The final dimensions of the robot were 45.7 × 45.7 × 62.23 cm. The size constraints were 50 × 50 × 80 cm. The reason the height being much lower than the maximum was to ensure that there was enough room left for a vision system to be added to the top of the robot at a later time. The final weight of the robot was 31.1 kg, which is about 9 kg under the 40 kg weight constraint. Again, the 9 kg difference left enough room for the future vision system.

4. Electrical Design

The electrical design of the robot includes the eZdspF28335 development board that holds the TMS320F28335 Digital Signal Controller (DSC), a printed circuit board (PCB) that contains the solenoid driving unit, the proximity sensor interface, the Hall-effect sensors interface, and the motor driving circuit. Figure 10 shows a block diagram for the whole electrical system. The eZdspF28335 development board is used to program the TMS320F28335 DSC and to get easy access to the I/O pins of the DSC. The PCB was completely designed using Multisim and Ultiboard software. Figure 11 shows the PCB layout and the functions of the various components. A solenoid driver was used to drive the solenoid valves for the pneumatic cylinders. The proximity sensor was implemented to detect the soccer ball when it is close enough to be kicked. A 25.9 V Lithium Polymer Ion battery rated at 326.34 Wh (watt-hour) was selected to power the whole robot. The detailed design of all the sensor interfacing circuits and a complete power analysis for the whole system can be found in [10].

The motor drivers (controllers) are the Syren10 manufactured by Dimension Engineering. The motor driver can support up to 10 A continuous and 15 A peak and support voltages from 6 V to 24 V with the absolute maximum of 30 V. The driver has regenerative current, which means that it will return power to the batteries when the robot is decelerating or changing direction. Regenerative current returns the inductive energy from the motor to the battery instead of dissipating it as heat. The motor controller operated at a simplified serial mode that uses TTL level single

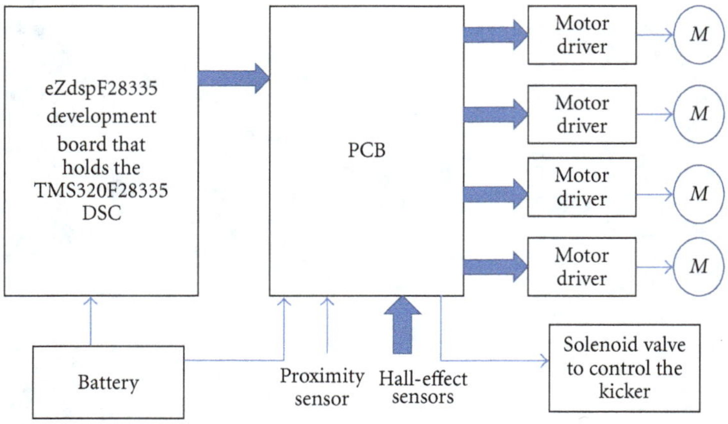

Figure 10: Block diagram for the electrical system.

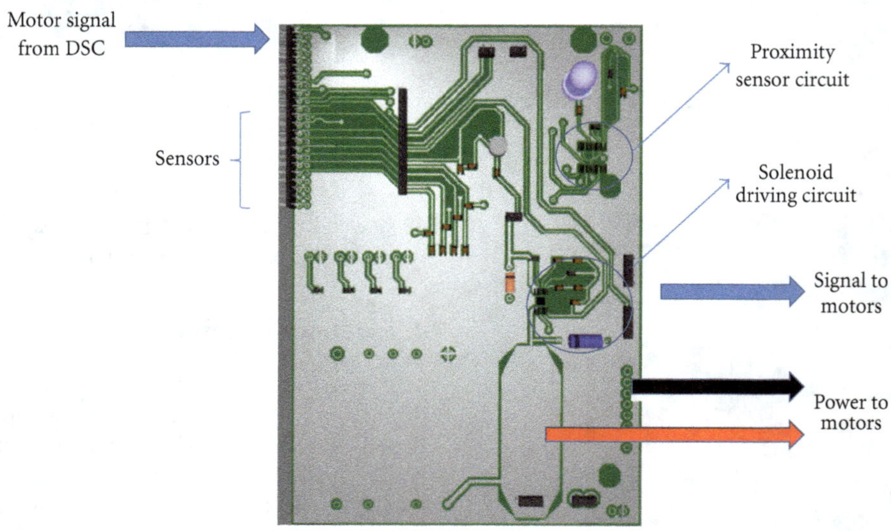

Figure 11: The PCB layout.

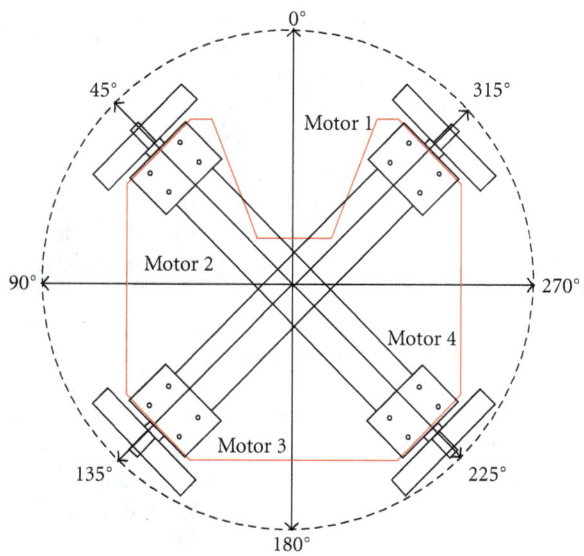

Figure 12: Eight different directions.

byte serial commands. The driver outputs a pulse width modulated signal at 32 kHz.

5. Testing Procedures and Results

The design requirement and specification of this project were set such that the robot should be capable of moving omnidirectionally with a speed of at least 2 m/s and able to kick a size 5 FIFA soccer ball for a distance of at least 5 meters.

The testing for the kicking distance started with the robot in a stationary position. Then the soccer ball was placed in front of the robot, and the kicking mechanism was actuated. The total air and rolling kicking distance was measured and recorded. The experiments were repeated 10 times. The rolling distance was well beyond the 5 meter goal. The horizontal kicking distance in the air was measured as approximately 3 meters, as shown in Table 3.

The speed of the robot movement was measured in eight different directions. The eight different directions are defined in angles with 0° being the direction that goes straightly

Table 3: Kicking distance in Air.

Trial	1	2	3	4	5	6	7	8	9	10	Avg.
Distance (m)	3.07	3.22	2.72	2.72	2.94	2.89	3.25	2.86	2.86	3.30	2.98

Table 4: Average speed in 8 different directions.

Direction (angle in degree)	0	45	90	135	180	225	270	315
Speed (m/s)	2.16	2.06	2.24	1.81	2.00	2.15	2.10	1.83

forward. The arrangement of the eight directions is shown in Figure 12. The robot was programmed to move and accelerate for one second, and then the distance travelled in 1 second was measured. The experiments were run 10 times on each direction. The average speeds for each direction are shown in Table 4. As shown in the table, the robot achieved the 2 m/s goal set by the problem statement for all but two of the angles, which are 135° and 315° along the same line but in opposite directions. This is due to the fact that the four motors were not mounted at the same level. The team mounted the four motors on the bottom plate using only visual inspection without the aid of any alignment tools. As consequence, one or two of the four motors, motors 2 and 4, are slightly off the plane with respect to the other two motors. Therefore the two wheels attached to these two motors have less traction with the floor. Motor 2 and 4 are the ones that are engaged for the movements in the directions of 135° and 315°. That is the reason why the robot moves slower in these two directions when compared with the others. In the future, using proper alignment tools, the four motors will be positioned to ensure that they are at the same level, and a more uniformly distributed speed would be achieved in all directions. Also note that the 1 second test includes the acceleration of the robot. For a linear acceleration process, the final speed is twice as much as the average speed. To verify the final steady speed that the robot can achieve, an 8-second test was done to allow the robot to move in 8 seconds and the distance measured. The average speed calculated to be nearly 3 m/s which represents how fast the robot can go.

Some other requirements were also tested based on specific components involved. The power supply was designed to supply full power to the robot for 45 minutes. The capacity of the power supply was tested by turning on all four motors and allowing them to run at full speed when the robot was set on a bench. After allowing the motors to run for a total of three (3) hours, it was determined, by visual inspection, that the performance of the motors was not at all affected after this length of running time. The number of kicks a full nitrogen tank can provide was also determined through testing. The nitrogen tank used in the robot was 48 cubic inch in size, which was filled with nitrogen gas to a pressure of 3000 psi. The manual pressure regulator was set to 100 psi. The team ran the experiments three times. The results showed that on the average 126 kicks were produced per tank.

As mentioned in the introduction, the initiative to design and build a new robot is to overcome the shortcomings of the Pioneer robot. The maximum speed that the Pioneer robot can reach is 1.6 m/s not considering the time required for turning. Our robot can reach a speed of more than 2 m/s in most of the directions without the need of turning. The kicker of the Pioneer robot cannot kick the ball into the air, while our robot can kick the ball into the air with an average distance of ~3 meters in the air.

6. Discussions and Conclusions

In this paper, a multidisciplinary engineering capstone design project is described. This project is about the design, building and testing of an omni-directional robot designed specifically for the Robocup competition. The robot was designed to meet requirements of moving omnidirectionally with speed of 2 m/s, kicking a size 5 soccer ball 5 meters. Various design choices were studied and best combination for functional capability, easiness of implementation, modularity for interface with other components, reliability, and cost were selected. A prototype was built according to the final design. Testing results of the prototype show that the delivered system meets the requirements specified at the start of this project.

The current design was also done with further improvements and functional additions in mind. The mass and height of the robot was well under the Robocup limit, allowing room for vision system. For the 2009-2010 academic year another group of engineering students worked at adding an omni-directional vision system to this robot platform. The team finished the design of the vision system in December 2009. In the spring 2010 semester, the team built the system and conducted experimental testing. The detailed description of the system and the experimental testing results can be found in [11]. In 2011, this robot including the vision system will be modified as needed to make it as reliable as possible. Then a couple of more robots will be replicated such that cooperative behaviors and team play strategy can be experimented.

Acknowledgment

The authors would like to thank Raytheon Fort Wayne for their support in this work.

References

[1] Robocup, 2009, http://www.robocup.org/.

[2] Y. Liu and J. Zhao, "A kicking mechanism for a soccer playing robot—a multidisciplinary senior design project," in *American*

Society of Engineering Education (ASEE) Annual Conference & Exposition, Austin, Tex, USA, 2009.

[3] D. Bruijnen, J. van Helvoort, and R. van de Molengraft, "Realtime motion path generation using subtargets in a rapidly changing environment," *Robotics and Autonomous Systems*, vol. 55, no. 6, pp. 470–479, 2007.

[4] J. Silva, N. Lau, J. Rodrigues, J. L. Azevedo, and A. J. R. Neves, "Sensor and Information fusion applied to a Robotic Soccer Team," in *RoboCup Symposium*, Graz, Austria, 2009.

[5] Y. Takemura, A. Sanada, T. Ichinose et al., "Development of "Hibikino-Musashi" omni-directional mobile robot," *International Congress Series*, vol. 1301, pp. 201–205, 2007.

[6] F. Santos, L. Almeida, L. S. Lopes, J. L. Azevedo, and M. B. Cunha, "Communicating among robots in the RoboCup Middle-Size League," in *RoboCup Symposium*, Graz, Austria, 2009.

[7] Y. Li, W. I. Lei, and X. Li, "Multi-agent control structure for a vision based robot soccer system," in *Proceedings of the 11th IEEE International conference on Mechatronics and Machine Vision in Practice(M2VIP '04)*, pp. 151–159, Macao SAR, China, November-December 2004.

[8] W. H. T. M. Aangenent, J. J. T. H. de Best, B. H. M. Bukkems et al., "Tech United Eindhoven Team Description 2009," http://www.techunited.nl/index.php?p=28.

[9] J. Apple, T. Frank, M. Saylor, and T. Siegel, "A soccer playing robot," Capstone Senior Design Report I, Department of Engineering, Indiana University—Purdue University Fort Wayne, 2008.

[10] J. Apple, T. Frank, M. Saylor, and T. Siegel, "A soccer playing robot," Capstone Senior Design Report II, Department of Engineering, Indiana University—Purdue University Fort Waynee, 2010, http://www.engr.ipfw.edu/capstone/sd_2_spring09/soccer_playing_robot.pdf.

[11] B. Abels, M. Akers, and G. Harden, "Omni-directional vision system for a soccer playing robot," Capstone Senior Design Report II, Department of Engineering, Indiana University—Purdue University Fort Wayne, 2010, http://www.engr.ipfw.edu/capstone/sd_2_spring10/Omnidirectional_Vision_System_Final_Report.pdf.

Permissions

All chapters in this book were first published in JR, by Hindawi Publishing Corporation; hereby published with permission under the Creative Commons Attribution License or equivalent. Every chapter published in this book has been scrutinized by our experts. Their significance has been extensively debated. The topics covered herein carry significant findings which will fuel the growth of the discipline. They may even be implemented as practical applications or may be referred to as a beginning point for another development.

The contributors of this book come from diverse backgrounds, making this book a truly international effort. This book will bring forth new frontiers with its revolutionizing research information and detailed analysis of the nascent developments around the world.

We would like to thank all the contributing authors for lending their expertise to make the book truly unique. They have played a crucial role in the development of this book. Without their invaluable contributions this book wouldn't have been possible. They have made vital efforts to compile up to date information on the varied aspects of this subject to make this book a valuable addition to the collection of many professionals and students.

This book was conceptualized with the vision of imparting up-to-date information and advanced data in this field. To ensure the same, a matchless editorial board was set up. Every individual on the board went through rigorous rounds of assessment to prove their worth. After which they invested a large part of their time researching and compiling the most relevant data for our readers.

The editorial board has been involved in producing this book since its inception. They have spent rigorous hours researching and exploring the diverse topics which have resulted in the successful publishing of this book. They have passed on their knowledge of decades through this book. To expedite this challenging task, the publisher supported the team at every step. A small team of assistant editors was also appointed to further simplify the editing procedure and attain best results for the readers.

Apart from the editorial board, the designing team has also invested a significant amount of their time in understanding the subject and creating the most relevant covers. They scrutinized every image to scout for the most suitable representation of the subject and create an appropriate cover for the book.

The publishing team has been an ardent support to the editorial, designing and production team. Their endless efforts to recruit the best for this project, has resulted in the accomplishment of this book. They are a veteran in the field of academics and their pool of knowledge is as vast as their experience in printing. Their expertise and guidance has proved useful at every step. Their uncompromising quality standards have made this book an exceptional effort. Their encouragement from time to time has been an inspiration for everyone.

The publisher and the editorial board hope that this book will prove to be a valuable piece of knowledge for researchers, students, practitioners and scholars across the globe.

List of Contributors

Jing Bai, Xiaoyu Zhang, Xiansheng Qin and Xiaoqun Tan
School of Mechanical Engineering, Northwestern Polytechnical University, Xi'an 710072, China

Zhanxi Wang
School of Mechanical Engineering, Northwestern Polytechnical University, Xi'an 710072, China
Xinlian College, Henan Normal University, Zhengzhou 451400, China

Yali Zhao
Xinlian College, Henan Normal University, Zhengzhou 451400, China

Shaobin Lan and Zhibin Song
Key Laboratory of Mechanism Theory and Equipment Design of Ministry of Education, Tianjin University, Tianjin 300072, China

Xingcheng Li, Qilin Li and Yuan Chao
College of Mechanical Engineering, Jiangsu University of Technology, Changzhou 213001, China

Li Ding
College of Mechanical Engineering, Jiangsu University of Technology, Changzhou 213001, China
Department of Industrial and System Engineering, The Hong Kong Polytechnic University, Kowloon 999077, Hong Kong

Bayu Erfianto and Riyanto
School of Computing, Telkom University, Bandung, Indonesia
School of Electrical Engineering & Informatics, Institut Teknologi Bandung, Bandung, Indonesia

T. Bambang and Hilwadi Hindersah
School of Electrical Engineering & Informatics, Institut Teknologi Bandung, Bandung, Indonesia

Intan Muchtadi-Alamsyah
Faculty of Mathematical Sciences, Institut Teknologi Bandung, Bandung, Indonesia

Zhenwei Wang
College of Mechanical Engineering, Jiangsu University of Technology, Changzhou 213001, China

Li Ding
College of Mechanical Engineering, Jiangsu University of Technology, Changzhou 213001, China
Department of Industrial and System Engineering, The Hong Kong Polytechnic University, 999077 Hung Hom, Hong Kong

Zhongwei Li, Yanping Bi, Shan Zhao and Liming Zhou
School of Biomedical Engineering, Xinxiang Medical University, Xinxiang, Henan 453003, China

Wu Ren
School of Biomedical Engineering, Xinxiang Medical University, Xinxiang, Henan 453003, China
State Key Laboratory for High Performance Complex Manufacturing, Central South University, Changsha, Hunan 410083, China

Bo Peng
Hunan Special Equipment Inspection and Testing Research Institute, Zhangjiajie Branch, Zhangjiajie, Hunan 427000, China

Yong Wang
College of Mechanical and Electrical Engineering, China University of Mining and Technology, 221000, China
Xuzhou Products Quality Supervision and Inspection Center, 221000, China

Yu Zhou and Qing Chen
Xuzhou Products Quality Supervision and Inspection Center, 221000, China

Peng Tian
Xuzhou Products Quality Supervision and Inspection Center, 221000, China
Institute of Environment an+d Surveying and Mapping, China University of Mining and Technology, 221000, China

Fernando Reyes-Cortés
Facultad de Ciencias de la Electrónica, Benemèrita Universidad Autónoma de Puebla, 72570 Puebla, PUE, Mexico

César Chávez-Olivares
Departamento de Ingeniería Robótica, Universidad Autónoma de Aguascalientes, Campus Sur, Av. Prolongación Mahatma Gandhi No. 6601, 20340 Aguascalientes, AGS, Mexico

Emilio J. González-Galván
Facultad de Ingeniería, Universidad Autónoma de San Luis Potosí, Av. Manuel Nava 8, Zona Universitaria, 78290 San Luis Potosí, SLP, Mexico

Muhammad Latif Anjum
School of Electrical Engineering and Computer Science (SEECS), National University of Sciences and Technology (NUST), H-12, Islamabad, Pakistan

Stefano Rosa and Basilio Bona
Department of Control and Computer Engineering (DAUIN), Politecnico di Torino, Corso Duca degli Abruzzi 24, 10129 Torino, Italy

Huanzhao Chen and Guohui Tian
School of Control Science and Engineering, Shandong University, Jinan, Shandong 250061, China

Yunwang Li, Yuwei Zheng, Feng Tian and Xucong Yan
School of Mechatronic Engineering, China University of Mining and Technology, Xuzhou 221116, China
Sumei Dai School of Mechanical and Electrical Engineering, Xuzhou Institute of Technology, Xuzhou 221111, China

Shou-yan Chen, Tie Zhang and Yan-biao Zou
School of Mechanical and Automotive Engineering, South China University of Technology, Guangzhou, Guangdong 510640, China

Xiliang Ma and Hua Zhu
School of Mechanical and Electrical Engineering, China University of Mining and Technology, Xuzhou, Jiangsu 221116, China

Zhiwei Kong and Qiang Lu
School of Automation, Hangzhou Dianzi University, Hangzhou, China

Sharon Rabinovich, Renwick E. Curry and Gabriel H. Elkaim
Computer Engineering, University of California Santa Cruz, Santa Cruz, California, USA

Enhong Xing and Xinhua Zhao
Tianjin Key Laboratory for Advanced Mechatronic System Design and Intelligent Control, School of Mechanical Engineering, Tianjin University of Technology, Tianjin 300384, China

Bin Li
Tianjin Key Laboratory for Advanced Mechatronic System Design and Intelligent Control, School of Mechanical Engineering, Tianjin University of Technology, Tianjin 300384, China
National Demonstration Center for Experimental Mechanical and Electrical Engineering Education, Tianjin University of Technology, Tianjin, China

Yaguang Zhu and Tong Guo
Key Laboratory of Road Construction Technology and Equipment of MOE, Changán University, Xi'an 710064, China

Rui Ma and Hongtao Wu
College of Mechanical and Electrical Engineering, Nanjing University of Aeronautics and Astronautics, Nanjing 210016, China

Li Ding
College of Mechanical Engineering, Jiangsu University of Technology, Changzhou 213001, China

Gilbert Tang and Phil Webb
School of Aerospace, Transport and Manufacturing, Cranfield University, Building 83, Cranfield, Bedfordshire MK43 0AL, UK

Index

www.ingramcontent.com/pod-product-compliance
Lightning Source LLC
LaVergne TN
LVHW081039210926
848485LV00004B/1058